Rock Engineering Risk

Frontispiece

Rock: Large, rugged mass of stone
Engineering: Contriving, designing, inventing and plotting
Risk: Hazard, danger, exposure to mischance or peril

Oxford English Dictionary

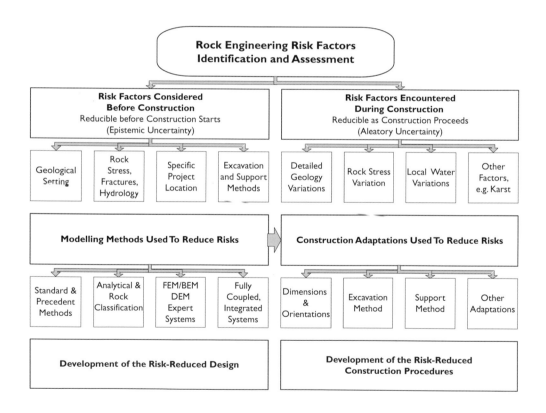

Rock Engineering Risk

John A. Hudson
Department of Earth Science and Engineering,
Imperial College London, UK

Xia-Ting Feng
Institute of Rock and Soil Mechanics,
Chinese Academy of Sciences, Wuhan, China

CRC Press
Taylor & Francis Group
Boca Raton London New York

CRC Press is an imprint of the
Taylor & Francis Group, an **informa** business

A BALKEMA BOOK

Published by:
CRC Press/Balkema
P.O. Box 447, 2300 AK Leiden, The Netherlands
e-mail: Pub.NL@taylorandfrancis.com
www.crcpress.com – www.taylorandfrancis.com

First issued in paperback 2021

Library of Congress Cataloging-in-Publication Data

Hudson, J. A. (John A.), 1940-
  Rock engineering risk / John A. Hudson, Department of Earth Science and
  Engineering, Imperial College London, UK, Xia-Ting Feng, Institute of Rock and
  Soil Mechanics, Chinese Academy of Sciences, Wuhan, China.
    p. cm
  Includes bibliographical references and index.
  ISBN 978-1-138-02701-5 (hardcover : alk. paper) – ISBN 978-1-315-73857-4
  (ebook : alk. paper) 1. Rock mechanics–Risk assessment. 2. Structural failures.
  I. Feng, Xia-Ting. II. Title.

  TA706.H79 2015
  624.1'5132–dc23

                                                          2015008976

ISBN-13: 978-1-03-209867-8 (pbk)
ISBN-13: 978-1-138-02701-5 (hbk)

Dedication

ISRM

This book is dedicated to the International Society for Rock Mechanics

Contents

Preface

The purpose of this book is to describe the subject of risk as it relates to the design and construction of engineering projects located on or within rock masses. Traditionally, such projects include facilities such as building foundations, dams, slopes, tunnels, caverns and mines; but, more recently, other increasingly complex rock engineering projects are being developed, constructed and operated, such as geothermal energy, radioactive waste disposal, CO_2 storage and hydraulic fracturing for shale gas. In all these projects, there are risks involved in the separate aspects of site investigation, modelling, design and construction—together with their cumulative effect on the operation of the whole project.

Our previous 2011 book, "Rock Engineering Design", published by CRC Press/Balkema (Taylor & Francis Group), was also concerned with the design of projects constructed on or in rock masses, i.e., slopes, dams, hydroelectric schemes, mines, and repositories for radioactive waste disposal. That book covered rock engineering design methodologies, associated flowcharts, the information required, technical auditing of design, a rock slope case example, an underground hydroelectric powerhouse case example, Protocol Sheets for auditing rock engineering design, and examples of the use of Protocol Sheets. This new book, "Rock Engineering Risk", covers the related and important subject of risk using the Frontispiece flowchart in which the risks are considered in terms of the uncertainties associated with 'before construction' and 'during construction' factors.

The emphasis in the book is on the physical aspects of these subjects, the rock mechanics and the rock engineering, rather than the financial aspects, although of course there are financial ramifications associated with the mitigation of the physical risks. Chapters 1–5 provide information on the subject of risk and the approaches to reducing risk, especially in the context of design and construction for underground rock engineering, although the general principles apply also to surface rock engineering. Chapters 6 and 7 contain two detailed, major case examples from China relating to long tunnels at great depth and a hydropower cavern complex. These two Chapters contain a wealth of information relating to the practical risk reduction methods described in the earlier chapters and the experiences of their application during both tunnelling and cavern construction.

Both our previously published book on 'rock engineering design' and this book on 'rock engineering risk' are outputs from the International Society for Rock Mechanics (ISRM) Commission on Design Methodology. The earlier 'rock engineering design' book was generated when John A. Hudson was President of the ISRM (2007–2011), and this new book on 'rock engineering risk' was generated while Xia-Ting Feng was President of the ISRM (2011–2015).

We are pleased to report that this book is the first in the newly established CRC Press/Balkema ISRM Book Series.

John A. Hudson and Xia-Ting Feng,
2015

Acknowledgements

ISRM *International Society for Rock Mechanics*

We are especially grateful to Professor E.T. Brown of Golder Associates in Australia who originally recommended that the ISRM Design Methodology Commission should focus on 'rock engineering risk' in the 2011–2015 ISRM Presidential tenure period—following the earlier 2007–2011 research on 'rock engineering design'. The risk subject certainly required attention and we hope that now, four years later, the Commission has indeed made a worthwhile contribution through the publication of this book.

Thus, the ISRM Design Methodology Commission was active in the period 2011–2015 and the authors, as respectively Commission President and Commission Co-President, are significantly indebted to the ISRM Commission Members listed below who actively took part in discussions held in association with ISRM symposia, made many suggestions and provided reference material.

Dr Conrad Felice, Mr Erik Johansson, Prof. Frederic Pellet, Prof. Wulf Schubert, Prof. Alexandros Sofianos, Prof. Ove Stephansson, Prof. Leslie G. Tham, Dr Antonio Samaniego, Dr Mostafa Sharifzadeh, Prof. Resat Ulusay, Mr Lauri Uotinen, Dr Philippe Vaskou, Dr Christophe Vibert, Dr Thierry You, Dr Yingxin Zhou.

The authors are additionally grateful to Professor Qian Qihu, Academician of the Chinese Academy of Engineering and President of the Chinese Society for Rock Mechanics and Engineering for his continuing support of the ISRM Design Methodology Commission's work.

Also, the authors are indebted to Dr Chin-Fu Tsang (former DECOVALEX Project Chairman) and Dr Lanru Jing (DECOVALEX Project Secretary) for their assistance in providing some of the DECOVALEX historical material in Chapter 5. The authors appreciate and thank the DECOVALEX2015 Funding Organisations for their financial and technical support of the DECOVALEX project work described in that Chapter. The statements made in this book are, however, solely those of the authors and do not necessarily reflect those of the DECOVALEX Funding Organisations.

In addition, we thank the following personnel who contributed to the content of Chapters 6 and 7 describing the major tunnel and hydropower case examples; their help has significantly enhanced the content and value of the book. Dr Qiu Shili and Dr Zhang Yongjie wrote the first drafts of Chapters 6 and 7, respectively, and Dr Qiu Shili also assisted with some of the diagram preparation. Professor Jiang Quan provided ideas and considerable information relating to risk assessment of underground cavern groups. Professors Zhang Chunsheng, Hou Jing and Chen Xiangrong were involved with the geological conditions and design information for the Jinping II Project. Professor Chen Bingrui, Dr Xiao Yaxun and Mr Feng Guangliang took part in the rockburst monitoring and warning system in the headrace tunnels and water drainage tunnel at the Jinping II Project site, and Professors Wu Shiyong, Wang Jimin and Zeng Xionghui also provided support for that aspect of the work.

* * * * *

Lastly, we express our profound thanks to Carol Hudson for her meticulous checking of all the details of not only the original manuscript, but also both the initial and final proofs of the book. We may not have eliminated all the errors but, through Carol's help, there are far fewer.

About the authors

John A. Hudson and Xia-Ting Feng at an International Society for Rock Mechanics (ISRM)
Task Force meeting of the Commission on Design Methodology
held at the Institute of Rock and Soil Mechanics, Chinese Academy of Sciences, Wuhan, China.

ISRM President 2007–2011: John A. Hudson
ISRM President 2011–2015: Xia-Ting Feng

PROFESSOR JOHN A. HUDSON

John A. Hudson graduated from the Heriot-Watt University, UK, and obtained his PhD at the University of Minnesota, USA. He has spent his professional career in consulting, research, teaching and publishing in engineering rock mechanics, and was awarded the DSc. degree by the Heriot-Watt University for his contributions to the subject. He has authored many scientific papers and books, and was the editor of the 1993 five-volume "Comprehensive Rock Engineering" compendium, and from 1983–2006 editor of the International Journal of Rock Mechanics and Mining Sciences. Since 1983, he has been affiliated with Imperial College London as Reader, Professor and now Emeritus Professor. In 1998, he became a Fellow of the UK Royal Academy of Engineering and was President of the International Society for Rock Mechanics (ISRM) for the period 2007–2011. In 2015, the 7th ISRM Müller Award was conferred on Professor Hudson in recognition of "an outstanding career that combines theoretical and applied rock engineering with a profound understanding of the basic sciences of geology and mechanics".

PROFESSOR XIA-TING FENG

Xia-Ting Feng graduated in 1986 from the Northeast University of Technology and obtained his PhD in 1992 at the Northeastern University, China. He was then appointed and acted as Lecturer, Associate Professor and Professor at the same university. In 1998, he was admitted by the Hundred Talents Programme to the Chinese Academy of Sciences (CAS). Subsequently, he permanently joined CAS's Institute of Rock and Soil Mechanics at Wuhan, China. In 2003, he obtained the support of the China National Funds for Distinguished Young Scientists; in 2010, he became a Chair Professor of the Cheung Kong Scholars' Programme, Ministry of Education, China; and, in 2009, he was elected as President of the International Society for Rock Mechanics for the period 2011–2015. He is currently Director of the State Key Laboratory of Geomechanics and Geotechnical Engineering in Wuhan. Additionally, in 2012, Professor Feng became the Co-President of the Chinese Society for Rock Mechanics and Engineering. He has made original contributions to the subject of 'intelligent rock mechanics' and his methods have been applied to large rock engineering projects in China and other countries.

Chapter 1

Introduction and background

It is said that it took Plato ten books to provide a definition of Justice and even then the definition was only approximate. In the case of risk, and in particular, rock engineering risk, there is certainly enough material to fill ten books. However, our objective for this book is to present the salient information in reasonably compact form, including the relevant background material, current approaches to the subject, risks before and during construction, plus two major case examples of long, deep tunnels and hydropower cavern groups—with our Frontispiece flowchart providing the overall conceptual structure.

1.1 THE PREVIOUS BOOK "ROCK ENGINEERING DESIGN" AND THIS BOOK "ROCK ENGINEERING RISK"

This book, "Rock Engineering Risk", is a product of the Design Methodology Commission of the International Society for Rock Mechanics (ISRM) and results from the work of the Commission during the years 2011 to 2015. It follows the production of the previous book "Rock Engineering Design" (Feng & Hudson, 2011) which also resulted from the work of the Commission, but in the earlier period 2007 to 2011. In the introduction to our previous book, we noted that the term 'Rock Engineering Design' encompasses a wide variety of design aspects relating to the planning and the construction of civil and mining projects on and in rock masses—from dam foundations to underground tunnels and caverns, although our emphasis was on rock engineering for underground projects. The term also covers a variety of scales from the length and spacing of rockbolts used to stabilise a rock mass to the overall design of an underground hydroelectric scheme, a large mine, or an underground radioactive waste repository, all of which can have footprints of several square kilometres.

We also noted that the subjects of rock mechanics and rock engineering became recognised as disciplines in their own right in the 1960s with the establishment of the International Society for Rock Mechanics (ISRM) in 1962. Over the 50+ years since the ISRM was formed, rock mechanics and rock engineering have developed considerably. There are now sophisticated methods of site investigation, advanced computer modelling techniques, and much improved construction methods. Moreover, we anticipate many exciting developments in the future, such as computer perception of monitoring results and virtual underground laboratories. Thus, the primary purpose of our previous book was to report on the design developments and to prepare the

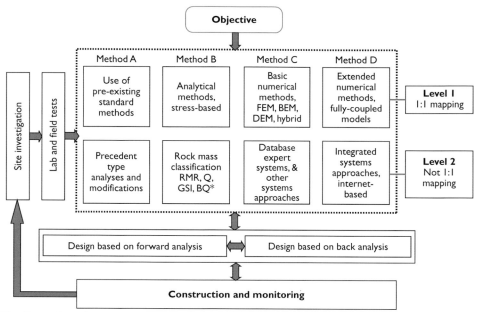

*The Chinese Basic Quality, BQ, classification system is described in Appendix B.

Figure 1.1 Flowchart of rock mechanics modelling and rock engineering design approaches (Feng & Hudson, 2011).

way for modern rock engineering design in a world where computing capability will continue to increase rapidly with all the associated advantages, there will be more emphasis on the use of the Earth's resources and sustainability, and more accountability of engineers' decisions.

A governing flowchart (Figure 1.1) relating to design methods was included in the previous book. The two *rows* of boxes in the dotted main box in Figure 1.1 represent the two main ways in which the mechanisms are mapped in modelling. In the top row, 'Level 1, 1:1 mapping', there is an explicit attempt in the modelling to represent the geometry and operating mechanisms on a 1:1 basis. For example, if there is a major fracture in the vicinity of a tunnel, this can be incorporated explicitly in a numerical model. However, in the lower row, 'Level 2, not-1:1 mapping', the modelling techniques are not designed to include the geometry and mechanisms directly. For example, in the neural network technique, there are nodes in the neural network used but a specific node cannot necessarily be directly associated with either a location or a physical variable, as in the finite element approach. The previous book contained five chapters on the background and foundations for the future methodologies and five further chapters on illustrations of the modern approach to rock engineering design. Also included was a set of Protocol Sheets for the technical auditing of rock engineering design. Thus, the previous book set the scene for this current book in which we now focus specifically on the risk aspects of rock engineering, noting that the book's emphasis is on the physical aspects of rock engineering and not on the associated financial risks.

1.2   ROCK ENGINEERING RISK

As in the previous book, we concentrate here on underground rock engineering—because risk in surface rock engineering is already well covered in the book "Guidelines for Open Pit Slope Design" by Read and Stacey (2009) and specifically in their Chapter 13 where, with reference to risk management, it is stated that, "To be effective it requires perception and detailed understanding of the range of risks involved in an undertaking, the development and implementation of risk assessment and risk management procedures, the leadership of management in developing a risk culture, documented procedures, high levels of communication and consultation, and personnel training and commitment." Their Chapter 13 has the following sections: Introduction, Overview of Risk Management, Geotechnical Risk Management for Open Pit Slopes, Risk Assessment Methodologies and Risk Mitigation.

The photograph (Figure 1.2) on this book's cover illustrates one important aspect of rock masses in relation to engineering risk: the non-ideal nature of *in situ* rock as an engineering material. In particular, rock masses are invariably discontinuous because of the presence of joints, faults and bedding planes (except for a few special cases) which causes problems in the excavation of tunnels and caverns, for civil and mining engineering projects. A particular problem is the formation of rock blocks around an excavation which can fall or slide into the excavation—this being a function of the scale and shape of the excavation as related to the rock mass fracturing, as illustrated

Figure 1.2 Book cover photograph: well-bedded and jointed, alternating limestones and thin mudstones, South Wales, UK.

a Limestone and mudstone inter-bedded strata, with conceptual tunnel.

b Limestone strata, with conceptual tunnel.

c Chalk strata with horizontal, vertical and inclined, conjugate fractures.

Figure 1.3 What would be the effect of excavating a tunnel in these laminated, fractured rock types?

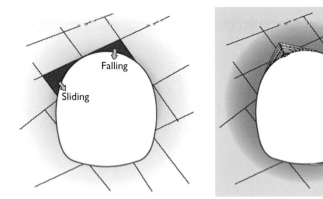

Figure 1.4 Two of the main risk factors in underground rock engineering: a) rock blocks sliding or falling into the excavation; and b) concentrated *in situ* rock stresses damaging the excavation-peripheral rock (from Derek Martin).

in Figure 1.3(a) & (b). Rock masses can have complex fracturing patterns, as illustrated in Figure 1.3(c), in which a tunnel would be very susceptible to rock blocks falling into the excavation in this chalk rock sequence.

Not only are rock masses discontinuous as illustrated in Figure 1.3(a), (b) & (c), but they are also inhomogeneous (e.g., consisting of limestones and mudstones in the case illustrated in Figure 1.2) and hence there is different rock material at different locations in the rock mass. They can also be anisotropic, having different properties in different directions, noting that we would expect different mechanical properties in the vertical and horizontal directions for the rock mass in Figure 1.2. Moreover, the details of these characteristics can vary with the specific location in the rock mass being considered. Often the rock mass will be beneath the water table and, depending on the *in situ* rock mass secondary hydraulic conductivity (which is a function of the fracturing), water inflow can add to the hazards and hence the risks. Additionally, rock masses contain a natural stress field caused by the movement of tectonic plates, the weight of the overlying rock, and other local factors; in other words, *in situ* rock masses are also pre-loaded. Two of the main risks associated with these factors are illustrated in Figure 1.4: rock blocks falling or sliding into an excavation and the concentration of the *in situ* stress damaging the excavation-peripheral rock.

These factors mean that engineers are faced with excavating tunnels and caverns in pre-loaded, discontinuous, water filled, natural, rock masses with complex geometrical and mechanical properties which can vary from location to location and in different directions. Needless to say, this causes the approach to rock engineering to be very different to other forms of engineering using man-made materials, indicating that the risks involved need to be specifically identified within the particular rock engineering environment.

1.3 GOVERNING FLOWCHART FOR THE BOOK

The risks in rock engineering occur throughout the overall process summarised by the flowchart in Figure 1.1, i.e., through site investigation, the associated laboratory and field tests, the design process using one or more of the eight A→D methods, before and during construction, interpreting and acting on back analysis information, and short or long term monitoring. All of these activities involve risks in the sense of generating invalid information, making errors in the modelling and hence the design, recommending inappropriate construction procedures, and misinterpreting back analysis. This situation led us to construct the book's governing flowchart, included both as the Frontispiece and in Figure 1.5, which has the two streams of risk factors, those considered before construction and those encountered during construction. In the lowest two boxes of the flowchart, the term 'risk-reduced' is used: it is possible to reduce the risks, but it is usually not possible to eliminate all the risks—because of the pre-loaded, discontinuous, water-filled, inhomogeneous and anisotropic nature of rock masses.

The terms 'epistemic uncertainty' and 'aleatory uncertainty' have been used in the Figure 1.5 flowchart to characterise respectively those uncertainties which are due to lack of knowledge and those which are due more to chance. The terms are clearly defined in Section 2.3.

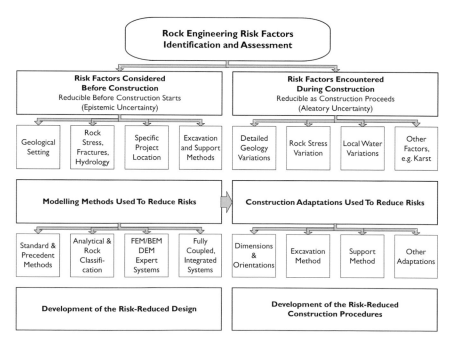

Figure 1.5 Flowchart of risk factors: those considered before construction and those encountered during construction, enabling risk-reduced design and risk-reduced construction procedures.

1.4   STRUCTURE AND CONTENT OF THE BOOK

The book has been structured with seven further chapters, the titles and a brief summary of which are listed below.

Chapter 2—Uncertainty and Risk: The nature of uncertainty and risk in rock engineering is described, together with an explanation of the concepts of epistemic uncertainty (lack of knowledge) and aleatory uncertainty (lack of precise predictability)—with illustrative examples.

Chapter 3—Rock Engineering Systems (RES), Technical Auditing and Protocol Sheets: The RES method provides a systematic approach to studying the main project variables, their interactions and hence the engineering options as related to risk, i.e., the left-hand stream in Figure 1.5. The associated subjects of technical auditing of rock engineering projects and Protocol Sheets are explained; these assist in formalising the risks for any particular project.

Chapter 4—Geological Factors and *In Situ* Rock Stress: The 'before construction' risk factors include, *inter alia*, the geological setting and the rock stress. The need to place sufficient emphasis on these subjects is explained, together with a brief review of *in situ* rock stress.

Chapter 5—Radioactive Waste Disposal: One of the most difficult, if not the most difficult, rock engineering challenges today is the design of a repository for the disposal of radioactive waste which will isolate the waste for thousands of years. The factors involved in reducing the risk of radionuclides migrating back to the biosphere at an unacceptable level are discussed in the rock engineering context.

Chapter 6—Major Case Example 1, Risks Associated with Long, Deep Tunnels: To illustrate the application of the information, philosophies, flowcharts and techniques described in the previous chapters, we provide two major and extensive case examples in Chapters 6 and 7; this first one explains the risks and the techniques for their mitigation involved in the construction of long, deep tunnels.

Chapter 7—Major Case Example 2, Risks Associated with Hydropower Cavern Groups: The second major case example explains the risks and their mitigation associated with hydropower cavern groups (and, by extension, any such caverns), i.e., the construction of the generator and transformer chambers which have large dimensions compared to the headrace tunnel example in Chapter 6.

Chapter 8—Concluding Remarks: Original inspiration, the journey through the book, and Protocol Sheets.

In addition, included at the end of the book are an eight page compilation of colour illustrations, two Appendices (the first on cavern risk events during construction; the second outlining the Chinese BQ, Basic Quality, rock mass classification system), the References and Bibliography section, and an index.

1.5 CHAPTER SUMMARY

This book on "Rock Engineering Risk" follows the previous 2011 book (Feng & Hudson, 2011) on "Rock Engineering Design" by the same authors, both books being outputs of the International Society for Rock Mechanics (ISRM) Commission on Design Methodology. The content of the previous book was explained, noting that the Figure 1.1 flowchart was the governing flowchart for that book with its emphasis on rock mechanics modelling and the associated design approaches.

The emphasis in this book has been placed on the rock engineering difficulties and risks inherent in creating excavations in material that

– is natural,
– often has a complex geological history,
– is discontinuous because of its formation and subsequent stress history causing faults and joints, collectively termed fractures,
– is inhomogeneous, having different properties in different locations,
– is anisotropic, having different properties in different directions,
– is 'pre-loaded' through the presence of a three-dimensional *in situ* stress field,
– is water-filled below the water table,

and thus requires special consideration of the risks involved in modelling and construction aspects. The rock mass illustrated on the book's front cover and in Figure 1.2 illustrates most of these rock mass features. The particular cases of rock blocks sliding or falling into an excavation and concentration of the *in situ* rock stress damaging the excavation-peripheral rock were highlighted. Given these factors, the governing flowchart for this book was then included in Figure 1.5, with its two main components of risk factors considered before construction (leading to risk-reduced design) and risk factors encountered during construction (leading to risk-reduced construction procedures). The remaining contents of the chapters in the book were then summarised with their explanations of uncertainty and risk, Rock Engineering Systems (RES), geological factors and *in situ* rock stress, radioactive waste disposal, and two major case examples (one dealing with long, deep tunnels and the other with large cavern groups).

Chapter 2

Uncertainty and risk

"The banana skin is waiting for you around the corner and you don't know where it is."

Ari Vatanen, 1980s Finnish Rally Driver

2.1 INTRODUCTION

The study of rock engineering risk by the International Society for Rock Mechanics (ISRM) Commission on Design Methodology in the years 2011–2015 was stimulated a) because it is a logical follow-on from the previous book "Rock Engineering Design" (Feng & Hudson, 2011) as explained in Chapter 1, and b) because a major meeting on risk was held in Wuhan, China, in 2012, hosted by the Chinese Academy of Engineering (CAE): "International Summit Forum on Safe Construction and Risk Management of Major Underground Construction" at which there were presentations on many aspects of rock engineering risk. The topics listed for consideration by the Forum were:

1   Mechanism, understanding, prediction theory and warning systems of rockburst, collapse, water inrush, or large deformation of major underground engineering;
2   Optimal design methodology for major underground engineering under conditions of high stress, karst, high water pressure, or weak rocks; and
3   Risk management methods and strategies for safe construction of major underground engineering under conditions of high stress, karst, high water pressure, or weak rocks.

In discussing the "Challenges and countermeasures of underground construction safety", Professor Qian Qihu explained that the safety situation related to civil engineering construction in China should be improved—accident casualties being just below the number of road traffic accidents and coal mine accidents. The vast majority of accidents were caused by rock collapse, gushing sand, and rockbursts; so the complex hydrogeological environment was the main cause, especially sudden water inrushes, with rock 'explosions' being the most difficult and hence the most challenging. [Note that this latter subject is discussed in detail in Chapter 6 in relation to the headrace tunnels at the Jinping II hydropower project.]

Of the many presentations given at this Forum, we highlight the seminal presentation by E.T. Brown of Golder Associates in Australia on "Risk Assessment and Management in Underground Rock Engineering–An Overview", the associated paper being published in the Journal of Rock Mechanics and Geotechnical Engineering (Brown, 2012). So, we begin this Chapter by noting some of the key points in that presentation which was included in Forum Topic 3.

The following definitions were included (from the Australian source AS/NZS ISO 31000: 2009):

- **Risk:** the effect of uncertainty on objectives
- **Risk source:** an element which alone or in combination has the potential to give rise to risk
- **Level of risk:** magnitude of a risk or combination of risks, expressed in terms of the combination of consequences and their likelihood
- **Event:** an occurrence or change of a particular set of circumstances
- **Consequence of an event:** the outcome of an event affecting objectives
- **Likelihood:** the chance that something will happen
- **Epistemic uncertainty:** arises from limitations in fundamental knowledge or understanding of the problem, alternatively, conceptual uncertainty
- **Aleatory uncertainty:** the irreducible randomness or variability associated with phenomena that are naturally variable in time or space.

There can be confusion concerning the meaning of words used when discussing risk, but definitions are available from the 2009 ISO (International Standardisation Organisation) Guide 73–Risk Management Vocabulary. We discuss epistemic and aleatory uncertainties in the following Section 2.3.

Brown (2102) also noted geomechanics-related risk sources as follows:

- Geological structures including dykes, faults and shear zones, possibly containing low shear strength minerals,
- Orientations, spacings, persistences and shear strengths of joint sets in the rock mass,
- Values and distributions of the compressive strengths and elastic properties of the rock materials,
- Rock mass classification values and their use in estimating rock mass properties,
- Lack of knowledge of the behaviour of the rock and rock mass under high stress,
- Weathered or otherwise weakened rock,
- High water pressures,
- High water inflows, including from karst features,
- Estimated pre-excavation stresses and the effects of geological structures and rock mass anisotropy and heterogeneity on those estimates,
- High horizontal pre-excavation stresses, even at relatively shallow depths, and
- Excavation-induced seismicity, including the effects of geological structures.

Most of these play a part in the major case examples that are included in this book, Chapters 6 and 7.

Brown (2012) also listed some risk analysis and evaluation tools as follows.

– Fault tree analysis
– Event tree analysis
– Consequence or cause-consequence analysis
– Bowtie diagrams
– Decision and decision tree analysis
– Probabilistic risk analysis
– Bayesian Networks
– Analytic Hierarchy Process
– Fuzzy logic and other artificial intelligence methods.

Brown (2012) notes that, "For the last 30 or 40 years, Probabilistic Risk Analysis (PRA) has probably been the most widely used method of quantitative risk analysis, generally of the risk of failure." With reference to Figure 2.1, he explains that, "As we progress from the preliminary to the detailed and then the final design stage, our knowledge of the design parameters improves, or is refined, as a result of further investigation, and the probability of failure reduces." An example of a risk register matrix is included as Figure 2.2.

In his conclusions, Brown (2012) notes that, "The literature contains a number of examples of the advanced use of risk assessment and management methods in

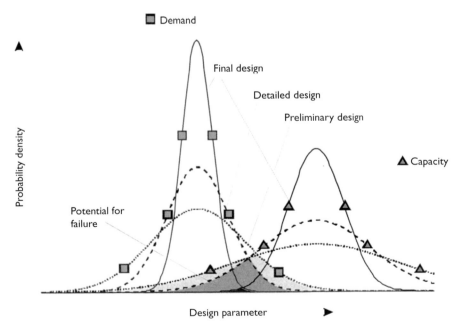

Figure 2.1 Illustration of uncertainty reduction during the development of a project until the potential for failure is minimised to an acceptable level (Valley *et al.*, 2010, after Hoek, 1991).

Likelihood	Consequences				
	Insignificant	Minor	Moderate	Major	Catastrophic
A	H	H	E	E	E
B	M	H	H	E	E
C	L	M	H	E	E
D	L	L	M	H	E
E	L	L	M	H	H

E: Extreme risk – immediate action required; unacceptable risk
H: High risk – Senior management attention required; unacceptable risk without action
M: Moderate risk – management responsibility; acceptable with control measures
L: Low risk – manage by routine procedures; acceptable risk

Figure 2.2 Qualitative risk determination matrix (Brown & Booth 2009).

underground rock engineering, in geotechnical engineering more broadly, and in construction management in China. It is suspected that, as is generally the case elsewhere in the world, greater use could be made in China of both the more routine qualitative and semi-qualitative and the more formal quantitative methods of risk assessment summarised in this paper ... a study reported by Tang *et al.* (2007) identified some advances made, and some scope for improvement, in the use of risk management in the Chinese construction industry. They suggested that, *inter alia*, 'current risk management systems are inadequate to manage project risks, and the lack of joint risk management mechanisms is the key barrier to adequate risk management'. They further suggested that, 'future studies should be conducted to systematically improve the risk management in construction by different approaches that facilitate equitable sharing of rewards through effective risk management among participants. Such studies should also consider the establishment of an open communication risk management process to permit the corporate experience of all participants, as well as their personal knowledge and judgment, to be effectively utilised.'" To address the suggestion in this last sentence, in Chapter 3 we present the Rock Engineering Systems (RES) approach which enables, not only a systematic approach to all the relevant factors, but also enables input from all members involved in the risk assessment process.

In the introduction to the "Catalogue of Notable Tunnel Failure Case Histories"* produced by the Geotechnical Engineering Office in Hong Kong (GEO, 2012), it is stated that the purpose is, "to disseminate information and promote awareness on tunnel failures which could pose a danger to life and property. The possible causes of the failures, the geotechnical problems and the lessons learnt, where these are known, are outlined in the catalogue." Many of these tunnel failures have occurred in soil, but some of those listed have been in rock and a selection is presented below in abbreviated form.

*http://www.cedd.gov.hk/eng/publications/geo/doc/HK%20NotableTunnel%20Cat.pdf

Example Case 1. Orange-Fish Tunnel, South Africa, 1970. Tunnel designed to carry irrigation water from the Orange River (80 km long and 5.3 m in diameter, 1,200 m above sea level). Tunnelling using the rail-mounted drill and blast method and lined with *in situ* concrete. Failure—Heavy water inflow, Water inflow of about 55,000 litres/min into the tunnel at 14 bars. Entire 1.6 km tunnel section flooded within 24 hours. Cause of failure: the tunnel passed through a shallow anticline and intersected a fissure, about 75 mm wide, almost perpendicularly.

Example Case 2. Holmestrand Road Tunnel, Norway, 1981. Road tunnel 1.78 km long and 10 m wide tunnel constructed by the drilland-blast method. The failure was a minor cave-in from the face and partly from the crown which occurred during the process of moving the steel formwork for cast concrete lining forward to the face. Cause of failure: a weak fault zone was encountered, no spiling bolts ahead of the face to support the weak ground.

Example Case 3. Seoul Metro Line 5–Phase 2, Korea, 1991. After blasting, daylight collapse up to ground surface, involving the embankment of a river, 20 m × 15 m × 4 m deep crater at the ground surface, water from river flowed into the tunnel. Cause of failure: thin weathered rock cover, inflow of soil and groundwater (Figure 2.3).

Example Case 4. Lærdal Road Tunnel on European Highway E 16, Norway, 1999. Road tunnel at 1,100 m depth, 24.5 km long and 9 m wide constructed by the drill-and-blast method. The failure was a cave-in involving a 17 m length of tunnel and extending up to about 11–12 m above the crown. The volume of the failed rock mass was estimated to be 1,200–1,500 m³. Possible causes of failure: poor communication—the driller did not inform the engineer about abnormal drilling rate encountered; expansion of the swelling clay under high stress to water during drilling of the rock bolts; the combination of the swelling of the clay and high stress produced a squeezing effect, which resulted in gradual weakening of the rock mass in the tunnel. Reinforced

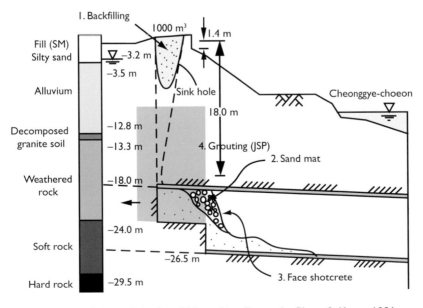

Figure 2.3 Failure of the Seoul Metro Line 5 tunnel—Phase 2, Korea, 1991.

ribs of sprayed concrete in addition to layers of sprayed concrete and rockbolts were installed just behind the cave-in zone. Rock material was hauled into the tunnel building up a barrier up to 2 m below the crown and concrete was pumped through a steel pipe to fill the void above the debris. Debris was gradually hauled out with stepwise installation of rock anchors and sprayed fibre reinforced concrete.

Example Case 5. Tunnel Failure in Japan, 2003. Ground collapse of an avalanche type containing cobbles, gravels and water took place at the point 900 m away from the tunnel portal. A large crater was observed at the ground surface about 130 m above the tunnel. Possible causes of failure: existence of high groundwater pressure, decrease in cover of the mudstone layer, water path created by the investigation drillhole. Remedial measures: filling the caved-in area with foam concrete; grouting under the collapse area; boring for drainage from the tunnel.

Example Case 6. Singapore MRT, 20 April 2004. This tunnel was excavated in soil but is included here because of the lessons learnt (Figure 2.4).

Excavated trench, 15 m wide and 33 m deep, mainly in marine clay with some fluvial clay supported by 0.8–1.0 m thick diaphragm wall which was 35–45 m deep without rock socket. Steel struts: 4–5 m horizontal and 3 m vertical spacing, bottom-up construction, jet grouted base slabs, layer 1–1.5 m thick at 28.5 m below ground. Layer 2–3 m thick at 33.5 m below ground (Layer 2 not yet constructed when collapse occurred). Causes of failure: under-design of the strutting system, incorrect use of Finite Element Method, no proper design reviews, disregard of different warnings, for example, excessive wall deflections and surging inclinometer readings, poor construction quality, ineffective instrumentation and monitoring system, failure to implement risk management. Lessons learnt: there is a need for robust design, risk management, design review and independent checking, purposeful back analysis, effective instrumentation, monitoring and interpretation regime, effective system of management

Figure 2.4 Collapse during construction of the Singapore MRT, 2004 (Government of Singapore).

of uncertainties and quality during construction, corporate competencies and safety management.

Example Case 7. Oslo Metro Tunnel, Norway, 2004. At the junction where the two tunnels met in an acute angle, tunnel cave-in after removal of most of the rock pillar between the tunnels. Possible cause of failure: unfavourable direction of the bedding planes in relation to the geometry and span of the tunnels; over excavation of the rock pillar and the removal of the remaining rock pillar and old concrete wall before the planned concrete pillar was constructed. Lessons learnt: the importance of adequate ground investigation; the need to follow the sequence of rock support installation in accordance with the design plans during construction.

Example Case 8. Hsuehshan Tunnel, Taiwan, 1991–2004. Construction of two 12.9 km long and 4.8 m diameter tunnels, plus a pilot tunnel. Eastbound 28 collapses occurred; Westbound TBM badly damaged due to tunnel collapse and groundwater inflow of 45,000 litres/min into the tunnel; Pilot Tunnel eight collapses occurred. Unexpected difficult geology with fractured rock and massive inflows of water; six major faults found along the tunnel alignment.

Example Case 9. Hanekleiv Road Tunnel, Norway, 2006. The tunnel was supported with a combination of rockbolts and steel fibre reinforced concrete. A section of tunnel caved in 10–11 years after excavation. Possible cause of failure: unfavourable geometry with joints almost parallel to the tunnel axis (Figure 2.5); the rockbolts were installed mainly parallel to the rock joints and with limited influence on the stability.

These nine case examples of tunnel failure have demonstrated key points that we address in this book: the need for more comprehensive understanding of the site

Figure 2.5 Hanekleiv Road Tunnel, Norway, where a section of tunnel caved in 10–11 years after excavation having unfavourable rock geometry with joints almost parallel to the tunnel axis.

conditions, especially the geological and hydrogeological settings; the need to understand rock fracturing; and the need for auditing all activities.

2.2 APPROACHES TO RISK MANAGEMENT

"If you don't understand the geology, you don't understand the problem, and it is unlikely that you will be able to provide a good solution."

D.P. Richards

In Chapter 1, we discussed the overall subject of risk, but how is a risk procedure to be implemented in a given project? There is a variety of approaches to risk via flow-charts, see for example the overview flowcharts in Figures 2.6 and 2.7, but for our specific subject of rock engineering risk, these flowcharts need to be expanded and presented in a form that can be utilised directly. In the previous book "Rock Engineering Design" (Feng & Hudson, 2011), we included a suite of Protocol Sheets which enabled the modelling prior to the design *per se* to be audited and hence risks to be

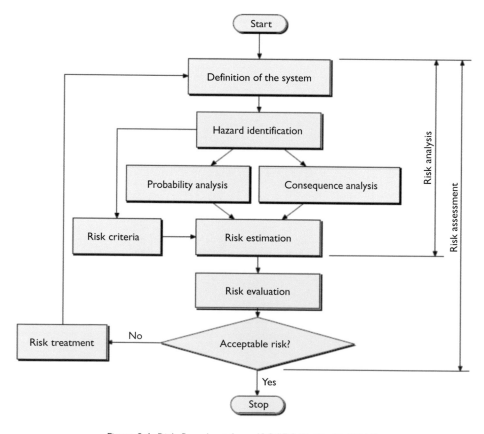

Figure 2.6 Risk flowchart from ISO/IEC Guide 51 (2014).

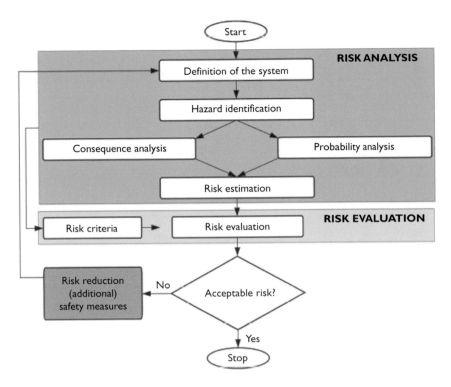

Figure 2.7 Risk flowchart from PIARC (World Road Association).

identified to some extent, but how is the risk to be minimised through the whole rock engineering process?

In a paper on "Sources of geotechnical risk", Baynes (2010) notes that the engineering properties used in design may not be reliable and may not be representative—and that the source of this risk is usually the project staff who are responsible for the choice—but that risks associated with geology are related more directly with uncertainty. He explains that three sources of geological uncertainty can be identified, creating three sources of geotechnical risk.

1   Variability in the 3D distribution of geological units and variability in the geological characteristics of each of the units. The hazard results from the presence of geological detail that is unforeseeable within the practical limits of an investigation, and differences between the scale of the project, the scale of the investigation and the scale of the geological features will have a profound influence on the extent to which this source of risk is allowed to develop. [Epistemic uncertainty, see Section 2.3]

2   The occurrence of actual hazardous geological conditions or processes within the ground. There may be uncertainty about the temporal occurrence of some hazards; for example, a landslide may be known to exist at a certain location or an active fault might have been identified, but the precise timing of the next

movement of the landslide or the fault will not be known. [Aleatory uncertainty, see Section 2.3]

3   When there is an absence of knowledge of what might be in the ground in which the project is being built, a hazard to the project is created because of the possibility of encountering an unforeseen ground condition that might adversely affect the project. [Epistemic uncertainty, see Section 2.3]

A particularly insightful and helpful paper is the one by Ross-Brown and Richards (2014) which discusses the empirical design of long-life underground excavations (such as mine ramps and long tunnels) which are expected to have a life of 20 or more years and which are often located in remote mountainous regions. The term 'empirical design' means using past data to guide the design based on, for example, rock mass class systems. The authors state that "The overall methodology of tunnel design is reviewed with an emphasis on the empirical methods used, and the challenges that designers face when geological, geotechnical, and hydrogeological data are sparse and expensive to obtain, especially when projected to the proposed tunnel alignment." The paper discusses why the preliminary design is so often inadequate, and why estimated costs are frequently poorly estimated. It discusses approaches that can be used to improve the design and the expectations of Owners, based on recent industry experience and the use of risk management tools. One conclusion is that "the empirical methods now being used are generally adequate, but that the data input into these models is often insufficient and that this deficiency is not fully recognised."

The paper deals with excavations which feature, in addition to a long life, a relative lack of geological data (per 100 m length of excavation, or per 10,000 m³ of excavated volume), requiring much judgment in projecting the geology and estimating the percentage of tunnel in each 'rock class' using rock mass classification systems. The authors note that the worse ground conditions to be encountered usually occur in faulted ground, that fault orientation with respect to the proposed tunnel makes a difference in tunnel excavation, and that faults are often associated with weak shear zones, soft and squeezing ground, high stress, and high water flows—which we will describe in Chapter 6. It is noted that, as pointed out by Carter (2011) *inter alia*, almost all collapses are associated with faults and that "the worst conditions are caused by some combination of faulting (on either a large regional scale or a local scale), high rock stress, and high groundwater pressure."

In discussing why the preliminary design is so often inadequate, leading to a difference between the estimated cost and the final cost, Ross-Brown and Richards (2014) note that the problem lies more with the limited quality of the data input into the models, and how the geologists and geotechnical staff are handling the data. What is needed ideally is

–   good quality data, implying good core recovery, competent core logging, using an experienced geologist on site and good empirical correlations,
–   a sufficient quantity of data with sufficient number of boreholes to characterise the regional and local geology,
–   boreholes to sufficient depth and to the tunnel depth or near the tunnel alignment,

– good knowledge of the regional geology and hydrogeology which requires sufficient surface mapping, boreholes and geophysics to build a realistic 3D geological model,

– the ability to project the geology, hydrogeology and geotechnics to the initial tunnel alignment and to subsequent adjustments to that alignment,

– knowledge and experience within the design team of working in similar ground conditions, and

– the team's ability to relate to similar design and construction projects.

Ross-Brown and Richards (2014) go on to say that, "The authors have encountered mistakes made by geologists and geotechs in applying empirical methods that have led to disputes. Although there are many excellent papers written about the different rock classification systems and their application to empirical models relating to excavation behaviour and support requirements, these are not skills easily learned in the classroom or from reading papers. Practice is required under field conditions, especially with someone who has a long experience in judging the values to be assigned to the rock mass parameters and applying them in the empirical models." The authors also point out that Carter (2011) has made generic systematic recommendations that apply to geological and/or geotechnical evaluations of such remote, deep, difficult sites. These include:

– in addition to the evaluation of rock mass classification indices, for deep tunnel alignments several other key geological factors need consideration: structural geological regime, current regional tectonic state, and likely paleo-stress history and that, for deep tunnels, the key geological risks include high rock stress, high temperature, and high groundwater pressure. [Note that explanations of rock fractures and rock stress are included in this book in Chapter 4.]

– when applying any rock mass characterisation technique (empirical design method), several prime factors must be considered and evaluated: suitability in terms of the rock mass in question, potential inaccuracy because empirical methods yield only mean values of specific parameters, while actual values may be substantially different; validity in terms of the actual mechanisms of rock mass behaviour, and reliability.

Ross-Brown and Richards (2014) also recommend that large projects should develop a **Risk Register** as a management tool for the planning, design, construction, and in-service conditions for long, semi-permanent tunnels and similar infrastructure. The risks would include, *inter alia*, the facts that the site investigation data may be insufficient, the presence of unanticipated geological features, and lack of experience of personnel in assessing the parameters for the various rock mass classification schemes. [Note that this relates strongly to epistemic (lack of knowledge) and aleatory (unpredictable uncertainties) which we discuss in the next Section.]

However, the authors (Ross-Brown & Richards, 2014) point out that, "The development of a Risk Register is only the first step in the risk evaluation process. Following this, each identified risk must be evaluated in terms of its Frequency of Occurrence (F) and the Severity of Consequences (C), if it should actually develop." They then provide general guidelines for performing such an exercise on a new project

(Pennington & Richards, 2010) via Risk (R) being considered as the product of frequency (F) and consequence (C): R = F × C.

Ross-Brown and Richards (2014) summarise their paper saying that the risk issue is less about the quality of the empirical methods and more about the quality of the data and the way geologists and engineers are handling the data. In particular, they note that, "All rock mass classification systems (and, hence, empirical design methods) are less reliable as the rock mass conditions become worse. This is due to the fact that poor quality rock is more difficult to core, often resulting in poor core recovery with the worst of the rock disintegrated during the drilling process. Thus the remaining drill core, albeit of poor quality, with a low RQD (Rock Quality Designation) value, is what the rock mass quality is based upon, knowing that some percentage of the rock was not even recovered. This is accounted for somewhat by variations in the RQD value, but no assessment can be made of the missing core for other input parameters for the assessment of rock mass quality". They also state that, "Geological and other risks should be evaluated in a formal manner using, as a minimum, the Risk Register approach."

Finally, the authors make their key statement as follows, "Sometimes we feel that empirical methods are often wrongly blamed for unanticipated geological conditions, tunnelling difficulties, and large cost overruns. But it is not the empirical methods that are to blame. The blame is mostly due to inadequate site investigation and our inability to understand the geological details at the site. As a result we often put oversimplified or incorrect data into our empirical models which yield poor results. This takes us back to the dictum 'If you don't understand the geology, you don't understand the problem, and it is unlikely that you will be able to provide a good solution'. The problem is magnified when engineers and owners do not recognise the likely extent of these 'geological' unknowns in their project."

Consideration of the subjects discussed in this Section led the current authors (Hudson & Feng) to reach three conclusions.

1   It is helpful to have a flowchart to provide a structure for thinking about the risk factors before and during construction and a useful addition would be the generation of associated Protocol Sheets to support risk management, of a similar type to those provided in the previous book on "Rock Engineering Design" for auditing design aspects. The risk subject has too many specific factors and idiosyncrasies to be included in even a significantly extended flowchart, unless this were to be supported by Risk Protocol Sheets covering all the manifold subjects involved.

2   Two subjects that have prime importance are the rock fractures and the rock stress. Accordingly, we have included explanations and discussions on these topics in Chapter 4. [More detailed information is included in Cosgrove & Hudson (2015)].

3   As a result of the first conclusion (that it would be difficult to compile a flowchart which accommodated all the possibilities) in the next ISRM Presidential period, 2015–2019, the ISRM Design Methodology Commission may consider how to generate a set of Risk Protocol Sheets similar to the ones included in the earlier Rock Engineering Design book—in order to be able to audit (both contemporaneously and subsequently) the ability to identify and accommodate risks in rock engineering projects. These will be based on

Figure 2.8 Construction of large caverns for hydropower projects.

i A study of experiences reported in the literature,
ii The experiences gained in the driving of the long tunnels described in Chapter 6, and
iii The experiences gained in the construction of major hydropower projects, as illustrated in Figure 2.8 and as described in Chapter 7.

2.3 EPISTEMIC AND ALEATORY UNCERTAINTIES

"There are some geological conditions that are 'unforeseeable', and when those conditions are encountered there will inevitably be some undetectable variations in the geology that can never be completely investigated within practical limits; for example, cavernous ground as a result of karst may be recognised but it may be impractical to attempt to investigate the details of every single cavity; the details are 'unforeseeable'."

Peter Fookes

2.3.1 Explanation of the terms 'epistemic' and 'aleatory'

The adjectives 'epistemic' and 'aleatory' provide a convenient and useful distinction between the two types of uncertainty that can contribute to rock engineering risk. Their characteristics are as follows:

Epistemic uncertainty

- Relating to knowledge, from the Greek '*episteme*' for knowledge
- Due to lack of knowledge of processes or quantities
- Reducible through further investigation
- Examples: What type of geological structure is present?; lack of understanding of coupled processes, or lack of data.

Aleatory uncertainty

- Dependent on chance, from the Latin '*alea*' for game of chance, dice
- Due to chance, intrinsic randomness
- Conceptually not resolvable
- Examples: At exactly what chainages will water bearing fractures be encountered in the tunnel?

So, the process of site investigation reduces epistemic uncertainty because knowledge of the rock mass is increased by the site investigation process. On the other hand, there will always remain uncertainties that cannot be fully resolved until the underground facility is being excavated. As Ralph Peck said, "No matter how much subsurface exploration is done on a tunnel project, you don't find out what is really there until you drive the tunnel".

This consideration of the two types of uncertainty is why our governing flowchart (Figure 2.9) has the two streams of

- Risk Factors Considered Before Construction, relating to epistemic uncertainties which are reducible to some extent before construction starts, and
- Risk Factors Encountered During Construction, relating to aleatory uncertainties which are reducible to some extent as the excavation proceeds and the features are encountered.

In other words, there is the possibility of increasing knowledge about some factors before construction, such as the geological setting, the hydrogeological setting, the overall rock stress field, the types of rock fractures, the facility location and the appropriate excavation and support methods—and this information can be used to reduce risks. But there are other factors that cannot be determined before construction: they can only be established through encountering them directly, such as the severity of faults and local variations of the rock stress magnitudes (see Chapter 4), the location and severity of water inrushes, the chainage where karst caverns will be intersected. These are characteristics of the local rock mass details, and can be some of the greatest risks. The two major case examples in Chapters 6 and 7 of long, deep tunnels in a highly stressed rock mass and the construction of caverns in faulted rock are presented following the Figure 2.9 flowchart structure.

It should be noted that we have used the terms 'epistemic uncertainty' and 'aleatory uncertainty' in line with the previous text in this Section and for convenience, but it is argued more strictly that aleatory uncertainty is irreducible—in the sense that one cannot predict the heads or tails outcome of the toss of a coin. So, the right-

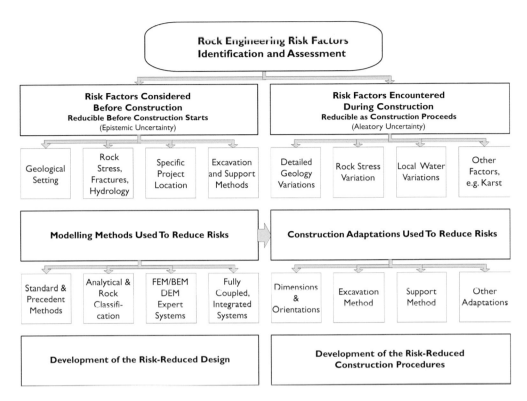

Figure 2.9 The two streams of the book's governing flowchart: risk factors considered before construction, and risk factors encountered during construction, relating respectively to epistemic and aleatory uncertainties.

hand stream of Figure 2.9 is reducible as excavation proceeds because knowledge of the ground conditions is incrementally gained, i.e., there is an increase in epistemic knowledge which does help to reduce the aleatory uncertainty (equivalent to learning about the bias on a die). The same applies to the prediction of rockbursts through microseismic monitoring as described in Chapter 6, although the precise location of water inrushes remains essentially an aleatory characteristic. Similarly, the locations where *in situ* stresses that have been significantly concentrated and therefore represent a hazard for tunnelling also remain essentially as an aleatory uncertainty, considering that a tunnel could be driven through the rock mass shown in Figure 2.10, where the local stress magnitudes and directions have been significantly perturbed by fractures. In other words, although the aleatory uncertainty is technically irreducible, the decrease in epistemic uncertainty during excavation reduces the types and level of aleatory uncertainty that can occur—in the same way that knowing the bias of a coin enables a better prediction of the heads and tails outcomes.

If microseismic emissions were being recorded as a tunnel was being driven through the rock mass modelled in Figure 2.10, the modelling might predict the occurrence of the emissions, but not their precise magnitudes and locations, see Figure 2.11.

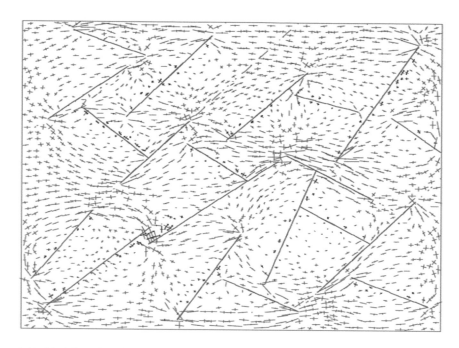

Figure 2.10 Modelling the effect of fractures on the magnitude and direction of the principal stresses (from C. Fairhurst).

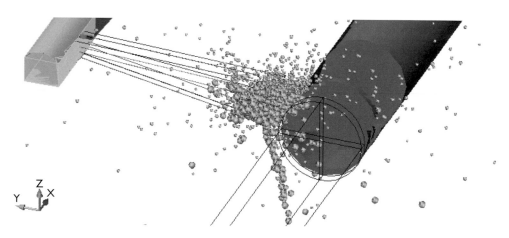

Figure 2.11 Microseismic emissions recorded in boreholes driven from a test tunnel to the periphery of Headrace Tunnel No. 3 in the Jinping II hydropower project described in Chapter 6.

Another surface example of aleatory uncertainty relates to the prediction of natural cliff falls. In Chapter 3, we will discuss how the Rock Engineering Systems (RES) interaction matrix approach can reduce the epistemic uncertainty of natural cliff falls, but there is also an aleatory component, as illustrated in the cliff fall shown in Figure 2.12. The likelihood of such cliff falls is predictable, but the precise location of

Figure 2.12 Beach cliff fall in horizontal sandstone strata with alternating weak and soft layers, plus sub-vertical fractures. Failure is caused by undercutting erosion and then sliding on the sub-vertical planar fractures (Jurassic Bridport Sands, UK).

such an event is an aleatory uncertainty because it depends on the specific location of the major vertical fractures in the rock mass which cause the large scale failure.

2.3.2   Procedures for dealing with epistemic/aleatory uncertainties and Eurocode 7

The flowchart in Figure 2.9 provides an overall strategic guide to coping with epistemic and aleatory uncertainties which has been used by the authors in reporting on the two major case examples in Chapters 6 and 7. But it is helpful to have more formal and specific diagrams and flowcharts to guide our approach to the subjects. Harrison (2012) notes that the dependence of the form of modelling on the state of knowledge: part (a) of Figure 2.13 indicates "how complete ignorance is the extreme of epistemic uncertainty, and that, as knowledge increases, so it may be possible to recognise that aleatory variability exists; and that part (b) shows this in a

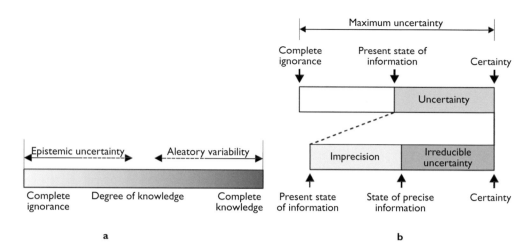

Figure 2.13 (a) Uncertainty, variability and degree of knowledge, after Guo & Du (2007), (b) Uncertainty and information states, after Aughenbaugh & Paredis (2006), both from Harrison (2012).

slightly different way, demonstrating that, as knowledge (or information) increases, a threshold—the state of precise information—may be crossed. Beyond this threshold only irreducible uncertainty remains, and additional knowledge will serve only to improve the characterisation of this." Harrison (2012) goes on to say that, "… it is critical to:

– recognise the presence of epistemic and aleatory phenomena when characterising rock mechanics properties;
– for each property in question, assess these phenomena appropriately in terms of the degree of knowledge; and then
– apply appropriate mathematical techniques to handle the uncertainty and variability in our analyses and designs."

Bedi (2013) provides the necessary detailed guidance in his PhD thesis, "A proposed framework for characterising uncertainty and variability in rock mechanics and rock engineering" which was supervised by J.P. Harrison at Imperial College London. In the thesis abstract, Bedi explains that, "… a novel taxonomy is developed and presented that facilitates characterisation of epistemic uncertainty and aleatory variability in the context of rock mechanics and rock engineering. Using this taxonomy, a new framework is developed that gives a protocol for correctly propagating uncertainty and variability through engineering calculations." Bedi provides a series of diagrams and flowcharts, of which Figure 2.14 is one example.

Given all that has been described and discussed in this book so far, it is evident that following existing codes of practice, e.g., Eurocode 7 (the Eurocode for Geotechnical Design known as EC7), will not enable the necessary subtleties of rock masses to be utilised successfully for engineering purposes—indicating that significant revisions

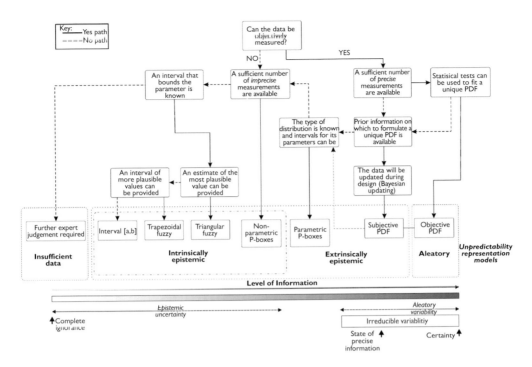

Figure 2.14 Data characterisation strategy sub-chart, from Bedi (2013), and after Aughenbaugh and Paredis (2006), Dubois and Guyonnet (2011), Guo and Du (2007) and Wenner and Harrison (1996). PDF ≡ probability density/distribution function.

need to be made to existing codes, or new codes need to be developed from scratch. In 2014 and in association with the ISRM-sponsored symposium EUROCK2014, there was a Workshop on "Applicability and Application of Eurocode 7 to Rock Engineering Design" which was chaired by Prof. J.P. Harrison. He explained that there are many reasons why EC7 is inappropriate for rock engineering design, but especially because EC7 embraces limit state design principles. It is anticipated major modifications will be made to EC7 during the period 2015–2018, as recommended by the Evolution Group 13 under the chairmanship of J.P. Harrison.

2.4   CHAPTER SUMMARY

In this Chapter on uncertainty and risk, we first explained how the book benefitted from a Chinese Academy of Engineering Forum meeting on 'safe construction and risk management of major underground construction'. We highlighted the paper presented there by E.T. Brown (2012) which provides a good deal of information on risk, risk sources and risk analysis methods. We then highlighted nine illustrative case examples of tunnel failures and explained the approaches to risk management.

The terms 'epistemic uncertainty' and 'aleatory uncertainty' were also explained— the word 'epistemic' relating to knowledge, and the word 'aleatory' relating to

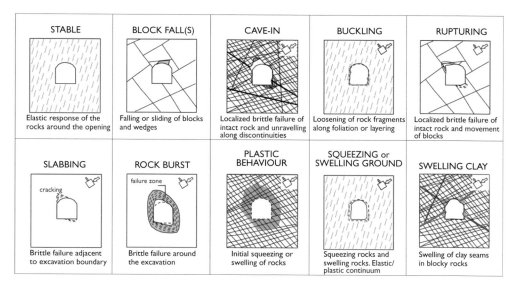

Figure 2.15 Some types of behaviour in underground openings (from Palmström and Stille (2007) and partly from Martin *et al.*, 1999 and Hoek *et al.*, 1995).

chance—together with the link to the governing flowchart for this book and the structure of the presentations of the two major case examples in Chapters 6 and 7. The paper by Harrison (2012) and the thesis by Bedi (2013) are particularly helpful with regard to the need for revision of Eurocode 7 to accommodate both the idiosyncrasies of rock mechanics and the related uncertainty, and the development of methods and associated flowcharts to incorporate uncertainty and risk analyses into engineering practice, although we suggest that these should be supplemented by new Protocol Sheets which can be used contemporaneously for risk analysis and subsequently for auditing purposes.

We conclude with the diagram in Figure 2.15 from Palmstrom and Stille (2007) illustrating the manifold types of rock behaviour caused by underground rock excavation. The rock fractures and the *in situ* stress play a dominant role in these types of behaviour, two subjects we cover in Chapter 4. Clearly, the analyses of uncertainty and risk associated with the different types of rock movement and failure in Figure 2.15 initially require understanding of the failure mode in question at a particular site, leading to the method of risk assessment, e.g., using the methods of Goricki and Schubert based on rock mass types and behavioural types (Goricki *et al.*, 2002).

Chapter 3

Rock Engineering Systems (RES), auditing and Protocol Sheets

"All things by immortal power,
Near or far
Hiddenly
To each other linkèd are,
That thou canst not stir a flower
Without troubling of a star."

Francis Thompson, English Victorian poet, 1859–1907

"We need to better understand the interconnectivity and interrelationship
of risk in the modern construction project."

From the Keynote Lecture on Risk Management by Professor Roger Flanagan,
School of Construction Management and Engineering, University of Reading, UK,
given at the ISRM EUROCK Symposium held in Stockholm in 2012.

3.1 INTRODUCTION TO THE SYSTEMS APPROACH AND AUDITING CONCEPTS

In the previous Chapter, we discussed the nature of uncertainty and risk and the concepts of epistemic and aleatory uncertainties, the latter being useful because they relate respectively and mainly to uncertainties before and during construction. In this Chapter, we discuss how the epistemic uncertainty (i.e., the knowledge available before construction) needs to be extracted in a coherent way, i.e., through a systems approach. A group of people working on a particular project often have a wealth of information between them, but no method of integrating this knowledge into a coherent whole. Also, when modelling and designing a rock engineering structure, it is necessary to ensure as far as practicable that the work has been done correctly; this leads to the necessity for technical auditing and a range of Protocol Sheets which are used to check the variety of detailed operations in the modelling and design work.

In the 2014 UK Royal Academy of Engineering Report* "Thinking Like an Engineer", the following points are made in the context of education and needing to think in a systems way for engineering problems as follows:

---

*http://www.raeng.org.uk/publications/reports/thinking-like-an-engineer-implications-full-report

- to recognize essential interconnections and to appreciate that systems may have unexpected effects that cannot be predicted from the behaviour of individual sub-systems;
- seeing whole systems and parts and how they connect, pattern-sniffing, recognising interdependencies, synthesising.

This way of thinking is definitely required for rock engineering problems and is the motivation for the systems approach described in this Chapter.

3.2 REDUCING EPISTEMIC UNCERTAINTY USING THE ROCK ENGINEERING SYSTEMS APPROACH

Throughout this book we use the flowchart in Figure 3.1 as a structural guide to the reduction of rock engineering risk, recalling that it is necessary to consider both types of uncertainty—epistemic and aleatory—the former being concerned with lack of knowledge about a process or model and the latter being concerned with the inherent randomness of a process or model. We now review the Rock Engineering Systems (RES) approach as a method for extracting knowledge about the particular rock engineering system in question, thus reducing the epistemic uncertainty for a given project and hence the risk. In particular, we should like to establish which factors/variables/parameters are the most interactive and which are the most dominant in a project. This then enables consideration of the emphasis required in site investigation, the focus of the modelling, and what can be done to 'damp down' the adverse effects of dominant factors, e.g., by adjustment of the location and/or geometry of the facility, by grouting fractures, by altering the excavation procedures, and so on. The RES procedure is explained first and then selected published applications are described to demonstrate the value of the approach.

As its basis, the RES approach uses an interaction matrix in which the main parameters governing a particular circumstance (e.g., slope failure, use of blasting or tunnel boring machines, underground stability and support) are selected and the interactions between them are considered. This involves a comprehensive assessment of the factors and interactions, the advantage being that all potential influencing factors can be included initially. The reason why the RES approach reduces epistemic uncertainty is because study of the interactions between the factors indicates which of these are most interactive in the system being considered, which are dominant, and which have a lesser or insignificant contribution.

Also, once the structure of the problem is developed in this way, the acceptability of simplifying assumptions in the project or model is clarified, again reducing the epistemic uncertainty. Moreover, the likelihood of a major hazard being overlooked is reduced. In addition, the subjectivity introduced into a project or model when it is studied by a single person is reduced through a group approach. Although epistemic uncertainty is emphasised here, the reduction of aleatory uncertainty can also be incorporated into the RES approach through the use of probability distributions and other strategies.

The senior author of this book developed the Rock Engineering Systems (RES) approach and authored the earlier book "Rock Engineering Systems: Theory and Practice" (Hudson, 1992). Since 1992, i.e., in the 20+ years since then, the approach has been used for a variety of rock engineering problems, including, *inter alia*, the stability of natural and artificial slopes, excavation methods, the stability of underground

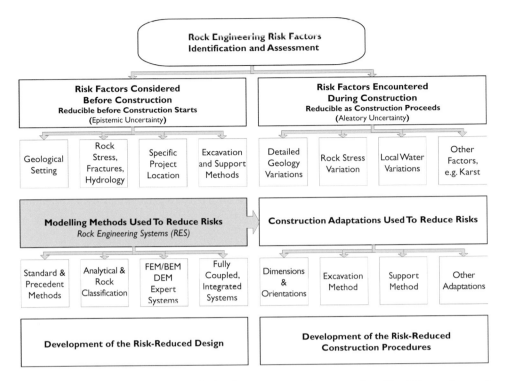

Figure 3.1 Flowchart used throughout the book to structure the approach to identifying and assessing rock engineering risk factors. The Rock Engineering Systems (RES) approach is one of the 'modelling methods used to reduce risks', i.e., the central box in the left-hand column of the flowchart.

openings, and organising the many factors (features, events and processes) in radioactive waste disposal. Additionally, the RES methodology has been extended to problems outside the sphere of rock engineering, such as air pollution and agricultural innovation. Moreover, there are now many case studies available from the application of RES to engineering problems in a variety of countries, i.e., Bangladesh, China, Greece, Iran, Italy, Korea, Spain, Sweden, Turkey, UK, and the USA. Firstly, the RES structured approach will be explained, together with the consequential developments that follow naturally from the use of an interaction matrix. Then, a series of RES applications described in the literature will be reviewed.

3.3   A REVIEW AND EXPLANATION OF THE ROCK ENGINEERING SYSTEMS (RES) METHODOLOGY

The RES methodology is an analytic approach, rather than a synthetic approach. In other words, the rock mechanics/engineering model is not built up by assuming certain variables should automatically be included (synthesis), but by studying the problem, breaking the problem down into its constituent variables (analysis), and assessing their significance so that their relative importance can be established and an appropriate model then constructed. One of the key aspects is firstly to establish the objective of the rock engineering project—because the importance of the respective variables depends on the project objective. An automatic output from the analysis process is the ability to assess the risk of inappropriate modelling and/or design.

3.3.1 The interaction matrix

The basic tool of the RES methodology is the interaction matrix, as illustrated in Figure 3.2. This enables a compilation of the manner in which each particular system factor affects all other system factors, together with the complementary manner in which all system factors affect that particular factor. When these effects are established for all the factors being considered, many operations can be conducted on the interaction matrix for a variety of purposes, in particular and in the context of this book, the evaluation of risk.

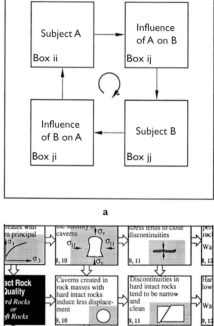

Figure 3.2 The interaction matrix, the basic tool of the RES methodology. (a) a 2 × 2 interaction matrix with leading diagonal terms Subject A and Subject B, from Hudson (1992); (b) a portion of a larger illustrated interaction matrix with 12 leading diagonal terms.

A simple 2×2 matrix is illustrated in Figure 3.2(a) having the two factors/variables Subject A and Subject B. These are placed in the diagonal boxes from top left to bottom right, this being known as the 'leading diagonal' of the matrix. The influence of Subject A on Subject B is contained in the top right box, and the influence of Subject B on Subject A is contained in the bottom left box. In this way, a clockwise influence convention is used. If the influence of Subject A on Subject B is the same as the influence of Subject B on Subject A, i.e., the two off-diagonal terms in Figure 3.2(a) are the same, the matrix is termed 'symmetrical'. In the problems we will discuss, the off-diagonal terms will be different: e.g., in the case of Subject A being the rock stress and Subject B being a rock fracture, the influence of the rock stress on a fracture is not the same as the influence of a fracture on the rock stress. The boxes in the square interaction matrix are indexed with the rows being i ($i = 1$ to N) and the columns being j ($j = 1$ to N). In Figure 3.2(a), we have just Boxes 1,1; 1,2; 2,1; 2,2 but, in Figure 3.2(b)—which is a portion of a 12×12 matrix—there are higher numbered boxes.

Let us say that we are interested in the interactions within a rock mass having the three main variables of rock structure, rock stress and water flow, all critically important in many rock engineering problems, e.g., the long, deep tunnels described in Chapter 6. The six separate binary interactions are shown in Figure 3.3. So, including the extra variable, construction, the resulting 4×4 interaction matrix is shown in Figure 3.4.

Note that each row in the interaction matrix contains the influences that the leading diagonal variable in that row has on all the other variables. For example, Row 2 in Figure 3.4 contains the three binary influences of the *in situ* stress on rock

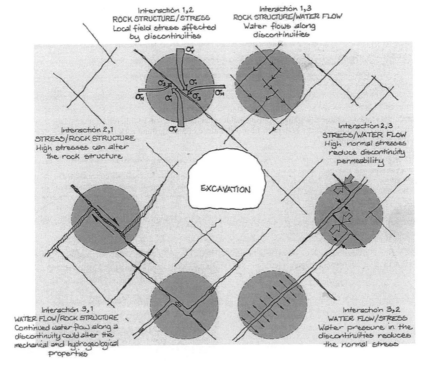

Figure 3.3 Consideration of the interactions in a rock mass with the three variables: rock structure, rock stress and water flow (from Hudson, 1989).

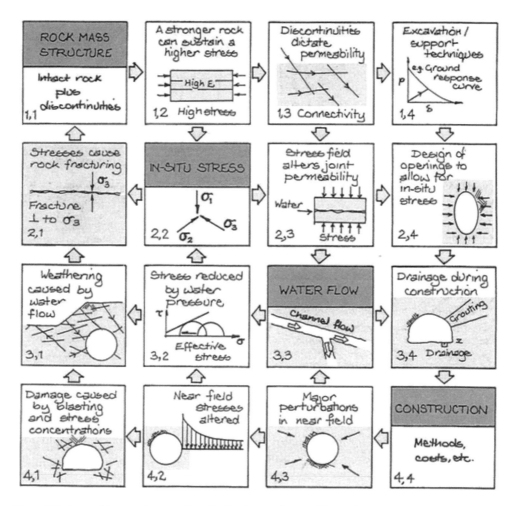

Figure 3.4 A 4 × 4 interaction matrix with the four main variables rock mass structure, *in situ* stress, water flow and construction along the leading diagonal, and the 12 binary interactions between the pairs of main variables in the off-diagonal boxes (from Hudson, 1989).

mass structure, water flow and construction. Conversely, each column in the matrix indicates how the other variables affect the variable in that column. For example, Column 2 in Figure 3.4 contains the three binary influences of the three variables rock mass structure, water flow and construction on *in situ* stress. Note especially that, by locating the construction variable in the lower right corner of the interaction matrix, i.e., Box 4,4 in Figure 3.4, Row 4 indicates the way in which construction affects the other variables and Column 4 indicates the way in which the other variables affect construction. This splits the matrix into those boxes related to rock mechanics and those boxes related to rock engineering, as illustrated in Figure 3.5. The interaction matrices described and illustrated in the later sub-Sections of this Chapter will follow the convention of locating the key engineering variable in the bottom right hand box of the matrix, i.e., Box *N,N*. In the case of studying the safety of natural rock slopes, the variable of interest, potential instability, would thus be located in Box *N,N*.

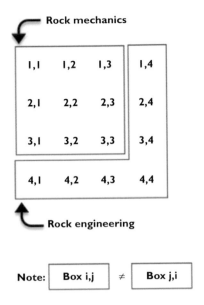

Figure 3.5 By locating construction in Box 4,4, the interaction matrix shown in Figure 3.4 is conveniently split into the interactions relating to rock mechanics and those relating to rock engineering (from Hudson, 1989).

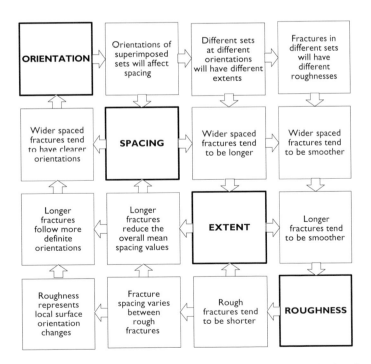

Figure 3.6 Example interaction matrix for rock fracture characteristics with the four leading diagonal terms orientation, spacing, extent and roughness (from Harrison & Hudson, 2000).

When any rock engineering problem is being considered, an interaction matrix can be constructed by firstly establishing the primary variables for the leading diagonal terms and then identifying the interactions for the off-diagonal boxes. In Figure 3.6, we present a simple example for rock fractures. When studying natural fractures in a rock mass, assume that we are interested in the 'variables'[1] of fracture orientation, spacing, extent (or persistence) and roughness. Using these four parameters as the leading diagonal terms of a 4 × 4 interaction matrix, we can identify the content of the 12 off-diagonal boxes and hence show that these variables are likely to be related.

The interactions listed in the 12 off-diagonal boxes of the Figure 3.6 matrix are of several forms: direct cause and effect, commonly observed correlations, and effects caused by sampling. The existence of the interactions indicates that there will be relations between the orientations, spacings, extents and roughnesses of fractures in a given rock mass.

An example of a generic rock mechanics/rock engineering 12 × 12 interaction matrix can be found in Hudson (1991). This matrix has the 12 leading diagonal terms: excavation dimensions, rock support, depth of excavations, excavation methods, rock mass quality, discontinuity geometry, rock mass structure, *in situ* stress, intact rock quality, rock behaviour, discontinuity aperture and hydraulic conditions. Because there are 144 boxes, there are 132 off-diagonal boxes (with brief sketches illustrating the interactions). A similar 12 × 12 interaction matrix has been constructed for rock slopes (Hudson, 1992) and an interesting 15 × 15 interaction matrix also containing sketches was compiled by Cancelli and Crosta (1994).

An interaction matrix of this type can be made for any rock engineering problem by first establishing the leading diagonal variables and then filling in the off-diagonal boxes. The off-diagonal boxes can then be allocated values according to their significance enabling further analyses, as will be explained in the following sub-Sections. Even before further analysis, the compilation of an interaction matrix for any rock engineering problem will always clarify the situation and assist in progressing down the left-hand column of the flowchart in Figure 3.1. Moreover, it enables the composite knowledge of a group of people to be organised and recorded. In other words, the interaction matrix provides the structure for eliciting and presenting the information relevant to a particular problem—and hence reducing the epistemic uncertainty.

3.3.2   Coding the interaction matrix, and the Cause–Effect plot

Having constructed an interaction matrix, the next step is to 'code' the off-diagonal components in order to express their importance or to enable mathematical manipulation of the matrix. There are five main methods to accomplish this coding for the off-diagonal boxes in an interaction matrix (Hudson 1992):

---

1 The term 'variable' has been used to describe the leading diagonal terms—because of the potential use of mathematics to analyse the interaction matrices. However, for Figure 3.6, it may be felt that the term 'parameter' or 'factor' would be more appropriate. The correct use of the terms 'parameter' and 'variable' is as follows: the term 'parameter' refers to the a, b and c in an expression such as $ax + by = c$, i.e., the coefficients making the relation specific; whereas, the term 'variable' refers to the unspecified, unknown x and y in the equation. Any of the terms variables/parameters/factors will be used in this Chapter for the leading terms of the interaction matrix, depending on the context.

1   Binary: the mechanisms in the off-diagonal boxes are either switched on or off, so the coding is either as 1 or 0.

2   Expert Semi-Quantitative: a number from 0 to 4 is allocated as follows:

0 – No interaction
1 – Weak interaction
2 – Medium interaction
3 – Strong interaction
4 – Critical interaction

alternatively this scale could be 1–5 as in the following example, Figures 3.7 & 3.8.

3   According to the slope of an assumed linear relation.
4   More numerically via a partial differential relation.
5   Explicitly via complete numerical analysis of the mechanism.

By far the most widely used of the five coding methods is Method 2, emboldened above. This is because Method 1 does not provide enough discrimination, and the information for Methods 3–5 is rarely available. Method 2 provides the necessary discrimination and the 0–4 (or 1–5) interaction values can be established by one person, or preferably by discussion within a group of persons familiar with the project being considered.

Consider the coded simple interaction matrix in Figure 3.7—which is the Figure 3.4 matrix coded using the Expert Semi-Quantitative method just described

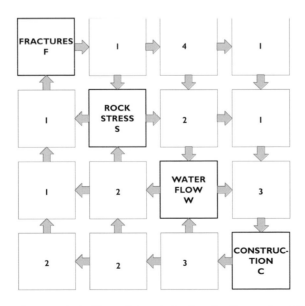

Figure 3.7 The interaction matrix in Figure 3.4 coded using the Expert Semi-Quantitative method. Summing the values in a row gives the *C* ordinate for a leading diagonal variable; summing the values in the column through the same variable gives the associated *E* ordinate. For example, the *C–E* co-ordinates for the leading diagonal term Fractures are (6,4) (from Harrison & Hudson, 2000). **Note that this simplified example is only included to illustrate how the C–E co-ordinates are obtained; a larger matrix is required for real applications.**

(Harrison & Hudson, 2000). In the first row of the matrix, for example, it is considered that the ESQ values for the influence of fractures on rock stress, water flow and construction are 1,4,1 respectively. So the boxes in Row 1 of the Figure 3.7 matrix contain the influences of fractures on all the other leading diagonal variables in the matrix, which we can sum to a value of 6. In a complementary way, we note that Column 1 contains the influences of all the other variables on fractures, i.e., 1,1,2 which totals 4. We term the total of Row 1 as the *Cause* (because this is the total way in which the first variable influences the system) and we term the total of Column 1 as the *Effect* (because this is the total way in which the system affects the first variable. So the *Cause–Effect* (C,E) co-ordinates for the first variable are (6,4). Performing the same operation on the other three variables gives the C,E co-ordinates for the four variables in Figure 3.7 as (6,4), (4,5), (6,9), (7,5). This ESQ method of matrix coding can be undertaken for any sized interaction matrix. If the matrix is 12 × 12, then there will be 12 pairs of C,E co-ordinates, one for each variable in the leading diagonal. The next step is to prepare a table of the ESQ values, as in Table 3.1, list the sum and difference and plot the pairs of co-ordinates, as in Figure 3.8.

Table 3.1 List of the C,E co-ordinates from Figure 3.7, together with the sum and difference for each leading diagonal variable.

Leading diagonal variable	C	E	C + E interactive intensity	C − E dominance/ subordinacy
Fractures, F	6	4	10	2
Rock Stress, S	4	5	9	−1
Water Flow, W	6	9	15	−3
Construction, C	7	5	12	2
Sum	23	23		
Mean	5.75	5.75		

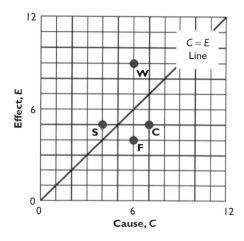

Figure 3.8 Cause–Effect plot for the leading diagonal variables in Figures 3.4 and 3.7: F–Fractures; S–Rock Stress; W–Water Flow; C–Construction (from Harrison & Hudson, 2000).

Note that, in Table 3.1, the $C + E$ value represents how active that particular variable is within the matrix system: the higher the value, the more active the variable. The $C - E$ value represents how dominant the variable is within the system: a positive value indicates that the variable is affecting the system to a greater extent than the system is affecting the variable. So, positive values of $C - E$ represent a dominant variable, whereas negative values of $C - E$ represent a subordinate variable—the system is affecting the variable more than the variable is affecting the system. Another point with reference to Table 3.1 is that the sum of the C values (ΣC) for the whole matrix is in fact the sum of all the values in the off-diagonal boxes, but the same applies to the sum of all the E values (ΣE). Thus, the mean of the C values equals the mean of the E values. The next step is to plot the (C,E) co-ordinates on a *Cause–Effect* plot, as in Figure 3.8.

With reference to Figure 3.8, more interactive variables, i.e., with larger $C + E$ values, will plot further away from the origin, e.g., W with a $C + E$ value of 15. The more dominant factors, $C > E$, plot to the right of the $C = E$ line—as is the case for F and C. The overall conclusion is that we have a moderately interactive system structure in which Water Flow has the strongest interaction. Fractures and Construction slightly dominate the system (being slightly to the right of the $C = E$ line) and Rock Stress and Water Flow are slightly dominated by the system (being slightly to the left of the $C = E$ line). These conclusions depend, of course, on the values assigned to the interactions. Note that the centre of gravity of the C,E co-ordinates will always lie on the (0,0) to (12,12) diagonal whatever coding values are assigned to the off-diagonal interactions in the matrix.

In this sub-Section, the intention has been to explain the interaction matrix coding, the *Cause–Effect* co-ordinates and the *Cause–Effect* plot together with its implications. For this reason, a simple 4×4 matrix has been used in Figure 3.8 for illustrative purposes. However, the method used for this demonstration applies in exactly the same way regardless of the number of variables, i.e., the dimension of the interaction matrix. Thus, we emphasise that this example is only to explain how the C,E co-ordinates are obtained and utilised; practical examples will be represented by larger interaction matrices with typically between nine and fifteen leading diagonal terms.

In the general case and for the larger interaction matrices included later in this Chapter, the type of constellation formed by the positions of the leading diagonal variables in the *Cause–Effect* plot indicates the type of system being investigated. There are four main types of *Cause–Effect* plot:

Type 1: Clustered around the centre of gravity of the points
Type 2: Dispersed around the centre of gravity of the points
Type 3: In an elliptical zone around the $C = E$ diagonal
Type 4: In an elliptical zone around the other diagonal.

These four types are illustrated in Figure 3.9(a) for the case of four different 12×12 interaction matrices.

In Figure 3.9(a), C,E co-ordinate points plotting further along the $C = E$ diagonal represent system variables with a greater interactive intensity because $C + E$ is greater; whereas those plotting significantly to the right of the $C = E$ diagonal represent dominant variables because $C \gg E$. This is illustrated in Figure 3.9(b).

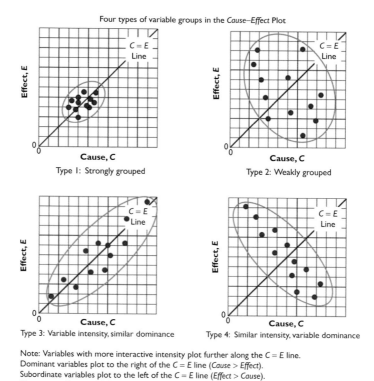

Four types of variable groups in the *Cause–Effect* Plot

Type 1: Strongly grouped

Type 2: Weakly grouped

Type 3: Variable intensity, similar dominance

Type 4: Similar intensity, variable dominance

Note: Variables with more interactive intensity plot further along the *C* = *E* line.
Dominant variables plot to the right of the *C* = *E* line (*Cause* > *Effect*).
Subordinate variables plot to the left of the *C* = *E* line (*Effect* > *Cause*).
The centre of gravity of the points always lies on the *C* = *E* line (Sum of C_i/n = Sum of E_i/n).

Figure 3.9a Four different types of variable clustering in the *Cause–Effect* plot.

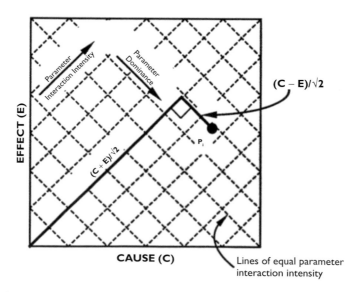

Figure 3.9b Interpreting the position of a parameter in *Cause–Effect* space (from Hudson, 1992).

Meanwhile, we can consider a *C,E* plot for the generic underground excavation matrix (Hudson, 1991—this matrix is too large to include in the book). The 12 variables are as follows:

1	Excavation Dimensions	7	Rock Mass Structure
2	Rock Support	8	*In Situ* Rock Stress
3	Depth of Excavations	9	Intact Rock Quality
4	Excavation Methods	10	Rock Behaviour
5	Rock Mass Quality	11	Discontinuity Aperture
6	Discontinuity Geometry	12	Hydraulic Conditions.

Using the ESQ coding for the significance of the off-diagonal interactions given earlier, one group of engineers coded the matrix with the values given in Figure 3.10(a). Summing the values in the row and column through each leading diagonal term and plotting the *C,E* co-ordinates for each variable results in the plot shown in Figure 3.10(b).

Interpreting the constellation of points in Figure 3.10(b) with reference to the guidance in Figures 3.9(a) and (b), we can make the following observations.

–   The system is significantly interactive because all the points are on or above the 50% interactivity line ($E = 44 - C$). Also, many of the parameters have a similar interactivity.

–   Parameters 7, 6 and 9, Rock Mass Structure, Discontinuity Geometry and Intact Rock Quality are the most dominant, being the furthest to the right of the $C = E$ line.

–   Parameters 10, 1, 2 and 8, Rock Behaviour, Excavation Dimensions, Rock Support and In Situ Rock Stress are the most subordinate, these being the furthest to the left of the $C = E$ line.

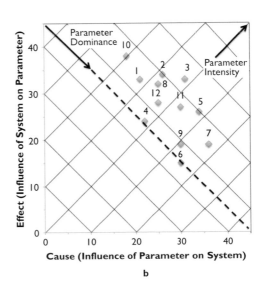

Figure 3.10 Matrix ESQ coding and the resultant *Cause–Effect* plot for a 12 × 12 interaction matrix for underground excavations.

A similar 12 × 12 matrix has been developed and ESQ coded for rock slopes (Hudson, 1992). In this case, the 12 leading diagonal terms are

1	Overall Environment	7	Hydraulic Conditions
2	Intact Rock Quality	8	Slope Orientation and Location
3	Discontinuity Geometry	9	Slope Dimensions
4	Discontinuity Mechanical Properties	10	Proximate Engineering Disturbance
5	Rock Mass Properties	11	Support/Maintenance
6	*In Situ* Rock Stress	12	Construction.

The ESQ coding and the resultant *Cause–Effect* plot are shown in Figure 3.11.

We note that in a similar way to the *Cause–Effect* plot in Figure 3.10(b) for underground excavations, the parameters in Figure 3.11 for rock slopes have similar interactive intensities BUT the interactive intensity is less than for the underground excavations case. This is because the underground rock mass system is more closely linked than the surface rock mass system. In this case, Parameter 7, Hydraulic Conditions, is the most interactive and Parameter 3, Discontinuity Geometry, is the most dominant.

* * * * *

The procedure described above is the one by which researchers and engineers have evaluated the interactivity and dominance/subordinacy of the different systems described in the various case studies presented later in this Chapter. But next, mechanism pathways through the interaction matrix will be described to indicate the consequences of a sequence of mechanisms occurring, rather than just each separate off-diagonal mechanism occurring in isolation. This will, in turn, be followed by an explanation of what happens when the whole matrix system is 'turned on', either as a step-by-step process or as a fully-coupled system.

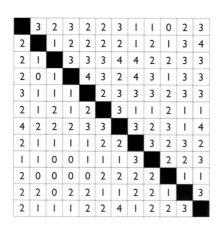

a

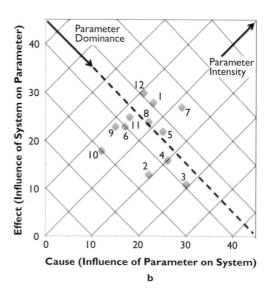

b

Figure 3.11 Matrix ESQ coding and the resultant *Cause–Effect* plot for the 12 × 12 interaction matrix for rock slopes.

3.3.3   Mechanism pathways

Each element of the interaction matrix represents one interaction between two of the leading diagonal variables. Let us now consider a sequence of interactions, i.e., one interaction, followed by another, followed by another, etc., thus forming a pathway of interactions through the matrix. A study of the subject was conducted by Harrison and Hudson (2006) in the context of hazard and risk, and a summary of that work is provided here.

There are many sequential mechanism pathways through an interaction matrix. The minimum length of a pathway is a single mechanism, e.g., fractures increase water flow in Figure 3.4. The number of these single-stage interactions is simply the number of off-diagonal terms in the interaction matrix, i.e., the number of terms in the matrix minus the number of leading diagonal terms. For a 4×4 matrix (e.g., Figure 3.4), the number of paths is $(4 \times 4) - 4 = 12$. For a 7×7 matrix, the number of single-stage interactions is $(7 \times 7) - 7 = 42$. However, there are many more multiple-stage pathways as illustrated by the number of longer length mechanism pathways in Table 3.2 and Figure 3.12.

Table 3.2 Number of mechanism pathways of various lengths within a 4×4 interaction matrix.

Mechanism path length	Number of pathways
1	12
2	36
3	96
4	240
5	528
6	1080
7	1872

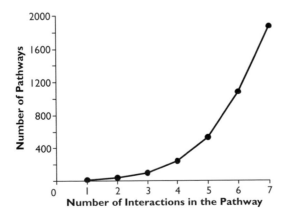

Figure 3.12 Relation between the number of interactions, i.e., off-diagonal mechanisms, in a pathway through the interaction matrix and the associated number of such pathways for a 4×4 matrix.

By associating the interactions between the leading diagonal parameters, i.e., the off-diagonal terms, with a coding system that indicates whether the interaction is one of amplification or attenuation, the consequences of any mechanism pathway through the matrix can be assessed. All the pathways can be identified by using a straightforward search algorithm once the matrix has been constructed. After giving the interactions appropriate values, the overall amplification or attenuation of a pathway can be determined as the product of the interaction values that make up the path. As an example of this procedure, we will consider the design of one type of rock engineering project for which the issue of risk identification and evaluation is supremely important: an underground repository for radioactive waste disposal. The design of such a repository has a number of characteristics which, taken together, make it a unique type of rock engineering project.

1   The repository has a very long design life (in the order of thousands of years, compared to the typical 120 years for conventional civil engineering projects).
2   The function of the repository is that an unacceptable quantity of radionuclides should not migrate to the biosphere. The main migration pathway is water flow through the network of rock fractures.
3   The repository is located at a significant depth, e.g., 500 m, in comparison to other shallower civil engineering works.
4   The rock mass is the protection for the inner barriers around the waste canisters, and so maintaining its integrity is important.
5   The access and repository tunnels could be left open for periods up to 50 years during the pre-closure phase of the operation.

The hazards associated with such a repository fall into three categories: during construction; during the depositional phase; and post-closure. It follows that there has to be a consideration of the impact of construction on rock mass properties so that the construction does not adversely affect the operational and post-closure circumstances. In other words, it is the geomechanical conditions *after construction* that should be suitable, not those *before construction* as determined in the site investigation. There-fore, in terms of geomechanical hazards associated with construction there are two issues: is the construction process itself hazardous?; and will construction adversely affect the rock mass integrity for the repository operation and function? This leads to four evolutionary stages in the hazard assessments related to the rock mass and the repository: 1. the baseline pre-construction natural rock mass system; 2. the construction phase, which has inevitable and additional construction-specific effects on the stability of the rock mass; 3. the operational period, which will include the emplacement of some significantly heat generating waste, backfilled with bentonite; and 4. the long-term phase, and the corresponding performance and safety assessments.

For the identification and evaluation of hazards, the first three of these can usefully be considered via sequential application of the interaction matrix approach, i.e., the development of the baseline matrix, then the effect of construction, and finally the effect of operation. The first step is to generate a basic interaction matrix with the key subjects on the leading diagonal, and to code it according to the importance of the interactions. This coding could be performed on the basis of expert knowledge, the results of specific numerical analysis, or the results of field studies performed, for example, in a dedicated

Underground Research Laboratory. Other forms of coding could be to accept a degree of variability or uncertainty in the individual significance values, and then perform the corresponding probabilistic (i.e., Monte Carlo) or fuzzy analyses.

For the illustrative *and simplified* case example here, consider the seven leading diagonal variables of stress, intact rock, fractures, water flow, temperature, ground water chemistry and introduced perturbations, thus extending the four variable matrix in Figures 3.4 and 3.7. The two terms of temperature and ground water chemistry have been included because they are critical for this particular engineering application. In Figure 3.13, the interaction matrix is shown together with some key words highlighting the interactions. The matrix is coded with the importance of the interactions in Figure 3.14.

ROCK STRESS	Rock fracturing	Normal and shear displacement	No direct effect	No direct effect	Minimal direct effect	No direct effect
Stiff rocks attracts stress	**INTACT ROCK**	Fracture type	Primary permeability	No effect	Mineral content	No direct effect
Perturbation of stress state	No direct effect	**FRACTURES**	Secondary permeability	Minimal direct effect	Fracture filling	No direct effect
Effective stress	Erosion	No direct effect	**WATER FLOW**	Heating or cooling	Dilution or concentration	No direct effect
Coefficient of linear expansion	Increased alteration rate	No direct effect	Convection	**TEMPER-ATURE**	Solubility	No direct effect
No direct effect	Solubility, alteration, replacement	No direct effect	No direct effect	No direct effect	**GROUND-WATER CHEMISTRY**	No direct effect
Significant changes	EDZ	EDZ	Alteration of hydraulic head	Heat source or sink	Ionic changes	**PERTUR-BATIONS**

Figure 3.13 An illustrative *and simplified* 7 × 7 interaction matrix for underground radioactive waste disposal. The values of the off-diagonal terms in Figure 3.14 show a coding for the importance of the interactions—on a scale from 1 (minimally important), through 2 (low), 3 (medium), 4 (high) to 5 (extremely important). These values are appropriate for the demonstration here: further study could reveal either that they need amending, or that a scale with greater resolution is required.

ROCK STRESS	5	5	I	I	I	I
2	**INTACT ROCK**	2	2	I	3	I
4	I	**FRACT-URES**	5	I	2	I
3	2	I	**WATER FLOW**	4	5	I
2	2	I	3	**TEMPER-ATURE**	4	I
I	2	I	I	I	**GROUND-WATER-CHEM-ISTRY**	I
5	2	3	4	5	2	**PERTUR-BATIONS**

Figure 3.14 Illustrative 7×7 interaction matrix (from Figure 3.13) with coded interactions.

Higher stage pathways, i.e., greater than just the single interaction via an off-diagonal box, can take many forms, but the main defining issues are whether

- pathways should begin and end on the same leading diagonal variable (e.g., and with reference to a 4×4 matrix as in Figure 3.4, $x_4 \rightarrow x_2 \rightarrow x_3 \rightarrow x_4$) or
- whether different leading diagonal variables can be used (e.g., $x_1 \rightarrow x_2 \rightarrow x_4 \rightarrow x_3$) for the beginning and end of a path, and
- whether multiple instances of a given interaction are allowed (e.g., $x_1 \rightarrow x_2 \rightarrow x_1 \rightarrow x_2$).

Here, we have elected to consider pathways that begin and end on the same leading diagonal variable, and that pass through only one instance of a given interaction. As a first analysis, pathways comprising just two interactions are considered: the histogram of pathway intensities is shown in Figure 3.15. In this case of a pathway with only two interactions, the pathways with maximum intensity could have been established by inspection, since they are simply given by the product of two interactions each with a value of 5. In general, however, the pathways corresponding to a particular pathway intensity value are not so easily identified.

The pathways in Figure 3.15 with intensity values of 25 and 20 are listed in Table 3.3. A study of this Table reveals that the temperature and groundwater

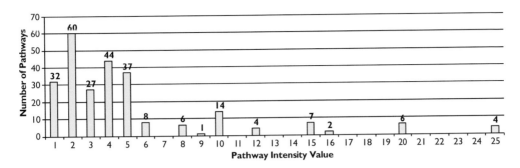

Figure 3.15 Number of pathways for each intensity value for all 2-interaction pathways in Figure 3.14.

Table 3.3 Two-interaction pathways with intensities of 25 and 20 from the matrix in Figure 3.14.

Pathway intensity = 25

1-3-4	Rock stress affects the Fractures, which affects the Water Flow
3-4-6	Fractures affects the Water Flow, which affects the Groundwater Chemistry
7-1-2	Perturbations affects the Rock Stress, which affects the Intact Rock
7-1-3	Perturbations affects the Rock Stress, which affects the Fractures

Pathway intensity = 20

1-3-1	Rock Stress affects the Fractures, which in turn affects the Rock Stress
3-1-2	Fractures affects the Rock Stress, which affects the Intact Rock
3-1-3	Fractures affects the Rock Stress, which affects the Fractures
3-4-5	Fractures affects the Water Flow, which affects the Temperature
7-4-6	Perturbations affects the Water Flow, which affects the Groundwater Chemistry
7-5-6	Perturbations affects the Temperature, which affects the Groundwater Chemistry

chemistry appear in two of the pathways. *Arguably it is these mechanisms—with which most geotechnical engineers have less familiarity—that might be omitted in the usual forms of civil engineering hazard analysis.* Similar tables can be produced for all the pathways associated with the specific pathway intensity values illustrated in Figure 3.15.

We now consider pathways with three interactions. In the histogram shown in Figure 3.16, the pathway intensity values have been plotted with class intervals of 5. The highest values are 2 pathways with intensity value of 125, and 4 pathways with intensity value of 100. The distribution of pathway intensities in this histogram is somewhat similar to those shown in Figure 3.15, with a large number of small values and a small number of high values. It is tempting to say that the distribution tends towards a negative exponential distribution as the number of pathway interactions increases, but this observation requires formal testing.

In Table 3.4, the six pathways with intensity values of 100 and 125 are listed, noting that these are the highest values of the 1470 pathways. The mean of the 1470 pathway intensity values is 10.5, with 1041 pathways having intensities below the

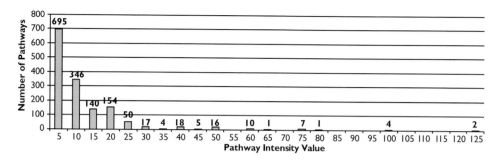

Figure 3.16 Number of pathways for each intensity value for all 3-interaction pathways in Figure 3.14.

Table 3.4 Three-interaction pathways with intensities of 100 and above from the matrix in Figure 3.14.

Intensity = 125

1-3-4-6	Rock Stress affects Fractures, which affects Water Flow, which affects Groundwater Chemistry
7-1-3-4	Perturbations affects Rock Stress, which affects Fractures, which affects Water Flow

Intensity = 100

1-3-1-2	Rock Stress affects Fractures, which affects Rock Stress, which affects Intact Rock
1-3-4-5	Rock Stress affects Fractures, which affects Water Flow, which affects Temperature
3-1-3-4	Fractures affects Rock Stress, which affects Fractures, which affects Water Flow
7-1-3-1	Perturbations affects Rock Stress, which affects Fractures, which affects Rock Stress

mean value. However, for an engineering scheme to be stable and successful, **all** these pathways must be acceptable within the engineering design, construction and operation processes. This leads to the concept that the many pathways with small intensity value are automatically accommodated within the engineering, but that the fewer pathways with high intensity values are the potential hazards.

By introducing the concept of importance, or significance, of the mechanism interactions, we have shown how sequences of interactions, represented by pathways through an interaction matrix, can be quantified in terms of their overall significance. By using a search algorithm to comprehensively identify all pathways of a given length, the distribution of pathway intensity values can be produced and the most critical pathways identified. These can then be subjected to further, more rigorous, analysis. The subject of the application of RES to radioactive waste disposal is further explained through the case examples presented later in this Chapter, and the validation of computer codes in the radioactive waste context is discussed in Chapter 5.

3.3.4   Step-by-step evolution of the interaction matrix

We have now seen how the RES interaction matrix is established, how the off-diagonal mechanisms can be coded for significance, how a *Cause–Effect* plot can be constructed, and how mechanism pathways can be constructed through the matrix. We now demonstrate what happens when the matrix is 'switched on', i.e., when all the matrix mechanisms are simultaneously operating, and how this leads via a step-by-step evolution to the concept of a fully-coupled model in which an off-diagonal mechanism between two variables is then no longer just the binary mechanism in isolation, but the global mechanism taking into account that all the mechanisms in the matrix are simultaneously operating.

To establish the matrix behavioural modes when all the mechanisms are operating simultaneously, a computer simulation can be developed (Hudson & Hudson, 1993). Three computer case examples are presented in this sub-Section; these illustrate a decay of energy peaks as energy is dispersed throughout the system, to more complex modes involving energy patterns, energy oscillations and energy instabilities.

The approach considers that each of the leading diagonal boxes in the interaction matrix is associated with an energy potential E_i. (This E_i refers to energy and is not the Effect discussed earlier). Via the mechanisms in the off-diagonal boxes, energy is transferred from the ith leading diagonal box, having energy E_i, to the jth leading diagonal box, having energy E_j (i refers to the matrix row number; j refers to the matrix column number). For a full matrix, this means that energy can be transferred from all leading diagonal boxes to all other boxes. However, energy is only transferred from a higher potential to a lower potential. One can think of an analogy where there are vertical pipes arising from the leading diagonal terms (with the height of water in them indicating the potentials) and horizontal pipes with one-way flow connecting the bases of all vertical pipes to each other. These horizontal pipes have different diameters so that the energy transfer rates between the leading diagonal boxes are different. Thus, for a matrix with N leading diagonal terms, or N vertical pipes, there are $N(N-1)$ off-diagonal terms, or $N(N-1)$ horizontal one-way flow pipes with different diameters.

Given a set of initialised potentials along the leading diagonal, the transfer of energy from one leading diagonal box to another, i.e., ΔE_{ij}, in an increment of time is given by:

$$\Delta E_{ij} = q_{ij} \cdot k_{ij}(E_i - E_j) \quad \text{or} \quad \Delta E_{ji} = q_{ji} \cdot k_{ji}(E_j - E_i), \text{ whichever is positive} \tag{3.1}$$

where the q_{ij} are damping coefficients and the k_{ij} are the energy transfer coefficients (or related to the diameters of the pipes in the analogy). Equation 3.1 represents the increment of energy being transferred from one leading diagonal box to another and that this depends on the difference in potentials, the ease with which the energy can be transferred, and the damping—through which energy is 'lost'.

There are many modifications that could be made to this basic model; however, the intent here is to explain and demonstrate the concept of 'switching on' the whole matrix and the engineering implications of construction. One aspect that is particularly interesting is to consider the two separate but complementary effects: the effect of a leading diagonal term on the system; and the effect of the system on the leading diagonal term (analogous to the C and E ordinates described earlier). Considering the row and column through a leading diagonal parameter P_i in the interaction matrix, for a given time increment of energy transfer, the sum of all the energy increments in the **row** through P_i represents all the energy leaving P_i, whereas the sum of all the energy increments in the **column** through P_i represents all the energy being transferred to P_i. These provide (C_i, E_i) energy transfer co-ordinates that allow one to plot the system changes. The value of $C_i + E_i$ is a measure of the total energy transfer occurring via a particular P_i. $C_i - E_i$ is a measure of the parameter's energy effect. When $C_i - E_j$ is positive, P_i is acting as a source; when $C_i - E_j$ is negative, P_i is acting as a sink. These (C_i, E_j) co-ordinates can be plotted as in Figures 3.10 and 3.11.

A computer program was developed not only to calculate the energy changes that could occur within the interaction matrix but also to generate displays that would help in the understanding of the resulting matrix behavioural modes. There are five main components to the display, Figure 3.17. Component A of the screen display (the matrix at the top left) shows the basic interaction matrix (in all three of the case examples the matrix has dimension 12 × 12). The parameters are located along the leading diagonal of the matrix from the top left to the bottom right. For the purposes of the simulation, the values of the parameters have been initialised at values between 0 and 100 units.

The actual values of the parameters at any particular time are indicated by the colour scheme—in equal increments from blue, representing a value between 0 and 10, through various shades of purple, to red for parameter values between 90 and 100 (see the Plates in the colour section at the end of the book). This scheme is self-scaling if the 0–100 range is exceeded by the activity of the matrix. The off-diagonal terms represent the energy transmission coefficients, the k_{ij} in Equation 3.1 and are colour-coded with 10 green shades: dark shades of green represent low energy transfer coefficients; light shades of green represent high coefficients. Thus, the leading diagonal colours will change as the simulation proceeds but the off-diagonal terms will not change because they represent the fixed k_{ij} energy transfer values (the diameters of the pipes).

Component B of the screen display (the vertical bar graph at the bottom left in Figure 3.17) also shows the energy values of the 12 leading diagonal terms. The colour-coding of these bars is exactly the same as those in the diagonal of the matrix.

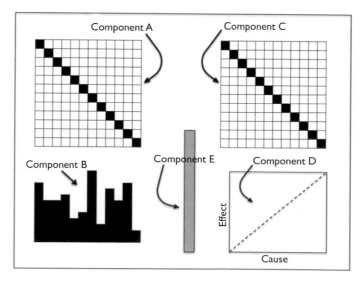

Figure 3.17 Components of the computer simulation output display (see text for explanations of the components).

This vertical bar chart display has been included to provide an easier interpretation of the dynamic distribution of parameter energy values. Note that in Cases A and B of the simulation, shown in Figures 3.18 and 3.19, the leading diagonal values have been initialised with a peak at the centre of the matrix and a linear drop off to the two corners. However, in Case C, shown in Figure 3.20, all the leading diagonal parameters have been initialised with a value of 20 except for the 12th parameter which has been initialised with a value of 100, representing a sudden large energy source in the rock engineering system. This is evident from the bar graph display. Many of the simulations involve quite complex patterns and 'eddy currents' in the matrix and the

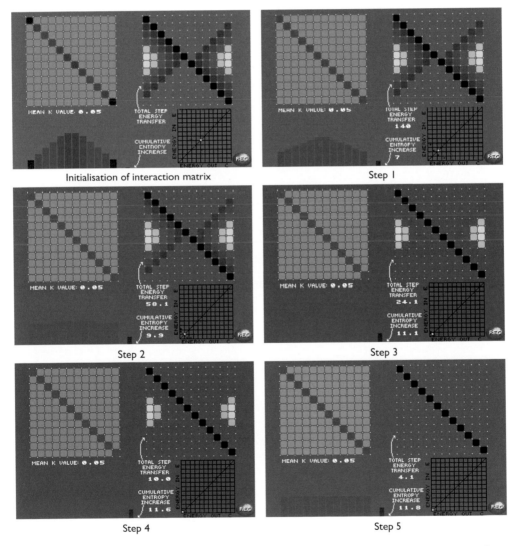

Figure 3.18 Computer simulation of step-by-step energy transfer within the interaction matrix—Case A: regular attenuation.

vertical bar graph display provides a clearer overview of the overall energy transmission pattern.

Component C of the screen display (the matrix at the top right in Figures 3.18–3.20) represents the energy increments being transferred from one leading diagonal parameter to another during each time step. It represents all the energy transmission increments that are about to take place in the next step and therefore can be directly correlated with the values of the leading diagonal parameters in the matrix to the left. Because of Equation 3.1 and the fact that energy is only transmitted from a high-value

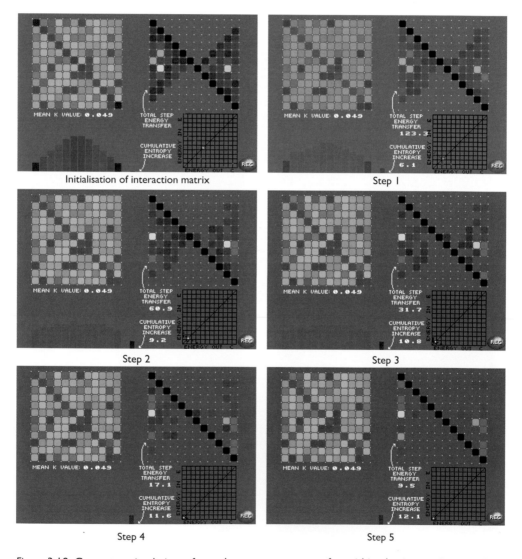

Figure 3.19 Computer simulation of step-by-step energy transfer within the interaction matrix—Case B: irregular attenuation. Note that colour versions of Figures 3.19 and 3.20 can be found in the colour plate section at the end of the book.

parameter to a low-value parameter, half of the off-diagonal terms will be coloured to represent energy flow: if energy is flowing from parameter P_i to P_j, it will not then be flowing from P_j to P_i. Thus, only one of the complementary pairs of the off-diagonal boxes, P_{ij} and P_{ji} will be occupied. The leading diagonal boxes are coloured black to indicate that energy is not considered to be flowing within any leading diagonal box itself. The off-diagonal boxes are coloured from deep brown to bright yellow on a self-scaling equal increment basis (between the highest and lowest energy transfer values) as the simulation proceeds, i.e., separately scaled for each increment of energy transfer. Below a very low energy transfer value, the boxes cease to be coloured.

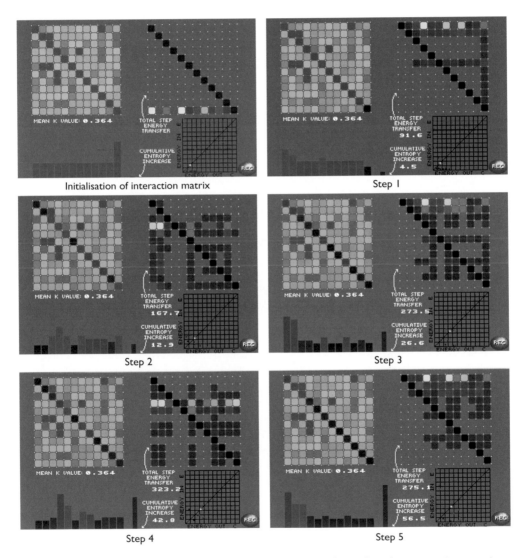

Initialisation of interaction matrix

Step 1

Step 2

Step 3

Step 4

Step 5

Figure 3.20 Computer simulation of step-by-step energy transfer within the interaction matrix—Case C: chaotic behaviour. Note that colour versions of Figures 3.19 and 3.20 can be found in the colour plate section at the end of the book.

Component D of the screen display (the graph in the bottom right of the display in Figures 3.18–3.20) is a graph of the 'energy out' *vs.* 'energy in' for each parameter, i.e., the energy *Cause–Effect* plot. As described earlier, from the matrix C above this graph, the total energy leaving a parameter is found by the sum of the increments in the matrix row through the parameter. The total energy arriving at a parameter is found from the sum of the increments in the matrix column through the parameter. These points have also been colour coded on an absolute system according to $C_i + E_i$, i.e., the parameter's total energy interaction intensity for that particular increment. The colour coding is from bright red for the highest value of $C_i + E_i$ to blue for the lowest values of $C_i + E_i$. Also, the diagonal line representing $C_i = E_i$ is shown. The yellow spot on this line is the mean of all the C_i values and the mean of all the E_i values—which are equal through the conservation of energy. Thus, the position of the brown spot along the $C_i = E_i$ line is a direct indication of how much energy in total is being transferred throughout the matrix during a particular time increment.

Finally, component E of the display (the blue vertical bar originating at the lowest centre of the display in Figures 3.18–3.20) is termed the entropy bar. In these simulations, all the damping coefficients, q_{ij}, have been given a value of 0.95 for illustrative purposes, i.e., energy transmission occurs with an efficiency of 95%. The remaining 5% of energy is put into the entropy bar and is lost in what could be stress waves, hysteresis, heat losses, etc. What was previously useable energy has now been converted into a form that can no longer be used directly: this process is termed entropy and occurs with all real mechanisms. The more energy that is being moved around the matrix, the higher the entropy bar will rise. In Cases A and B of the simulation (Figures 3.18 and 3.19), there is fairly rapid attenuation of the system and little rise in the entropy bar. However, in Case C (Figure 3.20), chaotic behaviour is exhibited and the entropy bar can be seen to rise to a significantly higher level. The numerical values of the total energy transfer for each matrix increment and the entropy level are also given on the screen displays.

Thus, with these five main components, together with the two numerical values, the behaviour of the matrix, given its fundamental off-diagonal structure and the initialisation of the leading diagonal terms, can be followed clearly. Three examples have been chosen to demonstrate both the operation of the simulation and the modes of engineering behaviour that can be exhibited, especially when the mean energy transfer coefficient is significantly increased.

In Case A, shown in Figure 3.18, all the energy transfer coefficients, the k_{ij}s in Equation 3.1, have the same relatively low value of 0.05 and the leading diagonal values have been initialised linearly from 0 at the top left of the matrix to a peak of 100 in the central two boxes and linearly back to 0 at the lower right of the matrix. Note that, because all the energy transmission coefficients are the same, the off-diagonal boxes have a uniform shade of green in the top left matrix. Case A illustrates the steady redistribution of the energy to the final stage when the leading diagonal parameters are the same. This is demonstrated by all five of the output screen display components as the simulation proceeds from Step 1 to Step 5 in Figure 3.18.

In Case B, Figure 3.19, the values of the leading diagonal terms have been initialised at the same values as in Case A but, instead of the energy transmission coefficients being the same, they have random values between 0 and 0.1 with a mean of 0.05 (which is the same value for all the k_{ij}s in Case A). Thus, they have the same mean value as in Case A, but they are statistically variable following a uniform distribution. As might be expected, the behavioural mode of attenuation is essentially the same as for Case A, except for the perturbations introduced by the random values of the energy transmission coefficients. The latter are clearly visible as the mosaic of green colours in the top left of the displays in Figure 3.19 as compared to the uniform shade of green in the top left matrix of Figure 3.18. Also, comparison of all the other components of the displays in Figure 3.19 vs. Figure 3.18 shows the symmetry of the displays broken up because of the varying k_{ij}s.

Case C, shown in the displays in Figure 3.20, has a totally different behavioural mode. Here the k_{ij} values are also uniformly random but ranging from zero to 0.7 with a mean of 0.35, a much higher value than the previous two cases. With these values of k_{ij}, larger amounts of energy can be transferred. The values of the leading diagonal parameters were initialised, all with a value of 20 except for the 12th parameter which has a value of 100. The energy transmission from this 'spike' in the 12th box does not attenuate gradually into the matrix, as was the case with the central peak in Cases A and B. As the simulation proceeds, it can be seen that there are oscillatory and chaotic modes. Within the context of the water analogy, one is reminded of the behaviour of the Lorenz waterwheel (Gleick, 1987) which has a steady rotational behaviour when the water flow is low but becomes chaotic, oscillating backwards and forwards, when the water flow is at a higher level. The model is also similar to chemical reaction rate models discussed in Prigogine and Stengers (1985) in which all is peaceful until certain reaction rates are exceeded.

One of the keys in interpreting the energy transfer behaviour is to study the yellow spot on the $C = E$ line in the graph of the incremental energy transfers at the lower right of the displays in Figures 3.18–3.20. The yellow spot location is proportional to the total energy transfer that is occurring with each step. Note that, through the five steps in the Case C simulation shown in Figure 3.20, the yellow spot does not monotonically move to the origin of the 'energy out–energy in' graph. Thus, this is one example of chaotic behaviour being manifested when the energy transfers are high and is reminiscent of rapidly changing weather in a meteorological system where the energy transfer values are also high.

The three step-by-step simulation cases in Figures 3.18–3.20 represent gradual attenuation, irregular attenuation and chaotic behaviour. The chaotic behaviour arises from a simple set of rules which then lead to an unpredictable outcome and the possibility of major system changes. It is instructive to study such behavioural modes through computer simulation to understand the range of activity that one could expect and also to learn how to control the system in order to achieve the engineering objectives, e.g., by grouting to reduce a hydrogeological leading diagonal parameter effect.

* * * * *

In the first part of this Chapter we have reviewed the Rock Engineering Systems (RES) methodology including the generation of a binary interaction matrix for a particular project, how the off-diagonal components of the matrix can be coded using the expert semi-quantitative method, the development of a *Cause–Effect* plot to indicate the dominance/subordinacy of the main factors, the concept of mechanism pathways through the matrix, how the matrix can evolve through energy transfer steps and how a fully-coupled interaction matrix can be developed to represent the matrix being 'switched on'. The main use of RES in the risk context is the production of the

Cause–Effect plot which indicates the significance of the different factors and how engineering actions can mitigate the effects induced by adverse factors—in other words studying the ways in which engineering activities can be adjusted to reduce risk.

In the second part of this Chapter, we present the essences of a variety of published RES applications in the period 1996–2014, all of which, with one exception, use the basic binary interaction matrix outlined in Sections 3.3.1 and 3.3.2. These examples demonstrate the value of the RES approach, especially when a risk index is required for a particular rock engineering situation.

3.4 EXAMPLES OF ROCK ENGINEERING SYSTEMS (RES) APPLIED TO ROCK MECHANICS AND ROCK ENGINEERING DESIGN

In the 20+ years since the publication of the RES book (Hudson, 1992), the approach has been used for studying a wide variety of systems, not only in rock engineering but also in other subjects. These applications, from the 1990s to 2014 are now described in sections covering surface blasting, natural and artificial slope stability, underground blasting, tunnel boring machines, underground support, siting facilities and radioactive waste disposal. Additionally, some applications in other fields are briefly described because of the relevance of the techniques developed to rock engineering. The issue of risk is relevant in all of these applications.

3.4.1 Natural and artificial surface rock slopes

In this sub-Section, the applications of RES to surface blasting and both natural and artificial slope stability are described.

3.4.1.1 Surface blasting

Latham and Lu (1999) developed a RES application for the quantitative assessment of the blastability of rock masses for surface mass blasting. The authors point out that, "The failure to promote blast design tools beyond rules of thumb might have resulted from the fact that the influence of *in situ* rock properties, discontinuity structures and their interactions are often too difficult to be quantitatively isolated and identified." They also note that, "The problem of obtaining a satisfactory measure of blastability from an assessment of numerous potentially influential factors has at least three features which have often been neglected in early attempts to investigate blastability. One is the interactions between factors. Another is the degree of influence (or the weighing) to be attributed to each factor or coupled factors. A third is the need to treat subjective data, a situation often encountered in geotechnical engineering with systems of soils, rocks, fluids and discontinuities."

The parameters and associated units that the authors used for their 12×12 interaction matrix are given in Table 3.5. The resultant *Cause–Effect* plot indicated that the blasting system has a low/medium intensity (because the parameters were found to be nearly half way along the $C = E$ line). Also, all the parameters had essentially equal dominance/subordinacy (because they were clustered around the $C = E$ line. By analysing the relative contributions of the parameters, they created a Blastability Index (BD).

Table 3.5 List of parameters and their units used by Latham and Lu (1999) for the RES development of a Blastability Index.

Parameters and units for RES Blastability Index

1   Strength (Uniaxial compressive strength, *UCS*, MPa, also via Point Load Index)
2   Resistance to fracturing (Uniaxial tensile strength, σ_t, MPa)
3   Sturdiness of the rock (Density, ρ, t/m^3)
4   Elasticity of rock (Young's modulus, *E*, GPa)
5   Resistance of rock to dynamic loading (P-wave velocity, V_p, km/s)
6   Hardness of rock (Schmidt rebound hardness value, *SHV*, rebound height scale)
7   Deformability (Poisson's ratio, *v*, dimensionless)
8   Resistance of rock to breaking (Fracture toughness, K_{Ic}, MPa-m$^{1/2}$)
9   *In situ* block sizes (Mean of block size distribution, mean)
10 Fragility of rock mass (Fractal dimension of rock block sizes, *D*)
11 Integrity of rock mass (Ratio of field: lab P-wave velocities, R_v, dimensionless)
12 Fracture plane's strength (Cohesion, *c*, MPa and friction angle, ϕ, degrees)

The greater the value of *BD*, the more difficult the rock is to blast. This blastability assessment system was applied to a case study at a highway improvement cutting site in North Wales, UK. The range in parameter interaction intensity was found to be quite wide, so only those factors contributing to a total of 72.5% of the (*C* + *E*) in the ordered histogram, that is, the eight parameters, P1, P2, P3, P4, P5, P6, P9, P10 in Table 3.5 were chosen as the main contributory factors to the blastability of the rock masses at the site. Applying the measured values of the parameters on site, the *BD* indicated that the rock masses are, in general, difficult or moderately difficult to blast.

In their article, Latham and Lu point out that their work is related to "uncontrollable factors governed by *in situ* geological conditions and the term 'blastability' has been deliberately restricted to quantify this intrinsic resistance of the rock mass". In their scheme, the emphasis is on the intact rock properties, which would be expected because the purpose of blasting is to reduce the natural block size distribution to the required fragment size distribution. Note that in Table 3.5 Parameters 3, 4, 5 and 7 are intact rock properties not directly related to failure, but are included because they characterise the quality of the rock. Parameter 6 is more strongly correlated with the failure properties, and Parameters 1, 2 and 8 are direct measurements of rock strength. In terms of the pre-existing fracturing in the rock mass, Parameters 9, 10 and 11 are indicators of the degree of fracturing present in the rock mass, but the only parameter explicitly representing the fracture failure properties is Parameter 12, the Mohr–Coulomb values. However, the art in applying engineering rock mechanics principles and the RES approach to rock engineering design is to adopt a pragmatic approach by successfully capturing the essence of the problem without introducing unnecessary complications—which the authors have done. Moreover, the use of a blastability index such as the one described in their paper, reduces the risk of adopting an inappropriate blasting scheme. For a full description of this RES application, the reader is referred to the philosophy and full case study presented in Latham and Lu (1999).

In a later publication (Faramarzi *et al.*, 2012) describe the development of a RES-based model for risk assessment and prediction applied to backbreak in ANFO (Ammonium Nitrate–Fuel Oil) bench blasting. The term 'backbreak' refers here to the extension of damaged rocks beyond the last row of production holes in a surface mine. The authors explain that such "backbreak is an adverse phenomenon in rock blasting operations, which causes safety reduction due to the instability of mine walls, high dilution, increasing loading and hauling costs, poor fragmentation, increasing water inflow due to fractured rock, and uneven burden in subsequent blasts".

Based on the work by Benardos and Kaliampakos (2004a), which is described later in this Chapter, a Vulnerability Index (*VI*) was developed for predicting backbreak and to analyse the associated risk encountered during surface blasting. Faramarzi *et al.* (2012) explain that the first step is to identify the parameters that are responsible for the occurrence of risk in the case of backbreak, analyse their behaviour, and evaluate the significance (weight) that each one has in the overall risk conditions. They chose the 16 parameters listed in Table 3.6. In the second step, a Vulnerability Index, *VI*, can be determined, using Equation 3.2.

$$VI = 100 - \sum_{i=1} a_i \frac{Q_i}{Q_{max}} \qquad (3.2)$$

where a_i is the weighting of the *i*th parameter, Q_i the value (rating) of the *i*th parameter, and Q_{max} is the maximum value assigned to the *i*th parameter (as a normalisation factor). Based upon the estimated *VI* (expressed on a 0–100 scale), the level of backbreak risk can be identified. The authors also use a Backbreak Index, $BBI = (1 - VI)$.

Using the parameters in Table 3.6, Faramarzi *et al.* (2012) constructed a 16×16 interaction matrix and coded the off-diagonal terms using the ESQ method. The resultant *Cause–Effect* plot is shown in Figure 3.21. Note that, because the mean of all the *Cause* values is the mean of all the coded off-diagonal terms in the interaction matrix and the mean of all the *Effect* values is also the mean of all the coded off-diagonal terms in the interaction matrix, the mean *Cause* equals the mean *Effect*, i.e., the centre of gravity of all the parameter points in Figure 3.21 must lie on the $C = E$ diagonal line, which is the dashed line in Figure 3.21.

Interpreting this *Cause–Effect* plot, we note that the system has a relatively low total interactivity (the mean of the parameter points is less than a third of the way along the dashed diagonal. The three most dominant parameters (*Cause* >> *Effect*)

Table 3.6 The 16 parameters used by Faramarzi et al. (2012) for the RES evaluation of blasting breakback at a surface mine. The reader is referred to the authors' paper for a detailed description of these properties.

Parameter			
P_1	Burden	P_9	Time delay
P_2	Maximum instantaneous charge	P_{10}	Discontinuities orientation to face
P_3	Last row powder factor to total powder factor	P_{11}	Velocity of detonation
P_4	Powder factor	P_{12}	Blasthole deviation
P_5	S/B ratio	P_{13}	RMR
P_6	S_T/B ratio	P_{14}	Blasthole inclination
P_7	Number of rows	P_{15}	Hole diameter
P_8	Stiffness ratio (H/B)	P_{16}	B/D ratio

are 13, 10, 15, i.e., RMR (Rock Mass Rating), Discontinuity Orientation, and Hole Diameter, respectively. The three most subordinate parameters are (*Effect >> Cause*) are 3, 2, 9, i.e., Powder Factor Ratio, Maximum Instantaneous Charge, and Time Delay, respectively. This is a significant conclusion—because it indicates that the breakback phenomenon itself is mainly dominated by the rock conditions, with the blasting parameters having less significance. In other words, in this application, the engineer has less control on the (breakback) outcome.

Faramarzi *et al.* (2012) used the method described to predict backbreak and the level of risk corresponding to each blast for 30 blasts carried out at Sungun copper mine, western Iran. They state that, "the results obtained were compared with the backbreak measured for each blast, which showed that the level of risk achieved is consistent with the backbreak measured.", see Figure 3.22. Thus, the RES methodology

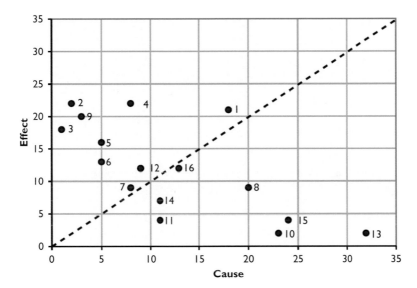

Figure 3.21 RES *Cause–Effect* plot used for estimating blasting breakback, from Faramarzi *et al.* (2012), see Table 3.6 for parameter identification.

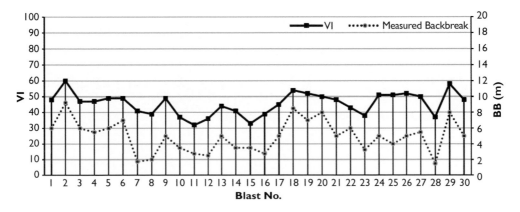

Figure 3.22 Variation in backbreak measured and Vulnerability Index (*VI*) for different blasts at Sungun copper mine.

and the vulnerability index, VI, can be used for blasting breakback risk assessment at surface mines using the technique described in which 16 parameters were used—a mixture of rock conditions and blasting factors.

3.4.1.2 Natural slopes

One of the earlier RES papers on natural rock slope stability set the scene for a variety of subsequent rock slope stability papers based on the same approach: this paper was "A comprehensive method of rock mass characterisation for indicating natural slope instability" by Mazzoccola and Hudson (1996). The purpose of this paper was to use the RES methodology to develop a new rock mass classification sensitive to large-scale instabilities in natural slopes and hence suitable for indicating unstable slopes, the area under study being in the Italian Central Alps. The possibility of the occurrence of large landslides in these regions limits land-use and threatens existing urban centres.

For natural slopes, analysis is often difficult because of a lack of data, geological complexity, the scale of the instability phenomena and the high number of interacting factors. So, in order to be able to have a structured approach to such complexity, the RES approach was adopted. Following this approach, 19 parameters relating to the general environment and to the rock mass characteristics were chosen, see Figure 3.23. Their causes and effects were analysed in order to weight each parameter according to its degree of interactivity in the system. The rock mass instability index developed takes into account the variability of the parameter values for different slopes when assigning ratings to different classes of the parameter values. In parallel, a predictability rating was computed, according to the presence in the field of a number of 'indicators of instability'. Both indices allow discrimination of critical slopes, and were found to be in good agreement with field evidence.

The rock types in the study area are mainly represented by two lithologies: a porphyritic, massive granitic gneiss and different rock types of sedimentary origin. The high degree of landslide hazard in the area results mainly from the interaction between two major factors: the morphogenetic agents and the nature of the structural features. The morphogenetic agents result from the combined action of uplifting and erosion. The structural features are linked to both high tectonic disturbance and post-glacial stress relief which cause the opening of wide and persistent fractures, frequently already sheared, which in turn induce deep-seated gravitational movements.

A 20×20 interaction matrix was compiled and coded using the ESQ method described earlier in this Chapter. The 20 leading diagonal factors chosen to characterise this natural slope stability circumstance were: 1–geology; 2–folds; 3–faults; 4–rainfall; 5–freeze and thaw cycles; 6–previous instability; 7–intact rock strength; 8–weathering; 9–number of fracture sets; 10–fracture orientation; 11–fracture aperture; 12–fracture persistence; 13–fracture spacing; 14–fracture mechanical properties; 15–rock mass strength; 16–hydraulic conditions; 17–slope orientation; 18–slope dimensions; 19–*in situ* stress; and 20–potential instability (this last one being the factor being studied). The 20 parameters are plotted in Figure 3.23

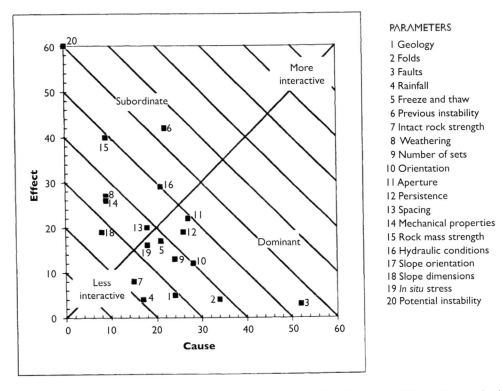

Figure 3.23 The *Cause-Effect* plot using the co-ordinates established from the ESQ coding method described earlier in the Chapter.

according to their resultant (*Cause, Effect*) co-ordinates using the method described earlier.

At this stage, having defined the relative interactive intensity as a measure of the significance of the parameters, the actual parameter values must come into play and a more detailed data input is needed from the field. The parameter values were chosen from a 'pulldown menu'. The list of the 19 relevant parameters was used in the field to collect data on 20 slopes, located in the Cimaganda rockslide area. Note that the parameter 'potential instability' is, of course, not used and so the number of indicator parameters is 19. Some parameters were described qualitatively; others were described quantitatively. For this reason, it was not possible to utilise the actual parameter values directly to compute an instability index, but a rating was assigned to different classes of parameter descriptions and values. Three classes of parameter values were set, with ratings of 0 for 'low contribution', 1 for 'contributory' and 2 for 'strongly contributing'.

The Rock Mass Instability Index was then defined in Equation 3.3 as

$$RMII_j = \sum_{i=1}^{19} a_i P_{ij} \qquad (3.3)$$

where i refers to parameters 1 to 19, j refers to the slopes (from 1 to 20), a_i is the $C + E$ scaled value for each parameter and P_j is the rating assigned to different classes of parameter values and is different for different slopes (i.e., the jth slope). A Predictability Rating (PR) was also developed—more detail being given in Mazzoccola and Hudson (1996). This method of establishing a RES 'vulnerability' index, which is explained in detail in Hudson (1992) and Mazzoccola and Hudson (1996), has been used in many of the case examples for different applications in this Chapter.

Rozos *et al.* (2008) used RES to rank the instability potential of natural slopes in Karditsa County, Greece, and hence to develop a method of zoning landslide risk in the area. The authors already had general data on 388 case studies and specific data for 224 of these failure sites on the main dimensions, width and length, of the area affected by landslide activity. A predictive index was required because the landslides affect both urban and cultivated areas, as well as engineered structures. By far the greatest frequency of landslides at this location occurs in flysch and molasses formations, with the remainder occurring in schist–cherts and in transition zone beds, with most landslides being of the rotational type.

Interaction matrix														
P1	0	3	3	0	1	3	1	3	2	4	4	3	27	
1	**P2**	2	0	0	3	3	0	4	0	3	4	4	24	
0	0	**P3**	1	0	2	1	0	1	0	2	2	4	13	
0	1	2	**P4**	1	2	2	1	0	0	2	2	3	16	
1	0	3	3	**P5**	0	4	0	3	1	4	0	3	22	
2	0	3	1	0	**P6**	2	0	2	0	0	3	3	16	Cause-C
0	0	1	1	0	1	**P7**	0	1	0	1	1	3	9	
3	0	3	1	2	0	2	**P8**	2	2	4	2	4	25	
1	0	2	0	0	3	2	1	**P9**	0	4	2	2	17	
1	3	2	0	0	1	2	0	2	**P10**	1	3	1	16	
0	0	3	3	3	2	3	1	3	0	**P11**	3	3	24	
0	0	2	2	0	2	3	2	3	2	1	**P12**	4	19	
0	0	0	0	0	0	0	0	0	0	0	0	**P13**	0	
9	4	26	15	6	17	27	6	24	5	26	26	37	228	
Effect-E														

P1 = Lithology	P2 = Rainfall	P3 = Slope inclination	P4 = Slope orientation
P5 = Geometry of main discontinuities	P6 = Human intervention on vegetation	P7 = Human intervention on slope geometry	P8 = Tectonic regime
P9 = Geomechanical action of water	P10 = Altitude	P11 = Geological structure	P12 = Thickness of weathering mantle
			P13 = Potential instability

Figure 3.24 Interaction matrix for evaluating natural rock slope instability with Expert Semi-Quantitative (ESQ) coding of the off-diagonal terms, from Rozos *et al.* (2008).

The thirteen parameters selected for the interaction matrix leading diagonal by Rozos *et al.* (2008) were lithology, rainfall, slope inclination, slope orientation, geometry of discontinuities, tectonic regime, altitude, geological structure, geomechanical action of water, thickness of weathering mantle, human intervention on slope geometry, human intervention on vegetation—plus potential instability of the slope. Each of the first 12 parameters was separated into five categories representing specific conditions and a number ranging from zero to four was assigned for each category, with the 0 category representing the most stable conditions and the 4 category the most potentially unstable conditions. The details of these categories are given in Table 1 of Rozos *et al.* (2008).

The matrix interactions were then coded (Figure 3.24) using the Expert Semi-Quantitative (ESQ) method with values of 0 to 4. The *Cause + Effect* values were used as weighting coefficients, which express the proportional share of each parameter (as a failure-causing factor) in slope failure. The resultant *Cause–Effect* plot is shown in Figure 3.25. Note that this plot is similar to the Type 4 case in Figure 3.9, i.e., having similar parameter intensities but variable parameter dominances.

It can be seen from Figure 3.25, that the most dominant parameters are 1, 2, 5, 8 and 10, i.e., lithology, rainfall, geometry of discontinuities, tectonic regime and altitude (where 'altitude' here means whether the condition is 'plain', 'semi-hilly', 'hilly', 'semi-mountainous', and 'mountainous'). Based on the *Cause* and *Effect* values, Rozos *et al.* (2008) created a slope Instability Index. The authors found that, not only did the Instability Index provide an indication of a slope's instability potential *per se*, but it also indicated the size of the area likely to be affected by the consequential landslide—leading to the related implications for land use and development planning processes in landslide susceptible areas.

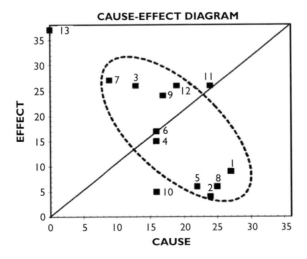

Figure 3.25 The *Cause–Effect* plot for the coded matrix in Figure 3.24, from Rozos *et al.* (2008).

In a paper on analysing earthquake-induced slope instability for the purpose of risk assessment, Castaldini *et al.* (1998) used the RES approach in a case study from the Northern Apennines in Italy. They used a multidisciplinary approach incorporating tectonics, seismology, geology, hydrogeology, geomorphology, and soil/rock mechanics. Their study area was Montese in the Garfagnana region of the Appennines which has been subjected to large earthquakes from historical times to the present. Castaldini *et al.* (1998) provide considerable information on the regional geology, the historical earthquake record, plus the rock mass and landslide characteristics. They used a comprehensive 21 × 21 interaction matrix with the following leading diagonal factors: lithology, active faults, inactive faults, *in situ* stress, seismic magnitude, seismic ground response, slope orientation, slope dimension, previous instability, rainfall, intact rock strength, weathering, number of discontinuity sets, orientation of discontinuities, aperture of discontinuities, persistence of discontinuities, spacing of discontinuities, mechanical properties of discontinuities, rock mass strength, hydraulic conditions, and potential instability. Their coded matrix is shown in Figure 3.26.

From their *Cause–Effect* plot for this RES case described by Castaldini *et al.* (1998), they found that the dominant variables, i.e., those furthest to the right of the $C = E$ diagonal with $C \gg E$, are active faults, inactive faults, lithology, intact rock strength and the orientation of discontinuities. In order to assess the level of potential instability of the slopes as induced by earthquakes, Castaldini *et al.* (1998) explain that they "defined a Rock Mass Instability Index (*RMII*), in such a manner that the higher the index value is, the more critical the slope will be. The relative importance of each parameter is expressed by the sum of *Cause* and *Effect*, which, in its turn, is expressed in terms of the percentage of the total $(C + E)$ and scaled in a way that, when all ratings are equal to a maximum value of 2, the maximum possible *RMII* is 100. Having scaled both the value of this sum a_j for each parameter and the rating of each parameter for each slope, the *RMII* can be computed according to the formula: $RMII_i = \Sigma a_j \times P_{ij}$ where *i* refers to the slope number, *j* refers to the parameter number (1 to 20); a_j is the scaled sum (*Cause + Effect*) for each parameter; and P_{ij} is the rating assigned to each parameter."

They also note that, "The distribution of *RMII* values usually seems to show three main portions: an upper part which groups slopes with similar values; a middle part where the values more or less gradually decrease; and a lower portion where the values stabilised around minimum values." They found that, for the non-seismic case, there were the three classes of low (L), intermediate (I) and high (H) relative proneness to instability with the *RMII* values: 0–42; 43–59; >60, and that, with the consideration of seismicity in the area, the values are similar but some slopes then become classified in the H region.

The authors comment in their paper that the parameters in the interaction matrix need to be easily detectable through field work or simple laboratory tests and that the ESQ coding method may be too subjective, saying that it would be improved by using mathematical relations in the off-diagonal boxes of the matrix. Indeed, this latter suggestion has been considered by the current authors, possibly by including the necessary relations and then taking the Laplace transform of these in order to eliminate time from all the off-diagonal relations. Another alternative is to use a neural network approach.

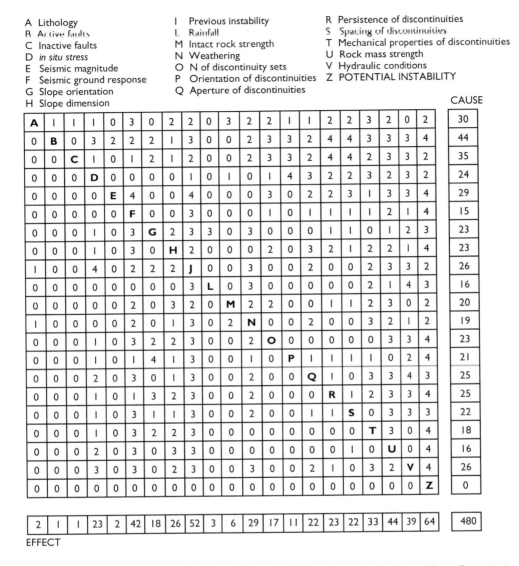

A Lithology
B Active faults
C Inactive faults
D *in situ* stress
E Seismic magnitude
F Seismic ground response
G Slope orientation
H Slope dimension

I Previous instability
L Rainfall
M Intact rock strength
N Weathering
O N of discontinuity sets
P Orientation of discontinuities
Q Aperture of discontinuities

R Persistence of discontinuities
S Spacing of discontinuities
T Mechanical properties of discontinuities
U Rock mass strength
V Hydraulic conditions
Z POTENTIAL INSTABILITY

CAUSE

EFFECT

Figure 3.26 The interaction matrix of Castaldini *et al.* (1998) coded using the Expert Semi-Quantitative (ESQ) method with the range 0–4, and indicating the *Cause* and *Effect* co-ordinates.

Zhang *et al.* (2004) describe a RES application to rockfall hazard assessment on the Chengdu-Lhasa highway in China. The authors explain that, "In the area under investigation, rockfall hazard analysis is complicated by the extensive distribution of blocks and boulders, lack of data, complex geological structure, difficult site conditions and the high number of interactive factors" and that, "the highway section is characterised by deep valleys, steep slopes, fractured and weathered rocks, extensive areas of debris accumulation, and high rates of geomorphological evolvement ... as a consequence, rockfall events in this area are usually unexpected, uncertain and frequent ... rockfall hazard assessment is of important significance not only for rockfall hazard mitigations along the highway connecting inland China and Tibet but also for the planning of the Chengdu–Lhasa Railway line...".

The authors used three indices to characterise the rockfall hazard: RHI, the Rockfall Hazard Index; RII, the Rockfall Intensity Index; and RFI, the Rockfall Frequency Index—noting that $RHI = RII \times RFI$ and that the RII and RFI values (from 0–1) were obtained via the RES methodology. For the RII interaction matrix, 10 factors were chosen: block geometry, block strength, slope geometry, slope materials, slope vegetation, motion mode, coefficient of restitution, resistance coefficient, rockfall velocity and maximum travel distance. Using the ESQ (0–4) coding method, they established that the five most dominant parameters in their system are slope geometry, block geometry, slope materials, block strength and slope vegetation. For the RFI interaction matrix, four factors were chosen: number of blocks, stability of blocks, rockfall triggering events, and rockfall history. This matrix was also coded using the ESQ (0–4) method.

The method of developing these two indices is further explained in Zhang et al. (2004). Their case study involved studying 19 slopes along the Highway which were divided into five classes according to their Rockfall Hazard Indices from extremely hazardous ($RHI > 0.7$) down to basically not hazardous ($RHI < 0.3$). The authors conclude their paper by indicating that these rockfall hazard indices will provide the scientific basis for optimisation of mitigation costs along the Chengdu–Lhasa Highway.

A paper by Shang et al. (2005) was published on a related topic: an engineering geological zonation for one section of the Sichuan–Tibet Highway in China. The authors explain that, "In linear engineering projects, such as those of the highways and railways in the northern Yarlu-Tsangpu Grand Canyon of Tibet that cross various geological and geomorphological units, engineering geological zonation must be carried out in advance because of complicated and diverse engineering geological conditions." A photograph of adverse slope conditions is shown in Figure 3.27 from Shang et al. (2005). They used the interaction matrix approach to develop an Engineering Geological Zonation Index (EZI) from semi-quantitative analysis of the data from the Basu–Linzhi section of the Sichuan–Tibet Highway and the formula $EZI_j = \sum_{i=1}^{n} a_i p_{ij}$ where j refers to the jth zone or sub-zone, n is the number of factors, a_i is the weighting ratio for factor i and p_{ij} is the rating value of parameter i in zone or sub-zone j. This follows the formula developed by Mazzoccola and Hudson (1996) for rock slope instability.

They used 20 factors in their EZI index approach as the main influential factors for 19 sub-zonations: igneous rock, metamorphic rock, sedimentary rock, talus, gully, valley, gorge, basin, weathering, stream scouring, slope erosion, cut slope, mining activity, cut trees, joint, fault, earthquake, groundwater cycle, water head difference, and discharge. The most dominant factors in their study were fault, igneous rock, metamorphic rock, sedimentary rock and earthquake.

Budetta et al. (2008) also used the RES methodology for landslide hazard zonation along the coastal slopes and cliffs (Figure 3.28), about 118 km in length, of the Cilento region between Agropoli and Sapri in Italy. The major geomorphological,

Figure 3.27 Adverse highway landslide conditions along the Sichuan–Tibet Highway (northern Yarlu-Tsangpu Grand Canyon of Tibet), from Shang *et al.* (2005).

Figure 3.28 Coastal stretch near Agropoli, Italy, affected by recurrent rock-falls (from Budetta *et al.* (2008).

geological and structural features of about 154 slopes and cliffs were studied and the interaction approach used to develop an Instability Index, *II*, (1–100), which was linked to a Geographic Information System. They found that rapid, but small, rock-falls can cause more casualties than moderate speed, but large, slides.

With reference to the type of instability occurring, and Figure 3.29, the authors explain that, "Normally, there is a cyclic instability, which starts itself with early

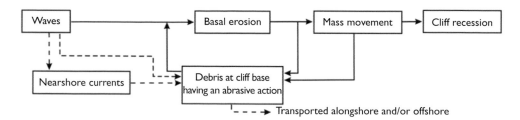

Figure 3.29 Coastal cliff recession system, after Sunamura (1992), modified by Budetta *et al.* (2008).

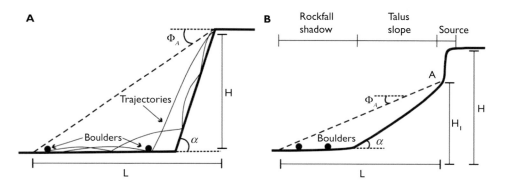

Figure 3.30 Rockfall mechanisms for a cliff (A) and a talus slope (B), from Budetta *et al.* (2008).

detachment, with the landslide debris moving away from the foot of the cliff because of the action of the sea and at times with reactivations: this sequence is strongly influenced by climatic variations even in short times and by the lithological and strength features of the rock masses." The authors used 11 parameters for the cliffs and 12 parameters for the slopes. The items for the cliffs were: cliff height (P1), cliff slope (P2), cliff orientation (P3), attitude of bedding planes (P4), jointing (P5), vegetation (P6), rainfall intensity (P7), groundwater (P8), wave motion (P9), pre-existent instability (P10), man-made structures (P11). For the stony flysch slopes the items are: slope height (P1), slope inclination (P2), slope orientation (P3), attitude of bedding planes (P4), clay fraction (P5), structural complexity (P6), vegetation (P7), rainfall intensity (P8), groundwater (P9), wave-motion (P10), pre-existent instability (P11), man-made structures (P12).

Their instability index, *II*, is calculated using the formula $II_j = \sum_{i=1}^{n} a_i P_{ij}$ expressed as a percentage, where *i* refers to the parameters, *j* to the examined cliffs or slopes (from 1 to *n*); a_i is the C + E value for each parameter, P_{ij} is the code allocated to different classes of values of the parameters which is different for different cliffs (*i*th cliff or slope). Values of the *II* are grouped into three classes indicating low, medium and high landslide hazard. The authors also considered the rockfall travel distance for the cases of a cliff and a talus slope (Figure 3.30).

The coded interaction matrices for the rocky cliffs and the stony flysch slopes are included as Figures 3.31 and 3.32, respectively. The explanation of the 'weights'

Weights	Wi												Cause
5.28	P1	3	0	0	0	0	0	0	0	3	0	3	9
6.22	3	P2	0	0	0	3	0	1	0	3	0	3	13
1.32	0	0	P3	0	0	2	0	0	1	2	0	2	7
3.21	3	2	0	P4	1	1	0	2	0	4	0	3	16
5.28	3	4	0	1	P5	2	0	4	1	4	0	4	23
5.47	0	1	0	0	1	P6	0	2	0	2	0	2	8
2.27	0	0	0	0	0	3	P7	3	0	3	0	3	12
4.72	0	0	0	0	0	3	0	P8	0	2	0	2	7
3.96	4	4	0	0	2	1	0	1	P9	3	0	3	18
9.06	4	4	0	0	1	3	0	2	1	P10	0	4	19
3.21	2	2	0	0	0	3	0	3	0	3	P11	4	17
-	0	0	0	0	0	0	0	0	0	0	0	P12	0
Effect	19	20	0	1	5	21	0	18	3	29	0	33	149

Figure 3.31 Matrix of binary interactions ESQ coded (0–4) for the rocky cliffs (limestones, dolomites, sandstones) of the Cilento area. Key: P1 = cliff height: P2 = cliff slope; P3 = cliff orientation; P4 = attitude of bedding planes; P5 = jointing; P6 = vegetation; P7 = rainfall intensity; P8 = groundwater; P9 = wave motion; P10 = previous instability: P11 = man-made works; P12 = potential instability. From Budetta *et al.* (2008).

Weights	Wi													Cause
5.30	P1	3	0	0	0	0	0	0	0	0	3	0	3	9
5.79	3	P2	0	0	0	0	3	0	1	0	3	0	3	13
1.16	0	0	P3	0	0	0	2	0	0	1	2	0	2	7
3.15	3	2	0	P4	0	3	1	0	2	0	4	0	3	18
4.48	4	3	0	1	P5	4	2	0	4	0	4	0	4	26
4.15	3	3	0	0	0	P6	0	0	3	0	3	0	3	15
4.63	0	1	0	0	0	0	P7	0	2	0	2	0	2	7
1.98	0	0	0	0	0	0	3	P8	3	0	3	0	3	12
4.63	0	0	0	0	0	0	3	0	P9	0	2	0	2	7
2.98	4	4	0	0	0	0	1	0	1	P10	3	0	3	16
8.94	4	4	0	0	1	3	3	0	2	1	P11	0	4	22
2.81	2	2	0	0	0	0	3	0	3	0	3	P12	4	17
-	0	0	0	0	0	0	0	0	0	0	0	0	P13	0
Effects	23	22	0	1	1	10	21	0	21	2	32	0	36	169

Figure 3.32 Matrix of binary interactions coded for the stony flysch slopes (meso-cenozoic flysch deposits) of the Cilento area. Key: P1 = slope height; P2 = slope inclination; P3 = slope orientation; P4 = attitude of bedding planes; P5 = clay fraction; P6 = structural complexity; P7 = vegetation; P8 = rainfall intensity; P9 = groundwater; P10 = wave-motion; P11 = previous instability: P12 = man-made works; P13 = potential instability—from Budetta *et al.* (2008).

is given in Budetta *et al.* (2008). The *Cause–Effect* plots resulting from the coded matrices in Figure 3.31 and 3.32 are shown in Figure 3.33.

The five most dominant parameters ($C \gg E$) for the rocky cliffs are found from the values in Figure 3.31 as jointing, man-made works, wave motion, attitude of bedding planes and rainfall intensity. Similarly, the five most dominant parameters for the instability of the stony flysch slopes are found from the values in Figure 3.32 as clay fraction, man-made works, attitude of bedding planes, wave motion and rainfall intensity. It can be seen that these dominant parameters are the same in the two cases except for the jointing for the rocky cliffs and the clay fraction for the stony flysch slopes—reflecting the material involved.

The method described enabled the calculation of the instability index for the 154 cliffs and slopes being investigated. The authors explain that, "The database was connected to a GIS (ArcView) to obtain the landslide hazard for the coast. Using this system it is possible to check the whole available geological data and the parameters that led to the calculation of the instability index and the relative hazard class ... with GIS it is possible to modify the input parameters, if necessary, in order to obtain, as an output, a map showing the cliff and its varied hazard level." The authors found that, "The study revealed that almost 56% of the coastal area displays high landslide susceptibility, 27% is characterised by fair landslide susceptibility, whereas only 17% is characterised by low landslide susceptibility".

Ceryan and Ceryan, (2008) describe an application of the interaction matrices method for slope failure susceptibility zoning in N.E. Turkey which enabled a slope failure susceptibility map to be created for the Dogankent area. They used eight parameters for the interaction matrix: P1: water conditions, P2: weathering, P3: shear strength parameters, P4: slope angle, P5: vegetation density, P6: distance from faults and shear zones well-developed, P7: discontinuity frequency, P8: previous stability. Of these, they found three dominant parameters: water conditions, fault distance, discontinuity frequency.

Rozos *et al.* (2011) continued their work using RES as described in "Comparison of the implementation of rock engineering system and analytic hierarchy process

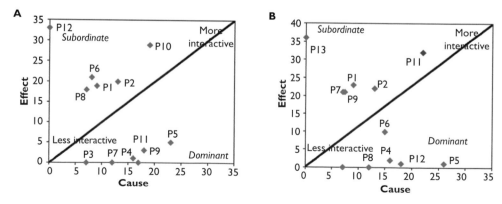

Figure 3.33 Cause–Effect plots resulting from the interaction matrices in Figures 3.31 and 3.32. (A) rocky cliffs, (B) stony flysch slopes (from Budetta *et al.* (2008)).

methods, upon landslide susceptibility mapping, using GIS: a case study from the Eastern Achaia County of Peloponnesus, Greece". In this paper, they compared the Rock Engineering System (RES) and the Analytic Hierarchy Process (AHP). This latter process is a semi-quantitative, multi-objective and multi-criteria decision making methodology (Saaty, 1990, 2006) comprising "the analytical hierarchy of involved parameters and the comparison between the various pairs of them for the assignment of a relevant ratio for each parameter ... it can estimate the weight of each parameter according to their preference, through the linear correlation of each one relative to the others. This is achieved by means of relevant correlation of them in pairs, as they are shown in a relative matrix, regarding the landslide vulnerability of the area."

For the study, landslides in the North Eastern part of Achaia County were examined. Rozos *et al.* (2011) describe the work as follows. "Ten parameters were used in both methodologies, and each one was separated into five categories ranging from 0 to 4, representing their specific conditions derived from the investigation of the landslides in the western part of the study area (ranking area). A layer map was generated for each parameter, using GIS, while the weighting coefficients of each methodology were used for the compilation of RES and AHP final maps of the eastern part of the study area (validating area). By examining these two maps, it is revealed that even though both correctly show the landslide status of the second site, the RES map reveals a better behaviour in the spatial distribution of the various landslide susceptibility zones." In the ranking area there are 277 recorded landslides; in the validating area there are 270 recorded landslides.

The two sites are shown in Figure 3.34; the ten principal parameters in Figure 3.35; and the ESQ coded (0–4) RES interaction matrix in Figure 3.36. The five most dominant parameters in this system ($C \gg E$) are lithology, geometry of main discontinuities, altitude, rainfall, and slope aspect.

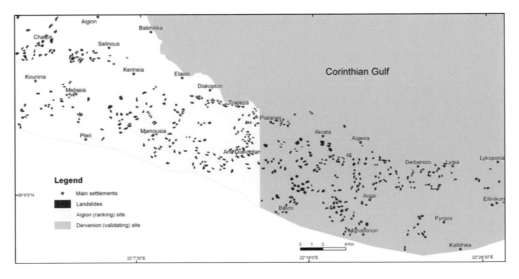

Figure 3.34 Map showing the landslide distribution in the separated two parts of the study area: the Aigion (ranking) site and the Dervenion (validating) site, Greece, from Rozos *et al.* (2011).

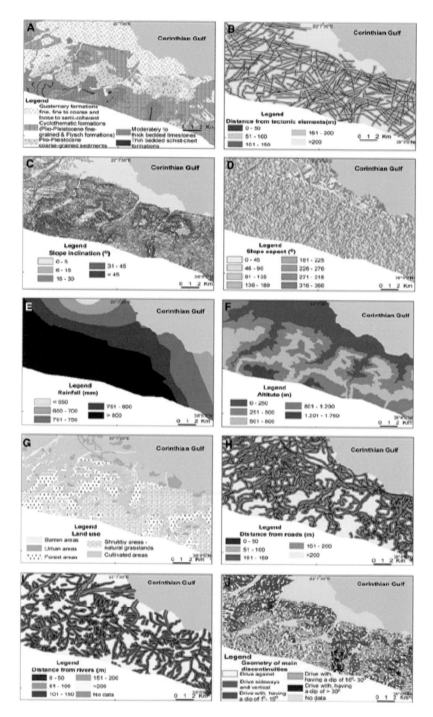

Figure 3.35 The thematic layers of the ten principal parameters involved in the study by Rozos *et al.* (2011): A: lithology, B: distance from tectonic lineaments, C: slope angle, D: slope aspect, E: rainfall, F: altitude, G: land use, H: distance from roads, I: distance from rivers, J: geometry of main discontinuities.

INTERACTION MATRIX											
P1	2	3	3	0	2	4	1	2	2	3	22
0	P2	3	1	0	1	0	2	3	2	3	15
1	2	P3	1	0	0	4	3	2	2	4	19
0	0	2	P4	1	0	2	1	1	1	3	11
2	2	3	0	P5	0	1	3	3	1	4	19
1	1	2	0	4	P6	3	2	1	0	1	15
0	1	0	0	0	1	P7	1	1	1	3	8
0	1	0	1	1	0	0	P8	0	1	3	7
1	1	0	2	3	0	2	0	P9	1	3	13
1	3	3	3	1	1	2	3	3	P10	3	23
0	0	0	0	0	0	0	0	0	0	P11	0
6	13	16	11	10	5	18	16	16	11	30	152

CAUSE-C

EFFECT-E

P1 = LITHOLOGY	P2 = DISTANCE FROM TECTONIC	P3 = SLOPE	P4 = SLOPE ASPECT
P5 = RAINFALL	P6 = ALTITUDE	P7 = LANDUSE	P8 = DISTANCE FROM ROADS
P9 = DISTANCE FROM RIVERS	P10 = GEOMETRY OF MAIN DISCONTINUITIES		P11 = POTENTIAL INSTABILITY

Figure 3.36 The ESQ coded interaction matrix used by Rozos *et al.* (2011).

Regarding the spatial development of the landslide susceptibility zones, their percentages of the total area from the RES map (Figure 3.37) are: 7% for the 'very low' zone, 23% for the 'low' zone, 39% for the 'medium' zone, 25% for the 'high' zone and 6% for the 'very high' zone. Although the relevant percentages are similar for the corresponding AHP map, as previously indicated the authors state that, "The comparison of these maps revealed that the RES method gives better results regarding the spatial distribution and the concentration of the most important susceptibility zones, i.e., depicting better the various zones in the most prone to landslide sites as its approximation helps in the elimination of any false judgment."

The work described in KhaloKakaie and Zare Naghadehi's (2012) paper ranks the rock slope instability potential of the Khosh-Yeylagh Main Road, in Iran using a 17 parameter interaction matrix. Figure 3.38 shows the nature of the roadside rock mass.

The authors' interaction matrix is shown in Figure 3.39 and the levels of their Instability Index in Figure 3.40, this index having been generated using the Mazzoccola and Hudson (1996) method. The five most dominant parameters in this 17-parameter system are faults and folds, geology and lithology, discontinuity orientation, rainfall, and number of discontinuity sets.

In their paper, "A probabilistic systems methodology to analyze the importance of factors affecting the stability of rock slopes", Naghadehi *et al.*, (2011) describe a PESQ coding methodology in which uncertainties in the assignments of the ESQ coding for the interaction matrix are expressed using probabilities, together with the related analysis of rock slope stability in the Khosh-Yeylagh region of Iran (Figure 3.41). They used nine parameters and found that the existence of previous instabilities was the most important parameter, directly indicating the importance of a site survey of

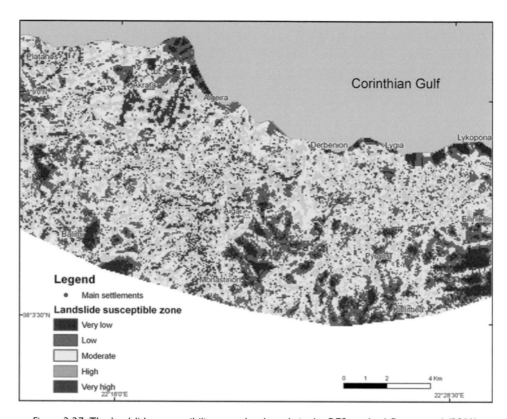

Figure 3.37 The landslide susceptibility map developed via the RES method, Rozos *et al.* (2011).

Figure 3.38 Nature of the rock mass along the Khosh-Yeylagh main road in Iran (from KhaloKakaie & Zare Naghadehi, 2012).

P1	2	3	3	3	2	0	0	2	2	2	2	2	3	0	0	3	29	Cause
2	P2	3	0	1	4	4	4	4	4	3	3	2	2	0	0	3	39	
0	0	P3	0	1	0	0	2	0	1	2	1	3	3	0	1	2	16	
0	0	2	P4	3	2	0	0	0	1	2	0	2	2	0	0	2	16	
0	0	3	2	P5	0	0	0	0	0	2	0	0	0	0	0	3	10	
0	0	3	0	2	P6	0	0	0	0	0	3	2	2	0	4	4	20	
0	0	3	0	0	0	P7	2	2	2	1	2	2	3	0	3	4	24	
0	0	3	0	3	0	0	P8	2	2	3	4	0	1	0	0	4	22	
0	0	3	0	2	0	0	1	P9	1	2	3	3	2	0	2	4	23	
0	0	3	0	1	0	0	1	1	P10	0	3	1	0	0	0	4	14	
0	0	2	0	0	0	0	0	0	0	P11	0	0	2	0	0	4	8	
0	0	3	0	3	0	0	2	1	0	3	P12	0	0	0	2	3	17	
0	0	3	0	0	0	0	0	1	1	0	2	P13	2	0	0	3	12	
0	0	4	0	0	0	0	0	0	0	0	0	0	P14	0	0	4	8	
0	0	3	0	2	0	0	0	0	0	2	4	0	0	P15	2	3	16	
0	0	3	0	3	0	0	3	2	0	1	2	0	0	0	P16	3	17	
0	0	0	0	0	0	0	0	0	0	0	0	0	0	0	0	P17	0	
2	2	44	5	24	8	4	15	15	14	23	29	17	22	0	14	53		
Effect																		

P_1: Geology and lithology; P_2: Faults and folds; P_3: Previous instability; P_4: Intact rock strength; P_5: Weathering; P_6: Number of sets; P_7: Orientation; P_8: Aperture; P_9: Persistence; P_{10}: Spacing; P_{11}: Mechanical properties; P_{12}: Hydraulic conditions; P_{13}: Slope height; P_{14}: Slope inclination; P_{15}: Rainfall; P_{16}: Freeze and thaw cycles; P_{17}: Potential instability

Figure 3.39 The 17 × 17 interaction matrix developed by KhaloKakaie and Zare Naghadehi (2012) for studying rock slope instability potential along the Khosh-Yeylagh Main Road, Iran.

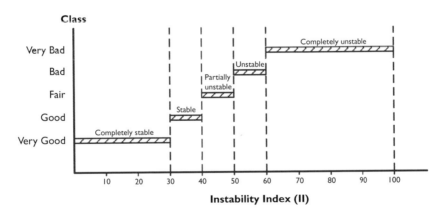

Figure 3.40 Levels of the slope Instability Index *II* used by KhaloKakaie and Zare Naghadehi (2012).

similar slopes in the area. The authors also interpreted probabilistically the degree of parameter dominance or subordinancy.

Ten parameters were used: P1: Geology and lithology; P2: Faults and folds; P3: Previous instabilities; P4: Intact rock strength; P5: Weathering; P6: Mechanical properties of discontinuities; P7: Hydraulic conditions; P8: Slope height; P9: Slope inclination; P10: Potential instability. The Expert Semi-Quantitative (ESQ) coding method was extended by including coding probabilities, i.e., to the PESQ coding method. The authors explain that, "...instead of assigning a unique (and deterministic) coding

Figure 3.41 Example of the roadside rock mass along the Khosh-Yeylagh Main Road in a mountainous area approximately 90 km north of Shahrood City, north-eastern Iran (from Naghadehi *et al.*, 2011).

value to each interaction, probabilities are assigned for each interaction to represent the likelihood of each possible coding value considered (from 0 to 4 in this case)". They continue, "This can be expressed by five matrices (M0 to M4, one for each code value from 0 to 4), where the off-diagonal elements of each matrix contain the probabilities for occurrence of that particular code for that particular interaction." This means that a parameter does not plot as a single point in the *Cause–Effect* plot, but as a distribution, as the example in Figure 3.42. Also, the values of dominance and subordinacy can be expressed as a mean and standard deviation, Figure 3.43. The five most dominant parameters in this system are geology and lithology, faults and folds, intact rock strength, weathering, and slope height.

A method similar in purpose to the RES approach but using a somewhat different approach has been presented by Wang *et al.* (2012), together with a case study conducted in Guizhou Province, China. Their system is, "a weighting method, integrating subjective weight with objective weight, for landslides susceptibility mapping based on geographical information system (GIS)". For parameters, the causal factors they use are: the landslide inventory, aspect, slope, proximity to streams of drainage network, proximity to railway, proximity to road, topography, elevation, lithology, tectonic activity and annual precipitation. The authors use the objective weights of these calculated according to the landslide area density based on an entropy weighting method. A fuzzy number weighting approach was also used to assess the sub-classes of each key factor. A map was then created, classifying the study area into the four categories of landslide susceptibility: low, moderate, moderate-high, and high.

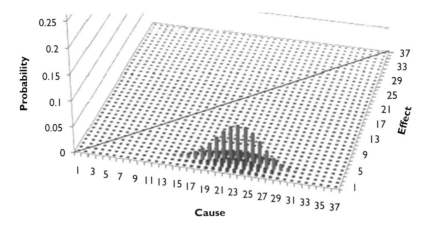

Figure 3.42 Use of the probability assignments causes a parameter to plot as a distribution, rather than a point, in the *Cause–Effect* plot. This example is for the dominant parameter P1, geology and lithology, in the RES analyses by Naghadehi *et al.*, (2011).

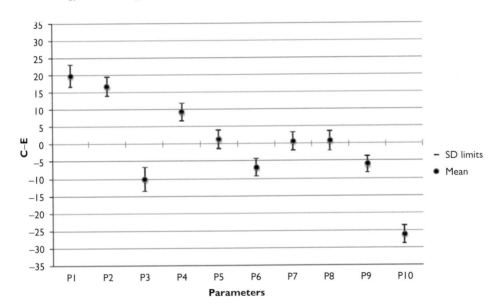

Figure 3.43 Probabilistic plot of parameter dominance, above the *C–E* zero line, and parameter subordinacy, below the *C–E* zero line, from Naghadehi *et al.* (2011). P1: Geology and lithology; P2: Faults and folds; P3: Previous instabilities; P4: Intact rock strength; P5 Weathering; P6: Mechanical properties of discontinuities; P7: Hydraulic conditions; P8: Slope height; P9: Slope inclination; P10: Potential instability.

As an alternative to RES, Sabatakakis *et al.* (2012) developed an ambitious landslide susceptibility zonation for Greece based on historical archives (1365 landslides) and using GIS aided mapping techniques. They used the ten landslide 'predisposing factors (predictors)': lithology, slope angle, elevation, hydrographic network-drainage density, rainfall, climate, seismicity, land use, road network density and population density. The authors explain that, "the correlation between the landslide locations and predictor classes was analysed by using the Landslide Relative Frequency, and R-mode factor analysis was applied to study the inter-relations between predictors (independent variables) while weighting coefficients were determined. The resulting landslide susceptibility map was verified using a data set of 375 new landslide locations." Sabatakakis *et al.* (2012) found that the five most significant predictors for the landslide phenomena are rainfall, climate, elevation, land use and lithology.

3.4.1.3   Instability of artificial rock slopes

An early use of a method similar to RES was used by Nathanail *et al.* (1992) to assess the stability of open-cast coal mine slopes at the Ffos Las site in Wales, UK. This was an exercise in the design and management of excavated slopes in highly deformed Coal Measures strata located in the severely tectonically deformed western part of the South Wales coal basin. Uncertainty in ground conditions was due to the geological complexity caused by Variscan deformation. A set of indicators, capable of providing an early warning of impending adverse slope behaviour, was developed—leading to a 'Ffos Las Alarm for Instability Hazard'. Eight slope failures were examined to test the success of the predictive indicators.

The work utilised the 12×12 generic interaction matrix for slopes (Hudson, 1992). The back analysis of previous major mass movement events at the site for comparison with the 'Alarm' prediction was analysed in terms of the following indicators: rainfall, old workings, geological structure, rate of geological change, geological complexity, major shear zone/fault, tight syncline, joint persistence, joint friction, joint waviness, post-peak behaviour, adverse discontinuities, pore pressure, rock mass strength, tension cracks, depth, slope orientation, slope extent, failure type, whether progressive failure, volume, coal lost and remedial action. Thus, by using this set of indicators, each of eight previous slope failures was examined to test the success of the 'Alarm' as a predictive indicator for proposed and newly excavated slopes; and by linking the complex geological circumstances with engineering activities, the 'Alarm' system provided an early warning of impending adverse slope behaviour.

In their paper, "Rock mass characterisation to indicate slope instability at Bandarban, Bangladesh: a Rock Engineering Systems approach" by Ali and Hasan (2002) the authors used the Rock Mass Instability Index (*RMII*) method with the 14 parameters of lithology, folding, rainfall, previous instability, rock strength, weathering, slope orientation, slope height, slope angle, compaction, rock discontinuities, vicinity to faults, and hydraulic conditions in the interaction matrix. This *RMII* was then applied to 25 road-cut slopes in the hilly areas of south-eastern Bangladesh where landslides have become more common because of anthropogenic activities such as road-cuts, quarries and mass scale cultivation.

The authors' 14×14 interaction matrix is included as Figure 3.44 and the resulting *Cause–Effect* plot as Figure 3.45. It can be seen from Figure 3.45 that the five most dominant parameters for this system (*Cause >> Effect*) are rainfall, lithology, faults, folds and slope orientation. They used the $RMII_i = \Sigma a_j \times P_{ij}$ Rock Mass Instability Index and found that slopes with a *RMII* value of <40 were stable and those with a value >40 were unstable.

1	2	3	4	5	6	7	8	9	10	11	12	13	14	
Lithology	2	0	2	3	2	0	0	0	2	2	1	1	3	18
0	Folds	0	3	0	0	0	0	0	0	3	2	1	3	12
0	0	Rainfall	4	2	3	0	0	0	2	1	0	4	4	20
0	0	0	Prev. inst.	0	2	0	2	2	3	2	0	1	4	16
0	0	0	2	Rock str.	3	0	0	0	3	2	0	0	2	12
0	0	0	1	2	Weath.	0	0	0	4	3	0	1	3	14
0	0	0	3	0	1	Slope or.	0	0	0	0	0	0	3	7
0	0	0	3	0	2	0	Slope ht	0	0	0	0	0	2	7
0	0	0	1	0	0	0	0	Slp. ang.	0	0	0	1	2	4
0	0	0	3	4	3	0	0	0	Compac.	1	0	2	2	15
0	0	0	4	3	3	0	0	0	3	Discont.	0	3	3	19
0	0	0	2	2	2	2	2	2	1	4	Faults	2	2	21
0	0	0	3	1	3	0	0	0	1	1	0	Hydrau.	2	11
0	0	0	0	0	0	0	0	0	0	0	0	0	Pot. Inst.	0
0	2	0	31	17	24	2	4	4	19	19	3	16	35	176

Figure 3.44 The 14 × 14 interaction matrix for road-cut slopes in Bangladesh coded by the ESQ method with a 0–4 range, from Ali and Hasan (2002).

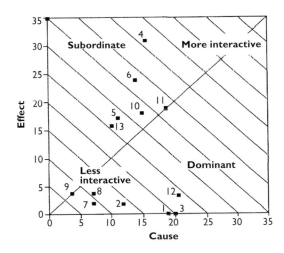

Figure 3.45 The *Cause–Effect* plot for the 14 parameters and ESQ coding shown in Figure 3.44, from Ali and Hasan (2002).

3.4.2   Underground rock engineering

The three subjects described in this sub-Section are underground blasting, tunnel boring machines and underground support.

3.4.2.1   *Underground blasting*

Andrieux and Hadjigeorgiou (2008) discuss a Destressability Index methodology for the assessment of the likelihood of success of a large-scale confined destress blast in an underground mine pillar. The work was aimed at reducing the rock stress in the particular context of large-scale choked destress blasts in mine pillars. The use of the RES approach has led to a destressability index which was applied to back-analyse a fully instrumented large-scale confined destress blast at Brunswick Mine, in Canada. The index indicates whether a given situation is conducive to being destressed by means of a large-scale confined destress blast, and, if so, whether the design of the blast is appropriate to achieve this goal.

The authors explain that, "Destress blasting can be defined as any attempt involving the usage of confined explosive charges (i.e., without free faces) to reduce the ground stresses in a particular region, and in which the blasted material is left in place ... it is the process of using confined explosive charges in order to damage the rock mass, for the purpose of softening its behaviour, reducing its capacity to carry high stresses and, hence, reducing the potential for it to undergo violent failure." For this RES approach to large-scale choked panel destress blasting, and based on much case study information, the authors used and explained in detail the following nine parameters for their interaction matrix: P1, the stiffness of the rock; P2, the brittleness of the rock; P3, the degree of fracturing of the rock mass; P4, the proximity of the rock mass to (static) stress-induced failure; P5, the orientation of the destress blast; P6, the width of the destress blast; P7, the unit explosive energy; P8, the confinement of the explosive charges; and P9, the result of the destress blast. The ESQ-coded interaction matrix is shown in Figure 3.46 and the associated *Cause–Effect* plot is shown in Figure 3.47.

Given that the dominant parameters $(C > E)$ are always below the $C = E$ line, it can be seen from Figure 3.47 that the dominant parameters are P7, P6, P8 and P5, i.e., the unit explosive energy, the width of the destress blast, the confinement of the explosive charges, and the orientation of the destress blast. The authors then developed a 'destressability index' rating, emphasising that this is, "...not as a direct design procedure, but, rather, an assessment of the likelihood of success of a proposed design in a given situation." The methodology was applied to a case study at the Brunswick Mine near Bathurst, NB, in Atlantic Canada. A cross-section through the case study instrumented pillar is included as Figure 3.48.

									Causes
P_1	2	2	2	1	1	2	1	2	13
2	P_2	1	1	1	2	2	2	2	13
1	1	P_3	2	1	2	2	2	3	14
1	1	2	P_4	1	1	3	1	4	14
2	2	2	1	P_5	1	1	1	3	13
3	2	4	2	2	P_6	1	1	3	18
4	3	4	4	1	1	P_7	3	4	24
2	2	3	2	1	2	2	P_8	3	17
3	3	4	4	2	1	2	1	P_9	20
Effects	18	16	22	18	10	11	15	12	24	146

Figure 3.46 The ESQ-coded (0–4) interaction matrix for large-scale choked pillar destress blasts, from Andrieux and Hadjigeorgiou (2008).

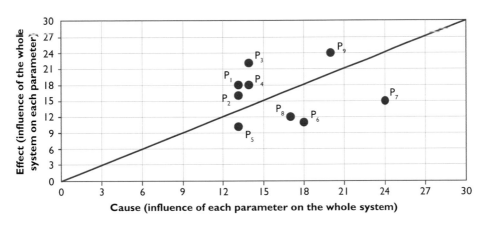

Figure 3.47 Cause–Effect plot for large-scale choked panel destress blasting, from Andrieux and Hadjigeorgiou (2008).

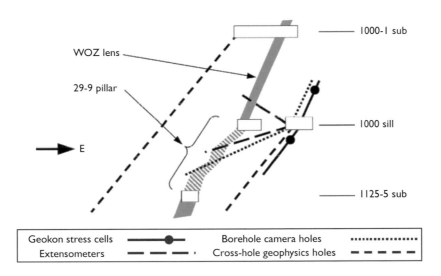

Figure 3.48 Cross-section through the 29-9 pillar at the Brunswick Mine, looking north and showing the projected location of the various instrumentation, from Andrieux *et al.* (2003).

The authors conclude that a particular appeal of the RES approach, "is that it provides a series of easily implemented steps that result in a rational assessment of the likelihood of success of a given destress blast design in a given situation of rock mass conditions and stress regime ... considering that (1) large-scale confined pillar destress blasts are usually a last resort endeavour with no possible second attempt, (2) the cost associated with their implementation is typically substantial and (3) the consequences of failure generally lead to significant ore losses and lost production (these blasts are only considered in the first place when large amounts of valuable ore are at risk, either directly or indirectly), this type of blast deserves sound engineering in order to maximise the likelihood of success." A detailed description of the approach and application is given in their paper.

3.4.2.2   Tunnel Boring Machines (TBMs)

The use of TBMs in rock has run the full spectrum: from high advance rates with no problems to becoming irretrievably stuck and having to be removed or abandoned. Thus, the rock engineering risk theme of this book has a special significance for tunnelling with the "expect the unexpected" refrain having a particular resonance.

The first RES example illustrated here in the tunnelling context has been reported in the paper, "A methodology for assessing geotechnical hazards for TBM tunnelling—illustrated by the Athens Metro, Greece" by Benardos and Kaliampakos (2004a). The authors state that, "The methodology presented in this paper aims at the identification of risk-prone areas, incorporating, at the same time, the uncertainty of ground conditions … the methodology assesses the hazards by introducing the concept of a vulnerability index [Figure 3.49], to identify the weighting of the parameters, and with probabilistic modelling to address the uncertainty in the parameters' values. The proposed model is illustrated via the Athens Metro case study, used also for validating its performance under actual construction conditions." Their risk analysis addressed: face instabilities/collapses and overbreaks; surface settlements; and water inflows. They used the RES approach together with probability assessments for the data values and a GIS application.

They used eight parameters: rock mass fracture degree as represented by RQD—(P1), weathering degree of the rock mass—(P2), overload factor—stability factor (N)—(P3), rock mass quality represented by RMR classification—(P4), uniaxial compressive strength of the rock—(P5), overburden, construction depth—(P6), hydrogeological conditions represented by the water table surface relative to the tunnel depth—(P7), rock mass permeability—(P8). These parameters were rated on a 0–3 scale for each of the 11 locations along the route being studied (note that these are not the interaction matrix ratings which are in Table 3.7), as illustrated in Figure 3.50.

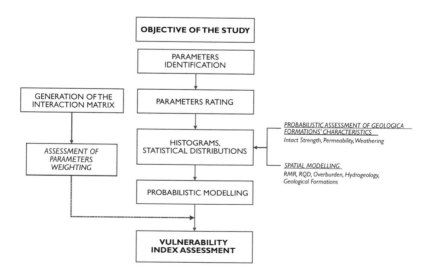

Figure 3.49 Flowchart of the proposed vulnerability assessment methodology, from Benardos and Kaliampakos (2004a).

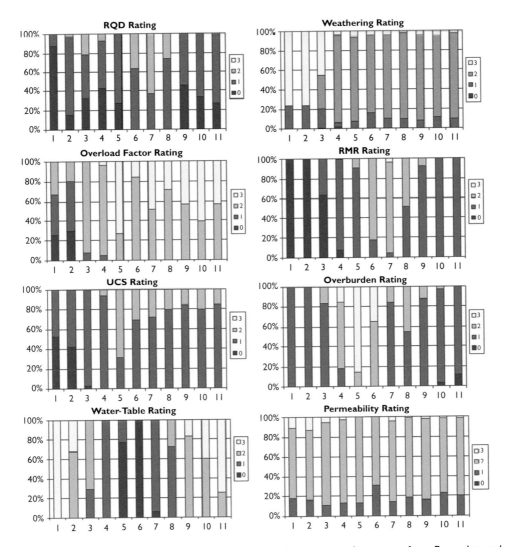

Figure 3.50 Rating of the principal parameters for each of the examined segments, from Benardos and Kaliampakos (2004a).

Each parameter's weighting factor, a_i, was then calculated from Equation 3.4

$$a_i = \frac{(C_i + E_i)}{(\Sigma_i C_i + \Sigma_i E_i)}(\%) \tag{3.4}$$

and the final weighting of the principal parameters are as in Table 3.8. This gives the dominant parameters (highest a_i values) as Rock Mass Rating (RMR), fracture degree (RQD), hydrogeological conditions, and rock mass weathering.

Table 3.7 The RES interaction matrix coded with ESQ ratings (0–4), from Benardos and Kaliampakos (2004a).

Fracture degree	4	1	4	1	1	4	3
2	Weathering	1	4	4	1	2	4
3	1	Overload Factor	2	2	3	0	2
1	2	2	Rock Mass Rating	3	0	2	4
3	1	3	4	UCS	2	0	1
3	1	4	2	1	Overburden	3	1
4	4	2	4	3	1	Hydrogeology	2
1	2	3	2	1	0	3	Permeability

Table 3.8 Final weightings of the principal parameters.

Principal parameter	a_i (%)
Fracture degree—RQD	14.11
Rockmass weathering	13.31
Overload factor—N	11.69
Rockmass rating—RMR	14.52
UCS	11.69
Overburden—construction depth	9.27
Hydrogeological conditions	13.71
Permeability	11.69

Finally, a Vulnerability Index (*VI*) is estimated via Equation 3.5

$$VI = \left(100 - \sum_{i=1}^{n} a_i \frac{P_i}{P_{max}} \right) \tag{3.5}$$

where *VI* (range 0–100) is in the form of a probability distribution and a_i is the weighting of the *i*th parameter in the system, P_i the value (rating) of the *i*th parameter, and P_{max} the maximum value a parameter can take (a normalisation factor). In terms of general categories, *VI* values 0–33 represent low vulnerability, 33–66 medium vulnerability, and 66–100 high vulnerability. So, for each of the 11 tunnel study lengths, the *VI* is expressed as a discrete probability distribution.

The 11 study lengths along the Athens metro tunnelling route are highlighted in Figure 3.51, the cumulative tunnelling progress for each study length is shown in Figure 3.52, and the correlation between the average advance rate and average vulnerability index demonstrated in Figure 3.53. The authors note that, "the mean values of *VI* and *AR* (Figure 3.53) for the 11 examined tunnel segments have a high negative correlation coefficient, about 0.92, signifying the coherent behaviour of the proposed vulnerability index methodology. A companion paper, Benardos and Kaliampakos (2004b), uses an alternative method of predicting tunnel progress for the Athens Metro, but making use of Artificial Neural Networks (ANNs).

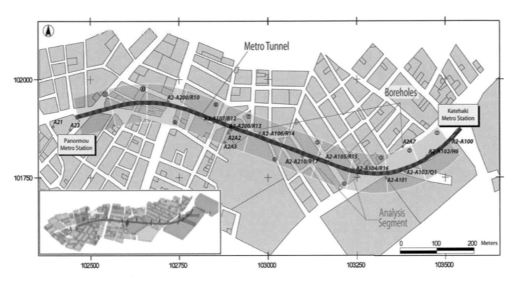

Figure 3.51 The 11 study lengths along the Athens metro tunnelling route, from Benardos and Kaliampakos (2004b).

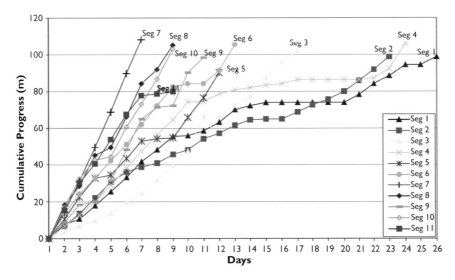

Figure 3.52 The cumulative tunnelling construction progress for each study segment of the Athens Metro, from Benardos and Kaliampakos (2004a).

3.4.2.3 *Tunnel stability*

Kim *et al.* (2008) explain that quantitatively identifying rock behaviour expected in excavating tunnels can assist engineers in selecting the best tunnelling method and support system and in evaluating tunnel stability through numerical analysis adjusted to rock behaviour. They used RES to develop a Rock Behaviour Index (*RBI*) for

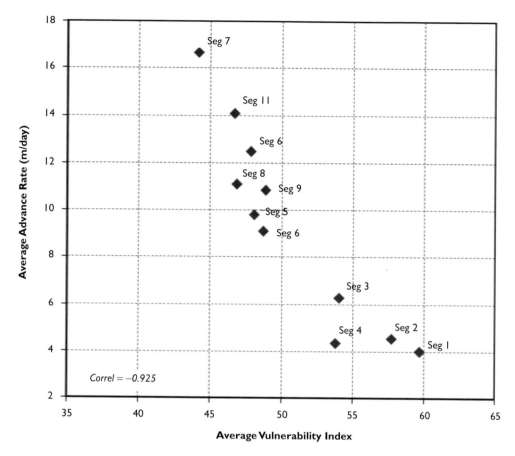

Figure 3.53 Relation between the average *VI* and *AR* values for each tunnel study length, from Benardos and Kaliampakos (2004a).

assessing plastic deformation and ground failure and illustrated the proposed model via a case study on the Seoul Metro Line 9.

Based on work by Cai *et al.* (2004) who provided a list of parameters that should be considered when describing a rock mass and using the results for design purposes, the authors settled on seven parameters influencing the rock behaviour: unconfined compressive strength, Rock Quality Designation (RQD), joint surface condition, stress, groundwater, earthquake, and tunnel span. They coded the interaction matrix by the ESQ method (0–4) and, using a similar method to the previous case examples, created three *RBI* indices: one for rock fall, one for cave-in and one for plasic deformation. These *RBIs* then indicated the potential for these cases according to the scale: 0–20, very low probability; 20–40, low probability; 40–60, moderate probability; 60–80, high probability; 80–100, very high probability.

A cross-section of the Seoul metro is shown in Figure 3.54 and the Rock Behaviour Indices (*RBIs*) for various tunnel sections are listed in Table 3.9. Further information on the development and use of these Indices is given in Kim *et al.* (2008).

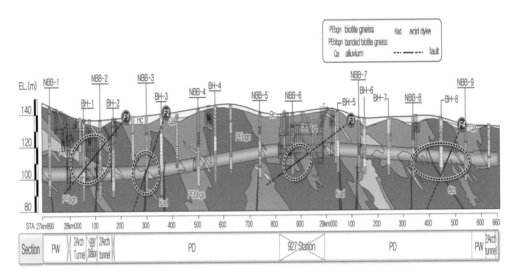

Figure 3.54 Seoul metro geology and tunnel longitudinal section (PD: double lane track section, PW: enlarged section), from Kim *et al.* (2008).

Table 3.9 Rating of parameters and Rock Behaviour Indices (*RBIs*) for various tunnel sections according to the support pattern of the double lane track section, from Kim *et al.* (2008).

Support pattern	Section	P1	P2	P3	P4	P5	P6	P7	RBI_1	RBI_2	RBI_3
PD-4	28,172 ~ 28,182	2	0	2	3	2	2	1	60.00	60.25	56.75
PD-5A	28,182 ~ 28,270	3	2	2	3	3	2	1	45.75	45.00	42.50
PD-4	28,270 ~ 28,344	2	0	1	3	4	2	0	61.75	61.00	58.25
PD-5B	28,344 ~ 28,414	1	1	1	3	4	2	1	55.25	54.75	54.00
PD-5A	28,414 ~ 28,445	2	0	1	3	3	2	1	61.25	61.00	57.50
PD-4	28,445 ~ 28,510	2	2	2	3	4	2	1	44.00	43.25	42.75
PD-5A	28,510 ~ 28,560	1	1	2	2	3	2	1	57.25	57.25	58.00
PD-5A	28,999 ~ 29,105	2	0	1	2	3	2	1	64.50	64.25	62.25
PD-5A	29,195 ~ 29,260	3	2	2	2	3	2	2	44.75	44.25	43.25
PD-3B	29,480 ~ 29,573	2	1	2	2	2	2	1	59.00	59.00	57.75

Shin *et al.* (2009) developed a methodology for quantitative hazard assessment for tunnel collapses based on case histories in Korea. They proposed a Tunnel Collapse Hazard Index (KTH-Index), an index system for assessing the hazard level of collapse at a tunnel face based on a sensitivity analysis of a database containing past collapse cases; 56 sets of such data were utilised. For the sensitivity analysis, the authors used a neural network based technique and RES (Figures 3.55–3.58). The assessment system was applied to the section of the SYK tunnel where large-scale collapses had already occurred and it is found that the predicted hazard levels were in good agreement with the field data already known.

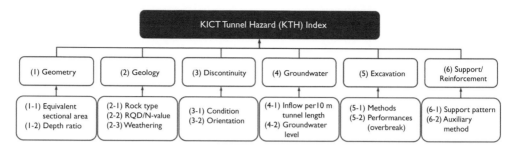

Figure 3.55 The parameters used for the RES interaction matrix in developing the Tunnel Hazard Index, from Shin *et al.* (2009).

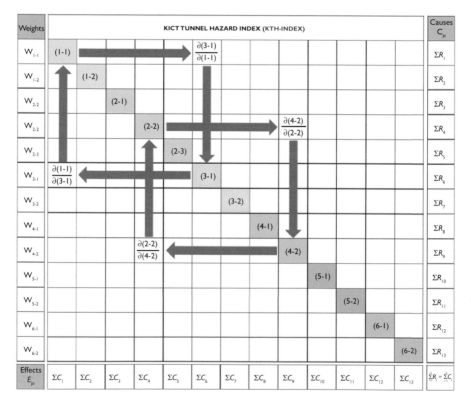

Figure 3.56 Configuration of the interaction matrix for the KTH-Index, from Shin *et al.* (2009). The large arrows indicate example off-diagonal interaction effects as partial derivatives, e.g., in the top row, the quantity $\partial(3\text{-}1)/\partial(1\text{-}1)$ is the change of the parameter (1-1), equivalent sectional area, with (3-1), discontinuity condition. See Figure 3.55 for the identity of the leading diagonal terms.

Weight	KICT TUNNEL HAZARD INDEX (KTH-INDEX)												
6.46E+00	1.00E+00	1.42E-06	2.97E-06	9.66E-06	9.79E-06	1.14E-05	1.30E-05	1.11E-05	8.08E-07	2.47E-06	4.78E-06	1.80E-05	3.40E-06
7.87E+00	1.84E-05	1.00E+00	5.03E-07	1.91E-05	1.02E-05	9.37E-06	2.87E-05	8.66E-06	1.76E-05	1.29E-05	5.66E-06	3.58E-05	9.83E-06
5.33E+00	1.15E-05	1.65E-05	1.00E+00	1.85E-05	9.08E-06	1.44E-06	3.01E-05	5.80E-06	1.07E-05	1.11E-05	1.28E-06	3.13E-05	9.62E-06
6.44E+00	5.81E-06	3.47E-07	9.50E-06	1.00E+00	1.32E-06	1.75E-05	3.22E-06	3.20E-06	3.07E-06	6.66E-07	4.13E-07	1.10E-05	4.18E-06
8.47E+00	7.23E-06	3.42E-05	1.39E-05	2.18E-05	1.00E+00	5.70E-06	2.98E-05	1.09E-05	2.48E-05	2.16E-05	4.85E-06	5.05E-05	1.14E-05
1.30E+01	5.95E-05	5.58E-05	4.72E-05	4.02E-05	4.12E-05	1.00E+00	8.19E-05	4.00E-05	4.94E-05	2.08E-05	4.88E-06	7.43E-05	1.66E-05
9.28E+00	1.98E-05	1.83E-06	8.64E-06	1.82E-05	1.41E-05	1.03E-05	1.00E+00	7.71E-06	9.32E-06	8.92E-06	2.07E-06	3.99E-06	2.10E-06
8.29E+00	1.93E-05	2.57E-06	1.14E-05	3.33E-05	3.66E-05	1.35E-05	4.02E-05	1.00E+00	3.57E-05	2.42E-05	2.07E-06	3.37E-05	1.75E-05
6.38E+00	9.79E-06	4.40E-06	6.42E-06	3.25E-06	5.64E-06	1.06E-06	1.12E-05	8.17E-06	1.00E+00	3.89E-07	1.69E-06	1.72E-05	1.18E-05
4.97E+00	2.34E-06	1.48E-05	1.96E-06	1.69E-05	1.06E-05	6.31E-08	1.11E-05	4.59E-06	1.35E-05	1.00E+00	1.28E-06	1.69E-05	8.03E-06
4.67E+00	4.21E-05	1.77E-05	9.13E-07	2.99E-05	9.80E-07	2.84E-05	2.77E-05	9.55E-06	1.28E-05	1.32E-05	1.00E+00	1.25E-05	4.35E-06
9.38E+00	1.36E-05	1.37E-05	1.46E-05	1.98E-05	1.82E-05	7.30E-06	2.96E-06	2.34E-06	1.97E-05	7.61E-06	1.47E-06	1.00E+00	5.59E-06
9.49E+00	3.55E-05	4.34E-05	2.69E-07	4.20E-05	3.49E-05	3.16E-05	6.54E-05	2.28E-05	5.10E-05	2.27E-05	1.05E-05	2.56E-05	1.00E+00
Effects, E_{pi}	2.45E-04	2.30E-04	1.18E-04	2.72E-04	2.01E-04	1.38E-04	3.72E-04	1.35E-04	2.48E-04	1.47E-04	4.09E-05	3.31E-04	1.04E-04

Figure 3.57 The Korean Tunnel Hazard (KTH) Index is calculated from these interaction matrix off-diagonal values using the method described earlier in this Chapter, from Shin *et al.* (2009). See Figure 3.55 for the identity of the leading diagonal terms.

The authors conclude by explaining that, "The weights calculated for the classification categories, which are key components in the calculation of the KTH-Index, were determined by the proposed methodology based on the interaction matrix and the ANN-based sensitivity analysis incorporated by a linear activation function. The linear activation leads to an assumption of a linear surface of analysis domain so that the values of the first order partial derivatives do not vary as the input values are changed. This means that ultimately determined weight values for major influence factors for tunnel collapse are unchangeable before recalculation with revised data. However, considering the actual situation of varying the major influence factors (e.g., groundwater for urban tunnels but discontinuity condition for mountainous rock tunnel), additional research needs to be performed by using non-linear activation functions. This will allow adaptable weights to be obtained for a given tunnel site."

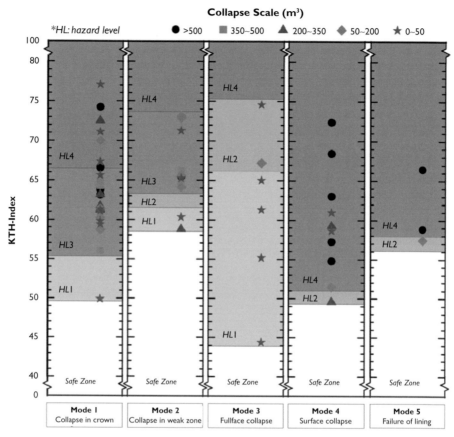

Figure 3.58 The potential tunnel collapse modes as a function of the Korean Tunnel Hazard (KTH) Index values together with an indication of the scale of collapse in m³, from Shin *et al.* (2009).

3.4.3   Underground radioactive waste disposal

The previous sections have covered a wide range of RES applications with a significant number of papers relating to natural rock slope stability which emphasise the large number of factors involved. Another subject with a large number of factors is radioactive waste disposal—because of the large number of features, events and processes (FEPs) that have to be studied during the compilation of a disposal licence application, a process which typically takes 20 years. Structuring the FEPs in a coherent way is most helpful and RES can provide the necessary capability. In this Section, several RES approaches are described which all assist in arranging the FEPs in a coherent manner. (The subject of underground radioactive waste disposal in the risk context and the use of RES for structuring the FEPs is further discussed in Chapter 5.)

Skagius *et al.* (1997) describe the process of structuring the FEPs in their paper, "Performance assessment of the geosphere barrier of a deep geological repository for spent fuel: the use of interaction matrices for identification, structuring and ranking of features, events and processes." The authors explain that, "the main purpose of the assessment is to identify the important issues affecting the long-term behaviour of, and the radionuclide migration within, the far-field rock of an underground repository for spent fuel." They used a 13 × 13 interaction matrix. Skagius *et al.* (1997) highlight an example content of one off-diagonal box as "Natural fracture system—transport of radionuclides, molecular diffusion of radionuclides in the natural fracture system. This will affect important transport parameters such as the surface area available for sorption and matrix diffusion. The sorption capacity of the rock is affected by the fracture minerals." Each such off-diagonal box has a reference to the Svensk Kärnbränslehantering AB (SKB) Swedish FEP database. The authors conclude by noting that the compiled information is valuable because it presents the fundamental background material in a structured and consistent manner. They also provide advice on the assembling

	1	2	3	4	5	6	7	8	9
1	Zone 4 and 5 Aquifer 2		Discharge through sediment						
2		Zone 4 and 5 Atmosphere				Deposition	Deposition		
3	Recharge		Zone 4 Riverbed Sediments	Resuspension	Uptake of water and nutrients, ingestion				
4		Evaporation, spray	Deposition of suspended sediment	Zone 4 River Water	Uptake of water, nutrients and suspended sediment	Irrigation	Irrigation	Ingestion	
5			Excreta and death	Excreta and death	Zone 4 River Biota				
6	Infiltration	Suspension				Zone 5 Soils	Root uptake and soil splash	Ingestion	Soil run-off
7						Weathering leaf litter	Zone 5 Plants	Ingestion	Detritus
8								Zone 5 Terrestrial animals	
9									Zone 5 Sinks

Figure 3.59 Interaction matrix representation of radionuclide transport and exposure pathways for the river and arable land—as part of the development of a biosphere model for underground radioactive waste disposal studies, from Agüero *et al.* (2008).

of interaction matrices in the radioactive waste context. However, in this application it should be noted that the interaction matrix is being used only for structuring information.

Agüero *et al.* (2008) in their paper "Application of the Spanish methodological approach for biosphere assessment to a generic high-level waste disposal site" also used the interaction matrix to structure the aspects of the biosphere. They explain that, "The biosphere in the context of high-level waste disposal is defined as the collection of various radionuclide transfer pathways that may result in releases into the surface environment, transport within and between the biosphere receptors, exposure of humans and biota, and the doses/risks associated with such exposures." The authors used an interaction matrix (e.g., Figure 3.59) for the development of their conceptual model as it relates to the Spanish generic disposal site which is within a landscape context 200×200 km located in Central-West Spain, along the Tajo river watershed. As with the work by Skagius *et al.* (1997), the construction of the conceptual model is complicated because of the many factors associated with the main components such as water, soil, vegetation and animals in and between which radionuclides may be transported or may accumulate.

The authors further explain that, "Based on climatological studies undertaken in an international framework ... narratives of environmental change were developed for the region of interest, covering a period from the present to 200,000 years AP. This narrative was decomposed into a series of biosphere states and transitions between them. These states and transitions were characterised in a structured way using an interaction-matrix-based approach ... the construction of a phenomenological interaction matrix provides a description of the intrinsic dynamics of the biosphere system, establishing the ways in which the principal component types are interrelated."

3.4.4 Use of the RES interaction matrix in other subject areas

Because of the utility of the RES interaction matrix in structuring a system's main factors and the inter-relations between them, the use of the matrix has been extended to subject areas beyond the specific rock engineering context. Some of these wider applications are highlighted in this Section.

Hill and Rosenbaum (1998) noted that, "The process of rock weathering may be regarded as a dynamic, multi-factorial system comprising the interactions between the rock mass itself, the agents of weathering and the environmental conditions." They used the RES interaction matrix and fuzzy sets to assess the significant factors in a rock weathering system considering the parameters Climate, Biological Activity, Mineralogy, Texture, Discontinuities, Permeability, Geomorphology and Time. Their paper provides a useful description of the eight parameters together with fuzzy set representations of these. The authors explain that, "An index of weathering (WI) may now be constructed for a rock mass at any site from a combination of the Fuzzy Set membership values assigned to each of the significant factors (FV_n), and their corresponding Factor Activity values (FA_n) derived from the Interaction Matrix. The site-specific value of each factor is weighted according to its activity within the rock weathering system operating at that site, and combined to produce a weathering index (WI) as follows: $WI = FV_1 \cdot FA_1 + FV_2 \cdot FA_2 + \cdots + FV_n \cdot FA_n$ where the FVs refer to the Factor Activities and the FAs refer to the Factor Values." More detail is given in Hill and Rosenbaum (1998).

Shang *et al.* (2000) describe a retrospective case example using RES and a comprehensive suitability index (*CSI*) for siting the pumped storage Shisan-Ling power station in China. The authors explain that, "The Shisan-Ling Pumped Storage Power Station was commenced to alleviate the power shortage in the northern area of China. Adjacent to the 13 Ming Dynasty tombs and a famous scenic location used by tourists near Beijing, the pumped storage power station has attracted attention because of its location and construction." There were two siting decisions: firstly, the location of the overall geological domain; and, secondly, the location of the powerhouse itself. At the stage of site selection, 1983–1985, the conglomerate of the Middle Jurassic Epoch was chosen and between 1985–1987 the decision-making was concentrated on locating the underground powerhouse, for which three schemes at the I, II and III positions arrayed over a distance of 1000 m were compared and postion II chosen. In 1995, the project was completed.

Thus, the work of Shang *et al.* (2000) relates to a retrospective analysis of the siting decision to establish if the power station is indeed located at the best choice of location. Eleven parameters were used for the comparison and selection of the three overall sites, i.e., the conglomerate, the andesite and the limestone. These were (1) lithology, (2) faults, (3) rock mass structure, (4) hydraulic conditions, (5) permeability, (6) joints, (7) geomorphology, (8) engineering layout, (9) environment, (10) cost and (11) construction. For locating the powerhouse, seven parameters mainly associated with faults and joints were considered: (1) spacing of faults, (2) width of fractures, (3) ratio of mud-bearing faults, (4) spacing of joints, (5) sets of

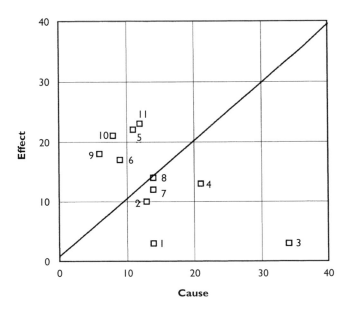

Figure 3.60 Cause–Effect plot for overall site selection. Legend: 1 lithology, 2 rock mass structure, 3 faults, 4 joints, 5 hydraulic conditions, 6 permeability, 7 geomorphology, 8 engineering layout, 9 environment, 10 cost, 11 construction, from Shang et al. (2000).

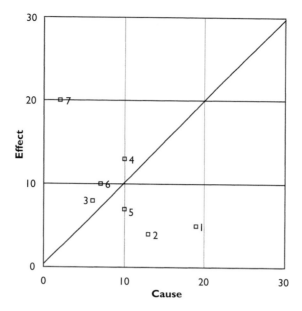

Figure 3.61 Cause–Effect plot for location of the powerhouse. *Legend:* 1 spacing of faults, 2 width of fractures, 3 ratio of mud-bearing faults, 4 spacing of joints, 5 sets of joints, 6 ancient weathered crust, and 7 waterflow, from Shang *et al.* (2000).

joints, (6) ancient weathered crust, and (7) waterflow. Unlike the overall siting objective, the parameters for the powerhouse location were orientated towards potential threats from deformation and failure within the rock masses.

The authors use the usual index scheme: $CSI_j = (100 - \sum_{i=1}^{n} \alpha_i P_{ij})\%$ where j indexes the schemes, i indexes the parameters, n refers to the number of parameters, a_i is the weighting factor for each parameter scaled from the C–E histogram, and P_{ij} is the rating assigned to different classes of parameter values—which is different for different schemes. The *Cause–Effect* plot for the overall siting options based on ESQ coding of the interaction matrix is shown in Figure 3.60. It is evident that the five most dominant parameters are faults, lithology, joints, rock mass structure and geomorphology. The *Cause–Effect* plot for the powerhouse locating options is shown in Figure 3.61. In this case, the three dominant parameters are spacing of faults, width of fractures and sets of joints.

At the actual design feasibility stage for the powerhouse, Shang *et al.* (2000) note that the conglomerate choice for the overall power station siting was originally made after comparison with the andesite and limestone formations, taking into account many aspects of geology, engineering, environment, cost, including the data from rock testing. For the powerhouse, position II was chosen because of detailed consideration of the local properties, proximity to faults, etc. The authors conclude, "The quantitative results from the *CSI* values are in agreement with the actual practical qualitative evaluations and engineering judgements that were made at the time. The demonstration retrospective case example has thus indicated that the RES-based *CSI* approach proposed … can provide a valuable supportive technique in the decision-making process at different stages in rock engineering design."

Avila and Moberg (1999) describe a systematic approach to the migration of the radionuclide ^{137}Cs in forest ecosystems using interaction matrices. They note that, "The migration of radionuclides in the environment is complex and involves multiple biotic and abiotic components and interactions. When developing conceptual and mathematical models of such processes, there is a risk that important components and interactions are omitted or underestimated."—and hence the use of interaction matrices. The authors used interaction matrices firstly to develop their conceptual model and afterwards as a diagram of the conceptual model itself. Their 9×9 matrix representing long-term migration of ^{137}Cs in a forest ecosystem is reproduced here in Figure 3.62, the interactions included having been established through a literature search. The authors also coded the matrix using the ESQ (0–4) method described earlier and generated the *Cause–Effect* plot, see Figure 3.63.

The *C–E* plot in Figure 3.63 is of Type 3 in Figure 3.9(a) in Section 3.3.2, i.e., factors with variable system intensity and similar dominance. The authors note that, "It can be seen that the components can be divided into three groups of different

Atmosphere (Air)	Interception rainfall snowfall	Interception	Interception rainfall snowfall			Interception	Interception rainfall snowfall	Interception inhalation
Transpiration burning	**Tree leaves**	Translocation	Leaf-fall weathering			Weathering interception	Weathering interception	Ingestion
Burning	Translocation	**Tree other**	Weathering interception	Fertilisation	Fertilisation	Mycorrhizae	Weathering interception	Ingestion
Resuspension		Rain splash	**Litter**	Decomposition Percolation	Percolation	Root uptake	Rain splash	Ingestion
		Root uptake		**Soil organic**	Percolation diffusion/advection	Root uptake	Root uptake	
		Root uptake			**Soil mineral**	Uptake	Uptake	
		Root uptake (mycorrhizae)	Fertilisation	Fertilisation	Fertilisation	**Fungi**	Root uptake (mycorrhizae)	Ingestion
Transpiration burning			Leaf-fall, weathering interception	Fertilisation	Fertilisation	Mycorrhizae	**Understorey**	Ingestion
			Fertilisation			Consumption	Consumption	**Wild animals**

Figure 3.62 Interaction matrix with nine diagonal elements describing the long-term migration of ^{137}Cs in a forest ecosystem, from Avila and Moberg (1999).

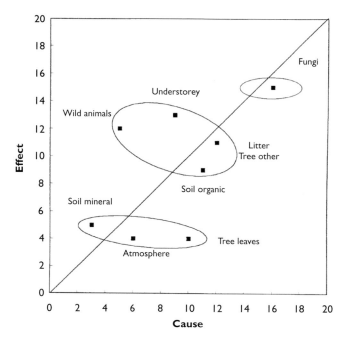

Figure 3.63 The *Cause–Effect* plot for ¹³⁷Cs migration in a forest ecosystem, from Avila and Moberg (1999).

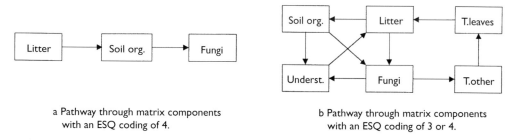

Figure 3.64 Pathways through the interaction matrix for ¹³⁷Cs migration in a forest ecosystem, from Avila and Moberg (1999). For the abbreviations, see Figure 3.63.

degrees of interactivity. Fungi show the highest values and mineral soil, atmosphere and tree leaves the lowest. Understorey vegetation, forest litter, organic soil, parts of trees (other than leaves) and wild animals have an intermediate position." The authors then draw an interesting conclusion: "Most components, with the exception of leaves and wild animals, are situated in the plot close to the main diagonal. This means that they influence the system approximately as much as the system influences them. This is an indication that the long-term behaviour of ¹³⁷Cs in forest ecosystems is characterised by near to steady-state conditions." Also, this application directly relates to isotope migration and hence to pathways through the matrix (Figure 3.64), as described earlier in Section 3.3.3.

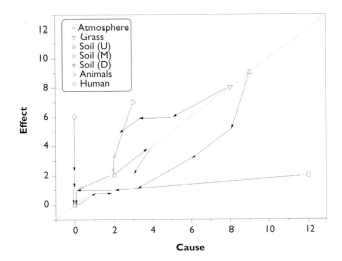

Figure 3.65 Temporal changes in *Cause–Effect* relations for ^{137}Cs migration pathways in a grassland ecosystem, from Velasco *et al.* (2006).

Velasco *et al.* (2006) also consider ^{137}Cs distribution and cycling in their paper, "Interaction matrices as a first step toward a general model of radionuclide cycling: Application to the ^{137}Cs behaviour in a grassland ecosystem". Following the Chernobyl accident, the Friuli Venezia Giulia Region in the northeastern part of Italy was subjected to heavier rainfall than other Italian regions and received the highest radioactive deposition. Through the pathways ideas described earlier in this Chapter, it is possible to identify the principal flux pathways in the system, these migration (or exposure) pathways being seen as a sequence of multiple interactions in the system.

An interesting aspect of the work by Velasco *et al.* (2006) is that they code their 7×7 interaction matrix for four successive time periods, thus enabling the trajectory of each parameter to be plotted in *Cause–Effect* space, as shown in Figure 3.65. The tendency for the parameters to become less interactive is evident in the Figure.

Another example of radionuclide pathways in the RES interaction matrix is included in the work by Smith *et al.* (1996) as part of their report "Biosphere Modeling and Dose Assessment for Yucca Mountain". [The Yucca Mountain area in the USA is being considered as a potential site for underground radioactive waste disposal.] In Figure 3.66, the radionuclide pathways within the interaction matrix are highlighted by the thin arrows.

In the paper by van Dorp *et al.* (1999) on "Biosphere modelling for the assessment of radioactive waste repositories; the development of a common basis by the BIOMOVS II reference biospheres working group", the authors explain that, "Performance criteria for radioactive waste repositories are often expressed in terms of dose or risk. If other criteria are used, they may well have been derived from dose or risk criteria. Some sort of biosphere modelling is therefore usually required to assess radionuclide migration and accumulation in the human environment, and to assess the associated radiation exposure." These authors also used RES to structure the supporting information. In the RES context, the authors conclude that, "The interactions or relationships identified

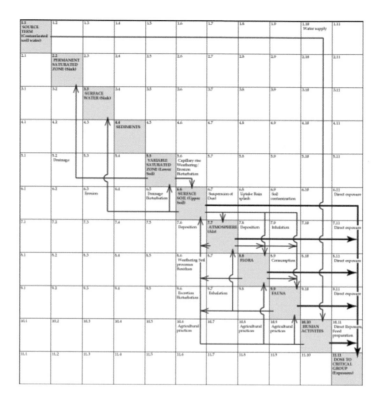

Figure 3.66 BIOMOVS II Reference Biosphere RES matrix modified for Yucca Mountain biosphere. No surface water body is assumed. The thin arrows refer to nuclide migration processes; the thick arrows refer to exposure pathways, from Smith *et al.* (1996). The leading diagonal terms (top left to bottom right) are Source Term, Permanent Saturated Zone, Surface Water, Sediments, Variable Saturated Zone, Surface Soil, Atmosphere, Flora, Fauna, Human Activities, Dose to Critical Group.

by the RES approach form a relational FEP [Features, Events and Processes] list. The relations identified in the relational FEP list form the textual basis (textual description or conceptual model) of the equations to be included in the mathematical description.

Condor and Asgharib (2009) utilise the Features, Events and Processes (FEPs) approach plus the RES interaction matrix to describe an alternative theoretical methodology for monitoring the risks of CO_2 leakage from sequestration wellbores. They use an 'Incident Potential Matrix (IPM)' instead of the ESQ coding, explaining that the inclusion of the IPM concept in the interaction matrix may allow stochastic modelling by the use of probabilistic density functions. In the IPM, the risk can be defined deterministically as *Risk = Exposure × Severity*. (Figure 3.67.)

The interaction matrix for this application is shown in Figure 3.68 and the associated *Cause–Effect* plot in Figure 3.69.

The form of Figure 3.69 corresponds to the Type 4 *Cause–Effect* plot in Figure 3.9, i.e., similar parameter intensities but with variable dominance. The four dominant parameters are Hydrogeology, Gas, Water Composition and Gas Composition. The authors conclude that, "This methodology has a great potential for being used in

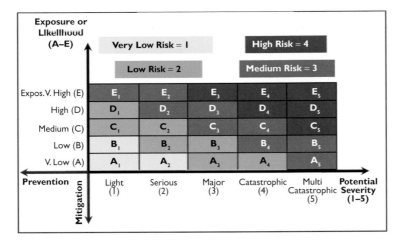

Figure 3.67 The Incident Potential Matrix, IPM, from Condor and Asgharib (2009). The colour coding for risk is similar to the RES ESQ coding (0–4).

[2,2] CASING	[2,3] a) Expansion/Contraction b)	[2,4] a) Sorption b) Colloid Filtering	[2,5] a) Water Flow b) Capillary Suction	[2,6] a) Corrosion b) Degradation Organic	[2,7] a) Gas Flow b) c)
[3,2] a) Expansion/Contraction b)	[3,3] CEMENT PLUG a) Recrystallisat/mineralizat	[3,4] a) Sorption b) Colloid Filtering c) Erosion	[3,5] a) Water Flow b) Capillary Suction	[3,6] a) Corrosion b) Degradation Organic	[3,7] a) Gas Flow b) c)
[4,2] a) Corrosion b) Colloid Filtering	[4,3] a) Dissolution/Precipitaion b) Degradation Organic c) Water Uptake	[4,4] WATER COMPOSITION	[4,5] a) Water Flow b) Convection c) Osmosis	[1,6] a) Gas Dissolution/Degassing b) Chemical Reaction c) Degradation Organic	[1,7] a) Corrosion b)
[5,2] a) Dissolution/Precipitation b) Corrosion c) Degradation Organic d) Erosion	[5,3] a) Dissolution/Precipitation b) Degradation Organic c) Erosion d) Diffusion	[5,4] a) Advection and Mixing b) Chemical Equilibrium c) Erosion	[5,5] HYDROLOGY	[5,6] a) Saturation b) Dissolution/ Degassification	[5,7] a) Two Phase Flow b) Expansion/Contraction
[6,2] a) Corrosion b)	[6,3] a) Carbonation b) Kinetics and Equilibria c) Cracking	[6,4] a) Gas Distribution/Degassing b) Colloid	[6,5] a) Saturation b)	[6,6] GAS COMPOSITION	[6,7] a) Gas Flow b)
[7,2] a) b)	[7,3] a) Diffusion b)	[7,4] a) b) c)	[7,5] a) Two Phase Flow b)	[7,6] a) b)	[7,7] GAS

Figure 3.68 Portion of an interaction matrix used by Condor and Asgharib (2009) for the analysis of the risks of CO_2 leakage from sequestration wellbores.

the probabilistic risk assessment and monitoring phases of wellbores and other components of geo-sequestration systems used for geological storage of CO_2. Its major advantage consists in its simplicity and practical results. Further studies may include the stochastic representation in the Incident Potential Matrix. In such a way, instead of having a fixed number for risk, probability density functions can incorporate the uncertainty."

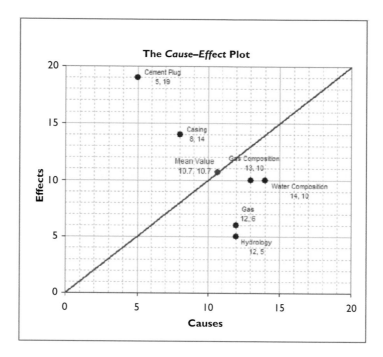

Figure 3.69 Cause–Effect plot for the coded interaction matrix in Figure 3.68, from Condor and Asgharib (2009).

In a different scientific area, Mavroulidou *et al.* (2004) described a qualitative tool combining an interaction matrix and a GIS to map vulnerability to traffic induced air pollution (carbon monoxide, CO) in the UK. The authors explain that, "The GIS results are presented in the form of a vulnerability map, which highlights the areas susceptible to poor air quality. This visual interpretation of the results is ideal for local authorities, who have to report to a wide range of non-specialists in the field, for example, planners, councillors and the public. The vulnerability map compares favourably with pollutant concentration patterns, obtained from an advanced dispersion model." The authors' matrix consisted of 20 parameters "from different disciplines including regional tectonics, geomorphological, chemical and physicochemical processes, and meteorology". The content of the interaction matrix and the GIS system procedures, together with their linkage are shown in Figure 3.70.

For the interaction matrix, the authors used the seven parameters: Traffic, Wind, Stability, Roughness, Topography, Buildings, and Air Quality; and the matrix was coded using the ESQ method (0–4). The resultant *Cause–Effect* plot is included as Figure 3.71 indicating that the most dominant parameters are Buildings, Topography, Roughness and Traffic. The final result of the procedure is the GIS traffic-induced air pollution vulnerability map for Guildford, UK, in Figure 3.72.

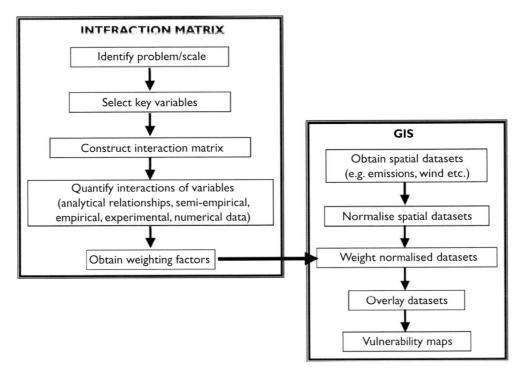

Figure 3.70 Schematic representation of the Mavroulidou et al. (2004) methodology.

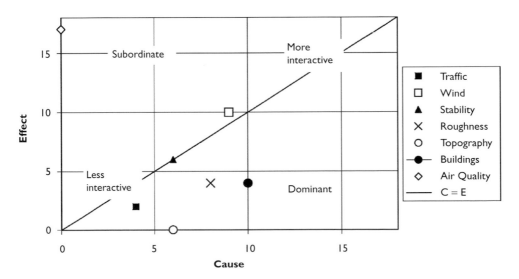

Figure 3.71 Cause–Effect plot for traffic-induced air pollution interaction matrix, from Mavroulidou et al. (2004).

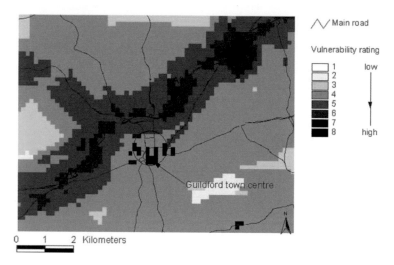

Figure 3.72 GIS traffic-induced air pollution vulnerability map for Guildford, UK., from Mavroulidou *et al.* (2004).

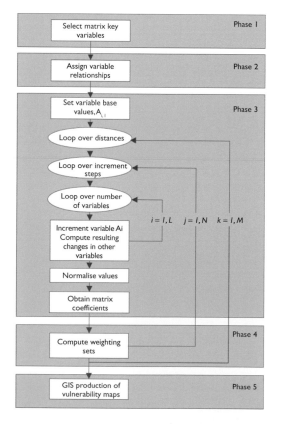

Figure 3.73 The Mavroulidou *et al.* (2007) procedure for assessing traffic-induced air pollution using variable RES matrix interactions, feedback loops to ensure convergence, plus a GIS system.

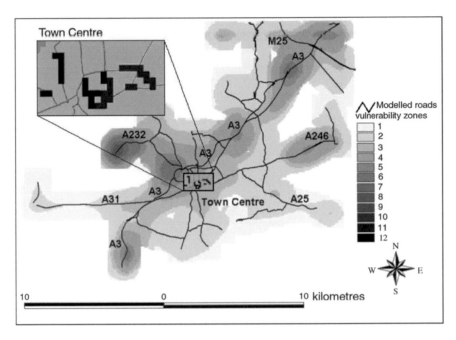

Figure 3.74 GIS vulnerability map for traffic-induced air pollution in the borough of Guildford, UK, based on the interaction matrix methodology, from Mavroulidou *et al.* (2007).

Policy Component (P) (Reorganization)	Formal & weak	Formal & weak	Formal & weak	Mixed & weak					Formal & medium Priority setting Programe devel. & review
Formal & weak	Research Component (R) (Reorganization)	Formal & weak		Mixed & weak	Mixed & medium	Informal & medium	Informal & medium Information sharing Problem diagnosis Technology diffusion Exchange of staff	Formal & weak Workshops/seminars Information sharing Personnel training
Formal & weak	Formal & weak	Education Component (E) (Reorganization)				Informal & weak	Informal & medium	Formal & weak Workshops/seminars Information sharing
Formal & weak			Credit Component (C) (Reorganization)					Formal & weak
Formal & medium Information sharing	Formal & medium Information sharing			Extension and Information Component (I)		Formal & medium Program development Problem diagnosis Priority setting Tech diffusion/demon. Training	Formal & medium Program development Tech. diffusion Info & finance sharing Workshops Seminars	Formal & weak Tech. diffusion & demonstration Information sharing
Formal & medium	Informal & medium				Private Enterprise Component (IPM)	Mixed & medium Tech. demonstration Training		Mixed & weak Program development Tech. development Workshops
Informal & medium Information sharing	Informal & medium Information sharing Problem diagnosis Technology diffusion Exchange of staff	Informal & weak Information sharing			Mixed & weak Tech. demonstration Training	Private Farm Component (F)		
Informal & medium Information sharing	Informal & medium Information sharing Problem diagnosis Technology diffusion Exchange of staff	Mixed & medium Information sharing		Formal & weak Program development Sharing of info. & finance Workshops	Informal & weak	Mixed & medium Problem diagnosis Priority setting Technology diff. & demonstration	Private Consultancy Component (CO)	Formal & weak Program development Tech. diffusion Info & finance sharing Workshops
Formal & medium Priority setting Program development Program review	Formal & weak			Formal & strong Priority setting Program development Technology development Technology diffusion and demonstration Information sharing	Mixed & weak	Formal & medium Problem diagnosis Program development Tech. demonstration Information sharing Training	Formal & medium Program development Tech. diffusion Info & finance sharing Workshops	External Assistance Component (EA)

Figure 3.75 Interaction matrix representing the structure of linkage mechanisms for an agricultural innovation system in Azerbaijan, from Temel *et al.*, 2002(a).

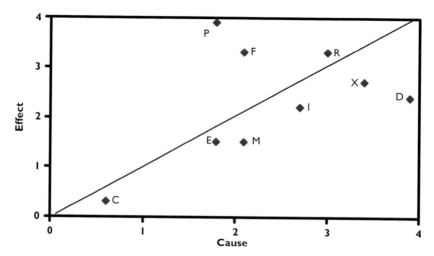

Figure 3.76 Cause–Effect plot relating to institutional linkages in the agricultural innovation system of Azerbaijan assessed by Temel *et al.*, 2002(a). *Legend*: P–Policy, R–Research, E–Education, C–Credit, I–Extension and Information, M–Private Enterprise, F–Private Farm, D–Private Consultancy, X–External Assistance.

The authors suggest that socio-economic factors (e.g., age, health, social status, professional or economic activity—obtained directly as digitised maps or extracted from geo-referenced databases, e.g., census data) could also be incorporated within the interaction matrix and hence enable socio-economic parameters to be included in the vulnerability assessment.

In a following paper on the same traffic-induced air pollution theme, Mavroulidou *et al.* (2007), introduce and assess a more versatile coding of the interaction matrix using *varying* off-diagonal matrix interactions. This enabled the introduction of non-linear relations between the leading diagonal primary variables. The six variables used for the matrix were emissions from traffic, wind speed, atmospheric stability, surface roughness, street canyons and air quality. The Mavroulidou *et al.* (2007) analysis scheme using non-constant interactions is shown in Figure 3.73.

After completing Phase 3 in Figure 3.73 and computing the weighting sets (as in creating the indices used in previously described applications), the authors constructed a base raster map showing the distance of each cell from the roads in the study area. The authors explain that, "This was then used to apply the appropriate interaction matrix and hence weighting value to each cell in the spatial dataset. Average values of the weighting factors for each distance interval were used. The resulting weighted spatial datasets for all selected variables were then overlaid using GIS map algebra, to produce pollution vulnerability maps." An example relating to the borough of Guildford in the UK is shown in Figure 3.74.

The authors' direct sequential method of introducing non-constant interactions between the primary variables in the interaction matrix is a most useful step forward and complements the indirect ANN method also described in this Chapter.

Widening the scope of the RES applications even further, Temel *et al.*, 2002(a), describe a "Systems analysis by graph-theoretic techniques: assessment of institutional linkages in the agricultural innovation system of Azerbaijan". The authors used the following nine leading diagonal components for their interaction matrix: Policy, Research, Education, Credit, Extension and Information, Private Enterprise, Private Farm, Private Consultancy and External Assistance. The matrix is shown in Figure 3.75.

One of their *Cause–Effect* plots generated from the coded matrix is included here as Figure 3.76. It can be seen directly from the plot that the five most dominant parameters are D–Private Consultancy, X–External Assistance, M–Private Enterprise, I–Extension and Information, and C–Credit. More information on this subject is given in Temel *et al.*, 2002(b).

3.5 FURTHER DEVELOPMENT OF THE RES METHODOLOGY

Yang and Zhang (1997), in noting that the Jiao and Hudson (1995) RES fully-coupled model was developed for linear relations between the variables, developed an hierarchical analysis for rock engineering using Artificial Neural Networks (ANNs) to overcome the limitation. They used the back propagation neural network architecture—which is an hierarchical design consisting of fully interconnected layers or rows of processing units. The authors explain that their interest was concentrated on "searching for a method of identifying how significant each factor is in the operation of ANN" and so the method of establishing the dominant factor is called an "hierarchical analysis". They used the Relative Strength of Effect (RSE) as the effect of an input on the output, thus determining the relative dominance of input variables (range –1 to 1) and hence the relative significance of all the input factors influencing rock behaviour.

The authors implemented the ANN RSE analysis using Sheorey's (1991) extensive data for coal mine roadways. Their initial inputs were: roadway span, roadway depth, uniaxial compressive strength, the six parameters in Barton's Q classification system (RQD, Jn, Jr, Ja, Jw, SRF), dry density, rock type and joint orientation. The output was the state of rock engineering, i.e., stable or unstable. The conclusion of Yang and Zhang (1997) was that the sequence of dominant parameters for the stability of Indian coalmine roadways is Rock Quality Designation, Barton's Stress Reduction Factor (SRF), density, rock type, span, uniaxial strength, number of joint sets, joint orientation, depth and joint roughness.

Yang and Zhang (1998) extended their 1997 work to the application of neural networks for RES and to provide an alternative method to graph theory analyses (Jiao & Hudson, 1995) for establishing the off-diagonal relations between the leading diagonal factors in the interaction matrix when the matrix is 'turned on', i.e., when the matrix acts dynamically, as in the earlier Figures 3.18–3.20 (also see Colour Plate section at the end of the book). They generated the Relative Strength Effect (RSE) and the Global Relative Strength Effect (GRSE) matrix.

Using a back propagation (BP) neural network architecture, they used the 17 factors overburden, intact rock strength, rock mass structure, average spacing/span, discontinuity tightness, discontinuity persistency, discontinuity type, discontinuity fillings, discontinuity planeness, discontinuity roughness, discontinuity dip, discontinuity strike, shear zoning or faulting, RQD, ground water condition, location of instability and the state of engineering. The authors state that their ANN procedure can in principle be extended to not only incorporate the purely rock mechanics factors but also the engineering factors, and even be able to link with environmental and financial aspects. Thus, Yang and Zhang (1998) developed a powerful extension of the RES methodology.

Cai *et al.* (1996) used neural networks with RES for cavern performance auditing consisting of a hybrid neural network and expert system. The authors explain that a knowledge-based expert system has a high symbolic reasoning capability and neural networks are powerful data processing tools in pattern recognition—so a hybrid system of the two utilises the advantages of both systems. The authors' hybrid system consists of five components for data acquisition and pre-processing (informing component), recording of engineering cases (case recording), learning or modelling of interaction mechanisms (matrix coding), prediction of altered parameters (matrix operating), and simulation of the dynamic process of rock engineering system (dynamics simulating). The interaction matrix consists of a rock mass quality subsystem (Q-value), excavation dimension (span or height, D) and rock support sub-system (bolt spacing S, bolt length L), thickness of shotcrete (T). The authors illustrated their scheme through the application of a computerised cavern performance auditing system and concluded that, "With the aid of the neural network's learning and the expert system's symbolic reasoning capabilities, the RES approach may be made 'intelligent' in solving the problems of rock engineering".

In a paper by Ferentinou and Sakellariou (2007), the authors describe computational intelligence tools for the prediction of slope performance under both static and dynamic conditions for landslide hazard assessment. They use a back-propagation algorithm, the theory of Bayesian neural networks and Kohonen self-organizing maps and estimate the slope stability controlling variables by combining computational intelligence tools with generic interaction matrix theory. They use the interaction matrix to establish the landslide causal factors rating to detect the dominance and interaction intensity of the principal factors. A later paper by Ferentinou *et al.* (2012) extends the work to the application of computational intelligence tools for the analysis of marine geotechnical properties in the head of Zakynthos canyon, Greece.

Both papers referenced in the preceding paragraph illustrate graphs in each inter-Action matrix box representing the relations between all pairs of leading diagonal variables, a 9×9 matrix in Ferentinou and Sakellariou (2007) and an 11×11 matrix in Ferentinou *et al.* (2012). It will be recalled that in Section 3.3.2, the potential matrix coding methods are listed as Binary, Expert Semi-Quantitative (the most widely used), According to the slope of an assumed linear relation, More numerically via a partial differential relation, and Explicitly via complete numerical analysis of the mechanism. In Ferentinou *et al.* (2012)., the authors extend the interaction matrix coding method to 'according to the slope of an assumed linear relation' and, together with further analysis techniques, have thus significantly advanced the RES analysis procedures.

The PhD thesis by Millar (2008) also addresses parallel distributed processing in rock engineering systems. In particular, the thesis text demonstrates how artificial neural networks can be directly coupled to finite difference numerical methods, thus providing a novel capability to embed the complexity of rock constitutive behaviour. The thesis explains how to "… associate the parameters describing environmental potentials and system material properties with identified modes of excavation behaviour … and parameter spaces and modes of behaviour are also visualised, such that the dynamic performance of rock engineering systems can be presented as a state trajectory on system performance maps."

Saeidi *et al.* (2014) extended the RES applications to rock groutability in the absence of a comprehensive ranking for this subject. They present an index to predict the groutability potential of rock masses, with the Bakhtiari dam in Iran selected as a case study where the proposed index was used to rank the seven geological units of the Sarvak formation at this dam.

Rafiee *et al.* (2014) applied the RES techniques to rock mass cavability in block caving mines using 'natural' parameters and 'induced' parameters and probabilistic coding. The probabilistic approach allows identification of "parameters with the highest probability of being dominant or subordinate, and also the parameters with the highest probability of being interactive, i.e., variability and/or uncertainties can be explicitly included in the analysis, and the effects of such uncertainties can be quantified.... results showed that the parameter related to the existence of *in situ* stress has the highest expected interaction." Rafiee (2014) also applied the RES approach to the development of a rock behaviour index for underground excavations.

There have been many other applications of RES to different rock engineering subjects. For example, Huang *et al.* (2013) applied RES to automated tunnel rock classification and used it for the Dazhushan tunnel in China. Frough and Torabi (2013) developed an application of rock engineering systems for estimating TBM downtimes. It is not easy to predict the directions that RES might travel in future research but the integration of RES with neural network computing is one of the most promising areas because RES provides the structure for the problem in hand and the neural network can be used to provide a dynamic learning procedure to refine the RES structure and provide a predictive capability.

3.6 AUDITING AND PROTOCOL SHEETS

Auditing is defined as a systematic and independent examination of data, statements, records, operations and performances of an enterprise for a stated purpose.

A protocol is a rule which describes how an activity should be performed.

A key aspect of reducing risk is ensuring that all the necessary procedures leading up to the design have been implemented properly. This leads to auditing procedures and the use of Protocol Sheets. The italicised text above explains the general meaning of these two words. Within the rock engineering risk context, it is of benefit to be able to formally audit the content of the rock mechanics modelling and rock engineering design of a project in order to ensure that all the necessary factors are included and that the technical work is correct. The term 'Technical Auditing' is used to describe this process and Protocol Sheets are the records made of its implementation. The subject has been described in detail in our previous book "Rock Engineering Design" (Feng & Hudson, 2011), so here we provide just a summary of the principles and discuss how the Design Protocol Sheets can be extended to Risk Protocol Sheets.

3.6.1 'Soft', 'semi-hard' and 'hard' technical audits and the audit evaluation

The technical auditing can be 'soft', 'semi-hard' or 'hard', with the characteristics as indicated in Figure 3.77.

For modelling, i.e., using one or more of the eight main types of modelling methods shown in Figure 3.78, any of the auditing types can be utilised, depending on the purpose of the auditing. The soft audit can be used initially to support the development of the modelling programme. However, the hard audit is necessary for the total audit evaluation and ability to state whether the modelling is adequate for the purpose. In the

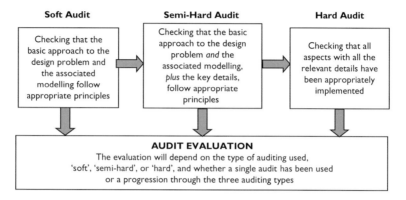

Figure 3.77 The 'soft', 'semi-soft' and 'hard' audits and the audit evaluation.

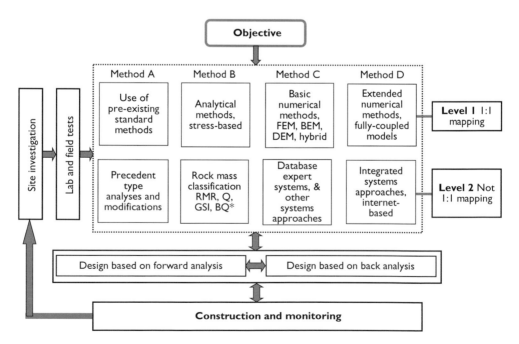

Figure 3.78 Flowchart of the eight main modelling methods within the context of rock engineering design and construction (from Feng & Hudson, 2011).

risk context of this book, the auditing is essential in order to be able to identify any deficiencies in the site investigation programme and in the modelling work supporting design, together with its implementation in the 'Design based on forward analysis' box. Similarly, auditing procedures are required for the 'Design based on back analysis' box.

Two technical auditing demonstration examples are included in our previous book (Feng & Hudson, 2011): one for auditing a site investigation measurement (*in situ* rock stress), and one for the modelling associated with the design of caverns for the Laxiwa hydropower project on the Yellow River in China. Also, Protocol Sheets for the auditing of modelling and design are included, together with completed example sheets for the design of the Jinping II underground powerhouse in China. Similar Protocol Sheets and demonstration examples are required for the evaluation and auditing of risk assessment procedures and these may be the subject of the ISRM Design Methodology Commission work in the period 2015–2019.

3.7   CHAPTER SUMMARY

We recall that the content of this Chapter has been a description of the reduction of epistemic uncertainty through the knowledge gained by utilising the Rock Engineering Systems (RES) approach and the reduction of errors through an auditing scheme supported by a portfolio of Protocol Sheets, i.e., the shaded area in the flowchart in Figure 3.79.

The RES system was outlined through the explanation of the interaction matrix with its leading diagonal of the key factors and off-diagonal boxes

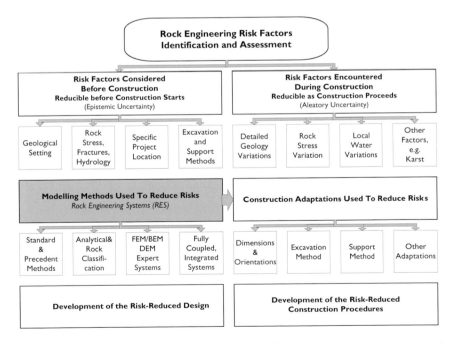

Figure 3.79 Rock Engineering Systems (RES) as a method of reducing epistemic uncertainty through the structuring of knowledge contributed from available sources.

containing the interactions between the key factors. The procedure involves coding the off-diagonal boxes according to their significance—the most popular coding method being the Expert Semi-Quantitative (ESQ) method, i.e., an integer from 0 to 4. Other potential coding methods are via the slope of an assumed linear relation, more numerically via a partial differential relation, and explicitly via complete numerical analysis of the mechanism. Once the interaction matrix has been coded, the *Cause–Effect* diagram can be constructed which indicates directly by eye the relative interactivity and dominance/subordinacy of the component factors. The concept of a concatenation of mechanisms is represented by pathways through the matrix which was illustrated by a radioactive waste disposal example. In reality, the mechanisms in the matrix system all operate simultaneously when a perturbation is introduced, such as blasting; and the dynamic models shown in Figures 3.18–3.20 demonstrate that, above a certain energy transmission value, the matrix can exhibit chaotic behaviour.

Then a series of examples was highlighted based on published papers describing applications in both the rock engineering context and the wider context. These included natural and artificial surface rock slopes, rockfalls, underground blasting, tunnel boring machines including the estimation of tunnel boring machine downtimes, tunnel stability, coal mine roadways, radioactive waste disposal, rock groutability, rock weathering, siting a hydroelectric project, migration of the radionuclide ^{137}Cs in forest ecosystems, CO_2 leakage, traffic induced air pollution and institutional linkages in an agricultural innovation system.

Recalling that the title of this Chapter is "Rock Engineering Systems (RES), auditing and Protocol Sheets", reducing the epistemic uncertainty using RES and hence the risk for any type of project is a sequential process involving a group of people undertaking the following activities.

- Consider the project objective,
- Choose the leading diagonal terms for the interaction matrix,
- Code the off-diagonal terms in the interaction matrix,
- Create the *Cause–Effect* plot,
- Determine the parameters with the most interactivity and dominance,
- Create an assessment index, e.g., an instability or vulnerability index (as in the case examples described),
- Calibrate the index via test cases, and then
- Use the index for the project being considered.

This procedure provides considerably more knowledge about the project, as evidenced by the suite of RES applications described in this Chapter. In fact, the knowledge may already have been present in the minds or computers of a project team; it is the RES procedures that enable structured elicitation of that knowledge for the risk assessment in hand—which, in turn, increases understanding concerning the project, reduces the epistemic uncertainty, and enables reduction of the risk. Moreover, in following the arrow from the shaded box in Figure 3.79 to the right-hand side of the diagram, the epistemic information is used to develop construction adaptations for reducing the risk. The aleatory uncertainty, which is a consequence of the inherent variability of the rock mass, will be discussed and illustrated in Chapters 6 and 7 describing major case examples relating to long, deep tunnels and hydropower cavern groups.

Chapter 4

Rock fractures and *in situ* rock stress

All rocks are traversed more or less distinctly by vertical or highly inclined divisional planes termed Joints.

—Archibald Geikie, p. 501,
Text Book of Geology, Macmillan, 1882

It may be noted that there are certain boundary conditions, which would apply in depth along any open fissure, and apply with uniformity along the surface of the ground. There can be no pressure or tension perpendicular to the surface, and no shearing force parallel to it, in its immediate vicinity.

—E.M. Anderson, p. 11,
The Dynamics of Faulting, Oliver and Boyd, 1942

4.1   INTRODUCTION

When characterising a rock mass for rock engineering design, it would be convenient if the rock were a CHILE material (Continuous, Homogeneous, Isotropic and Linearly Elastic). This would considerably simplify the site investigation and modelling procedures. However, rock masses are not CHILE materials: they are DIANE materials, i.e., Discontinuous, Inhomogeneous, Anisotropic and Not Elastic—which complicates the site investigation, modelling and design studies, and is a *raison d'être* for this book. Moreover, a rock mass is not like a specimen in a testing machine waiting to be loaded. Rock masses are pre-loaded with an *in situ* state of stress caused mainly by the movement of tectonic plates. Therefore, in our overall context of 'risk', it is critical that the two factors, rock fractures and *in situ* rock stress, are understood and taken into account during the modelling, design and construction for rock engineering structures.

The lack of continuity in a rock mass is caused by discontinuities, i.e., joints and faults, noting that joints are caused by a tensile stress and faults are caused by a shear stress. As evident from the Geikie (1882) quotation at the beginning of this Chapter, the role of fractures in affecting rock masses, such as in beach erosion, has long been recognised. Similarly, the role of *in situ* rock stress in the formation of faults was clearly outlined in the early book "The Dynamics of Faulting" by Anderson (1942). But note that, during underground construction, it is not just the roof that is affected: rather the state of the pre-existing stress field is altered by the excavation process. Also, there is a two-way interaction between the rock fractures and the *in situ* stress, as highlighted earlier in Chapter 3 in the Rock Engineering Systems (RES) explanations.

The rock stress can create and alter rock fractures which in turn can have a substantial effect on the magnitudes and orientations of the *in situ* principal stresses.

Because both the fractures and the *in situ* stress are two of the most important aspects for rock engineering construction, they must be considered in rock engineering risk evaluation. In particular and as illustrative examples, they influence repository design for the underground disposal of radioactive waste as discussed in Chapter 5 and are key components in the two major case examples reported in Chapters 6 and 7 on the construction of long, deep tunnels and hydropower caverns. We now provide explanations of the two subjects in the following Sections 4.2 and 4.3.

4.2   ROCK FRACTURES

In the early days of rock mechanics, solutions to the stress states around underground excavations were obtained via elastic solutions, i.e., by assuming that the rock mass is a continuous, homogeneous, isotropic, elastic continuum, and using rock properties obtained from rock samples in the laboratory. In the case of relatively unfractured rock masses, this assumption can lead to useful solutions; however, it has now long been recognised that, for most rock masses, the testing of intact rock specimens in the laboratory will not produce values representative of the *in situ* rock mass—because of the discontinuities in the rock mass, i.e., the joints and faults.

Indeed, this was the main reason why the International Society for Rock Mechanics (ISRM) was formed in 1962. As noted in the ISRM 50th Anniversary Commemorative Book (Hudson & Lamas, 2012), Professor Charles Fairhurst mentions that on 24 May 1962, i.e., the day before the constitutional meeting of the ISRM took place in Salzburg, a reporter asked Professor Müller, "Do we know the strength of rock?", to which Professor Müller replied, "For rock tested in the laboratory, yes. For a rock mass, no. This is what we need to determine. This is why we need an International Society for Rock Mechanics". Now, more than 50 years later, there is considerably more understanding of fractured rock masses—generated by enhanced structural geology knowledge, site investigation techniques and computer modelling.

4.2.1   The spectrum of brittle and ductile rock deformation

We begin by summarising the geological spectrum of deformational features that is structurally possible, from discrete fractures to pervasive fabrics, with descriptions of the geometrical characteristics of the main components. Each component is a member of the complete spectrum of possible structures, ranging from the extremes of individual tensile and shear fractures (which are examples of highly localised deformation) to a pervasive mineral fabric (which represents a uniform distribution of deformation). This enables the approach to the structural categorisation to be within several complementary contexts:

–   brittle to ductile deformation;
–   decreasing strain localisation or, conversely, increasing pervasiveness;
–   increasing pressure and temperature;

- decreasing deformational strain rate;
- from a continuum via a discontinuum back to a continuum;
- from homogeneity via inhomogeneity back to homogeneity, and
- from straightforward rock mechanics characterisation, through more problematical characterisation, back to straightforward characterisation.

Given this spectrum, it is important to be able to recognise multiple deformational sequences and, via the content of an interaction matrix (the interaction matrix device having been previously explained in Chapter 3), we illustrate examples of the possibilities when there have been two consecutive deformational events. Cases where there have been more than two such events can be represented by a pathway through the interaction matrix, a concept explained in Section 3.3.3.

Block (a) in Figure 4.1 represents a homogeneous, isotropic rock with a mechanically inert marker horizon. When stressed, this rock can form the variety of structures illustrated by blocks (b) to (f) depending primarily on the lithology, stress state, and the physical conditions operating during deformation (e.g., pressure, temperature and strain rate).

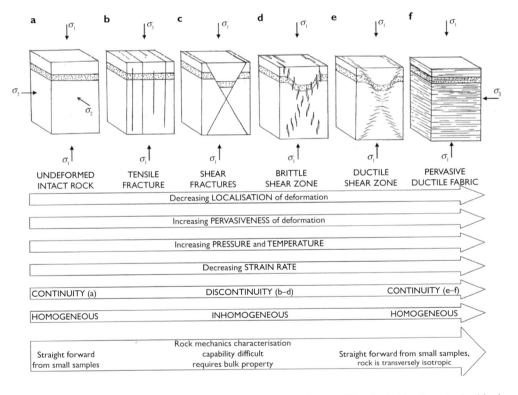

Figure 4.1 The main modes of structural deformation in a rock mass. The shaded horizon in the block diagrams is a mechanically inert marker band, included to indicate the deformation occurrence (diagram developed in association with Prof John Cosgrove of Imperial College London).

Block (b) contains tensile fractures which will form when the conditions under which the deformation is occurring are appropriate for brittle deformation (i.e., low pressure and temperature and high strain rate) and when the differential stress is low. The tensile fractures form perpendicular to the least principal stress and hence parallel to the major principal stress. The result is the formation of parallel fractures which form a fracture set. An exception to this uniform fracture orientation occurs when the intermediate principal stress is equal to the minimum principal stress; the fractures still form parallel to the major principal stress σ_1 but can open with equal ease in any direction, their strike orientation then being random with the generation of a polygonal fracture array.

In the case of Block (c) and like tensile fractures, shear fractures will form when the conditions under which the deformation is occurring are appropriate for brittle fracture (i.e., low pressure and temperature, and high strain rate) but when the differential stress is high. Shear failure occurs along two planes (both containing the intermediate principal stress, σ_2) and inclined at an angle of $(45 - (\phi/2))$ to σ_1 where ϕ is the angle of sliding friction, which for many rocks is about 30°. So, two sets of planes can develop in response to a single episode of shear deformation, each inclined at ~30° to σ_1. Shear fractures are characterised by displacements parallel to the fracture walls. If such fractures have little displacement, they can be termed joints; otherwise, they are termed faults. Fracture-parallel displacements tend to remove asperities on the fracture walls, resulting in them generally being less cohesive than tensile fractures.

Blocks (d) and (e) represent cases where the conditions under which deformation is occurring change from those appropriate for brittle deformation to those appropriate for ductile deformation, so the shear deformation becomes progressively less localised and is represented by a zone of shear deformation rather than a discrete shear plane. The varied expressions of localised shear failure, namely discrete conjugate shear fractures, brittle shear zones, and ductile shear zones often result in the formation of a network of highly deformed shear zones enclosing lenses of relatively undeformed rock.

Block (f) characterises the far end of the spectrum in Figure 4.1, noting that all the other structures described change the rock from an assumed originally homogeneous state into one which is inhomogeneous. Conversely, the formation of a pervasive ductile fabric does not affect the rock's homogeneity, but it does generate an important mechanical anisotropy in the rock mass.

4.2.2 Multiple deformational sequences

During their history, rocks have been subjected to a number of tectonic events, each of which may have generated structures of the type illustrated in Figure 4.1, but superimposed on a pre-existing structure, as shown via the interaction matrix in Figure 4.2 which illustrates the superposition of a new fracturing event on a pre-existing structure. Note the five basic deformational modes in the boxes along the leading diagonal (from top left to bottom right) in Figure 4.2. All the possibilities of one deformational mode superimposed on another are then illustrated in the 20 off-diagonal boxes. For example, the consequence of tensile fracturing being superimposed on shear fracturing is shown in Box (1,2), i.e., Row 1, Column 2 (using the clockwise convention previ-

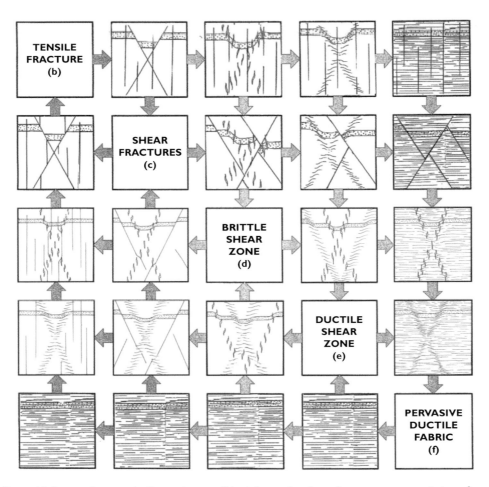

Figure 4.2 Interaction matrix illustrating possible deformational overlay sequences consisting of two deformational events. The leading diagonal terms (top left to bottom right) are the individual deformational modes illustrated in Figure 4.1. The off-diagonal terms indicate the superposition of one mode in the leading diagonal on another mode in the leading diagonal, following the clockwise direction of the arrows (the convention explained in Chapter 3).

ously described in Chapter 3). Conversely, the consequence of shear fracturing being superimposed on tensile fracturing is shown in the complementary off-diagonal box (2,1), i.e., Row 2, Column 1. This same matrix component locational principle is applied to all the other deformation mode superimpositions in Figure 4.2.

The Figure 4.2 matrix diagram considers just one deformational event superimposed on one other and assumes that the stress orientations in both cases are the same, i.e., with σ_1 acting from top to bottom of the page and σ_3 acting across the page. However, many episodes of ductile and brittle deformation may have affected the rock and the stress orientations linked to the different events are unrelated and therefore are likely to have been different and had different magnitudes. Such a change in the orientation of σ_1 would result in more complex deformation patterns than those shown in Figure 4.2. Similar matrices could be generated for considering more complex multiple superpositions according to other user interests, and hence provide a clearer appreciation of how such deformational structures are likely to impact on the rock mechanics properties of the rock mass. The key point, hopefully illustrated by this brief discussion of some of the potential rock mass structures that the engineer might encounter, is that it is essential in rock engineering design to have a good understanding of the structural geology setting, and hence to reduce the risk.

4.2.3   The risks associated with different types of rock mass

Rock masses will usually contain fractures—often introduced into the rock mass at different times and under different *in situ* stress conditions, as noted in the previous Section. The engineering risks associated with different types of fractured rock masses will vary therefore on the engineer's recognition of the specific local fracturing and the engineer's knowledge and skill in designing an appropriate engineering scheme that can cope with the fractures (and the intact rock if the rock stress–intact rock strength ratio is high). However, it is possible to give some general guidance on 'high risk' and 'low risk' rock masses in the context of rock engineering. Accordingly, the following Tables 4.1–4.5 have been included to provide a visual indication of the spectrum of rock masses and their engineering difficulty. The rock mass risks have been categorised in Tables 4.1–4.5 according to their very high, high, medium, low and very low engineering risk potential. The reader should note that this rock mass risk classification is only intended to indicate the general risk: all rock masses should be approached initially with engineering caution.

Exposure 1 in Table 4.1 is very high risk because of the significantly shattered and weathered nature of the rock mass and, in the photograph, there is a visibly open, large fracture, possibly forming a rock slab that can slide down into the road. Because of the topography and the long lengths of these roads in India, it is difficult to implement stability measures. The rock mass in Exposure 2 is a mass of shattered and unconnected blocks which have little resistance to movement if they were to be subjected to any engineering activity.

Exposure 3 is a magnified part of Exposure 2 with indications of dynamic breakage; note the fractography markings (fracture surface characteristics that indicate the mode of fracture formation) on the surface of the left central block. These markings occur when there has been high speed crack development. Exposure 4 is another very high risk case because the strata can easily separate and are significantly fractured, thus enabling rock blocks to form and fall out.

Exposure 5 is of a greywacke rock, a fractured, highly weathered, clayey, impure sandstone. This is a high risk rock mass because of the intensity of the fracturing and the weak bonding of the blocks. It would be easy for the blocks in this rock to loosen and fall/slide out in an underground excavation. 'Exposure' 6 in Table 4.2 is not of an

Table 4.1 Very high risk rock masses.

Very high risk

Exposure 1. Fractured and weathered roadside rock slope, India

Exposure 2. Shattered basaltic rock mass, Minnesota, USA

Exposure 3. Part of the shattered basaltic rock mass in Exposure 2, Minnesota, USA

Exposure 4. Coastal cave in a layered rock mass, UK

exposure *per se*, but of a highly fractured drillcore from the Jinping II tunnelling site in China. In fact, the fracturing of the drillcore, known as discing, is caused by the high *in situ* rock stress—the high 3-D *in situ* stresses are concentrated around the drill core as the hole is being drilled, causing discs of a more or less regular thickness to be broken off. So, the drillcore in Exposure 6 does not represent a pre-existing fractured rock mass, but one that can become fractured when an excavation is made as a result of the altered high stress field. The process of tunnelling and constructing caverns in this highly stressed, marble rock mass is presented in detail in the major case examples in Chapters 6 and 7, respectively.

The orientations of right-angled faces in Exposure 7 can be governed mainly by either the inherent anisotropy of the marble or by the pre-existing rock fracture network. Note where non-vertical fractures have caused block failure, thus leaving additional triangular faces on some of the blocks. In Exposure 8, the coastal rock mass is subject to progressive weathering and the increased possibility of block falls due to the pre-existing fractures.

The reason why Exposures 9 and 10 are medium risk rock masses is because of the propensity for rock blocks to form at excavation surfaces. The bedded strata in Exposure 9, see also the book cover, are vertically jointed so, together with the lack of significant bedding adhesion, it is easy for cuboid rock blocks to form at excavation surfaces. In Exposure 10, the upper chalk stratum also contains bedding fractures and vertical fractures *plus* inclined conjugate fractures, i.e., all the ingredients for the creation of both cuboid and tetrahedral rock blocks which can fall into an excavation made in this rock mass.

Exposure 11 represents a low risk rock mass because it has a high intact rock strength, is resistant to weathering, and the large fractures are relatively sparse (note the fracture traversing the rock face from top right to bottom left) so they do not easily form rock blocks. The bedded sandstone layers in Exposure 12 are similar: although parallelepiped blocks are evident, the rock mass is stable.

Table 4.2 High risk rock masses.

High risk

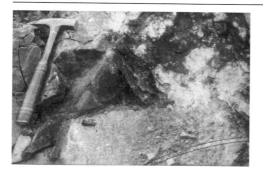

Exposure 5. Greywacke rock mass, Istanbul, Turkey

'Exposure' 6. Marble drillcore from the Jinping II tunnelling site, China

Table 4.3 Medium risk rock masses.

Medium risk

Exposure 7. Marble quarry in Portugal

Exposure 8. Coastal exposure in the UK

Exposure 9. Well-bedded and jointed, alternating limestones and thin mudstones, UK (see book cover)

Exposure 10. Fractured superincumbent grey-white Lower Chalk, above iron pigmented Red Chalk, above bioturbated ferruginous sandstone conglomerate, UK

Table 4.4 Low risk rock masses.

Low risk

Exposure 11. Pre-split, granitic basement rockface, Finland

Exposure 12. Sandstone rockface, Geopark, China

Exposure 13. Granite quarry, Poland

Exposure 14. Large basalt pillars. Note the rock climber in the upper centre of the photograph

Table 4.5 Very low risk rock masses.

Very low risk

Exposure 15. Machine excavated tunnel sidewall in the granitic Bohemian Massif, Czech Republic

Exposure 16. Fused volcanic layers, Jeju Island, South Korea

Exposure 17. Long drillcore stick from site investigation borehole in the Gobi Desert granite, China

Exposure 18. Tunnel in a disused salt mine, Colombia

In the cases of both Exposures 13 and 14 in granite and basalt, the intact rock strength is high—which means that, even if rock blocks form, they remain wedged in place, although failure of the type shown at the bottom left of Exposure 14 can occur.

In the case of Exposure 15, although the granitic rock mass is fractured, there is no coherent rock block formation and any blocks that do form are geometrically locked in place. In the case of Exposure 16, the rock is sufficiently strong and there are no fractures, so failure does not easily occur.

Both Exposures 17 and 18 are examples of rock masses, or at least portions of rock masses, where there are no pre-existing fractures, so failure will not occur unless the rock stresses are high compared to the intact rock strength. Note that in Exposure 18, the just discernible, curved excavation-peripheral fractures in the rock (upper right-hand side of the photograph) have been caused by the increase in the local stress state following excavation of the tunnel but, because there are no pre-existing fractures in the salt, large rock blocks do not form.

As has been indicated in this Section 4.2, not only the rock fractures but also the *in situ* rock stress needs to be studied in the context of risk in rock engineering. Accordingly, the next Section discusses the pre-existing state of stress in a rock mass, noting that it is the stress that causes many of the fractures, and, reciprocally, it is the fractures that can locally alter the *in situ* state of stress. Note that the individual hydroelectric powerhouse construction problems related to fractures are summarised in Appendix A.

4.3 *IN SITU* ROCK STRESS

There is a pre-existing stress in rock masses, partly because of the weight of the overburden and partly because of tectonic plate movement. Thus, underground rock engineering involves making excavations in a pre-loaded material. The construction process alters the rock stress components, increasing the magnitudes in some locations and reducing them in others. As noted in the quotation from Anderson (1942) at the beginning of the Chapter, the normal stress component perpendicular to a free surface is zero and the shear stress component parallel to the surface is also zero. The risk is related to the engineering activities causing an increase in the magnitude of other stress components and hence failure, especially if this is associated with the presence of fractures.

4.3.1 The stress state in a rock mass

The fundamental problem associated with the estimation of the *in situ* principal stress magnitudes and directions is that the stress, being a point property, is not uniform throughout a rock mass because of the perturbations introduced by the geological 'heterogeneities' on a variety of scales, as described in Section 4.2. The reasons for this *in situ* stress variability are now discussed in relation to the tectonic/regional scale, site scale, excavation scale, borehole/measurement scale, and microscopic scale, and we explain the influence of rock inhomogeneity, anisotropy, discontinuities and free faces. Note that, unlike force, stress is a tensor quantity requiring six pieces of information to define it. This is because stress is force per unit area. The stress quantity is expressed either as a symmetrical 3×3 matrix containing six independent terms (three normal stresses and three shear stresses) or as the magnitudes and directions of the three principal stresses (which are mutually orthogonal).

4.3.1.1   In situ *rock stress scales*

Tectonic plate scale—The main factor in generating a regional *in situ* stress state in rock masses is the movement of the Earth's tectonic plates. This results in the application of a stress field across large areas of a particular land mass, hundreds or thousands of kilometres, and hence creates a regional stress. Studies of the World Stress Map (Heidbach *et al.*, 2008) confirm the association between the tectonic plate movement and the directions of the major horizontal *in situ* principal stress.

Site scale—The site scale of interest for rock engineering purposes will depend on the purpose of the engineering facility. For example, a rail tunnel through a rock mass will generally have a relatively low volume of interest in terms of its cross-section, i.e., the rock mass immediately surrounding the tunnel. In special cases, this volume may be larger, as in the case of a tunnel in rock below the sea where there is the potential for water inflow. A cavern for storing a compressed gas will have a larger local volume of interest, not only because of its larger cross-sectional dimensions, but also because of the potential for gas leakage through fractures in the rock mass. An underground repository for storing radioactive waste will have a large rock mass volume of interest because the facility itself may be kilometres in extent and because of the over-riding safety criterion that unacceptable quantities of radionuclides should not escape to the biosphere, see Chapter 5. In the case of the long hydropower headrace tunnels discussed in Chapter 6, where rockbursts and water inflow were on-going major hazards during construction, the volume of interest was larger than usual because the mechanically stored strain energy causing the rockbursts and the rock mass volume supplying the water inflow were large. Similarly, the hydropower cavern complexes described in Chapter 7 have large dimensions for underground excavation schemes.

Excavation scale—Even though the project itself may occupy a large rock mass, as in the case of the radioactive waste repository or hydropower scheme mentioned above, the local *in situ* rock stress around each separate excavation (with dimensions less than, say, 100 m) in the project has to be considered, and this can vary from excavation to excavation—even though they are constructed within the same site scale stress field.

Borehole scale—*In situ* stress measurements are generally undertaken using overcoring or hydraulic methods in boreholes. Thus, we also have to consider the variability of the *in situ* stress on scales of the order of 0.1 m in order to interpret these measurements. Are there significant perturbations to the regional stress on the borehole scale? If so, we need some method of understanding these and being able to upscale the values to the excavation and site scale, as required.

Microscopic scale—In the overcoring method of stress measurement, strain gauges are used which are of the order of one centimetre long. Thus, we also need to understand the perturbations of rock stress on the microscopic, or at least grain size, scale, in order to interpret any anomalies in the strain gauge readings as the overcoring is undertaken.

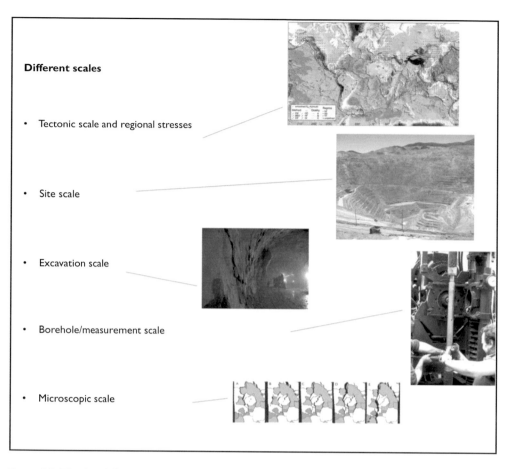

Different scales

- Tectonic scale and regional stresses

- Site scale

- Excavation scale

- Borehole/measurement scale

- Microscopic scale

Figure 4.3 The five different scales for considering the variation of *in situ* rock stress. The five illustrations, from the top down, are the global scale (the scale for the World Stress map), the site scale (illustrated by the Bingham Canyon Mine in Utah), the excavation scale (illustrated by a hydroelectric project tunnel in China where rock spalling and bursting occurs, see Chapter 6), the measurement scale (illustrated by a borehole straddle packer used in the hydraulic fracturing method of stress measurement), and the microscopic scale (illustrated by computer simulation of the microstructural breakdown of a crystalline rock).

These five scales, tectonic/regional scale, site scale, excavation scale, borehole/ measurement scale, and microscopic scale, are illustrated in Figure 4.3.

4.3.2   Stress perturbation factors

4.3.2.1   Rock inhomogeneity

On all the scales highlighted in the previous Section, the rock mass can be inhomogeneous. Hence, in order to understand fully the *in situ* stress variation, we have to also consider how rock stress is transmitted through an inhomogeneous medium.

In Figure 4.4(a), the Weibull probability density distribution is illustrated. This distribution is used here to show, via numerical modelling, how inhomogeneity can influence the internal stress state in a rock mass, see Figures 4.4(b) and (c).

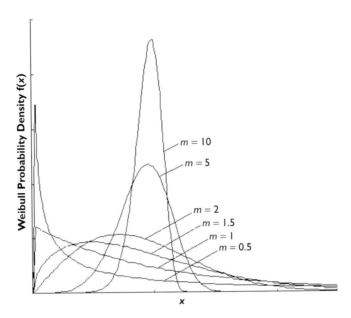

Figure 4.4a The probability density distribution, f(x) as a function of *x* for different values of the shape parameter, *m*, in the Weibull statistical distribution (from Tang & Hudson, 2010).

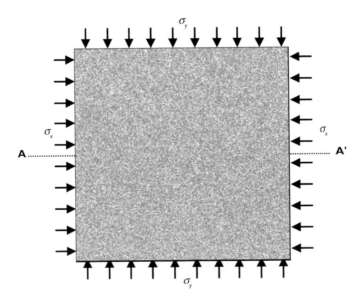

Figure 4.4b Numerical simulation model of an inhomogeneous rock block—with the grayscale indicating the variation in elastic moduli of the individual elements following the Weibull distribution shown in Figure 4.4(a).

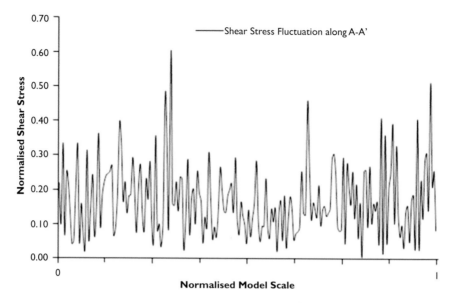

Figure 4.4c Shear stress fluctuations along the cross-sectional line AA′ shown in Figure 4.4(b).

The Weibull statistical distribution shown in Figure 4.4(a) is used to characterise the elastic modulus of the elements in a finite element model of an intact rock. As the homogeneity index, *m*, decreases, so the inhomogeneity of the elemental elastic moduli increases.

In Figure 4.4(b), a biaxially loaded 2-D inhomogeneous rock block is shown with 40,000 elements. In Figure 4.4(c), the normalised shear stress is plotted across the section AA′ shown in Figure 4.4(b). These Figures from Tang and Hudson (2010) are independent of absolute scale and illustrate the type of natural variation of stress that can occur within an inhomogeneous loaded rock mass. Considering that the overcoring method of stress measurement is based on the output of small strain gauges, it is not surprising that there can be a large variation in the *in situ* stress measurement results in, for example, a granitic rock with crystal sizes commensurate with the strain gauge dimensions.

4.3.2.2   Rock anisotropy

In a similar way to the variation in the stress as a function of inhomogeneity, the stress will also vary with the rock anisotropy—which may be coherent, as in the case of a set of regular rock strata, or more difficult to characterise when combined with large and small inhomogeneities. The rock strata in Figure 4.5 are from the Lower chalk formation in the south of England and comprised of alternating hard and soft chalk, the harder layers (the lighter ones in Figure 4.5) being able to sustain a greater shear stress than the softer layers.

Figure 4.5 Anisotropy in chalk strata composed of alternating hard and soft layers (dimension across the picture diagonal ~3 m, Lower chalk, Isle of Wight, UK).

4.3.2.3 Rock fractures

The other main cause of *in situ* stress variation is the presence of natural, pre-existing fractures in the rock mass. These occur on all scales, from microscopic flaws in rock grains to brittle deformation zones that can be kilometres in length. In Figure 4.6, a numerical finite element simulation indicates how the shear stress can vary in a granitic microstructure when it is subjected to uniaxial stress.

In Figure 4.7, there is an illustration of how the stresses can vary across a site several kilometres long containing a variety of extensive brittle deformation zones. This computer modelling example using the 3DEC program indicates how the presence of major faults in the rock mass can cause a large variation in the local principal stresses and hence why the collations of *in situ* stress measurement data can often show a large spread.

Such numerical modelling is helpful in explaining the trends and the types of variations that can occur. For example, from the modelling one can create a histogram of the magnitudes of the principal stresses occurring in a rock mass. To set up the numerical model, however, requires characterisation of the rock mass, requiring in particular the mechanical properties of the rock and major discontinuities: for the rock, the deformational elastic modulus, E, and the Poisson's ratio, v; and, for the discontinuities, the normal stiffness, k_n, the shear stiffness, k_s, the cohesion, c, and the angle of friction, ϕ. There are six main methods (Hudson *et al.*, 2008) to estimate the mechanical properties of fault zones, most of which are directed towards estimation of E and v.

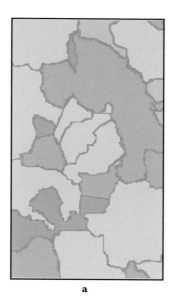 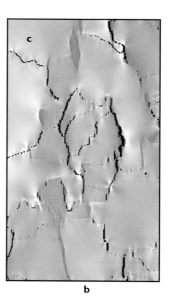

a   b

Figure 4.6 Illustration of stress variation on the small scale. (a) Simulation of a specific granite micro-structure with crystals of feldspar, mica and quartz (with uniform properties within each grain). (b) Shear stresses developed in the microstructure during failure when loaded in the vertical direction (higher stresses are a lighter grayscale).

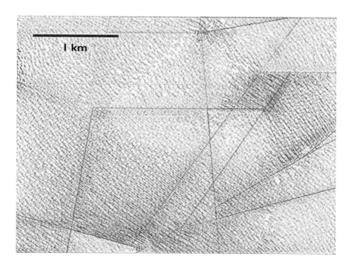

Figure 4.7 Portion of a numerical modelling (3DEC) output indicating the directions and magnitudes of the major principal stress for a horizontal section of a rock mass in Sweden at 450 m depth. The through-going lines are brittle deformation zones (faults) which have been incorporated in the modelling. The overall horizontal direction of the principal stress is NW–SE in line with the northern European regional tendency. The West–East direction is across the page. The shading indicates the magnitudes of the major principal stress, ranging from 18–60 MPa, from E. Hakami.

4.3.2.4 *The influence of a free surface*

A key aspect for rock engineering is the influence of a free surface on the *in situ* stress state, as shown in Figure 4.8. As noted by Anderson (1942), on a free rock surface, whether natural or engineered, there can be no normal or shear stresses; hence, by definition, the free surface is then a principal stress plane. This means that one principal stress is perpendicular to the free surface (assumed to be σ_3 in Figure 4.8) with a zero value; and the other two principal stresses must be parallel to the free surface. Thus, the orientations and magnitudes of these three local principal stresses will be different to those of the regional stress field. The Earth's surface is one example of a free surface, so the stress state must be as in Figure 4.8 with a zero value principal stress acting perpendicular to the surface and the other two principal stresses being parallel to the surface. If the Earth's surface is locally non-planar (e.g., as in Figure 4.9) there will be an associated perturbation in the near surface rock mass because a) the vertical stress (being generated by the weight of rock above) will vary from place to place, and b) the transmission of the horizontal stress will occur at a deeper level.

In the case of an engineered free surface, e.g., the unsupported periphery of a tunnel or cavern as in the tunnelling and cavern schemes described in Chapters 6 and 7, this redistribution of the pre-existing rock stress state occurs in conjunction with two other main effects: displacement of the periphery; and the creation of a sink for water inflow. The result is an Excavation Disturbed or Damaged Zone (EDZ)—being an inevitable consequence of removing part of the rock mass and creating the free surface. Such excavation not only removes the rock but reduces the mechanical and hydrogeological resistance of the region to effectively zero. Any additional disturbance above this inevitable threshold disturbance is caused by the particular mode of excavation, blasting or Tunnel Boring Machine (TBM). But the main effect in the current context is the significant alteration of the *in situ* stress state in the vicinity of the excavation periphery (Hudson *et al.*, 2009).

*   *   *   *   *

A summary of the main stress perturbation factors is given in Figure 4.10.

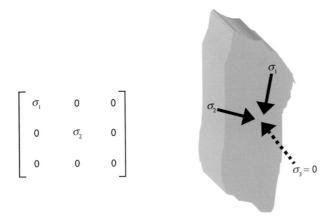

Figure 4.8 Principal stresses are parallel and perpendicular to open fractures and excavation surfaces.

Figure 4.9 Example of significant surface topography (in Switzerland) where the near-surface *in situ* stress is affected by the mountains and valleys.

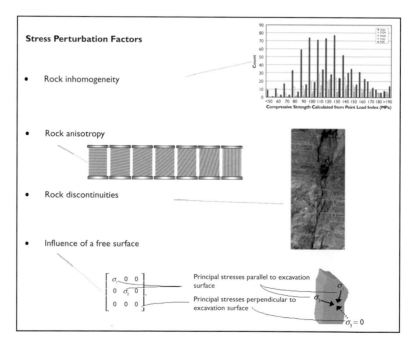

Figure 4.10 The four main stress perturbation factors of rock inhomogeneity, rock anisotropy, discontinuities in the rock mass (at all scales) and the influence of a free surface (whether natural or generated by engineering activities).

4.3.3   Evidence of *in situ* stress variability

The evidence of *in situ* stress variability comes not only from the variability encountered during individual stress measurement campaigns but also from the measured principal stresses compilations made over the years. Chief among these is the World Stress Map which contains the data from inferred stress states and direct stress measurements (e.g., Heidbach *et al.*, 2008), i.e., from

- earthquake focal mechanisms,
- wellbore breakouts and drilling-induced fractures,
- *in situ* stress measurements (overcoring, hydraulic fracturing, borehole slotter), and
- young geological data (from fault-slip analysis and volcanic vent alignments).

As useful overview information for any particular site, it is possible to custom create one's own stress map for any given area of the Earth's surface from this database.

4.3.3.1   *Stress vs. depth compilations*

Compilations of *in situ* stress components with depth vary from the early Hoek & Brown (1980) compilation to the relatively more recent ones of, for example, Lee *et al.* (2006). A problem with such compilations for the rock engineering designer is that the scatter of the data is too high for accurate specification of the stress state at any particular location and depth. Indeed, at first sight, one might conclude that many such compilations do not have any utility for rock engineering design because the scatter in the data is far too high, e.g., see Figure 4.11(a). There is a large scatter of points in Figure 4.11(a) for all three of the principal stresses—both for the horizontal and vertical directions. In other words, there is no clear stress state at any given depth level and only a general trend of the principal stresses increasing with depth.

However, an interesting approach aimed at reducing the scatter in the data was proposed by Lee *et al.* (2006) whereby the values of the principal stresses are plotted against the first stress invariant, $I_1 = \sigma_1 + \sigma_2 + \sigma_3$ (the sum of the three principal stresses), as in Figure 4.11(b), rather than against depth as in Figure 4.11(a), these data being from Western Australia. It appears from Figure 4.11(b) that in fact the measurement data as a stress vs. I_1 plot do appear more reliable than expected from the Figure 4.11(a) plot. Additionally, the ratios between the principal stresses are now evident: $\sigma_1/\sigma_2 = 1.5$; $\sigma_2/\sigma_3 = 1.5$; $\sigma_1/\sigma_3 = 2.3$. Lee *et al.* (2006) also collated the data for Eastern Australia and found the same principal stress ratios.

The existence of these ratios stimulated Harrison *et al.* (2007) to plot stress measurement data from the UK, Chile and Finland in the same way and to establish the principal stress ratios—as shown in Table 4.6. There is a remarkable consistency in the values in Table 4.6, given that they are from different countries and geological environments—indicating that the rock stresses are in fact more constrained than is indicated just by plots of the principal stresses versus depth. The reason for this (which is not that the ratios arise simply because the principal stresses are ordered triples) is that fractured rock masses can only sustain certain principal stress ratios. The mechanical basis is explored in Harrison *et al.* (2007).

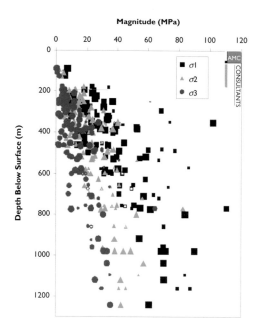

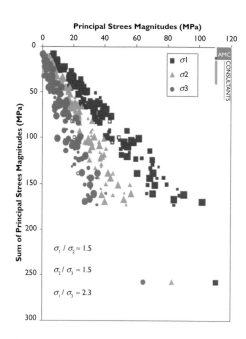

Figure 4.11a Variation of principal stress magnitudes plotted against depth for the Yilgarn Craton, Western Australia (from Lee *et al.*, 2006).

Figure 4.11b Variation of principal stress magnitudes plotted against the first stress invariant for the Yilgarn Craton, Western Australia, for the same data as in Figure 4.11(a) (from Lee *et al.*, 2006).

Table 4.6 Summary of principal stress mean ratios.

	σ_1/σ_2	σ_2/σ_3	σ_1/σ_3
Australia	1.5	1.5	2.3
Chile	1.6	1.6	2.6
Finland	1.7	1.7	3.0
U.K.	1.6	1.4	2.4

4.3.3.2   The ways ahead for improving the understanding of rock stress variability

The understanding of the variability of *in situ* rock stress can be achieved by the interaction of two approaches: enhanced measurement techniques and computer modelling incorporating the geological factors. The key to successful rock stress measurement and estimation lies in the understanding of the nature of rock stress and careful application of quality assurance procedures. This is why the four International

Society for Rock Mechanics (ISRM) Suggested Methods for Rock Stress Measurement (Ulusay & Hudson, 2007) begin in Part 1 with considerations relating to the strategy for rock stress estimation. Then, following the Part 2 and Part 3 Suggested Methods on overcoring and hydraulic methods, the Part 4 Suggested Method contains sets of tables relating to quality assurance and technical auditing. An additional Part 5 Suggested Method explains how to establish a model for the *in situ* stress at a given site (Stephansson & Zang, 2012, 2015). Needless to say, the measurement of *in situ* rock stress is not an easy task but the values will be more reliable if the guidance provided by these ISRM Suggested Methods is followed.

Geological knowledge and computer modelling are the keys to understanding rock stress variability in rock masses and there is currently considerable effort being allocated to the development of extended numerical codes which are able to incorporate more variables and more couplings of different types, e.g., thermo-hydro-mechanical-chemical codes, see Chapter 5 on radioactive waste disposal. Using computer modelling, it is possible to obtain an estimate of the complete principal stress variation across a site, to calibrate the model using *in situ* measurements mentioned in the previous paragraph, to incorporate the stress perturbation factors of rock inhomogeneity, anisotropy, the presence of major and minor discontinuities, to introduce free surfaces, and to explore the effect of variations in these factors, i.e., the sensitivity of the simulation of a particular site to the input variables. Moreover, virtual boreholes can be driven in the numerical models to compare with experimental data.

4.3.4 A case study of modelling *In situ* rock stress at the Olkiluoto site, western Finland

The Olkiluoto site in western Finland is proposed for the disposal of high-level radioactive waste by Posiva—which is the organisation in Finland responsible for research into the final disposal of spent nuclear fuel and for the construction, operation and eventual backfilling and closure of the disposal facility. Olkiluoto is an island (~10 km²) in the Baltic Sea and separated from the mainland by a narrow strait. The Olkiluoto nuclear power plant, with two reactors in operation, a third one under construction and a fourth in the planning stage, as well as the VLJ repository for low and intermediate waste, are located in the western part of the island. The repository for spent fuel will be constructed in the central and eastern part of the island. Site characterisation has been taking place for over 20 years with the ultimate aim of providing sufficient support for the selection of the most suitable locations for the first deposition tunnels and disposal holes for the spent fuel canisters. The case study included here concerns numerical modelling work related to understanding the *in situ* stress state.

The crystalline bedrock at Olkiluoto is mostly comprised of (Aaltonen *et al.*, 2010) high-grade metamorphic supracrustal rocks, the origins of which can be traced back to epiclastic and pyroclastic material, based on geochemical indications and the sporadic presence of relicts of primary bedding structures, such as graded bedding in metaturbidite sequences. These rocks were migmatised, with the development of abundant leucocratic granites, and are intruded by a few thin mafic dykes. In terms of field relations, modal composition, texture and migmatite structure, the rocks at

Olkiluoto can be divided into four major classes: 1) migmatitic gneisses with stromatic, veined and diatexitic varieties, 2) tonalitic-granodioritic-granitic gneisses or TGG gneisses, 3) other gneisses including mica gneisses, quartz gneisses and mafic gneisses, and 4) pegmatitic granites.

On the basis of refolding and cross-cutting relations, it is clear that the rocks at Olkiluoto have been subject to polyphase ductile deformation, including five stages (D1–D5). As the crust cooled, partly during or after D4 or D5, the thrusting that formed the low angle D4 shears was reactivated in the brittle regime during NW–SE contraction or extension, producing many of the site-scale fault zones at Olkiluoto. The faulting is likely to have initiated as semi-ductile to semi-brittle deformation, and only became truly brittle during the later phases.

The large-scale faults at the Olkiluoto site are thought to cause variability in the *in situ* stress field. It was therefore necessary for the supporting numerical modelling code to be able to successfully simulate these structures. Accordingly, the Itasca code 3DEC was utilised, which is a program based on the distinct element method and allows modelling of a site in three-dimensions. The distinct element method uses an explicit time-marching scheme to solve the equations of motion directly. The calculation cycle that is involved in solving the equations is the following: the law of motion and the constitutive equations are applied, sub-contact force-displacement relations are prescribed for both rigid and deformable blocks, integration of the law of motion provides the contact displacement velocities, and finally the sub-contact force-displacement law is then used to obtain the new sub-contact forces (Itasca 2010). Therefore the calculation cycle is not a measure of time as such, as it varies in each simulation according to how complex or extended the cycle is determined to be by the 3DEC program. Note that in the 3DEC code compressive stresses have negative values.

The 3DEC model is created by firstly introducing boundary blocks before creating the actual model. The function of the bounding blocks is to apply uniform boundary conditions unaffected by the Brittle Deformation Zones (BDZs). The initial block dimensions are $20 \times 16 \times 3$ km, after which successive cuts reduce it to $16 \times 12 \times 2$ km. This inner block includes the bounding lineaments which are also modelled as faults. The boundary conditions used have involved the bottom of the model being restricted in the y-axis (elevation) whilst otherwise free. This equates to a 'roller' condition. Force boundaries were applied from the NW–SE to simulate the overall current regional stress field, thought to be caused by the opening of the mid-Atlantic ridge having a 'push' of ~25 MPa. The rock mass is assumed to be homogeneous, isotropic and linearly elastic with the equivalent properties: density 2730 kg/m^3, Young's modulus 55 MPa and Poisson's ratio 0.25. The brittle deformation zones, (Hudson *et al.*, 2008), i.e., the large faults present at the site, were given a cohesion of 0.3 MPa, a friction angle of 20°, a tensile strength of 1 kPa, a normal stiffness of 50 GPa/m and a shear stiffness of 20 GPa/m. Here we just provide some highlights of the modelling results and implications. The full range of modelling simulations is given in Valli *et al.* (2011).

In the computer output graphic in Figure 4.12, we can see the influence of the faults (i.e., brittle deformation zones), indicated by the large grey lines in the diagram in affecting the local stress magnitudes and directions. This is a vertical section along the σ_1 plane in a thrust regime with σ_3 vertical. In this stress regime, the fractures most susceptible to shear movement are those dipping at 30° and striking at ~90° to

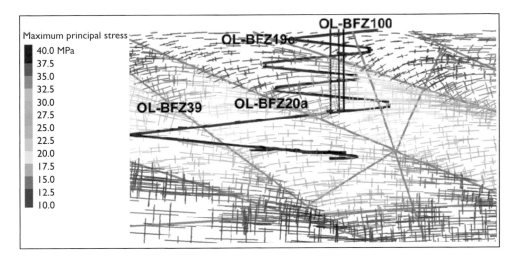

Figure 4.12 Maximum principal stress (magnitudes and directions) in a vertical cutting plane. The letters highlight some of the reference names of the BDZs (brittle deformation zones) and the zigzag line is the ~4 km spiral ONKALO tunnel leading to the proposed repository at ~400 m depth.

the direction of maximum compression, as shown in the upper part of the section in Figure 4.12. The fractures are clearly mechanically active because their influence on the local stress states is evident. At or near the surface it can be seen that the principal stresses are close to horizontal and vertical—a consequence of the free surface effect mentioned earlier. This effect attenuates with depth because the normal stress on the fractures increases and they are then more resistant to re-shear. Note how the changes of the stresses around the BDZs reduce with depth, as does the difference between the two principal stresses, i.e., the stress field becomes somewhat more hydrostatic. As also noted earlier, if the fractures are free surfaces, they could not sustain a shear stress and the stress trajectories would be rotated into an orientation parallel and normal to the fractures surfaces. This is not the case in Figure 4.12 and, despite being zones of weakness within the rock mass, the BDZs can sustain a shear stress.

The section in Figure 4.13 is the σ_2–σ_3 plane, i.e., orthogonal to the section in Figure 4.12. In this section, showing the maximum principal stresses in the plane, we can also see the BDZ perturbation effects.

In Figure 4.13, the applied regional stress field direction (NW–SE) is clearly evident from the principal stress in this horizontal plane, as is the impact of the BDZs on its local orientation and magnitude. There is considerable variability in the magnitude of the maximum principal stresses, from about 12 to 25 MPa. Note also the usefulness of this type of presentation in the general case for locating the position of an engineering structure with a view to minimising the effect of high stresses. The ability to determine a suitable project location cannot be established from point values of the stress state in a site investigation, but requires this type of numerical modelling—so that the whole stress field characteristics become apparent. There may be other stress criteria, such as the need to locate an unlined high pressure water tunnel in a region where the minimum principal stress was not too low.

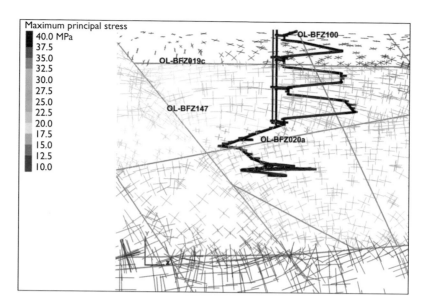

Figure 4.13 Vertical cutting plane, perpendicular to the applied forces. Scale: ~700 m depth represented.

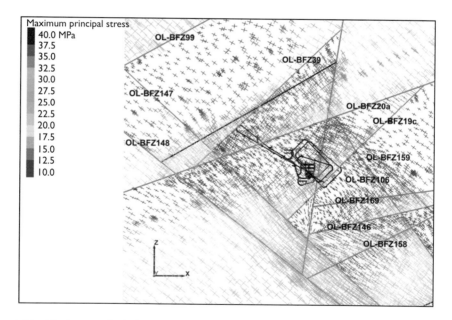

Figure 4.14a Maximum principal stresses in the horizontal cutting plane, depth level −150 m. (See colour plate section at the end of the book.) Scale: ~4 km laterally.

The computer outputs in Figures 4.14(a) & (b) are for horizontal sections, at, respectively, depths of 150 m and 420 m. Comparison of the two Figures indicates the tendency for homogenisation of the stress field (both in terms of magnitudes and directions) as the stress acting across the BDZs increases and hence inhibits their

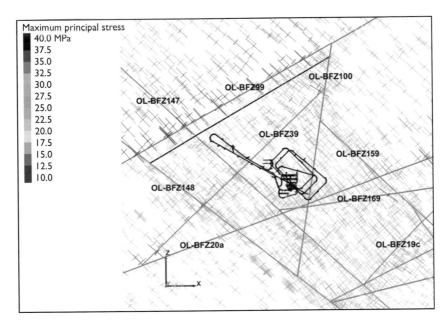

Figure 4.14b Horizontal cutting plane, depth level −420 m. (See colour plate section at the end of the book.) Scale: ~4 km laterally.

ability to deflect the stress field. In other words, the influence of the BDZs wanes with depth. Nevertheless, examination of the diagram shows that, even at this depth, there are still perturbations within the stress field. Taken together, Figures 4.12, 4.13, and 4.14 (a) & (b) indicate that the BDZs can cause compartmentalisation of the principal stress magnitudes and deflection of the principal stress orientations, together with the ramifications for the other components of the stress tensor. These and studies of other computer outputs (e.g., plots of the individual stress tensor components with depth) can provide a much enhanced understanding of rock stress variation, both in general and with reference to a specific site.

We have illustrated the types of output that can be obtained through numerical modelling and the value that these have in interpreting the influence of faults and potentially other structures on the distribution of principal stresses throughout a fractured rock mass. An understanding of the type of variation that can occur is most helpful in deciding on the content of a stress measurement campaign and the location of an engineering facility (assuming that there is a choice in its location). A further case study of modelling rock stress is included in the hydropower cavern major case example in Chapter 7.

4.4 CHAPTER SUMMARY

Rock fractures and *in situ* rock stress are two of the most important factors governing rock engineering risk. Indeed, we noted that it is the presence of rock fractures in a rock mass that stimulated the formation of the International Society for Rock Mechanics. Via Figure 4.1, the types of fracture that can occur were illustrated in the context of their mode of formation, noting that there is usually a sequence of geological deformation modes that create new fractures superimposed on an existing fracture array—which has implication for the way in which fractures can terminate against other fractures. In Tables 4.1–4.5, a spectrum of rock masses is shown, representing very high risk rock masses through to very low risk rock masses. The allocation of the type of risk is based on the types of fracture arrays and the rock types, the main key being whether it is easy or difficult for rock blocks to form and fall or slide into rock excavations.

The discussion in Section 4.3 intimated that the history of stress measurement is, firstly, one of difficulty in conducting the measurements themselves and, secondly, finding that there is a wide spread of stress values at a particular site. We have indicated how the local *in situ* rock stress field at a site can be influenced by a wide variety of factors, not least by the presence of fractures—so that we should not expect to find a constant state of stress across an engineering site of interest. The factors of both scale and perturbations to the stress field explain why a large spread of results is often experienced. However, to assist in the characterisation of the stress variability through a rock mass, we now have the ability to numerically model a rock mass with many of its idiosyncrasies, especially the presence of the major discontinuities, i.e., large faults. The case study of stress modelling for the Olkiluoto area in western Finland indicates that the rock mass fault parameters are critical because they can either allow stress–geology interaction to develop or deter it. In addition, the stress magnitude changes in the modelling example are mostly a result of brittle deformation zones that have a sufficiently gentle dip and are dipping in the same direction as the maximum principal regional stress field.

We recommend that stress measurement campaigns should begin by a study of the geology of the rock mass in question so that the possibility of a variable local stress state can be evaluated, bearing in mind the five size scales (tectonic/regional scale, site scale, excavation scale, borehole/measurement scale, and microscopic scale) and the four perturbation effects (rock inhomogeneity, rock anisotropy, discontinuities in the rock mass, and the influence of a free surface—whether natural or generated by engineering activities). Identifying especially the large rock fractures, as described in Section 4.2, enables a stress modelling procedure of the type described in Section 4.3 which will allow the main effects of the large fractures to be characterised and illustrated. This procedure will give a much deeper understanding of the factors affecting the stress field and their estimated magnitude, leading not only to more coherent stress measurement campaigns but also to a much improved method of interpreting the results. Although stress is somewhat of an abstract quantity, knowledge of the principal stress magnitudes and orientations is usually crucial to the design of safe rock engineering excavations, and the factors mentioned in this Chapter will assist in the estimation of the *in situ* stress field and its variations at a particular site.

Both rock fractures and rock stress play a large part in the major case studies described in Chapters 6 and 7.

Acknowledgement: The authors thank the Posiva Oy company in Finland for their support and for allowing the data in Section 4.3 to be used in this book. However, the viewpoints and conclusions presented in Section 4.3 are solely those of the authors and do not necessarily coincide with those of Posiva Oy.

Chapter 5

Radioactive waste disposal: overcoming complexity and reducing risk

5.1 THE DISPOSAL OBJECTIVE

Of all rock engineering projects, the design of an underground repository for high-level radioactive waste is the most difficult and requires the longest design time—of the order of 20 years. This protracted design period is the result of a unique combination of features, five of which are listed below.

1   The over-riding purpose of the repository is to ensure that unacceptable quantities of radionuclides do not escape to the biosphere.
2   Unlike other rock engineering projects, such as a metro tunnel, a hydroelectric project, or a mine, where there is a defining facility position and often orientation, the repository provides a static storage function rather than the active function of other rock engineering projects. This means that the repository can potentially be located in a variety of different rock masses and at different depths (apart from planning and regulatory restrictions).
3   Because of the required isolation function, the design involves understanding the coupled geological-thermal-hydrological-mechanical-chemical-biological processes to a significantly greater degree than for other rock engineering projects.
4   The regulators require that the design life of the repository is of the order of hundreds of thousands of years, as compared to a few days for the area required for the working face of a longwall coal mine and up to about 100 years for 'conventional' civil rock engineering projects.
5   There can be significant public involvement in the acceptability and location of a repository facility.

Ensuring that unacceptable quantities of radionuclides do not escape to the biosphere requires extensive research on protective canisters for the waste and wide-ranging site investigation at a chosen site to establish how and when such canisters could leak and how radionuclides could travel through rock fractures to the biosphere. The required very long design life may require consideration of the effect of, not one, but several future ice ages with the associated effects on the groundwater, and the subjects discussed in Chapter 4: i.e., the rock fractures and the *in situ* rock stress.

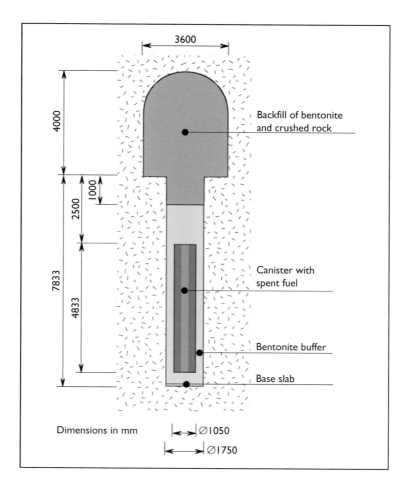

Figure 5.1 Schematic view of deposition hole with canister, buffer materials and backfill for radioactive waste disposal, dimensions in mm (from Pers *et al.*, 1999).

Moreover, the choice of location involves planning conditions and interaction with the general public. In order to reduce the risk to an acceptable level, detailed studies of all the relevant factors are required. A schematic of the emplacement of a canister containing radioactive waste in a deposition hole is shown in Figure 5.1 and a typical overall repository layout is shown in Figure 5.2.

5.1.1   An example of radioactive waste repository statistics

In 2013, Posiva (the implementer in Finland) produced a guide of disposal related key figures according to the situation in 2012 as information for the public. These statistics are listed below to provide an indication of the figures involved in designing and

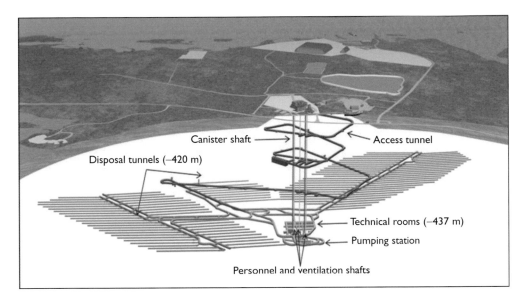

Figure 5.2 Schematic of a radioactive waste repository layout (from Posiva, Finland).

preparing for an underground radioactive waste repository (*cf*. Figure 5.2 and the spiral research tunnel, known as ONKALO) situated in Olkiluoto, western Finland.

- Number of site investigation boreholes: 57
- Total length of drill core samples: ~32 km
- Length of repository access tunnel: ~5 km
- Incline of access tunnel: 1:10
- Tunnel width × height: 5.5 × 6.3 m
- Combined length of shafts: ~1 km
- Volume of water leaking into ONKALO: 37 l/min
- Amount of spent nuclear fuel during entire life cycle of Finnish nuclear plants: 5,440 t
- Anticipated number of disposal canisters: ~2,800
- Anticipated repository volume: 1.3 million m^3
- Total length of disposal tunnels: ~35 km
- Distance between canisters in disposal tunnels: ~9 m
- Anticipated annual number of canisters for disposal: 36
- Canister dimensions: 1.05 m diameter. 3.55/4.75/5.22 m length
- The disposal project will require about 20,000 tonnes of copper and about 40,000 tonnes of iron and steel parts
- Multibarrier safety principle: canister, deposition hole bentonite, tunnel backfilling, bedrock
- Anticipated year start of disposal: 2022
- Anticipated year end of disposal and repository closure: 2120s.

Needless to say, this is a mammoth rock engineering design project, similar in magnitude to the Jinping II headrace tunnels and cavern group excavations described in detail later in Chapters 6 and 7. In order to ensure the safety of the repository, it is necessary to ensure that all aspects have been considered and that unacceptable quantities of radionuclides will not migrate to the biosphere. For this, the relevant Features, Events and Processes (FEPs) have to be identified and computer modelling is necessary. We explain the FEPs in Section 5.2, and ensuring the validity of computer modelling in Section 5.3.

CONSTRUC-TION/ LAYOUT (1.1)	Excavation method (1.2)	Excavation method Grouting Reinforcement (1.3)	(1.4)	Displacement effects (1.5)	Construction materials Stray materials (1.6)	(1.7)	Resaturation (1.8)	Repository depth Ventilation (1.9)	Tunnel dimension (1.10)	Ventilation Gas gener. Alt. of reinf. (1.11)	(1.12)	Environmental impact (1.13)
Swelling Temperature (2.1)	**BUFFER/ BACKFILL/ SOURCE** (2.2)	Swelling (2.3)	(2.4)	Swelling (2.5)	Colloid source Groundwater composition (2.6)	Changed flow, holes and tunnels (2.7)	Resaturation (2.8)	Heat transport (2.9)	Swelling pressure (2.10)	Gas transport (2.11)	RN transport (2.12)	(2.13)
Excavation method Amount of reinforcement (3.1)	Bentonite swelling Rock fallout (3.2)	**EDZ** (3.3)	(3.4)	(3.5)	Diss./prec. Colloid and particulate generation (3.6)	Changed permeability (3.7)	(3.8)	Heat transport (3.9)	Fractures affected (3.10)	Air diff. Gas transp. (3.11)	Matrix diff. Sorption (3.12)	(3.13)
Layout/ construction method (4.1)	(4.2)	Magnitude and geometrical extent (4.3)	**ROCK MATRIX/ MINERALOGY** (4.4)	Fracture characteristics and infilling mineralisation (4.5)	Rock-water interaction (4.6)	Matrix K Rock compressibility (4.7)	(4.8)	Heat transport (4.9)	Genesis, tectonic history and rock type (4.10)	Radon generation (4.11)	Sorption Matrix diffusion (4.12)	Land-use Potential human intrusion (4.13)
Avoid major fracture zones Constructability (5.1)	(5.2)	Mechanical properties and fracture frequency (5.3)	(5.4)	**NATURAL FRACTURE SYSTEM** (5.5)	Diss./prec. Colloid generation (5.6)	Flow paths Connectivity Channeling Storage capa. (5.7)	(5.8)	Heat transport (5.9)	Stress magnitude and orientation (5.10)	Transport path for gas (5.11)	Molecular diff. Matrix diff. Sorption (5.12)	Wells (5.13)
Depth affected by redox pot. Construction materials (6.1)	Chem. alt. Water chem. (6.2)	Precipitation/ bacterial growth (6.3)	Groundwater rock interaction (6.4)	Prec. and diss. of fracture minerals (6.5)	**GROUNDWATER CHEMISTRY** (6.6)	Density Viscosity (6.7)	Density affects groundwater head (6.8)	Heat transport (6.9)	(6.10)	Gas gener. Microb. act. (6.11)	Sorp. Prec./diss. Colloid transport (6.12)	Water-use Biotopes (6.13)
Canister positioning Construction methods (7.1)	Saturation Bentonite erosion (7.2)	Erosion (7.3)	(7.4)	Erosion and sedimentation (7.5)	Mixing (7.6)	**GROUNDWATER MOVEMENT** (7.7)	Equalisation of pressures (7.8)	Forced heat convection (7.9)	Two-phase flow (7.10)	Transport of diss. gas and RN Dispersion (7.11)	Recharge and discharge (7.12)	(7.13)
Construction methods (8.1)	(8.2)	(8.3)	(8.4)	(8.5)	Solubility (8.6)	Driving force due to pressure gradient (8.7)	**GROUNDWATER PRESSURE** (8.8)	(8.9)	Effective stress (8.10)	Gas solubility and exp./comp. (8.11)	(8.12)	Potential effect on vegetation (8.13)
Design/layout Construction methods (9.1)	Temperature in buffer/ backfill (9.2)	(9.3)	Thermal expansion and conductivity (9.4)	Permafrost (9.5)	Solubility, kinetics (9.6)	Heat convection (9.7)	Buoyancy eff. (9.8)	**TEMPERATURE/HEAT** (9.9)	Thermal expansion (9.10)	Gas solubility and exp./comp. (9.11)	Molecular diff. Matrix diff. Sorption (9.12)	(9.13)
Design/layout Construction methods (10.1)	Swelling Rock fallout (10.2)	Mechanical stability Fracture aperture (10.3)	Mechanical stability (10.4)	Mechanical stability Fracture aperture (10.5)	(10.6)	(10.7)	Confined aquifers (10.8)	(10.9)	**ROCK STRESSES** (10.10)	(10.11)	(10.12)	Mechanical stability (10.13)
Ventilation problems (11.1)	(11.2)	Opening of fractures Heat conduction (11.3)	Fracturing Thermal properties (11.4)	Fracture aperture (11.5)	Diss. of gas (11.6)	Two-phase flow (11.7)	Capillary forces (11.8)	Gas law (11.9)	(11.10)	**GAS GENERATION AND TRANSPORT** (11.11)	Colloid sorption on gas bubbles (11.12)	Rel. of radioactive gas (11.13)
RN release (12.1)	(12.2)	(12.3)	(12.4)	(12.5)	Radiolysis Redox front (12.6)	(12.7)	(12.8)	(12.9)	(12.10)	(12.11)	**TRANSPORT OF RADIONUCLIDES** (12.12)	RN release (12.13)
Siting Design/ layout (13.1)	(13.2)	(13.3)	(13.4)	(13.5)	Infiltrating water (13.6)	Surface water recharge & percolation (13.7)	Land use Climatic & tidal driving forces Hydraulic gradient (13.8)	Climatic driving forces (13.9)	Glaciation Erosion (13.10)	(13.11)	(13.12)	**BIOSPHERE** (13.13)

Figure 5.3 Far-field interaction matrix for radioactive waste repository design (from Pers *et al.*, SKB Report TR-99-20, *www.skb.se*). This interaction matrix device is explained fully in Chapter 3. The shading of the boxes indicates the importance of the subjects in the boxes: dark–important; grey–intermediate; white–not so important.

5.2 FEATURES, EVENTS AND PROCESSES

The Features, Events and Processes of the title of this Section are known as FEPs—which are all the characteristics of the repository project required to be understood in order to successfully achieve the isolation function. Use is made in Pers *et al.* (1999) of the RES interaction matrix device (explained in Chapter 3 of this book) for considering the interactions in the buffer, the near-field and the far-field, with Figure 5.3 being their far-field matrix. This matrix has 13 leading diagonal terms and hence potentially $13 \times 13 - 13 = 156$ interactions in the off-diagonal terms. The report by Pers *et al.* (1999) contains extensive Appendices with detailed listings of the factors for the three cases.

Pers *et al.* (1999) explain that, "safety assessments of radioactive waste repositories are based on predictive modelling of the performance of the engineered and natural barriers for very long time scales. To evaluate the performance of a repository, assumptions must be made on the future evolution of engineered barriers and natural conditions considering all relevant Features, Events and Processes, FEPs. There is therefore a need for systematic methods to make sure that all alternative future evolutions of the repository system relevant to a reliable assessment are considered." To a certain extent, this approach is required for all rock engineering projects, but is crucially required for radioactive waste disposal for the five reasons listed in Section 5.1.

Ensuring that unacceptable quantities of radionuclides do not escape to the biosphere requires extensive research on protective canisters for the waste and considerable site investigation at the chosen repository location to establish how and when such canisters could leak and how radionuclides can travel through rock fractures to the biosphere. The required very long design life may require consideration of the effect of, not one, but several future ice ages with the associated effects on the groundwater, the rock stress and the fractures—depending on the site location.

5.3 THERMO-HYDRO-MECHANICAL (THM+) PROCESSES

Because of the long design life resulting from the half-lives of the radioactive materials, a repository cannot be designed by precedent practice because optimal site selection criteria have not been established from engineering experience—nor can they be if the design life has to be hundreds of thousands of years. So, in the context of the FEPs discussed in Section 5.2, there has to be, *inter alia*, a systematic consideration of all Thermal, Hydrological and Mechanical effects (THM) that could prejudice the integrity of the repository and its man-made and natural barriers in the short and long terms; then an adequate model for the radionuclide migration can be developed in association with other key factors. We are concentrating here on the thermo-hydro-mechanical interactions and the associated computer modelling, but we emphasise that this in only one aspect of the repository design problem. Moreover, the THM subjects have to be linked with other subjects such as chemistry, geology, biology, etc.,—i.e., to THM+. The content of this Section has been updated from the paper by Hudson *et al.* (2001).

Within the 'risk' theme of the book, we also concentrate on auditing the computer programs and their use. THM+ effects occur over a variety of space and time scales, with additional complexity when the effects are coupled. Furthermore, the parameters associated with the THM+ processes and their significance are likely to be site specific. In order to design the repository to meet the performance and safety requirements, it is necessary to be able to adequately assess the role of THM+ processes and determine how they affect the performance and hence the safety case. Thus, THM+ processes have to be identified and modelled. However, it is anticipated that it will not be necessary to include all aspects of the coupled processes and so the identification of the key components is crucial.

When considering the THM+ mechanisms, it is important to show that a given process has relevance to the repository performance, or that increasing the complexity of characterisation and modelling is actually required. Given that there are first- second- and third-order THM+ processes, the modelling has to be developed to a useable practical scheme which captures the essence of the required processes. Some THM+ couplings will be concept, site and waste-type specific, e.g., whether high-, medium- or low-level waste is being considered.

A variety of numerical codes has been developed; some codes are relevant to repository design, and others have been developed specifically for disposal performance assessment. As intimated, it is essential to be able to audit these codes in the context of the THM+ processes to establish whether they are capable of capturing the essence of the THM+ issues in hand. This refers initially not to whether the codes are internally correct, but whether the variables and mechanisms that they represent are appropriate to a given analysis objective and associated system or sub-system.

5.3.1   The THM+ issues in context

To establish whether a particular coupled THM+ mechanism is important or not requires an appreciation of the role of the particular mechanism in each and all of the overall design criteria and assessments. A mechanism may have a strong effect on, for example, the extent of the Excavation Disturbed Zone (EDZ), but if the EDZ does not affect any of the construction, design and performance, and safety assessment criteria, then the coupled mechanism may be of limited significance in the overall repository performance, depending on the repository concept and site. Hence, there has to be some method for quantifying the effect of the coupled mechanism and evaluating the effect on the repository sub-systems and then on the whole system. Moreover, there also has to be a method of identifying and structuring all the THM+ coupled mechanisms.

A simple mechanism links two variables. A force applied to an elastic spring causes a displacement. This is an M mechanism linking the variable 'force' with the variable 'displacement' using the parameter governing the mechanism, the 'spring stiffness'. In the case of heat-generating radioactive wastes, it is possible to envisage a range of individual and coupled T, H and M processes that can operate. This is illustrated schematically in Figure 5.4, which reinforces the point that there are many different types of coupling; hence, there can be a variety of fully-coupled THM+ issues, which need attention at the conceptual, physical and modelling levels.

It is also possible to represent the THM couplings shown in Figure 5.4 using the interaction matrices of the Rock Engineering Systems (RES) methodology as explained fully in Chapter 3. In Figure 5.5, the separate thermal, hydrological and mechanical variables are placed along the leading diagonal of the matrix with the binary links as the off-diagonal terms. Note that the interaction matrix in Figure 5.5(a) is not symmetrical in the sense that TH, i.e., T → H, is not the same as HT, i.e., H → T. Taking the sequence into account, the T → H → M coupling is represented by a pathway through the interaction matrix, as shown in Figure 5.5(b) using the linking interactions of TH and HM. However, the three subjects can also be coupled as T → M → H, illustrated by the pathway in Figure 5.5(c) using the interactions TM and MH.

In fact, the shorthand 'THM' means the combination of the three processes—which includes all THM permutations, and indeed all possible partial couplings and

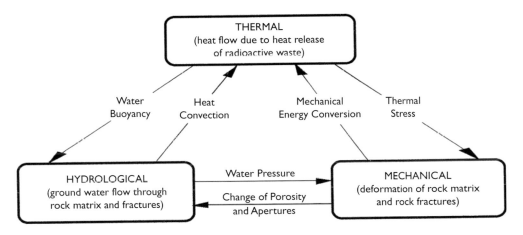

Figure 5.4 Examples of Thermo-Hydro-Mechanical (THM) couplings.

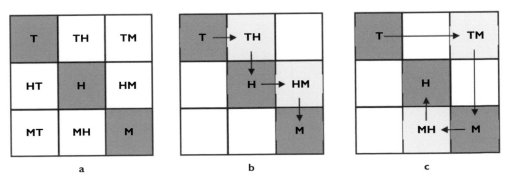

Figure 5.5 THM coupling illustrated using an interaction matrix: (a) represents the same network as that shown in Figure 5.4; (b) shows that THM coupling can be a specific pathway, such as T → TH → H → HM → M; (c) shows the alternative coupling as T → TM → M → MH → H, or TMH.

different pathways linking the three processes. This is an important aspect of THM modelling since the result will depend on the particular pathway used to incorporate all the interactions. Thus, a full THM coupling, if required, should invoke all the links represented by the network in Figure 5.4, which are the same as all the binary interactions in the off-diagonal terms of Figure 5.5(a). Considering now the whole disposal scheme, a systematic approach is required to categorise all the mechanisms potentially relevant to the repository design and disposal systems and hence to be able to identify the relevant THM and other issues, such as chemical effects, leading to THM+. In Figure 5.6, we show the five sub-systems defined by SKB, the Swedish implementer, noting that other radioactive waste agencies follow a similar classification.

In Figure 5.6, the shaded squares along the leading diagonal of each interaction matrix intimate the variables used to characterise each sub-system. The binary or pairwise interactions between all pairs of variables are found as the off-diagonal terms. Each off-diagonal term can be assessed in terms of whether the interaction is significant for the operation of the sub-system, as diagrammatically illustrated by the shaded off-diagonal terms in Figure 5.6. This is the RES method of structuring the information and hence studying the THM+ mechanisms and processes within the overall disposal context. As indicated earlier, a mechanism involving three (or more) variables is represented by a pathway through the matrix.

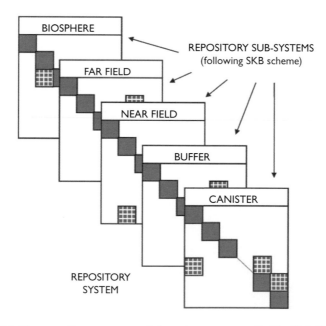

Figure 5.6 The repository system and the sub-systems studied by SKB (Sweden). Highlighted off-diagonal terms (illustrated here schematically) are the mechanisms identified by SKB as the components of the process system. Some of these mechanisms will be THM mechanisms; some will be other types in THM+, e.g., chemical mechanisms.

5.3.2   The excavation, operational and post-closure stages

It is useful to group the THM+ processes into several time periods, rather than according to the spatial distance from the near field to the far field of the repository. A convenient division of these time periods is as follows: the operational stage (up to permanent closure); the containment stage (from permanent closure to 300–1000 years, after the heat pulse has peaked); and the isolation stage (for long-term isolation up to, for example, 10,000+ years). However, for the consideration of coupled THM+ processes, we here emphasise the earlier time periods, which are when the most significant initial effects of the processes occur. Thus, in the following discussion, we group the processes according to (1) excavation stage, (2) operation stage and (3) post-closure stage.

5.3.2.1   The excavation stage

During the excavation stage, no radioactive waste is yet emplaced to provide thermal input, and the thermal effects of normal geothermal gradients and tunnel ventilation are expected to be small. Then a number of coupled MH processes are expected to occur. Here we have used MH, rather than HM, indicating that the direction of coupling at this stage is mostly from mechanical to hydrological effects. They are briefly described as follows.

A   The excavation of the repository causes a major perturbation of the rock formation by the creation of a large cavity. The impact depends on the initial stress field around the system, the nature of the excavation method and the repository design, plus the nature of the ground, e.g., clay or fractured rock. As we have discussed in Chapter 4, it is not easy to determine the *in situ* stress field of a region, especially in the presence of fracture sets forming a network, which could well be anisotropic. The excavation will concentrate stress changes around the cavity, which, in turn, will change the local fracture apertures and permeability. The question is how to determine the nature and extent of anisotropic change in hydraulic conductivity around the repository cavity. Since a repository will be excavated in stages, how does the first stage change the stress field and how do we calculate the MH effect of excavation at a later stage, which is conducted in a different stress field environment?

B   The excavation also represents a relatively sudden event and hence the normal and shear stress across nearby fractures may change in a short time, producing sudden aperture changes. This may cause the pore pressure to rise quickly before the water has a chance to move and equilibrate. Such a transient coupled MH effect may cause local failures, as well as local hydraulic conductivity changes.

C   The cavity at this early stage will be ventilated. This means that water will be taken out of the system and the pore pressure near the repository will be much reduced from the original condition, which may induce fracture closure (an HM coupling).

D   Reduction of pore pressure near the repository cavity may also result in degassing of the pore water. Gases that were in solution in water under pressure will be released. Thus, the flow in the rock surrounding the cavity will be two-phase flow (i.e., both water and gas are present), so that water permeability is much reduced due to gas interference.

5.3.2.2   Operational stage

During the operational stage, the nuclear waste and buffer/backfill or liner materials have been emplaced. Thus, at this stage, there is thermal input and a series of THM+ processes occur.

A   The thermal output from the nuclear waste will heat up the buffer/backfill and the rock. This will occur over several decades in the different repository rooms as they are successively filled with the waste. Thermally induced stresses will be created around the repository, which may change the hydraulic conductivity. The TM effects have been relatively well studied and much experience has been gained in their modelling and observation.

B   Of interest here is the heating up of the multiple media system, the waste canister, backfill materials (e.g., bentonite), and then the surrounding rock. These all have different expansion coefficients. How they move and compress each other and how the interfaces between them behave may cause a significant change in the hydraulic properties of these interfaces.

C   Since water and rock have different thermal expansivity, thermal input may also cause significant pore pressure changes. This is particularly the case for repositories in clay where the original hydraulic conductivity is low and much of the water is in closed pores. The thermal input will then cause changes in effective stress, possibly giving rise to local failures and hence increased local hydraulic conductivity.

D   For bentonite backfill, the incoming water from the rock will increase its saturation and cause it to swell. The imposition of the swelling pressure and the capability of the bentonite to fill in gaps and fractures will change the local hydrological properties and hence the water flow paths.

E   It is not obvious how to assess the behaviour of the gases, which could be either air from the open cavity migrating into the rock, or gas from degassing of the water during the first stage, or due to phase changes (steam). The gases will move into the system under buoyancy or temperature gradients, and then may redissolve into water or form gas pockets that expand with temperature and interfere with the groundwater pathways. The presence of gases will change significantly the local permeability to water flow, as represented by the relative permeability function.

F   The heat will induce convective flow in the rock and the local temperature near a waste canister may be high, and significant vaporisation will occur. The water vapour will move away from the repository and condense in cooler regions of the rock. This process forms a complex hydrological system that requires a fully multiphase code for its analysis. How this system will affect the effective stress field and thus the mechanical condition is an open question.

G   During this period, the repository may be kept open to allow for the option of retrievability. Thus, the system is ventilated and heat/moisture taken out of it. This, coupled with thermal evaporation, will cause dehydration in the near field.

5.3.2.3 Post-closure stage

This is the stage after permanent closure of the repository. The repository is now sealed and thus there is no ventilation or escape of moisture. The repository cavity is then resaturated (if below the water table) and repressurised to its original hydrostatic pressure corresponding to the depth. During this stage, a number of important changes take place.

A   Thermally, the temperature builds up to a certain level and then decreases. The temperature peak of the heating–cooling cycle is reached after 15–100 years near the waste canister, but may take 200–1000 years to be attained in the far field. The exact temporal and spatial distribution of the heating cycle depends on the waste inventory and repository design.

B   This is also the period when the hydraulic pressure is rebuilt in the backfilled and sealed repository opening. Thermally induced flow or convection (TH) depends on thermal energy imparted to the water, which will last much longer than the temperature pulse. The convective velocity could peak at around 10,000 years.

C   In the same way, the TM effect is also dependent on thermal energy imparted to the rock and is not directly dependent on temperature. Thus coupled TM processes will also last up to 10,000 years. One example is that the thermal expansion distribution may open fractures below the repository and close fractures above it, thus creating an under-pressure region below. The possible occurrence of this coupled TMH effect should be evaluated by modelling and be detected and studied by a repository monitoring programme.

Mechanical deformation during the resaturation and repressurisation is an irreversible coupled THM+ process. It is not expected that the system will return to pre-excavation conditions because of mechanical hysteresis. Similarly, the thermal strain during the cycle of temperature build-up and decrease may also be an irreversible process. The role of gases during this stage, in terms of their flow and dissolution in water, is also an open question. Other coupled processes have been envisaged away from this 'base case scenario'. These include links involving chemical effects, seismically induced HM, glacially induced THM+, erosion and asteroid/meteorite induced THM+.

5.3.2.4 Heterogeneity and multiple stage data needs

Along with the above discussions of coupled THM+ processes, there are many other factors but we highlight here just two of the most important ones. **The first is the subject of heterogeneity.** At least in the case of anisotropy and via the theory of elasticity, there is a method of characterising anisotropy through the elastic constants, i.e., Young's modulus and Poisson's ratio for a perfectly isotropic material, through transversely isotropic (five constants) and orthotropic (nine constants) materials, to a fully anisotropic material with 21 elastic constants. But, we do not yet have a system for characterising heterogeneity at a potential repository site, or how to design and perform a reasonable set of field measurements to obtain the basic heterogeneity parameters. Because of this, heterogeneity is a critical uncertainty that pervades all modelling input, the modelling sensitivities and output, and may well be one of the factors limiting the confidence level of THM+ predictions. The subject is important because the THM+ models support the performance assessment safety case for regulatory approval.

A special class of heterogeneity is the fracture network in the rock. This is well known in crystalline rocks. There is a suggestion that there are also fractures in clay formations (unless the clay is very plastic) and that fractures may be induced during dehydration of clay. All the coupled THM+ processes already mentioned will be present in fractured porous systems. In other words, both fractures (with their high permeability and low stiffness) and the porous block need to be considered. Furthermore, there can be strong heterogeneity even in the porous medium itself, and it is not extraordinary to find, for example, variations of hydraulic conductivity of one or two orders of magnitude or more in a porous medium. In the general case, heterogeneity means different THM+ properties at different points in space and time. Modelling the THM+ processes in a heterogeneous system will, however, be solved through the use of computer programs which are developing rapidly.

* * * * *

In terms of **multiple-stage data needs** the data required for each stage of the repository development and assessment should be built up from the previous stage. Each stage of repository development represents a major perturbation of the hydro-mechanical conditions of the rock formation. Responses to these perturbations can be utilised to characterise the system, and hence they need to be monitored in detail, both spatially and temporally.

Before the excavation starts, information is needed on the geological structures and stress distributions. The former include the geometry of faults and joints, and the presence of geological domains, as well as effective boundaries of the region to be studied. The stress distributions may require borehole measurements such as hydro-fracturing and profiling. Water injection and withdrawal pressure transient tests also need to be undertaken. The hydraulic and mechanical conditions at the chosen boundaries are of particular importance.

The period of the actual excavation of the repository represents an opportunity to obtain data at the scale of the repository and at the magnitude of its mechanical impact. After the excavation stage, the impact of the excavation on rock hydro-mechanical behaviour is useful to further characterise and understand the site. Deformations of the cavity profile, especially near fractures that intercept the cavity, are useful for estimating the mechanical condition near those locations (this is discussed in detail in Chapter 7 in the context of constructing hydropower caverns). Distribution of water emergence in the freshly excavated surfaces and changes in pore pressure distribution also will give information on major flow paths and how they are affected by mechanical changes because of the excavation. Careful monitoring design needs to be made prior to the excavation. Afterwards, the circumstances may be changed because any installation of rock support systems will alter local Excavation Damaged Zone (EDZ) conditions.

During the operational stage, assuming that the physical and chemical properties of the backfill materials are known from laboratory testing, monitoring of rock changes and hydro-mechanical conditions at the rock–backfill interface may be made at a number of representative locations. The objective of this monitoring programme would be to understand the rock responses with thermal input from the waste and under the swelling pressure of the backfill material (e.g., bentonite). The swelling is in general non-uniform at an early stage of deposition.

During the post-closure stage, drastic or abrupt changes in THM+ conditions are not expected but significant changes involving a temperature rise-and-fall cycle occur. Certain types of remote geophysical monitoring of these changes may still be possible. One school of thought is that the site should be abandoned after establishing monuments to warn future generations not to disturb the underground system. However, an alternative view is that monitoring should still be continued to detect any changes, and to ensure they are within model expectation ranges. Also, some countries may incorporate 'retrievability' into the design considerations.

5.3.2.5   Modelling phases and scaling

Three modelling phases can be carried out, corresponding to the three stages of repository development and based on data collected before the start of each stage. As the data and information build up for successive stages, so does the degree of sophistication of the model for successive phases. Each modelling phase should include:

1   model selection;
2   data evaluation and calibration of model structure/parameters;
3   modelling studies, including sensitivity study and uncertainty evaluation, with feedback to repository design and performance assessment; and
4   predictive modelling of the next stage.

To evaluate and understand the field observations and monitoring data at each stage, it is necessary to compare with predictive results of the models from the previous phase. Further, calibrations are particularly important to prevent within-phase errors and cumulative errors developing with successive phases of the modelling exercise. The degree of success with which this can be done depends on the design and execution of the monitoring and testing programme at each repository stage.

It is important to design a data-gathering strategy for each period and set it in place. The repository may otherwise be excavated without adequate preparation, and consequently proper data are not obtained for THM+ evaluation: if the unique opportunity to obtain large-scale THM+ responses is overlooked, it will be difficult to gain it again.

*   *   *   *   *

Considering the issues already discussed in this Chapter, how are the THM+ processes modelled and in what context should the THM+ issues be considered? The series of steps below indicates one method by which this could be achieved (noting that quantitative modelling may not be necessary, and that the modelling will be disposal concept- and site-specific).

Step 1: Model conceptualisation.
Step 2: Identification, definition and specification of THM+ mechanisms/processes in the model.
Step 3: Selection of those items identified in Step 2 which are considered to be important for the process system and hence performance assessment.
Step 4: Listing of analytical solutions and numerical codes able to simulate the selected THM+ processes.
Step 5: Auditing of codes to establish their capability to model sub-systems as defined by the performance assessment strategy.
Step 6: Choice of codes for numerical calculations.
Step 7: Obtaining all necessary supporting data for analyses.
Step 8: Conduct modelling and obtain results.
Step 9: Application of results for repository design and performance assessment.

This sequence of steps is a 'top-down' or analytical approach. To date, the modelling of the THM+ processes has been more of a 'bottom up' or synthetic approach—for the practical reason that the modelling has evolved by incrementally improving the numerical codes to include more and more components of the THM+ coupling, and by acquiring laboratory data on simple systems.

However, if the THM+ modelling is to reflect the nature of the THM+ coupling within the repository design and performance assessment contexts, the modelling must be tailored to the specific THM+ mechanisms and couplings required, as defined by the process system. Therefore, in the future it will become increasingly necessary to 'modularise' the numerical codes so that the necessary code components can be linked according to the anticipated sequence of engineering perturbations at the site and hence the sequence of THM+ links that require modelling. It may also involve the ability to link the interaction matrices presented earlier in this Chapter enabling the modelling output of one matrix to provide the modelling input of another, so that the THM+ aspects can be followed through the necessary space and time ranges.

Also, the auditing of the code components and the composite code will be necessary in order to facilitate the provision of an audit trail demonstrating that the modelling does indeed capture the THM+ essence of the problem, both in the structural components and in operational sequence.

5.3.3   The use of numerical computer codes

The necessary representation and modelling of the THM+ processes may be conducted qualitatively or quantitatively. In the event that numerical codes with THM+ coupled components are to be used for performance assessment purposes, the question of the level of complexity has to be addressed. Is a simple representation of the system adequate? Is it necessary to attempt to use fully-coupled THM+ codes? Which stage of the disposal is being supported by the modelling: operational or post-closure? It should also be noted that part of the use of the modelling and codes is to compare the results with the site information in the validation context. In this way, the numerical models and site data are complementary. However, given that experiments cannot be conducted over the long time scales of interest, i.e., thousands of years, there is probably only one acceptable method of including the combined THM+ (and additional) couplings for predictive purposes—the use of numerical codes containing algorithms for evaluating the coupling equations. Thus, it is important that the advantages and limitations of numerical codes are understood, and that contemporaneous auditing is included with their use.

5.3.3.1 The nature of numerical codes

Numerical codes have been applied widely in rock engineering for analysis of mechanical, hydraulic, thermal and coupled problems. They are commonly used for obtaining stress and displacement distributions, fluid inflow, fluid pressure distribution and temperature distribution in a rock structure for certain geometries and given boundary conditions. Numerical codes have been developed to a high level and are able to include complex geometries and some degree of coupled mechanical, hydraulic and thermal processes.

There are numerical codes that can be supplied ready for purchase and use. The question naturally arising then is 'Which numerical code should be used for the particular rock engineering problem in hand?' In the current waste disposal context, the question is 'Which codes can be used for which aspects of the problem?' noting that there are no off-the-shelf codes encompassing the whole problem. Referring to the sub-systems in Figure 5.6, which codes can be used for which aspects of the sub-systems? Moreover, because a transparent audit trail is required, follow-on questions are 'What is the formal procedure by which a particular code is chosen for use in this project?' and 'Has the code been adequately verified and validated?'

However, for the coupled mechanical-hydraulic-thermal rock engineering and performance assessment problem with complex rock geometry and additional potential factors such as chemistry and biology, the information involved is diverse and complicated. Many assumptions are made in numerical codes to simplify the information for the code's use. As a result, the outputs resulting from the use of different codes may be different due to the different ways in which the information is used and simplifications assumed. Because it is not immediately clear what information is relevant, it is not apparent which codes should be used.

5.3.3.2 Uncoupled and coupled codes

One of the important decisions will be to decide whether to use uncoupled or coupled codes, i.e., whether to use a set of codes, each dealing with one aspect of the problem, or codes which have algorithms representing the combined THM+ processes. Referring to Figures 5.4 and 5.5, there are codes that have been developed to study uncoupled problems, i.e., they solve problems in only one subject: thermal or hydraulic or mechanical processes. There are also codes that have been developed to study coupled problems, i.e., incorporating one or more of the network links in Figure 5.4 or the off-diagonal interactions in Figure 5.5. But can uncoupled codes capture sufficient essence of the problem for engineering purposes, which is a similar question to whether a continuum solution can be a sufficient approximation for a jointed rock mass?

The coupled codes have extra levels of complexity above uncoupled codes. Firstly, a coupled code has to contain more processes than a single-subject code. Secondly, the full THM coupling introduces two-way links between the separate T, H and M algorithms (*cf.*, the pathways in Figure 5.5(b) and (c)). However, we will never be able to characterise all the details of all the couplings with the characterising parameters, so work should be directed towards the optimal level of approximation.

5.3.3.3 Technical auditing of numerical codes

In our previous book, "Rock Engineering Design" (Feng & Hudson, 2011), we included a chapter (Chapter 5 in that book) on technical auditing of rock mechanics modelling and rock engineering design in general and we explained the overall motivation as follows.

"Within the context of the rock engineering design methodology, it is of benefit to be able to formally audit the content of the rock mechanics modelling and rock engineering design of a project in order to ensure that all the necessary factors are included and that the technical work is correct. The term 'Technical Auditing' is used to describe this process and the overall purposes of a Technical Audit (TA) are as follows:

1   to evaluate the logic of the work based on the stated objective;
2   to establish whether all the necessary physical mechanisms, variables, and parameters have been included in the relevant analyses;
3   to show that the supporting analyses are technically correct;
4   to consider whether conclusions are justified in terms of the project objectives and the work conducted; and
5   to provide an audit information, analysis and decision trail.

The key principles of an audit in general (Dunn, 2004) are that it is made according to evidence, known criteria and the current scientific framework. Auditing involves verification by evidence and the result is an opinion based on persuasive evidence. The audit should have an independent status, be free from investigatory and reporting constraints, produce a benefit, and result in a report. The audit result will always be an opinion and so the auditing must carry authority. These principles directly apply to the specific case of technical auditing for rock mechanics modelling and rock engineering design and have therefore been adopted here."

The Chapter in the previous book also included a diagram showing the content of the 'soft', 'semi-hard' and 'hard' audits—which is reproduced here as Figure 5.7.

These principles should be adopted in evaluating the appropriateness of given numerical codes for radioactive waste computer studies, one key aspect of the subject being matching a particular code's analysis capability with the problem in hand. Does the code capture the essence of the problem? Is this code the best one to use? Do we need to develop new codes?

The main issues associated with the application of numerical codes to the radioactive waste disposal problem are

–   defining the problem to be solved,
–   considering codes that could be used,
–   establishing that the codes can capture the essence of the problem,
–   checking that the code is internally correct, and
–   ensuring that the code is an adequate representation of reality.

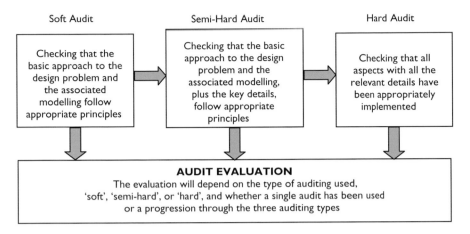

Figure 5.7 The 'soft', 'semi-soft' and 'hard' audits and the audit evaluation.

A formal method is needed to check that the codes do capture the essence of the problem. In the context of checking that the code is internally correct, there has to be some form of guarantee that the operations of the code are indeed internally correct. Methods, which have been developed in the nuclear and aerospace industries to approve numerical codes, will need to be applied to the codes used for modelling THM+ processes so that the audit trail is complete. In the performance assessment context, there are wider issues. Can a simpler representation of the system be shown to be conservative and hence be used in place of a full THM+ model? Finally, for checking that the code does model the real conditions acceptably well, fully-coupled *in situ* tests will be required and adequate 'acceptance' tests defined to demonstrate the code's adequacy. We discuss this in the following Sections 5.4 and 5.5.

5.3.3.4 Capturing the essence of the problem

In order to establish whether a code can capture the essence of the problem in hand, it is necessary to define the analysis capability of the code and compare this with the analysis requirements of the problem. Continued work around the world on the FEPs, the process system, performance assessment and safety studies will enable the components of the problem to be identified, but how can the analysis capability of a code be specified? We could start by compiling a table with the physical THM+ processes relating to the rock mass and rock joints. Also, this information and the information used in any numerical code can be presented by a Binary Interaction Matrix (BIM) similar to those presented in Figures 5.5 and 5.6. The state variables are those known to be required by the physics of the problem, and listed along the leading diagonal of an interaction matrix; the mechanisms linking the state variables form the off-diagonal components. The components of a binary interaction matrix necessary to model the problem would be established first in order to provide the reference requirements. The content of this reference interaction matrix in principle contains all the information required to solve the problem. For a specific code being considered, the matrix is also used to present the specific capability of the code in utilising information, i.e., a statement of the specific links between the state variables which are actually in the code.

An example of the type of interactions which one might like to have in a THM code is shown in Figure 5.8 and the components that might be in a code are shown in Figure 5.9. A comparison of the two diagrams will immediately indicate if the necessary code content is there. It is not sufficient, however, to only compare the content: it is also necessary to consider the code algorithms connecting the variables and mechanisms—because one missing algorithmic connection or an inappropriate coupling sequence could invalidate the whole code.

In numerical codes, the values of some variables can be directly obtained by solving the sets of equations with respect to certain boundary conditions. Examples are displacement in a mechanical equilibrium problem and fluid pressure (hydraulic head) in a fluid flow problem. In addition, some variables' values are obtained by physical laws. For example, stress can be directly obtained from strain (which can be directly obtained by differentiating displacement) via the stress–strain constitutive law. Fluid flow can be directly derived from hydraulic head via Darcy's law.

The first step in assessing the capability of the numerical codes is to determine the state variables that could be or are actually used in the codes. For example, the coupled THM+ problem will require a certain number of variables for a complete description of the problem, say n variables, and an associated number of binary relations, $(n^2 - n)$, included in the numerical codes to represent the maximal capability of utilising the information (this is the input information), see Figure 5.8.

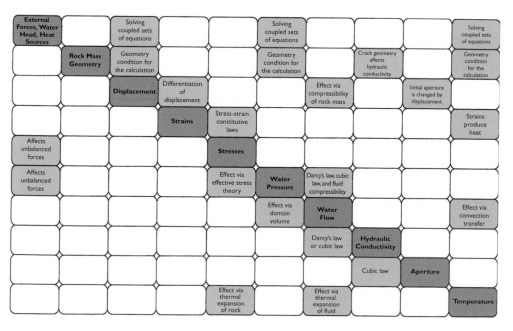

Figure 5.8 Using an interaction matrix to define the variables and interactions required to solve a THM problem (this matrix is illustrative of the technique and is not necessarily the one which would be used).

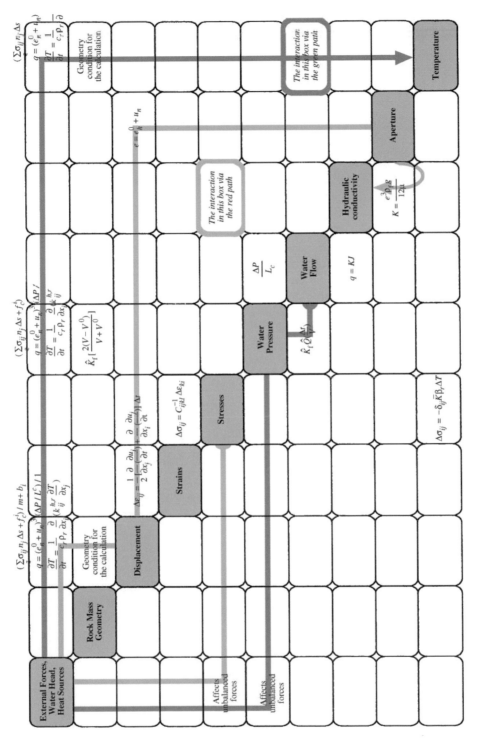

Figure 5.9 Presenting the physical variables and mechanisms that are actually included in a specific numerical code and how these are connected by the code's algorithms. (The content of this interaction matrix is then compared with the required code content as defined via Figure 5.8).

So, for a specific code in use, an information audit can be conducted directly by using the interaction matrix representation to compare the required variables (the leading diagonal components in the interaction matrix) and the required mechanisms (the off-diagonal components in the interaction matrix). Why is a particular box empty? Is it important? How does the coupled performance of this code compare to an optimal code? Answering these questions leads to the concept of assessing and auditing the analysis capability of numerical codes. Thus, it is possible to develop audit sheets for formalising this aspect of the THM+ modelling and hence to provide a way-marked audit trail that emphasises the rigour and transparency of the numerical modelling work.

5.3.3.5   The overall Technical Auditing (TA) procedure and risk

As noted, the term 'Technical Auditing' (TA) in the current context means examining the technical content of, for example, a thermo-hydro-mechanical numerical analysis to establish if it is adequate for the purpose. The reason for requiring such a capability is that, in order to be able to coherently design a repository in a rock mass, one must be able to predict the consequences of different design options, and hence reduce the risk. Currently, the main method of developing this predictive capability is to use THM+ and more extended numerical modelling techniques. However, there are problems relating to all such modelling—in terms of ensuring that all the relevant variables, parameters and mechanisms have been included in the modelling and that the model does indeed represent the rock reality. This includes both the basic modelling capability of the program and the fact that it has been used correctly.

Thus, questions have to be asked relating to the generation of the modelling output. Examples of such questions are listed below.

– What is the work/project objective?
– Have the relevant variables and mechanisms been identified?
– Is the model/code adequate?
– Which data are required?
– How should the data be obtained?
– Are the data adequate?
– Has the model been used properly?
– What are the prediction/back analysis protocols?

For modelling the underground environment, it is necessary to include the rock mass geometry, the *in situ* rock stress, the properties of the intact rock and fractures, the hydrogeology and the excavation process. For the design of a radioactive waste repository, other factors should also be included, e.g., thermal processes, geochemical processes, time-dependent processes. Establishing the basic subject information and the interactions between the subjects via the type of interaction matrix content illustrated in Figure 5.10 is the way to begin, followed by more detailed studies of the FEPs list described earlier and to what extent these can be accommodated in the program.

However, problems soon arise related to the:

– systematic evaluation of geological and engineering uncertainties;
– understanding and mathematical representation of large rock fractures;
– quantification of fracture shape, size, connectivity and effect of fracture intersections;

1,1 **Geology**	1,2 (High) Rock inhomogeneity and anisotropy, & fractures (geometry, etc.) affect rock properties/stresses	1,3 (High) Rock porosity, rock mass permeability, and water-rock interaction affect water flow	1,4 (High) Rock mineralogical composition and geometry of fractures affect hydrogeochemistry	1,5 (High) Mineralogical composition, porosity, textural and structural anisotropy effects
2,1 (Low) Stress data affecting the geo-interpretation of fracture systems and rock mass	2,2 **Rock Mechanics**	2,3 (Medium) Spatial distribution of *in situ* stress and EDZ influences the hydrogeological regime	2,4 (High) Stress changes near fracture zones affecting flow may change precipitation	2,5 (Low) Rocks & fracture zones subjected to higher stresses may be more thermally conductive
3,1 (Medium) Hydrogeological tests and measures can affect the geo-interpretation of permeable features	3,2 (Medium) Water pressure changes the effective stress	3,3 **Hydrogeology**	3,4 (Medium) The flow pattern affects dilution and mixing	3,5 (Low) The flow pattern will affect the temperature due to convection effects
4,1 (Low) Hydrogeochemistry affects fracture surface properties	4,2 (Medium) Precipitation in fractures affects the fracture stiffnesses, strengths and creep properties	4,3 (Medium) Effects of ground-water age, density, viscosity, and dissolution and precipitation	4,4 **Hydrogeo-chemistry**	4,5 (No effect)
5,1 (Medium) Thermal anisotropy and measurement affecting the geo-interpretation	5,2 (High) Change in temperature can change the local stress, possibly leading to failure	5,3 (Medium) Temperature gradients and thermal expansion in rock/fractures affect the water flow	5,4 (Medium) Dissolution and precipitation enhanced by thermal gradients	5,5 **Thermal Properties**

Figure 5.10 Establishing the content and significance of different subjects and their interactions as part of the Technical Auditing process.

- representation of rock mass properties and behaviour as an equivalent continuum and existence of the Representative Elemental Volume (REV);
- representation of interface behaviour;
- scale effects, homogenisation and upscaling methods;
- numerical representation of engineering processes, such as excavation sequence, grouting and reinforcement;
- time effects; and
- large-scale computational capacities.

Also, the most important step in numerical modelling is, not operating the computer code, but the earlier 'conceptualisation' of the problem in terms of the dominant processes, properties, parameters and perturbations, and their mathematical presentations—which are the coupled THM+ processes (GTHMCE processes in the Figure 5.10 case), as follows.

- G: Geological—site geometry, lithology, fractures;
- T: Thermal—heat loads, heat flow;
- H: Hydrological—water pressures, water flow;
- M: Mechanical—rock stress, stiffness, strength;
- C: Chemical—water chemistry, swelling rocks;
- E: Engineering—effects of excavation.

The 'soft' and 'hard' audits

The soft audit firstly establishes an overview of the THM+ modelling work and determines whether well-known issues of importance and difficulty in characterising and modelling rock masses have been addressed at the outset. Then, the purpose, style, features and content of the modelling are listed so that they can be presented in a compact manner.

Part 1 of the soft audit: 'Robustness questions'

These 'robustness questions' establish an overview of the modelling through a series of questions relating to the rock mass modelling issues of special importance and difficulty. The difficulties do not have to be fully overcome in the THM+ modelling, but there should be adequate awareness of them. They should already have been addressed in the modelling, at least in terms of explaining why the modelling is adequate given each difficulty. The suite of robustness questions is given in Table 5.1.

Part 2 of the soft audit: Specifying the components and features of the modelling

The components and features of the model are then specified through a suite of questions. These are listed in Table 5.2 under the four subject areas of:

- modelling objective;
- modelling concept;
- modelling technique; and
- modelling adequacy.

Developing from the soft audit to the hard audit

The hard audit covers the same subjects, but requires detailed justification of the answers to the questions. It is not enough in the hard audit to respond simply 'yes' or 'no' concerning the processes covered in the modelling: auditing work needs to cover all the basic items of processes, equations, properties, parameters and methods. The hard audit results should be presented in the same form as the soft audit results, but including the necessary hard audit details and justifications.

Table 5.1 Questions relating to overviewing the modelling and considering how well-known difficulties in modelling rock masses have been addressed.

1 What is the purpose of the modelling?
2 In what way is this work different to previous similar modelling work?
3 What is the scale of the rock mass being modelled?
4 What is the basic modelling geometry?
5 Has it been necessary to divide the rock mass into separate rock mass domains? (Rock mass domain: a region of the rock mass in which the rock properties are statistically similar, but different to the properties of the surrounding rock in other structural domains.)
6 Are the intact rock properties being specifically incorporated?
7 How are the fracture properties being incorporated?
8 Are features of the structural geology of the rock mass being incorporated?
9 Are the rock mass properties being input directly (as opposed to being a result of the input intact rock and fracture properties)?
10 How have the rock properties been estimated?
11 Is a constitutive law required for the rock mass? If so, how was it established?
12 Has the rock mass been modelled as a CHILE material? (CHILE: Continuous, homogeneous, isotropic, linearly elastic.) What has been done to account for the DIANE aspects of the rock reality? (DIANE: Discontinuous, inhomogeneous, anisotropic, not elastic.)
13 How have the stress boundary conditions been established?
14 Does the model include any failure criteria. If so, which one(s)?
15 Is the rock being modelled as a continuum, discontinuum, or combination of the two?
16 What are the hydrogeological conditions in the modelling?
17 How have the hydrological boundary conditions been established?
18 Are effective stresses being used?
19 How are the thermal properties being incorporated?
20 How are the THM+ components being included in the modelling: as uncoupled components, pairwise coupled components, fully-coupled components?
21 Are there any special boundary conditions, loading conditions, or rock mass features in the modelling?
22 Has physical rock testing been used to obtain any parameters supporting the model?
23 Has there been any study of potential adverse interactions that could lead to positive feedbacks and hence instabilities—in the rock mass and in the modelling?
24 Have all the potential failure mechanisms been identified?
25 Have modelling sensitivity studies been undertaken?
26 Have modelling protocols been used?
27 How will the modelling methods and results be presented?
28 Can the modelling be verified/validated?—in this study and in principle?
29 Are there any features of the model or modelling work not covered by the points above?

Presentation of the auditing results

There should be clear presentations of the auditing results. These can be in the form of a report, or an effective alternative is a poster type display. Three types of presentation should be made.

– What modelling work is being done or has been done in principle—the soft audit poster display?
– What is being done and why it is being done in detail—the hard audit poster display?
– Conclusions concerning whether the modelling is adequate for the purpose specified—the evaluation poster display.

Table 5.2 Soft auditing of components and features of the modelling and the associated questions.

Subject area 1: Modelling objective—establishing the purpose of the work

1 *The modelling objective*

1-1 Has the modelling objective been clearly established?

1-2 How will it be known when the modelling work is completed?

Subject area 2: Modelling concept—describing the modelling concept and content

2 *Conceptualisation of the processes being modelled; the sub-system(s) being isolated for study; the physical processes involved*

2-1 What rock mass systems are being considered?

2-2 What are the main physical processes being modelled?

2-3 What is the changing independent variable?

2-4 How is the system perturbed so that the mechanisms are initiated?

3 *Specification of the modelling content—what are the physical variables, connecting relations, parameters, boundary conditions, initial conditions, etc.?*

3-1 Listing of the physical variables

3-2 Listing of the THM+ couplings

3-3 Is the model 1D, 2D, 3D or some combination?

3-4 Is a continuum or a discontinuum being modelled?

3-5 Specification of the boundary conditions

3-6 Specification of the initial conditions

3-7 How is the final condition established?

4 *Modelling solution requirements: what type of model output is required, given the stated modelling purpose?*

4-1 What is the required model output?

4-2 Does the model output match the modelling objectives?

5 *Modelling solution technique*

5-1 In principle, how is the model output to be obtained: one code, one set of data, one run?—or a suite of numerical experiments? How is the required model output to be obtained?

5-2 Are any quality control checks in place? Checking the input data have been entered correctly, validation against known solutions, independent duplication of runs?

Subject area 3: Modelling technique

6 *Numerical code utilised: which numerical code is to be used? How is it known that the code is operating correctly?*

6-1 Which numerical code is to be used?

6-2 Why is that code being used?

6-3 Where did the code originate from?

6-4 How has the code been validated?

7 *Supporting model data & data input method. What are the necessary supporting data? How are they to be obtained? How are they to be input?*

7-1 Listing of type and justification of boundary conditions

7-2 Listing of input data with source of the data and justification

7-3 Do the data have to be adjusted before being input?

(Continued)

Table 5.2 (Continued)

8 *Model sensitivity analysis. How does the model output depend on the model input in terms of whether a sensitivity analysis is required?*

8-1 How does the model output depend on the input parameter values?

8-2 Is a sensitivity analysis being conducted? If so, what type of analysis? Processes, mechanisms, parameters, boundary conditions, couplings, etc.

8-3 How are the results of the sensitivity analysis to be summarised?

9 *Presentation of modelling results. Is it possible to demonstrate that the numerical code is operating correctly? Are the modelling results clearly presented?*

9-1 Is it possible to demonstrate that the numerical code is operating correctly?

9-2 Is it possible to show that the supporting data are reasonable assumptions for a rock mass?

9-3 How are the modelling results to be presented?

9-4 Does the presentation of the modelling results link with the modelling objective?

Subject area 4: Modelling adequacy

10 *Sources of errors. What are the main sources of errors?*

10-1 Have any errors been corrected?

10-2 List the sources of potentially significant errors

10-3 Do any of the potentially significant errors invalidate the modelling objective, concept and conclusions?

11 *Modelling adequacy. Does the modelling seem adequate for the purpose? Are there any problem areas? Is any corrective action required?*

11-1 Do all the previous questions indicate that in principle the model is adequate for the purpose?

11-2 If not, list the problem areas

11-3 What corrective action is required?

11-4 Does the soft audit have to be repeated after corrective action has been taken?

These three reports or poster displays (with supporting documentation) are then suitable for communicating the modelling information, not only to geoscientists but to clients, disposers, regulators, managers and the public. The results ensure that the modelling is transparent and traceable through the audit trail.

5.3.3.6 Validation

The term 'validation' means ensuring that a code does adequately represent the real conditions: the only way that this can be done is by comparing the modelled and actual effects of a perturbation under fully-coupled site conditions. (The way in which the validation of the total system could be developed from validation of sub-systems is a subject ripe for research and one that may be established in due course.) One adverse aspect of this subject is that increased study of coupled models simply increases the number of unsolved problems relating to their compilation and use—making an eventual solution more distant. Moreover, it will be difficult to validate complex numerical models. Both these aspects indicate the desirability of simpler models, if at all possible, if only so that validation is more practical. The subject of code validation is discussed further in Section 5.4 following which the motivation and Test Cases of the DECOVALEX project are presented.

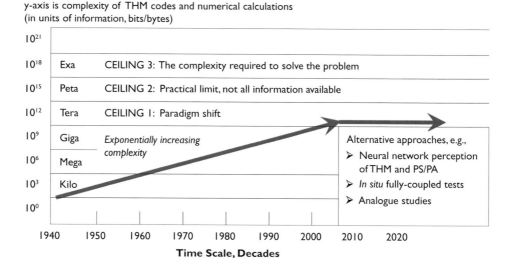

Figure 5.11 Conceptual diagram illustrating the increasing complexity of numerical codes with time and the modelling paradigm shift when the amount of information required to support the codes becomes excessive.

5.3.3.7 The future of numerical codes

A characteristic of numerical codes is that they have increased in complexity with time, as is illustrated by the conceptual diagram in Figure 5.11. Despite the possible preference for a simpler model for performance assessment validation, it is anticipated that increasing model and code complexity will continue, although the analysis techniques will change.

It is prudent to consider the ways in which the process system for radioactive waste disposal might be modelled in the future, given the inability for the increasing use of information illustrated in Figure 5.11 to be sustained. Are numerical codes the way to process the information and hence establish the repository design? It is likely that computers will be involved, but perhaps neural networks will be used rather than numerical codes *per se*. Perhaps computers using neural networks will enable us to have a 'perception' of the THM+ processes and the performance assessment—and indicate the way ahead themselves. Note that 'intelligent rock mechanics' approaches are used for both understanding and mitigating rockbursts (described in Chapter 6) and for learning the rock response when progressively deepening hydropower caverns (described in Chapter 7). Also, there is a view that the repository design problem should be solved by a procedure involving simplification rather than complication—for which the neural networks and related techniques are ideally suited.

5.4 THE DECOVALEX PROGRAMME

As discussed in the previous Section, computer modelling, together with *in situ* experimentation, is required to support radioactive waste disposal repository design: this is why the DECOVALEX work from 1992 to the present has focussed on computer modelling of benchmark cases and simulation of *in situ* experiments. It is essential to ensure that a) the coupled computer programs are operating correctly (verification) and b) that they do indeed represent the rock reality (validation). Computer modelling is now used for almost all rock engineering project design, but how do we know that the output from the computer modelling has anything to do with reality? Does the computer model contain the required physical processes? Does it contain the idiosyncrasies of the site being modelled? Have the data been input correctly? Needless to say, although verification studies can be helpful, it is validation that is required for the design of a radioactive waste repository to satisfy the regulators.

5.4.1   The development of the **DECOVALEX** programme

For both generic and specific radioactive waste repository design, *in situ* experimentation and project design needs to be linked with computer modelling and this is why the DECOVALEX program was initiated as early as 1992 with the focus on computer modelling of benchmark cases (verification) and simulation of *in situ* experiments (validation). It is essential to verify that the coupled computer programs are operating correctly and to validate that they do indeed represent the rock reality.

An international co-operative research project was launched in Stockholm, Sweden, managed by the Swedish Nuclear Power Inspectorate (SKI), for the period of 1992–1995. According to the agreement of the participating parties, including both Funding Organisations and their research teams, the overall objective of the DECOVALEX project was "to increase the understanding of various Thermo-Hydro-Mechanical (THM) processes of importance for radionuclide release and transport from a repository to the biosphere and how they could be described by mathematical models". Note that chemistry has since been included for THMC (Thermo-Hydro-Mechanical-Chemical) processes.

The work began with studies of fractured hard rock (because granite had been selected as the host rock by Canada, Finland, France, Japan, Sweden, and the UK, which were six out of the nine original Funding Organisations). There were three BMT (Bench Mark Test) problems and six TC (Test Case) problems. Due to the success of the initial and subsequent research phases, this project, with the acronymic name DECOVALEX (DEvelopment of COupled models and their VALidation against EXperiment) has been extended five times over more than two decades. The first DECOVALEX project was named DECOVALEX I and was followed by DECOVALEX II, DECOVALEX III, DECOVALEX-THMC, DECOVALEX-2011 and, at the time of writing, DECOVALEX-2015).

In line with the motivation for the DECOVALEX work, the organisation and research work have the following characteristics: simplicity and efficiency of project management; integrated modelling and experiments; aims of understanding and insight; aims also of method (testing and model) development; manageable task scopes; in-depth discussions; flexible agenda; relevance to Performance Assessment (PA) and Safety Assessment (SA); good achievements with many scientific publications; and a platform for training young scientists.

The results of this work during the DECOVALEX phases have been published in the international literature. There have been two books: "Coupled Thermo-Hydro-Mechanical Processes of Fractured Media" (Stephansson *et al.*, 1996) and "Coupled Thermo-Hydro-Mechanical-Chemical Processes in Geosystems: Fundamentals, Modelling, Experiments and Applications" (Stephansson *et al.*, 2004). The work has also been published in five Special Issues of international scientific journals as follows:

- Int. J. Rock Mech. Min. Sci. & Geomech. Abstr., 1995; 32 (5)
- Int. J. Rock Mech. Min. Sci., 2001; 38 (1)
- Int. J. Rock Mech. Min. Sci., 2005; 42 (5 & 6)
- Environmental Geology, 2009; 57 (6)
- Journal of Rock Mechanics and Geotechnical Engineering, 2013; 4 (4 & 5).

In addition, an overview paper by C.-F. Tsang and others was written on the Excavation Damaged Zone, EDZ (Tsang *et al.*, 2005). Chin-Fu Tsang was the DECOVALEX Chairman for the period 1992–2006; John A. Hudson is the Chairman for the period 2007–2015; and Jens Birkholzer is the Chairman from 2016 onwards. Lanru Jing has led the Secretariat for the whole duration of the DECOVALEX project. An important feature of the DECOVALEX project, given the extremely long repository design lead time, is that there has been a strong focus on the training of young researchers—with about 40 PhDs having been awarded since 1992 for research on the DECOVALEX tasks.

5.4.2 Research work in the current DECOVALEX phase: D-2015

The current phase of the DECOVALEX research (D-2015, i.e., the phase that ends in 2015) consists of research work on water flow through a rock mass, the responses of a rock mass and fractures to physical and chemical disturbances, and the effectiveness of sealing rock voids. Most of the work relates to *in situ* observations and hence to the *validation* of computer predictions. Moreover, the ten Funding Organisations from nine countries in the current phase represent significant international research co-operation: BGR/UFZ (Germany), CAS (China), DOE (USA), ENSI (Switzerland), IRSN (France), JAEA (Japan), KAERI (Korea), RWM formerly NDA (UK), NRC (USA) and SURAO formerly RAWRA (Czech Republic).

The DECOVALEX modelling tasks cover the range of argillaceous, sedimentary and crystalline rocks and are comprised of the following tasks:

- Task A: The Sealex *In Situ* Experiment, Tournemire Site, France
- Task B1: HE-E Heater Test, Mont Terri, Switzerland
- Task B2: EBS Experiment, Horonobe, Japan
- Task C1: THMC of single rock fractures
- Task C2: Water Inflow, Bedrichov Tunnel, Czech Republic

5.4.2.1 Task A: The Sealex in situ experiment, Tournemire site, France

The Task A Test Case involves study of large-diameter bentonite cores and bentonite–rock interfaces with the objective of evaluating the impact of the seals (bentonite and concrete plugs) on the performance and safety assessment functions (Figure 5.12). The knowledge gained from the experiments will be applicable to other host rocks, such as crystalline and sedimentary rocks (the main instrumented component is bentonite which is needed for repositories in crystalline rocks, such as granite). Also, the

Figure 5.12 Work on the *in situ* experimental studies at the Tournemire site in France.

experiment has a direct impact on design, implementation, evaluation and monitoring of the sealing systems (bentonite and concrete plug) of geological repositories, especially on the post-closure issues of safety assessment after the sealing system installation in different host rocks, for both near- and far-field safety cases.

The main testing features are: placement of pre-fabricated bentonite cores (seals) into horizontal boreholes (600 mm in diameter) in the wall of galleries of the URL; forced saturation of the bentonite cores with watertight sealing of the boreholes; intra-core instrumentation with non-disturbing wireless systems installed in the bentonite core to prevent potential preferential fluid flow pathways. The participating research teams are performing numerical simulations of the saturation phase of the Sealex *in situ* tests for different testing conditions and modelling the coupled hydro-mechanical behaviour of the bentonite–rock interfaces.

5.4.2.2   Task B1: The HE-E in situ *heater test, Mont Terri Underground Research Laboratory, Switzerland*

This work, linked to an experiment in the Opalinus clay at the Mont Terri Laboratory in Switzerland, is a Test Case based on the NAGRA PEBS (Long-term Performance of Engineered Barrier Systems) programme, Figure 5.13. The objective is the evaluation of the sealing and barrier performance of the EBS (Engineered Barrier System) with time during the heating phase of a repository. Similarly to Task A, the knowledge

Figure 5.13 The research tunnel at the Mont Terri Underground Research Laboratory, Switzerland.

gained from experiments will be applicable to other host rocks, such as crystalline and sedimentary rocks, since the main instrumented component is the EBS (bentonite and concrete plug) which is also needed for repositories in other rocks, such as granite. The work involves a combination of blind prediction, model calibration with measured data, and long-term prediction for Performance Assessment (PA) and Safety Assessment (SA) impact evaluation of the EBS.

The main scientific issues being considered are the thermal evolution, buffer (bentonite) resaturation process and *in situ* determination of thermal conductivity of the bentonite and its dependency on saturation, pore water pressure evolution in the near-field, swelling pressure evolution of the bentonite, and water input from rock to the EBS. The heating started in 2011 and was continued for a period of three years with a designed heater surface temperature of 135°C. Laboratory tests are also being performed to characterise the behaviour and parameters of bentonite blocks, granular bentonite particles and the Opalinus clay rock.

5.4.2.3 Task B2: The EBS experiment at Horonobe, Japan

This EBS (Engineered Barrier System) experiment is being conducted in sedimentary rock at the Horonobe Underground Laboratory in Japan with the main instrumented component being bentonite and sand, Figure 5.14. The objectives of the Test Case are to validate the coupled Thermo-Hydro-Mechanical-Chemical (THMC) model and to

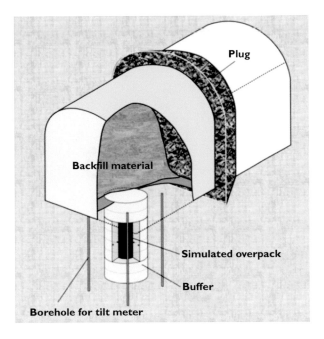

Figure 5.14 The Engineered Barrier System (EBS) at the Horonobe Underground Research Laboratory in Japan.

obtain wider data on an engineered barrier system and the associated THMC phenomena in the surrounding rock. Additional objectives are to confirm both the applicability of the measurement techniques for confirming the performance of an engineered barrier system and the 'set-up technology' of such an engineered barrier system, including the backfill of the tunnel using practical techniques given the *in situ* environment.

As with the Tasks A and B1, the knowledge gained from the experiments and modelling will be applicable to other host rocks such as crystalline and sedimentary rocks because the main instrumented components are the EBS (bentonite and plug) that are also needed for repositories in crystalline rocks such as granite. The work involves a combination of blind prediction, model calibration with measured data and long-term prediction for the impact of the EBS on PA/SA studies. The scientific issues concerning thermal evolution, buffer (bentonite) resaturation process, backfill effects, pore water pressure evolution in the near-field, swelling pressure evolution of the bentonite, water input from rock to the EBS (involving characterisation of rock saturation surrounding the EBS), and possible chemical issues, with model development and validation, and confidence building as one of the major objectives.

5.4.2.4 Task C1: THMC modelling of rock fractures

The work for Task C1 involves the study of laboratory sample-sized rock fractures for understanding and modelling the fully-coupled THMC processes as fluids flow through the fractures, Figure 5.15. It is based on data obtained from published papers and can thus be considered as both a Benchmark Test and a Test Case. Fully-coupled

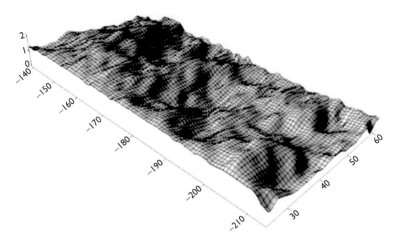

Figure 5.15 Illustrative example of an irregular rock fracture surface.

THMC processes of a single rock fracture have not been attempted before within the DECOVALEX Project phases but these processes are an issue of dominating importance for the PA/SA of repositories in fractured crystalline rocks such as granite and may also have important reference values for repositories in other types of host rocks.

The work is aimed at modelling the fully-coupled THMC processes of rock fractures based on data from the laboratory experiments on novaculite (Yasuhara *et al.*, 2006) and granite (Yasuhara *et al.*, 2011). The overall objective of the Task is to use the experiments to build and refine conceptual and physical process models for the single fracture system and to present this developed understanding in a way that can be useful input for underpinning science in radioactive waste disposal and safety case development.

The novaculite experiment is being studied first, together with considering the impacts of different approaches adopted by the research teams. The work will then move on to consider the granite case based on learning from the fracture surface evolution in the novaculite experiment—through moving from a well hydraulically constrained system (the novaculite) to a less constrained system (the granite).

5.4.2.5 Task C2: Modelling water flow into the Bedrichov Tunnel, Czech Republic

The work for this Task involves predicting the water flow into the Bedrichov water transfer tunnel located in the Bohemian granite massif in the Czech Republic, Figure 5.16. The objective is to evaluate the groundwater flow and tracer transport processes at the site scale and compare the result with the recorded data, and to consider the treatment of uncertainties for site characterisation. Thus Task C2 is also a Test Case. The experiment and research will assist with the site characterisation practices for radioactive waste repositories in crystalline rocks by considering the

Figure 5.16 The Bedrichov water tunnel in the Czech Republic.

challenging fracture system characterisation issues that play an important role in the reliability and uncertainty issues of PA/SA.

A major advantage of this Task is that considerable work has already been completed in the previous DECOVALEX phase in obtaining the basic information. A comprehensive database is already established containing the available data on site geology, fracture mapping (inside the tunnel), resistivity profiles, water inflow, water chemistry, stable isotope sampling and results, and fracture displacements.

The main scientific issues being considered are:

– hydrogeological characterisation of the test site;
– stress measurement and interpretation of results at the site scale;
– groundwater flow and reactive tracer transport study: measurements and numerical modelling, considering discontinuity and heterogeneity issues;
– borehole stability;
– impact of uncertainty of the fracture system geometry and hydro-mechanical behaviour on water flow and tracer transport.

The study considers the impacts of fluid flow pathways formed by the rock fracture system, heterogeneity of the rock mass and chemical composition, temperature, tracer mixing, and related water/tracer chemistry.

* * * * *

Further information on the history and current activities of the DECOVALEX programme can be found at www.decovalex.org.

The authors appreciate and thank the DECOVALEX2015 Funding Organisations for their financial and technical support of the DECOVALEX project work described in this Chapter. The statements made here are, however, solely those of the authors and do not necessarily reflect those of the DECOVALEX Funding Organisations.

5.5 UNDERGROUND RESEARCH LABORATORIES (URLs)

5.5.1 The purpose of URLs

For the reasons outlined in the previous sections of this Chapter, it is necessary to conduct *in situ* experiments, primarily in order to validate the computer modelling of processes but also for other reasons, such as establishing the rock mass response to excavation in different circumstances and the confirmation of *in situ* monitoring equipment. Such *in situ* experiments are not confined to the radioactive waste subject area, as exemplified some years ago by the Chinnor tunnel boring machine trials in the Lower chalk in Oxfordshire, UK, which were a precursor to the construction of the Channel Tunnel connecting England and France. Examples of other subjects requiring *in situ* experiments are the underground storage of CO_2 and geothermal energy production.

The OECD report on URLs (2013) in the radioactive waste context has listed their purposes as to:

– develop the technology and methodology required for underground experimentation;
– provide data to understand the behaviour and assess the performance of the repository system and of their interactions;
– demonstrate the robustness of the design and to show the potential areas of optimisation of engineering components and processes;
– train personnel for safe operation of a future repository; and
– build confidence with stakeholders for their understanding of the important processes governing repository performance.

There are now many URLs in different countries conducting *in situ* experiments on a variety of subjects relating to radioactive waste disposal. One of the most successful and well known was the Canadian URL which had a 420 m deep shaft and operated from 1985 to 2003 in the 2.6 billion year old granite at Pinawa, Canada. Particularly useful experiments conducted there helped to understand the *in situ* rock stress and the spalling that occurs on excavation in rock masses with high magnitude stress fields, (see Figure 5.17).

5.5.2 The Swedish Äspö URL

As highlighted in Table 5.3 in the previous Section, URLs have a variety of purposes, some of which are aimed at obtaining sufficient supporting information to ensure that radionuclides from the final repository will not travel to the biosphere in unacceptable quantities over the repository lifetime. Recalling the many FEPs discussed in Section 5.2 and the need to validate computer modelling and specific simulation studies, a URL is likely to host many different types of experiment. As an example, we will

Figure 5.17 'Discing' of a large rock core due to the effect of the *in situ* stress when coring takes place. Left to right: Neville Cook, John Hudson, Evert Hoek, Norbert Morgenstern, Charles Fairhurst in 1991.

Table 5.3 Roles of generic and site-specific URLs (from OECD, 2013).

Generic URL	Site-specific URL
Development and testing of technology and methodology–test methods for characterisation, construction techniques, monitoring.	*Evaluation of site and confirmation*–characterisation of geosphere immediately adjacent to repository and development of upscaling rules.
Development of understanding of processes and collection of generic data for safety assessment–sensitivity of rock mechanics, host rock-barrier properties and their interaction.	*Collection of site-specific data*–data required for performance assessment and for future optimisation of repository design, reduction in inherent conservatism in conceptual and safety assessment models.
Concept testing and demonstration–testing of disposal design concept and alternatives, operational options, demonstration of industrial-scale projects	*Demonstration of technology and techniques*–monitoring of near-field responses of the repository for regulatory purposes, address environmental impact assessment issues.
Building confidence and fostering international co-operation–experts from different disciplines interact to build technical confidence, develop experience among international professional communities, interaction between various stakeholders and interested public.	*Testing of final repository design as well as other operational aspects*–testing the robustness of the Engineered Barrier System (EBS) or other testing linked specifically to safety assessment requirements for licensing.
	Building confidence–demonstration of specific system design/techniques to regulators and the public.

consider the Swedish Äspö URL (SKB, 2014) run by the Swedish implementer Svensk Kärnbränslehantering AB.

In SKB's Annual report for 2013 (SKB, 2014), it is stated that, "The Äspö Hard Rock Laboratory (HRL) is an important part of SKB's work with the design and construction of a deep geological repository for the final disposal of spent nuclear fuel. Äspö HRL is located in the Simpevarp area in the municipality of Oskarshamn. One of the fundamental reasons behind SKB's decision to construct an underground laboratory was to create opportunities for research, development and demonstration in a realistic and undisturbed rock environment down to repository depth. The underground part of the laboratory consists of a main access tunnel from the Simpevarp peninsula to the southern part of Äspö where the tunnel continues in a spiral down to a depth of 460 m. Äspö HRL has been in operation since 1995 and considerable international interest has been shown in its research, as well as in the development and demonstration tasks." (See Figure 5.18).

The report SKB (2014) discusses geoscientific research as a basic activity, and the aim to increase the understanding of rock mass properties and measurement methods. The waste is isolated by both natural and engineered barriers: the natural barrier work involves studying sulphide in repository conditions, and modelling groundwater flow and transport of solutes; the engineered barrier studies involve a prototype repository, alternative buffer materials, horizontal emplacement, large scale gas injection test, canister corrosion testing, concrete and clay, low pH cementation products, deposition tunnel backfilling, tunnel end plugs, and several other projects. All these *in situ* experiments are geared to ensuring the successful isolation of the waste in the future repository located in the Swedish bedrock. Because research into radioactive waste isolation is being conducted in many countries, it is advantageous to have links

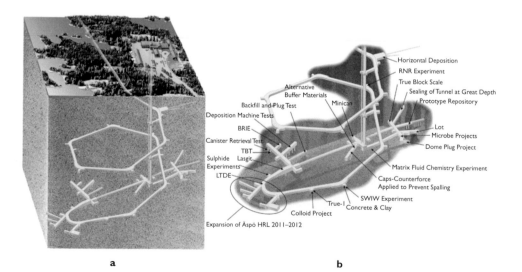

a   b

Figure 5.18 (a) Overview of the Äspö HRL facilities. (b) Allocation of experimental sites from −220 m to −460 level.

with other waste implementers and *in situ* laboratories, and the Äspö HRL is involved in co-operation with six other countries.

An interesting approach to the reproducibility of numerical modelling was conducted within the previous phase (i.e., 2007–2011) of the DECOVALEX research programme described in Section 5.4. In this Test Case example, the principal stress alterations caused by excavation of one of the small diameter shafts shown at the top right of Figure 5.19 at the Äspö Hard Rock Laboratory in Sweden were numerically modelled by separate teams from China, Czech Republic, Finland, Japan, Korea and Sweden, and compared with the actual results. As is evident from Figure 5.19, the stress path results, using different numerical models, are approaching sufficiently close reproducibility for the results to be considered adequate for engineering purposes.

The work conducted in such HRLs is wide ranging and extensive, with particular significance for rock mechanics and rock engineering, e.g., the stress path and rock spalling experimental results, see Figures 5.19 and 5.20. Moreover, many of the organisations involved produce research reports on the results of their work. These contain a wealth of scientific information and *in situ* test results which is freely available via downloadable pdfs, e.g., from www.skb.se (Sweden) and www.posiva.fi (Finland).

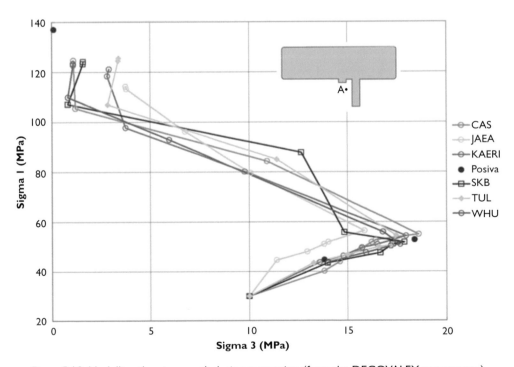

Figure 5.19 Modelling the stress path during excavation (from the DECOVALEX programme).

Figure 5.20 Rock spalling as a result of locally increasing the *in situ* rock stresses around a simulated deposition hole at the Äspö HRL (Andersson, 2007).

5.6   CHAPTER SUMMARY

Within the context of the book's theme of "Rock Engineering Risk", in this Chapter we have discussed the subject of radioactive waste disposal. This is an unusual project design and project risk subject because of the combination of a series of factors: the unique over-riding design criterion that unacceptable quantities of radionuclides must not escape to the biosphere; that the repository facility can be located any-where underground (subject to planning conditions); that complex coupled computer modelling is required, potentially geological-thermal-hydro-mechanical-chemical-biological coupled codes; and that the design life, rather than being around 100 years for conventional civil engineering projects, is of the order of hundreds of thousands of years.

To establish all the relevant aspects, lists of Features, Events and Processes (FEPs) have been prepared by the implementing organisations and we have shown how these can be structured using the Rock Engineering Systems (RES) interaction matrix approach described in detail in Chapter 3. To add to the complications, there are three main repository stages: excavation, operational, and post-closure. All this means that, in order to confirm the recommended site investigation, analysis and design activi-ties during these stages, a strict auditing procedure is required. Such auditing has been described in detail in our previous book "Rock Engineering Design" (Feng & Hudson, 2011) in which auditing Protocol Sheets were included. Here, in this chapter, in Tables 5.1 and 5.2 we provide questions related to the difficulties in rock mechanics modelling and the contents of a potential soft auditing procedure.

It is anticipated that, because of the complexities involved in direct one-to-one modelling of all the rock modelling and construction issues, that the modelling will move to 'intelligent' modelling via, for example, neural network perception approaches—as indeed we illustrate next in Chapter 6 on the risks associated with long, deep tunnels, and in Chapter 7 on the construction of caverns for hydropower projects. Furthermore, given the advances in computing capabilities, on the horizon is the possibility of developing virtual underground laboratories which, when 'calibrated', will enable virtual experiments to be conducted in the virtual rock mass.

For the validation of computer programs, *in situ* experiments are required, as illustrated via the work of the DECOVALEX project. We outlined the current work of this international project at the time of writing (2015) through the analyses of *in situ* experiments in France, Switzerland, Japan and the Czech Republic, plus consideration of the chemistry aspects of fluid flow through rock fractures. The general subject of Underground Research Laboratories was discussed with reference to the 2014 OECD report on the subject and illustrated by a specific rock spalling experiment conducted at the Äspö Hard Rock Laboratory in Sweden.

Although the design of a repository and consideration of all the associated risks—related to the over-riding design criterion that unacceptable quantities of radionuclides should not escape to the biosphere—are much more difficult and time consuming (typical site investigation and design work taking more than 20 years) than in 'conventional' rock engineering projects, there is one major redeeming feature: that the organisations involved generally have a policy of making all the scientific information available to the public. This means that, not only does the information provide the necessary evidence and credibility that the future repositories have been well designed, there is a wealth of scientific and engineering data immediately available through downloadable pdfs from the organisations' websites. Readers are therefore encouraged to take advantage of this facility because the information, data and related analyses can be used to advantage in research and to support analysis, design and construction in other areas of rock engineering.

Chapter 6

Risks associated with long deep tunnels

The entire physical world is most properly regarded as a great energy system: an enormous market-place in which one form of energy is for ever being traded for another form to set rules and values. That which is energetically advantageous is that which will sooner or later happen. In one sense, a structure is a device which exists in order to delay some event which is energetically favoured. It is energetically advantageous, for instance, for a weight to fall to the ground, for strain energy to be released, and so on. Sooner or later the weight will fall to the ground and the strain energy will be released; but it is the business of a structure to delay such events for a season, for a lifetime, or for thousands of years.

"Structures", J.E. Gordon, Penguin Books, 1978

6.1 INTRODUCTION

Recalling that the previous Chapters have included content on the introduction and background to the rock engineering risk subject, explanations of uncertainty and risk, the risk factors before, during and after construction, rock engineering systems and auditing, rock fractures and rock stress, we now present a major case example concerning the risks associated with long deep tunnels, both in general and specifically with regard to the Jinping II hydropower station in China. In this context, we begin by discussing both the epistemic and aleatory uncertainties and methods to assess and mitigate the risk for such long deep tunnels including rockbursts and water inrushes. Then, we explain in detail how the concepts have been applied for the specific case of the deep 17 km long tunnels at the Jinping II project.

6.1.1 Development of long deep tunnels

In order to satisfy the requirements for transportation, hydropower, water supply and other projects, long tunnels are being or will be constructed (see Table 6.1 for a list of some of the longest tunnels in the world). In particular and for our case example, there are seven long tunnels with a length of 17 km and overburden of 1900–2525 m at the Jinping II Hydropower Station in China. In line with the data in Table 6.1, it can be seen that there is a large number of long tunnels to be constructed over the next ten years throughout the world. Some of these will be excavated in difficult conditions which could well lead to difficulties during construction, such as the difficulty with rock support, instability caused by other hazards, increase in cost and duration of construction period.

Table 6.1 Some of the world's longest tunnels under construction or at advanced planning stage (some data are from en.wikipedia.org/wiki/List of longest tunnels in the world).

Name	Location	Length (m)	Type	Year	Comment
Qinling Tunnel, Han River Diversion to Wei River	Shaanxi, China	98,300	Water supply		
New York City Water Tunnel No. 3	New York State, USA	96,560	Water	2020	More water supply for New York City. Already in use; at completion it will be the world's third longest tunnel
Gotthard Base Tunnel	Lepontine Alps, Switzerland	57,072	Railway twin tube	2016	New Gotthard Railway; it will be the longest railway tunnel. Two tubes (East 57,091 m, West 56,978 m), 8.8–9.5 m cross section
Brenner Base Tunnel	Stubai Alps, Austria–Italy	55,000	Railway	2025	Pilot tunnel under construction
L9/L10 Barcelona Metro tunnel	Barcelona, Spain	43,710	Metro	2016 partially	The total system will have a length of 47.8 km, of which 43.71 underground and 4.09 on viaducts. North section of the line is in operation. In 2016 south section: airport El Prat/Collblanc will be in operation
Gaoligongshan Tunnel	Yunnan, China	39,600	Railway	2017	Railway between Dali and Ruili
Koralm Tunnel	Koralpe, Austria	32,900	Railway twin tube	2022	Boring of main tunnel started at the west portal May 2010
Neelum Jhelum HydroPower Tunnel	Muzaffarabad Azad, Pakistan	31,500	Hydro-electric	2016	
New Guanjiao Tunnel	Qinghai, China	32,645	Railway twin tube	2013	Longest tunnel on the upgraded dual-track Xining–Golmud section of Qinghai–Tibet Railway, longest railway tunnel in China, 3323–3380 m above sea level
Melamchi Water Supply Development Board	Melamchi to Kathmandu, Nepal	26,000	Water Tunnel	2014	Under construction, financed by Asian Development Bank
Musil Tunnel	Wonju–Jecheon (Jungang Line), South Korea	25,080	Railway	2018	
Pajares Base Tunnel	Principado de Asturias, Spain	24,667	Railway twin tube	2014	
Liyama Tunnel	Liyama, Japan	22,225	Railway	2015	For Hokuriku Shinkansen
Water transfer tunnel	Lambayeque, Peru	20,200	Water supply	2014	Under a 2000 m high mountain

(Continued)

Table 6.1 (Continued)

Name	Location	Length (m)	Type	Year	Comment
Headrace tunnels at Jinping II hydropower station	Sichuan, China	16,670	Water diversion	2016	Under a 1900–2525 m high mountain, four main plus two auxiliary tunnels
Ceneri Base Tunnel	Lepontine Alps, Switzerland	15,400	Railway twin tube	2019	New Gotthard Railway
Mount Ovit Tunnel	Erzurum Province–Rize Province, Turkey	14,700	Highway twin tube	2015	Groundbreaking 2012
Ryfast	Stavanger–Strand, Norway	14,300	Highway twin tube	2018	Groundbreaking 2012
Water diversion Tunnel	Qianhai, China	73,000	Water supply	2020	Under a 1100 m high mountain
Mont d'Ambin base tunnel	Cottian Alps, France–Italy	52,000	Railway	2020–2023	Access shafts under construction
Line 4 (Athens Metro)	Athens, Greece	33,000	Metro	2015–2025	
Follo Line	Oslo, Norway	19,000	Railway	2020–2021	
Fehmarn Belt Fixed Link	Germany–Denmark	17,600	Road & Railway	2020	Start of construction 2014
Förbifart Stockholm	Stockholm, Sweden	16,000	Road	2020	
Kresna Gorge Tunnel	Kresna, Bulgaria	15,000	Road	2020–2023	Part of Struma motorway connecting Sofia with Athens

Factors such as considerable depth, long distance and traverse through different geological units are likely to cause a series of special challenges during the construction of such deep tunnels, which are different from those in a 'conventional tunnel'. These challenges include rockbursts in hard rock and large deformations in weak rock, which are induced by high *in situ* stresses (e.g., Steiner, 1996; Hoek & Marinos, 2000; Dalgic, 2002; Meguid & Rowe, 2006; Jiao *et al.*, 2007; Zhang *et al.* 2012), as well as high-pressure mud flow and groundwater inrush when the tunnel passes through karst topography or fault zones (e.g., Tseng *et al.* 2001; Lin & Lee, 2009). Table 6.2 lists some of the major engineering problems that have occurred in deep and long tunnels over the past ten years. We note that the severe engineering hazards influencing tunnel construction are brittle failures invoked by high stresses, i.e., spalling and rockbursts, large deformations, groundwater inrush, collapse and tunnel boring machines becoming stuck.

Zhao *et al.* (2013) reviewed geological hazards which occurred during the construction of Chinese railway tunnels. Based on statistical analyses of the railway tunnel cases covering nearly 10,000 km in a variety of complex geological conditions in China, the major geological hazards and the potential risks are illustrated in Figure 6.1. It can be seen from the Figure that, in Chinese railway tunnel construction, water inrushes and large deformations are the challenging issues, accounting for about

Table 6.2 Major engineering hazards in deep and long tunnels.

Name	Location	Length (planned length) (m)	Maximum overburden (m)	Type	Excavation method	End of construction (year)	Major engineering problems					
							Overbreak and Spalling	Rock-bursts	Large deformation or squeezing (including floor heave)	Collapse	Ground-water inrush	TBM jamming or buried
Lyon-Turin Base Tunnel	Italy and France	2,329 (2000)	2500	Traffic tunnel	Mechanical method	2010			✓			
Headrace tunnels at Jinping II hydropower station	Sichuan, China	16,670	2525	Water diversion	TBM and D&B	2016	✓	✓	✓	✓	✓	✓
Karaj-Tehran tunnel	Tehran, Iran	32,000	800	Water supply	TBM				✓			
Headrace tunnel at the Parbati II hydro-electric project	Himachal, India	31,370	1500	Water Supply	TBM and D&B	2013	✓	✓				
The Syueshan tunnel	Taiwan	12,900	>700	Highway tunnel	TBM and D&B	2006			✓	✓	✓	✓
The Yacambú-Quíbor tunnel	Venezuela	25,000	1270	Water transmission tunnel	TBM to D&B	2008			✓	✓	✓	
The Kuhrang Tunnel project	Iran	23,409	>1100	Water conveyance tunnel	Open TBM and Blasting				✓	✓	✓	✓
The Zhegu mountain tunnel	Sichuan Province, China	4,400	1000	Highway tunnel	D&B				✓			
The Kaligandaki headrace tunnel	Nepal	5,950	600	Headrace tunnel					✓			
The Gotthard Base tunnel	Lepontine Alps, Switzerland	57,072 or 57,091		Railway twin tube		2016			✓			

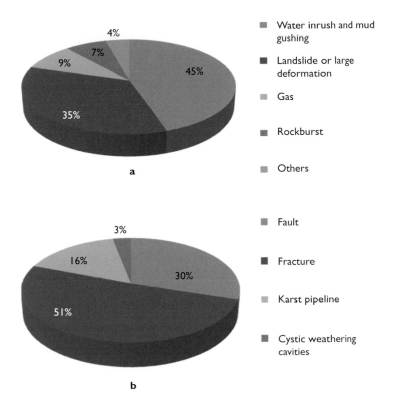

Figure 6.1 Types of geological problems in tunnels excavated by the drill-and-blast method during the construction of Chinese railway tunnels: (a) percentages of the phenomenon experienced, and (b) the percentages of mechanism type related to water inrush (Zhao *et al.*, 2013).

80% of the total. Note that the large book by Lu (2009) describes karst conditions in China. Furthermore, Figure. 6.1 also shows the percentage of cases of the different mechanisms which lead to water inrush; fractures, faults and karst conditions are three of the major mechanisms.

The above-mentioned hazards have resulted in loss of human life, economic loss and delays in construction. For example, Table 6.3 lists some case examples of rockbursts occurring during the construction of tunnels in China. Moreover, Figure 6.2 shows some field photographs of rockburst damage. It can be seen that these rockbursts led to severe failures of support systems, such as fractures and extreme deflection of steel arches (see (a), (b) and (c) in Figure 6.2), and crushed and broken rock masses near the excavation surfaces (see (d), (e) and (f) in Figure 6.2). Due to the obstruction caused by these hazards, it is necessary that their risks be assessed and the associated defence treatments established.

Since the 1970s, risk in tunnels has been considered. For example, Einstein (1994) has analysed the construction of the Adler tunnel and studied the potential risks and risk analysis methods in rock engineering. Sturk *et al.* (1996), Apeland *et al.* (2002), Likhitruangsilp and Ioannou (2004), Nyvlt *et al.* (2011), Huang (2006), and Sousa and Einstein (2012) have studied the risk of tunnels. The ITA (2009) has produced

Table 6.3 Some statistics of rockburst risks in Chinese tunnels.

Tunnels	Date of breakthrough	Max overburden (m)	Chainage at rockburst	Total length of rockburst zones (m)
Guanciba tunnel for Chenkun Railway	1966	1650		Sporadic
Water diversion tunnels for Ertan Hydropower Station	1993	200		315
Headrace tunnels for Taipingyi Hydropower Station	1993	600	>400	
Headrace tunnels for Tianshengqiao Hydropower Station II	1996	800	30	
Tunnels for Qinling Railway	1998	1615		1894
Erlangshan tunnels for Chuanzang Road	2001	760	>200	1252
Chongqing Tongyu Tunnels	2002	1050		655
Chongqing Lujialing Tunnel	2004	600	93	
Headrace tunnels for Futang Hydropower Station	2004	700	>400	900
Incoming traffic tunnel for Pupugou Hydropower Station	2005	420	183	
Qinling Zhongnanshan extra long highway tunnel	2007	1600		2664
Headrace tunnels, water drainage tunnel and auxiliary tunnels at Jinping II Hydropower Station	2011	2525	>750	
Headrace tunnels at Jiangbian Hydropower Station	2012	1678	>300	

guidelines for tunnel risk management. In addition, China has compiled the Railway Tunnel Risk Assessment Guidelines (2008), plus the highway, bridge and tunnel engineering construction safety risk assessment guide, and the specification of the risk management for the construction of city rail transit engineering (2011). However, the current risk analysis methods do not cover tunnels very thoroughly because these need more consideration of the risks of rockbursts, jamming of TBMs due to large rock deformations, and water inrushes at high pressure. Therefore, it is necessary to establish methods to assess and mitigate the risks for long tunnels at great depths.

6.1.2 Flowchart to develop risk management for long, deep tunnels

When considering the risk issues involved during tunnel design and construction and introducing a mechanism of risk management into the deep tunnelling process, research is required—especially identifying the associated uncertainties, including epistemic

Figure 6.2 Failure of the support system and damage to the rock mass due to rockburst and spalling occurrence. (a), (b) and (c): broken support system in TBM tunnel; (d), (e) and (f): crushed rock masses.

uncertainty and aleatory uncertainty, assessing the potential hazard types and their risks, as well as introducing acceptance criteria, designing treatment strategies and risk mitigation schemes before and during tunnelling, and optimising dynamic risk management approaches. We recall our 'master flowchart' in Figure 6.3 and follow this with the initial risk management flowchart for long, deep tunnels in Figure 6.4.

When this risk management is applied to an actual tunnel with long length and large overburden, the processes of risk management are divided into two aspects, as shown in Figure 6.4: (i) initial risk management and (ii) dynamic and final risk management. The initial risk management consists of epistemic uncertainty analyses,

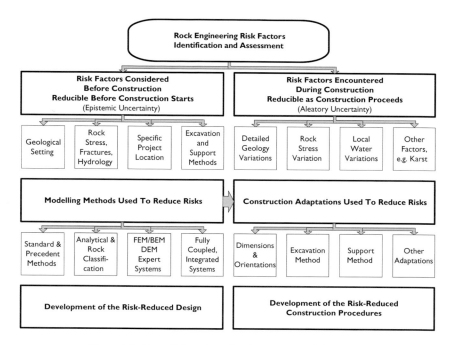

Figure 6.3 Overall flowchart for the content of this book.

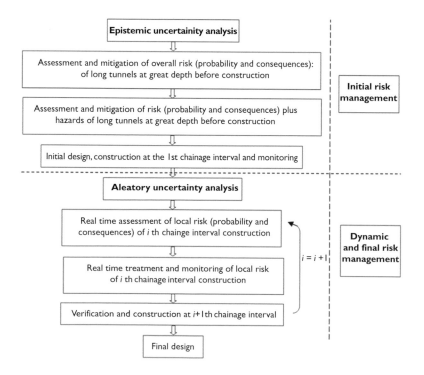

Figure 6.4 Flowchart to establish risk management for long, deep tunnels.

including the identification and analyses of the overall and local risks, as well as the designation of their treatment and mitigation, which must be conducted before the construction. Moreover, some initial designs, such as the support and monitoring schemes, must also be implemented in the context of the initial risk assessments. The latter assessments are a dynamic process to identify uncertainties and the factors inducing the engineering hazards, i.e., the uncertainties of geological structures, variable distributions of the *in situ* stress field, non-uniform mechanical characteristics of rock masses, etc. These uncertainties may be partially identified by borehole site investigation or by being exposed through excavation. Based on these identified uncertainties, an understanding of the hazards in such long tunnels can be obtained so that more detailed engineering designs may be established.

6.2 EPISTEMIC UNCERTAINTY ANALYSIS OF DESIGN AND CONSTRUCTION FOR LONG DEEP TUNNELS

Some key engineering geological factors directly influencing the stability of tunnels and their design process are (*cf.* Figure 6.3) geological setting, rock mechanical properties, *in situ* stress conditions, groundwater inflow through fractures and weakness/fault zones, and the procedures of excavation and support. Therefore, in the following Section, we discuss the epistemic uncertainties of the above-mentioned key factors.

6.2.1 Geological settings

6.2.1.1 Geological factors relating to rockbursts in deep tunnels

Rockburst mechanisms involve two aspects: the rockburst seismic source mechanism and the rockburst damage mechanism. Durrheim *et al.* (1995) identified these factors via back analyses of rockburst cases occurring in South African mines. It was found that one of the important factors is geological conditions, including the stress field, characteristics of rock masses, and their structural nature. In fact, the past geological activity leading to complicated geological settings may have caused complex rock mass structures and associated complex stress field distributions. More and more evidence has revealed that rockbursts exhibiting high intensities commonly occur in or near geological structures. For example, Castro *et al.* (2009) found that fault-slip is a crucial mechanism which may lead to large energy releases, evoking severe rockburst damage events during deep mining. This has also been recognised by other researchers (Ortlepp, 1978; 2000; Gay & Ortlepp, 1979; Reyes & Einstein, 1991). The effects induced by a fault have been described by Gay and Ortlepp (1979) in the following statement:

> "*The shear zones are themselves made up of smaller, en echelon, shear planes, which are connected by subsidiary conjugate shears and extension fractures. These shear planes become diffused and are replaced by extension and bedding-plane fractures at the boundaries of the fault zone.*"

Also, the occurrence of a rockburst event depends on both the regional and local geologies: the extent of the adverse geological structures provides an overall limit for the rockburst prone ground; and the local geology governs the rockburst location and rockburst intensity, as well as the rockburst mechanisms and associated types. Figure 6.5 shows several conceptual rockburst mechanisms given by Trifu and Suorineni (2009). In Figure 6.5, these mechanisms are shown to depend on the local geological structures and their failure modes under high stress conditions. Moreover, the local variability of rock mass structures and their mechanical properties also governs the occurrence of rockbursts/seismic events. For example, Figure 6.6 illustrates the difference between seismic events before and after a destress blast at Xstrata Zinc Brunswick Mine (Rose *et al.*, 2011). After destress blasting, the seismic events were clearly decreased. Such destress blasting changes the deformation and strength

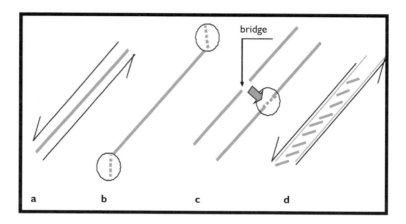

Figure 6.5 Conceptual rockburst mechanisms: (a) slip along a continuous fault, (b) fracture propagation, (c) damage of rock bridges co-linear with faults, and (d) damage of *en echelon* rock bridges within an overall continuous discrete fault (Trifu & Suorineni, 2009).

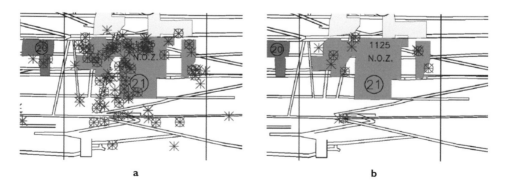

Figure 6.6 Difference of seismic events before and after a destress blast at Xstrata Zinc Brunswick Mine: (a) before destress blast (01/01/2009 to 05/24/2010); (b) after destress blast (05/25/2010 to 08/20/2010) (Rose *et al.*, 2011).

of rock masses in terms of increasing the length of fractures with better utilisation of explosive energy. Also, blast vibrations can reduce the friction of joint planes. Therefore, destress blasting alters the conditions, including the geological structures, their mechanical properties, and the stress field in the rock mass.

6.2.1.2 Geological conditions exhibiting squeezing or large deformation behaviour

As summarised in the review paper by Barla (2001), squeezing behaviour is associated with poor rock mass deformability and strength properties, and is encountered in ductile and altered rock complexes: gneiss, micaschists and calcschists (typical of contact and tectonised zones and faults), claystones, clay-shales, marly-clays, and similar rock types. Under squeezing ground conditions, one of the key factors in assessing the magnitude of the squeezing deformation is the geological environment, such as the rock mass structures, rock types, and also the geological history of the tectonic units. To illustrate the importance of geology, the squeezing deformation cases in the Sedrun section of the Gotthard Base Tunnel (Mezger et al., 2013) and the Lyon–Turin Base Tunnel (Bonini & Barla, 2012) are briefly reviewed. However, only problems invoked by geological conditions are analysed in this section and the detailed geological information for both case tunnels are excluded here, but may be obtained from key references listed in the References section, such as the information provided by Kovári and Staus (1996), Barla et al. (2007), Barla (2010), Bonini and Barla (2012) and. Mezger et al. (2013).

At the Sedrun section of the Gotthard Base Tunnel, it was found that the tunnelling-induced large convergences correlated reasonably well with the degree of shearing and schistosity orientation of the rock. As demonstrated by Mezger et al. (2013), two factors, namely the degree of shearing and the influence of schistosity, must be estimated by rock mass classification on the basis of advance core drilling and geological investigation. Therefore, two kinds of geological uncertainties occur during the above-mentioned estimation process, one of which is the subjective uncertainty of rock mass classification to determine the degree of shearing, and the other is the divergence of observed geological data, e.g., the schistosity orientation or dip angle and the angle between the strike direction and the tunnel axis, i.e., between those exposed in the excavated tunnel or engineering field, and those in the advancing tunnel faces. It also is worth noting that the non-uniform distribution of squeezing deformation over the cross-section of this tunnel was partially due to the overall anisotropy of the rock mass (Goricki et al., 2005), which was based on local structural rock features such as quartz inclusions or local schistosity orientation changes (Mezger et al., 2013). These local changes of geological condition can be complicated, as shown in Figure 6.7 and increase the uncertainties of the engineering geological settings. In Figure 6.7(a) and (b), it can be noted that the variation of the characteristics of the local schistosity plane is remarkable, such as the occurrence of local foldings and the change in the dip angle of the schistosity planes. Additionally, and considering the geological setting of the Gotthard Base Tunnel, as illustrated in Figure 6.7(c), heavy squeezing problems in this tunnel occurred in two tectonic units, namely the Clavaniev Zone (CZ) and intermediate Tavetsch-Massif (TZM), which consist predominantly of gneisses and schists (Kovári & Staus, 1996). According to the works published by Mezger et al. (2013), in these two tectonic units both the intensive shearing and schistosity govern the squeezing behaviour of the rock mass. However, the

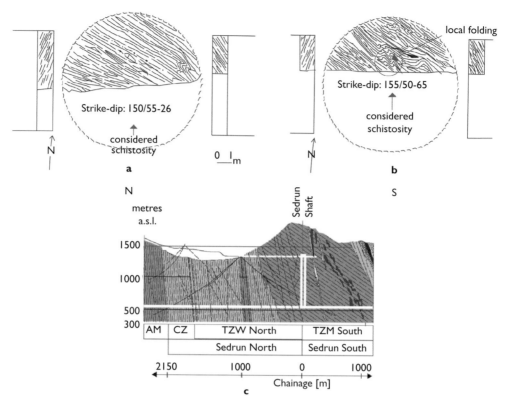

Figure 6.7 Geological conditions relating to the Gotthard Base Tunnel. (a) and (b) Geological mapping of the tunnel faces, chainage 1535 m in the NW tube and chainage 1202 m in the NE tube; (c) Geological profile (Guntli & Weber, 2009; Mezger *et al.*, 2013).

expressions of these in the two tectonic units are clearly different, due to the variation of local lithology and tectonic setting. Thus, the complexities of the geological conditions govern the uncertainties in such squeezing or large deformation behaviour.

A similar conclusion may also be obtained from the squeezing cases given by Bonini and Barla (2012), in the Lyon–Turin base tunnel, which is a key element of the Trans-European Network for passengers and freight, and is excavated between the portals in Italy and France (more information can be obtained from Barla *et al.*, 2007). One of the access adits of the Lyon-Turin base tunnel, i.e., Saint Martin La Porte, underwent large convergences between chainages 1200 and 1550 m. Metric convergences were encountered in the coal schists section (Houiller Brianconnais), associated with a large plastic zone around the gallery and a strong time-dependent behaviour, as the overburden reached 300 m. Figure 6.8 shows the partial geological condition of the Saint Martin La Porte area. Detailed mapping of the geological condition at the tunnel faces was performed systematically by Bonini and Barla (2012), and is illustrated in Figure 6.8(b). From these geological mappings, it may be noted that the rock mass gradually improves from a very disturbed rock mass (chainage 1325 m), including significant portions of weak rocks, to a relatively homogeneous condition, where the schist content gradually reduces and the presence of sandstone increases (chainages 2001 m and 2330 m). Due to the differences in the geological conditions along the tunnel's horizontal axis, both the variability of squeezing behaviour and the magnitude of tunnel convergence are remarkable, as shown in Figure 6.8(c), implying that the inhomogeneous distribution of the rock types and the presence of local tectonic features, such as shear zone, cracking zone and fault zone, can play essential roles in the squeezing behaviour of rock masses.

Thus, the distribution and presence of 'weak materials' may invoke large tunnel convergence (Bonini and Barla, 2012). These 'weak materials' include schist, shale, coal and cataclastic rock, which significantly alter the mechanical behaviour of rock masses, e.g. the tunnel section between chainage 1300 m and 1600 m in the Lyon-Turin

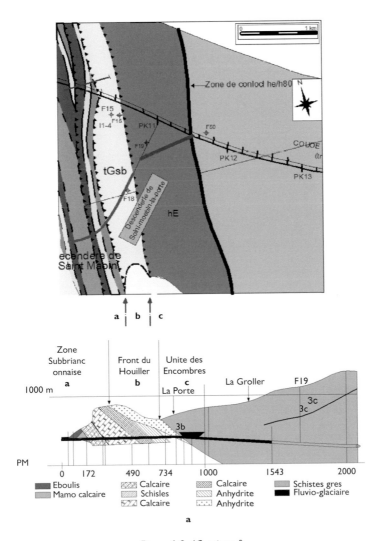

Figure 6.8 (Continued)

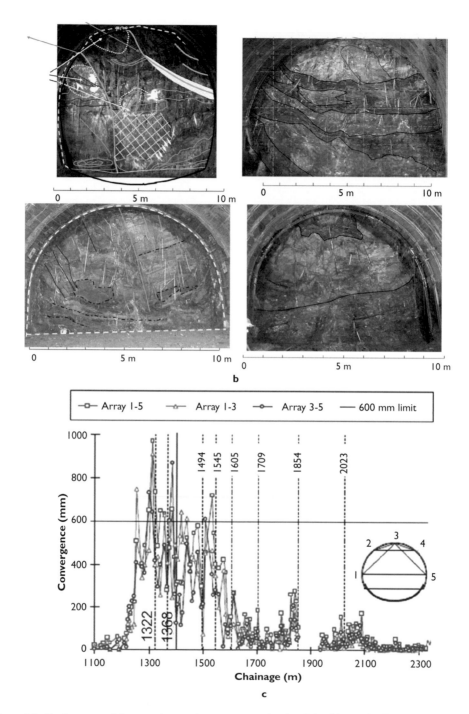

Figure 6.8 Geology conditions and tunnel convergence in the Saint Martin La Porte area: (a) Plan view at the base tunnel level and simplified geological longitudinal view; (b) Mapping of the geological conditions at the face at chainage (clockwise) 1325, 1670, 2001 and 2300 m. (c) Tunnel convergence along different arrays (Vu *et al.*, 2013, and Bonini & Barla, 2012).

base tunnel, where the 'weak materials' mentioned above are distributed extensively so that the convergence is larger, as shown in Figure 6.8(c). Importantly, the anisotropic character of the convergence measurements was also observed and analysed in the Saint Martin La Porte area by Pellet (2009), Bonini and Barla (2012), and Vu *et al.* (2012, 2013). This anisotropic trend can be interpreted by combining three main factors: lithology, orientation of the discontinuities and *in situ* stresses. Hoek and Marinos (2009, 2010) emphasised the fact that the magnitude and directions of *in situ* stresses are of great importance for squeezing behaviour, as well as the anisotropic closure interpretation. However, Vu *et al.* (2013) argued that the amount and the direction of anisotropic deformation depended on the lithology and some structural geological features described by the dominant discontinuity families. Bonini and Barla (2012) also pointed out the significance of the anisotropic geological structure on causing non-symmetrical strain distributions in the measurement cross-sections.

Based on the above information and the two case analyses, we can conclude that: (i) the variability of geological conditions can lead to squeezing behaviour; and (ii) the complexity of the geological setting increases the uncertainty of the prediction of the convergences during tunnel construction in squeezing ground, as well as that of an adequate temporary support and the excavation parameters. Thus, the key points are the lack of knowledge concerning the complexity of the geological setting and the different factors governing the squeezing behaviour. However, many tunnel cases have provided a wealth of knowledge relating to the effects of geological conditions on the squeezing behaviour of a rock mass (Aydan *et al.*, 1993, 1996; Steiner, 1996). Therefore, the uncertainty concerning the geological setting may be decreased greatly based on observations and knowledge, as follows.

i   **Identifying the lithology**—which is one of several important factors governing the squeezing behaviour of a rock mass. Weak materials refers to 'soft rocks' or 'squeezing rocks' and 'swelling rocks', such as shale, coal, schist, mudstone, clay, phyllite, cataclastic rock, siltstone, salt, potash or weathered and/or sheared metamorphic and igneous rocks. Proportional analysis of the various lithological facies plus rock classification, e.g., those methods indicated by Vu *et al.* (2013) and Bonini and Barla (2012), is an approach to classify the variability of potential squeezing conditions from the lithology at the tunnel face. Invoking mineralogy, Terzaghi (1946) pointed out that a prerequisite for squeezing is a high percentage of microscopic and sub-microscopic particles of micaceous minerals or of clay minerals with a low swelling capacity; the capacity to swell seems to be limited to those rocks which contain clay minerals, such as montmorillonite, which has a high swelling capacity. The main objective in identifying the lithology is to evaluate the mechanical characteristics of the rock mass, especially the strength and deformation properties, which are related to the squeezing potential.

ii **Identifying the tectonic conditions and local geological structures** such as faults, fractures and crushed zones (Kimura *et al.*, 1987; Schubert *et al.*, 1996; Dalgic, 2002; Khanlari *et al.*, 2012), intensive shearing zone (Mezger *et al.* 2013), lithologic transition zone, the igneous contact zone (Yassaghi & Salari-Rad, 2005), intensive folding zone, soft and hard rock contact zone (Feng *et al.*, 2013), discontinuity surfaces (bedding and foliation, Singh & Goel, 1999).

iii **Intensifying geological investigation and deformation analyses of excavated tunnel sections.** To predict the deformation of a rock mass in a to-be-excavated tunnel section under squeezing ground conditions, the understanding of the deformation mechanism in the excavated tunnel section is helpful. This consists of (a) determining the major geological factors governing the deformation behaviour of squeezing ground, which is commonly based on detailed geological investigation, including geology mapping and sampling; and (b) analysing the monitoring data to identify the deformation characteristics and understand the relation between the deformation and geological settings. The objective is to understand the characteristic features of the potentially anisotropic, highly heterogeneous, disrupted and fractured conditions of rock masses which may exhibit squeezing behaviour.

6.2.2   Rock stress

In situ stress is a key consideration for the design and construction aspects of rock engineering projects, as we have described in Chapter 4. The distribution of *in situ* rock stress results from a combination of the tectonic history, geological setting, and inherent variability of geological materials. It is also a major contributory cause of cracking, fracturing, spalling and rockbursting in a hard rock tunnel, and, if the rock is weak, it may give rise to squeezing behaviour, shearing and deformation. Moreover, there can be stress and hydraulic coupling (*cf.* Chapter 3), and a high induced rock stress after the tunnel is excavated. Thus, to consider the risk in geotechnical engineering problems, a profound understanding of the geological setting and the assessment of a geo-stress field is one of the most important analyses, including the estimation of the full *in situ* stress tensor and its variability across the site which can now be modelled using computer programs such as 3DEC (see the colour plate section at the end of the book).

Commonly, two types of analysis are conducted: (i) research on the macroscopic characteristics of the regional stress field in order to characterise the overall nature of the geo-stress field in the study area; and (ii) studies of the local stress field along the proposed tunnel line, which enables a deeper understanding of rock stress under specific engineering conditions, and as well as providing input for the mechanical analysis and engineering design of the tunnel. However, establishing the *in situ* stress field across a site is not easy and needs careful study. The basic principles of various methods have been summarised in publications: the basic principles by Ljunggren *et al.* (2003), Hudson *et al.* (2003) and Hudson (2010); *in situ* rock stress determinations in deep boreholes at the Underground Research Laboratory in Manitoba, Canada by Thompson and Chandler (2004); determination of stress orientation and magnitude in deep wells by Zoback *et al.* (2003); back regression analysis on the initial geo-stress field given by Hu *et al.* (2005), Xie *et al.* (2008), Liu *et al.* (2008), Gu *et al.* (2008) and He *et al.* (2009); the method for obtaining the local stress field based on the interpolation and numerical simulation method by Zhang *et al.* (2008); and estimation of *in situ* stress along deep tunnels buried in complex geological conditions by Zhang *et al.* (2012). The book by Zang and Stephansson (2010) "Stress Field of the Earth's Crust" provides an overview of the required subjects.

The epistemic uncertainties of rock stress in the current context stem from the variations of rock masses and the limitations of stress measurement, as well as the complexities of engineering in deep tunnels. The methods used to address the problem are of two types: (i) the direct approach, i.e., the measurement of *in situ* stress using the overcoring and hydraulic fracturing methods; and (ii) the indirect approach, in which the orientation and magnitude of *in situ* stress are determined using information such as borehole breakout, core discing, brittle failure of the surrounding rock mass and the Kaiser effect, back analysis and other information, or based on monitoring results. The overcoring method is used to obtain the rock stress under relatively low overburden conditions because, as the overburden and associated magnitude of the rock stress increases with depth, it is difficult to use the method due to effects such as core discing—a phenomenon indicative of high magnitude stresses which are normally encountered in deep tunnelling, as shown in Figure 6.9. This phenomenon stimulated the field measurement of the principal stresses at depth using the hydraulic fracturing technique, especially the Hydraulic Testing of Pre-Existing Fractures (HTPF) method developed by F. Cornet. However, under extremely high rock stress conditions, it may not be possible to conduct the hydraulic fracturing to measure the rock stress, due to the fact that the rock mass cannot be fractured by the hydraulic pressure, e.g., in the deepest tunnel section of the Jinping II tunnel (above 2300 m overburden). Moreover, due to the complexity of measurement methods and the mechanical properties of the rock involved in rock engineering, the test results are often found to have problems with respect to accuracy and representativeness. It is also worth noting that the selection and interpretation of measurement data, the rejection of unreasonable data, and back analyses of stress data are associated with subjective judgement. Computer modelling of stress fields, especially with the ability to incorporate major fracture zones, helps significantly in understanding the nature of *in situ* stress variability at a particular site.

The indirect method of determining rock stress is a back-analysis or back-calculation method based on observed information, such as the statistics of failure characteristics (e.g., brittle failures such as breakouts, spallings and rockbursts, large

Figure 6.9 Extreme rock core discing caused by high *in situ* rock stress and the process of core drilling (from Feng & Hudson, 2011).

deformation behaviour, damage zone characteristics), the phenomenon of rock core discing, analytical calculations based on mechanical models, and numerical analyses. But this is somewhat difficult to perform and ideal conditions are required to achieve useful and representative back-calculation results. As argued by Panthi (2011), the many parameters required to conduct back-analysis of rock stress must be available and assessed in advance, with assumptions required to simplify the analysis process.

For example, rock mechanics properties, in particular rock strength and deformability properties, are the key in the back-calculation of *in situ* stress magnitudes. However, the uncertainty of these parameters may be considerable and so it is difficult to obtain accurate magnitudes of these parameters. In fact, when determining these parameters, difficulties may be encountered due to the complexities of the geology and engineering conditions. Also, the numerical model used to conduct back-analysis is an ideal mathematical description—which decreases the complexities of the actual geological setting and mechanical behaviour of the rock mass, e.g., the simplification of geological structures such as faults and joints. Furthermore, to comprehensively consider the state of rock stress in the studied region, it is also noted that 'multi-source information' is required to assess the rock stress using the indirect method. For example, when Zhang *et al.* (2012) developed an approach to assess *in situ* stress along the deep Jinping II tunnels excavated in complex geological conditions, they adopted several approaches: multivariate regression, numerical simulation, stereographic projection, and employed large amounts of multi-source field data such as brittle rock failures, Excavation Disturbed Zone (EDZ) information, core discing, large-scale deformation of soft rock, and the World Stress Map. Although, of course, it is favourable to improve understanding and clarify the characteristics of the *in situ* rock stress, sometimes large amounts of decision-making data are a double-edged sword: by employing the multi-source data, we have to decide which data should be used and which removed when we conduct our analyses. Therefore, the accuracy of rock stress assessment via the indirect method also depends on the researchers' experiences and decisions.

6.2.3 Hydrogeology

Hydrogeological studies related to tunnels often focus on one or several of the following aspects (Vincenzi *et al.*, 2009): (i) from a technical perspective, groundwater is a problem for the construction of tunnels; (ii) from an ecological and hydrogeological perspective, tunnels represent a risk for groundwater integrity together with the connected surface waters and ecosystems; and (iii) from a scientific point of view, tunnels offer the opportunity to access and study underground environments, including aquifers and groundwater. In this sub-Section, we deal with the first of these aspects. Groundwater can lower the stability of the rock mass in a tunnel (Cesano *et al.*, 2000) and groundwater inrushes into tunnels are also a risk for the workers and machines, especially if these inrushes occur unexpectedly and at high pressures and/or flow rates.

According to the experiences summarized by Tseng *et al.* (2001) from the Syueshan tunnel in Taiwan, groundwater inflow at high pressure is mostly related to geological structures, such as faults and shear zones. Under such complicated geological structures, the fractured rock formation can be a strong groundwater bearing body, with fault zones providing the recharge routes. If these complicated geological structures are located in carbonate karstic rock regions, the potential and risk of groundwater inrush in the excavated tunnels will be much higher than those in non-karstic engineering areas (Mosayebi *et al.*, 2004; Day, 2004; Bakalowicz, 2005).

Zarei *et al.* (2012) presented a list of the recent tunnelling projects in Iran in karstic rocks, along with their respective geological hazards. The list illustrates the fact that, due to the particular hydrogeological features in karstic aquifers, the construction of tunnels faced a high risk of interference with groundwater and cavities or thick fill deposits. Zarei *et al.* (2012) reviewed several water inrush cases in the Kuhrang tunnel, where a large amount of water rushed in when a karstic cave was exposed in the tunnel floor, the water discharge exceeding 1,200 l/s after several hours. Therefore, the above-mentioned hydrogeological conditions can, not only threaten the safety of tunnelling projects as well as affecting the time and economic aspects, but they may also enhance the difficulty of tunnel design and engineering treatment. Tunnelling in karstic zones may present numerous challenges to the engineer, including the unpredictable occurrence of cavities and well-developed hydraulic conduits.

For the reasons outlined above, it is difficult to decrease the epistemic uncertainty caused by complicated hydrogeological conditions. Figure 6.10 illustrates a water inrush case in the Kuhrang tunnel described by Zarei *et al.* (2012). In this case, before the karstic channels were exposed, the rock mass was very strong, and no groundwater flows were experienced. The process of water inrush was very sudden and unexpected, thus highlighting the epistemic uncertainty about groundwater inrushes due to unclear hydrogeological conditions. However, some suggestions may be obtained from tunnel cases as described below.

i   Identifying major aquifers based on lithology and hydrogeological experiences and their groundwater types, such as the pore water, fracture water and karst groundwater. Aquifers are permeable and porous, and include rock types such

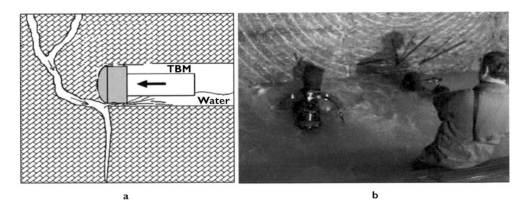

a   b

Figure 6.10 A water inrush case at the tunnel floor in the Kuhrang tunnel described by Zarei *et al.* (2012): (a) Schematic model of the geological condition and karst features; (b) Tunnel inundation and effort by divers to prevent inflow.

as sandstone, conglomerate, fractured limestone, and unconsolidated sand and gravel. Fractured volcanic rocks, e.g., columnar basalts, also make good aquifers. Rocks such as granite and schist are generally poor aquifers because they have a low porosity. However, if these rocks are highly fractured, they can make good aquifers. For deep tunnels, the types of groundwater which are major hazards for tunnelling are the fracture water and karst groundwater.

ii Identifying major hydrogeological structures and their conditions. When tunnel excavation proceeds to areas with faults, folds, shear zones and karstic zones, on-site geological investigation must be undertaken as part of the normal construction procedures and the hazards of such geological structures related to tunnelling must be identified in the standard risk analysis process. The potential engineering problems which may be encountered after the excavation of the tunnels must also be assessed, which includes the excavation schemes, support types and their parameters, monitoring and early-warning programme and systems, and the respective emergency measures.

iii Thoroughly studying the hydrogeological information obtained from the excavated tunnel section to comprehensively understand the characteristics of the hydrogeology along the tunnels and the associated influencing factors.

iv Conducting advanced geological forecasts and advanced drilling to identify hydrogeological structures and predicting the groundwater features, such as groundwater types, flow rates and flow pressures.

6.2.4 Properties of the rock mass

In deep and long tunnels, judgement of the quality of the rock mass is based on observations and tests. The investigations and procedures will vary according to the nature of the project, as well as the complexity of the geology, the background of the engineering company, and the experience of the individual geologists or rock mechanics engineers involved (Palmström, 1995). Due to the fact that geological formations are spatially variable, and that only a limited number of measurements or observations can be made, the determination of the properties of the rock mass along such tunnels inevitably involves a certain amount of extrapolation and guesswork. Therefore, initial epistemic uncertainty will be evident when we compare the properties assessed before tunnelling with those after tunnelling.

This is demonstrated by a rock mass classification case in the tunnels of the Baihetan Power Station. There are differences between the estimated rock mass classifications and the actual ones after the excavation process, as shown in Figure 6.11. The rock mass assessed as Class III before construction was reclassified as Class IV. This difference stems from the fact that the geological data are collected in stages such as collection before, during and after excavation. Therefore, the geological and mechanical information used to analyse rock mass properties is from less to more and from single-source to multi-source. Before construction, interpretations and extrapolations of rock mass properties are made based on observations and investigations performed on the surface, or those exposed in drilling holes or exploratory tunnels. During and after construction, the data for the characterisation of ground conditions can be collected either at the surfaces of excavations or at outcrops exposed as part

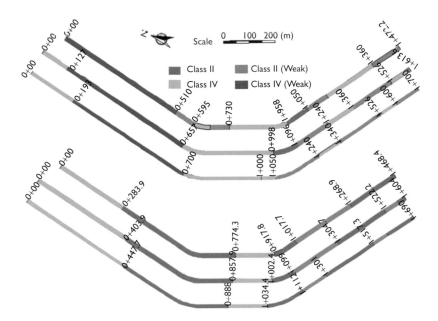

Figure 6.11 Difference between the estimated rock mass classification before construction, shown in the lowest three tunnels, and the actual rock mass classification for the same three tunnels, shown in the upper three tunnels (Baihetan Power Station, China). See the colour plate section at the end of the book.

of the construction work. Also, much monitoring, testing and evaluation can be performed directly within the rock mass. The uncertainty present during initial assessment of the rock properties is thus reduced.

When we predict the mechanical properties and behaviour of a rock mass and design an excavation scheme, as well as the associated support programme, rock mass characterisation is undertaken using the Q-system, RMR-system, GSI-system, and the Chinese BQ system (see Appendix for a description of the latter system) if the strength parameters of the rock mass can be evaluated and assessed. However, these classification systems have been established on the basis of tunnel cases or other underground project cases with relatively low overburdens compared to our current context of deep tunnels. As a result, most of these systems are only appropriate for conditions of low or moderate field stress and low water pressures. (Assume that, if the major principal stress is greater than 20 MPa, then the *in situ* stress is referred to as a high field stress.) High water pressure can be defined as water pressure which is greater than 1 MPa. Therefore, some engineers and researchers argue that the existing rock mass classification methods are not directly appropriate for deep tunnels, leading to proposed novel classification systems more applicable to deep tunnels: e.g., MRMR, proposed by Laubscher (1990); the HHQ-system, presented by Wang *et al.* (2006); and DBQ for deep engineering in a soft rock mass, established by Wang *et al.* (2013).

6.2.5   Project location

The determination of project location depends upon the project type, engineering cost and benefit, excavation difficulty, etc. However, some tunnels have to be excavated under complicated engineering conditions in order to satisfy particular tunnel functions. For example, headrace tunnels or intake tunnels are commonly excavated through mountains with a high overburden and/or in complex geological environments in order to obtain the high water heads. The seven Jinping II headrace tunnels in China involved seven tunnels which were excavated through the Jinping Mountains with a maximum overburden of 2525 m. Their east and west entrances were both located in the deep-cut valley of the Yalong River because the tunnels short-circuited a loop in the river, thus creating a 300 m water head.

An optimised design for choosing a tunnel location includes the layout of the tunnel line, selection of excavation depth, spacing between tunnels, and so on. As discussed, such designs are based on an understanding of the geological settings, rock stress, rock mass properties and behaviour, which can be obtained from site investigation, drilling exploration and laboratory and field testing. We note that empirical and analogy analyses from previously excavated tunnel cases are also crucial in the process of location design, if their conditions are similar to the new project. However, once a project location is determined and the tunnels begin to be excavated, uncertain risks must then be addressed during the tunnelling process itself.

6.2.6   Excavation and support methods

Excavation and support are naturally two of the important aspects affected by the deformation of squeezing ground or high rockburst risks in deep tunnels. However, considerable progress has been made in governing these hazards during the process of design and construction for such tunnels. In the early days of the rock mechanics subject and when considering the mining procedures in the gold mines of South Africa, Cook et al. (1966) determined the influence of the excavation sequence on the released energy during rockbursting, and then suggested an index to be used to estimate the released energy, together with suitably modified excavation geometries. For the rockburst issue in tunnels, modified and advanced designs for excavation and support have been suggested by Hoek and Brown (1980), Hoek et al. (1995; 2008), Kaiser et al. (1996; 2000), Kaiser and Cai (2012).

In order to systematically optimise the excavation parameters (such as tunnel size and shape, the excavation sequence and its rate, as well as tunnel excavation types), Feng et al. (2013) suggested a numerical index, namely the local energy release rate (LERR), for the design process of deep, hard rock tunnels. Using the LERR in conjunction with the excavation method, an effective rockburst control strategy has been developed which can decrease the released energy during tunnelling by means of an optimised excavation approach. Also, Zhang et al. (2012) illustrated the advantages of an advance small pilot tunnel ahead of a TBM face, which enables reduction of the rockburst intensities and seismic events.

Considering tunnel support in a rockburst prone region, Kaiser et al. (1996) summarised three key functions of the supports as follows: (1) to reinforce the rock mass to strengthen it and control bulking; (2) to retain broken rock to prevent fractured block failure and unravelling; and (3) to hold fractured blocks and securely tie back the retaining element(s) to stable ground (see Figure 6.12). The goal of reinforcing the rock mass using rockbolts is not only to strengthen it, thus enabling the rock mass to support itself (Hoek & Brown, 1980), but also to control the bulking process because rockbolts prevent fractures from propagating and opening up. Widely used retaining elements are wire mesh, reinforced shotcrete, strap, steel arch, or cast-in-place concrete. Shotcrete is reinforced by fibre or mesh to increase its tensile strength and toughness. As discussed by Kaiser and Cai (2012), mesh-reinforced shotcrete or mesh over shotcrete offers a superior retaining function under rockburst conditions. A holding function is required to tie retaining elements of the support system and the loose rock back to stable ground in order to dissipate dynamic energy due to rock ejection and rock movement, and to prevent gravity-driven falls of ground. In addition to these functions, the installed rock support system must be able to absorb dynamic energy while also accommodating large sudden rock deformations due to rock failure with associated bulking.

Although these excavation and support methods have been comprehensively applied in the construction of deep tunnels, it is difficult to select the most effective excavation and support methods to deal with rockburst hazards under complicated excavation conditions. One of many reasons for this is that the mechanics of rock support are complex, and no models exist which can fully explain the interaction of the various support components in a rock support system (Kaiser & Cai, 2012).

The same problems may occur during tunnelling in soft or weak ground. The goals of excavation optimisation and support designs are to control the large deformations, and to ensure the stability of both the rock mass and its support structures. Consequently, many methods of support techniques have been proposed. For example, the New Austrian Tunnelling Method (NATM) (Han, 1987), which is also known as the sequential excavation method (SEM), is a popular method used in modern tunnel design and construction.

Salamon (1970) studied the support system in terms of energy. The support structure and surrounding rock simultaneously generate compatible deformation, and the

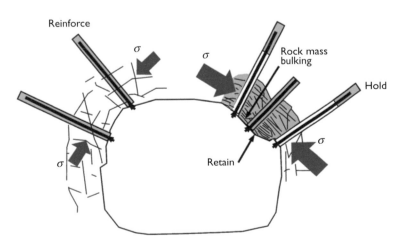

Figure 6.12 Three key functions (reinforce, retain, and hold) of rock support (Kaiser *et al.*, 1996).

support structure can absorb part of the dispersed energy from the surrounding rock mass. The combined support method (Feng, 1990) proposed that increasing the thickness of support was not the optimal method for tunnelling in soft rocks, and that the method of pre-flexibility and post-stiffness was a priority. He *et al.* (2007) suggested that the support technique for deep tunnelling should be carried out in two steps: first flexible support, followed by a coupling support for the critical parts. Dong *et al.* (1994) also proposed a support theory for 'loose circles' in the surrounding rock mass. More recently, several researchers have proposed a large number of novel excavation and support methods to control the squeezing behaviour of the rock mass, e.g., the DSM method adopted by Barla *et al.* (2007) and the (steel) fibre reinforced high-strength shotcrete method adopted by Hisatake (2003) and Sustersic *et al.* (2004).

In summary, these methods can be divided into two types: (i) the passive approach, and (ii) the active method. The so-called 'passive approach', which aims at accommodating large deformations, is preferred to the so-called 'active method', which aims at preventing rock deformation. The only feasible solution in heavily squeezing ground is a tunnel support that is able to deform without becoming damaged, in combination with a certain amount of over-excavation in order to accommodate the deformations. Supports which are based on this so-called 'yielding principle' can be structurally implemented in two main ways (Anagnostou & Cantieni, 2007): either by arranging a compressible layer between the excavation boundary and the extrados of a stiff lining, or through a suitable structural detailing of the lining that will allow a reduction in its circumference. In the first case, the ground undergoes convergences, while the clearance profile remains practically constant. This solution has been proposed particularly for shield tunnelling with stiff segmental linings (Billig *et al.*, 2008). The second solution is the one usually applied today. It involves steel sets with sliding connections in combination with shotcrete. Moreover, in order to achieve the required clearance profile after convergence, the tunnel is excavated to a certain size, which accommodates the convergence and support installation, including the inner lining. Consequently, a flexible support, over-excavation, longitudinal gaps in the shotcrete lining and yielding rock bolting are commonly applied to deal with large deformation problems in deep tunnels.

However, if the groundwater becomes a problem because of a large deformation during the excavation process, advance grouting may be necessary. The purpose of grouting is to solidify the rock formation and improve the overall watertightness in order to improve the stability of the excavation heading. Feedback from the results of such ground improvement should be reviewed carefully in order to improve understanding the geological characteristics of the rock formation. When groundwater ingress is encountered, the treatments mainly include water-stopping grouting and groundwater drainage for pressure reduction. These two concepts can be combined to increase the treatment effectiveness. Figure 6.13 shows a typical grouting design ahead of tunnel faces.

From the above, it is clear that a complicated design of tunnel excavation and support is required in complex engineering conditions. In such tunnel design, many design parameters, such as excavation method and sizes, bolt length and the thickness of lining, etc., will rely on experiences provided by engineers and designers, as well as analyses, assessments and calculations. The reliability of knowledge from experience,

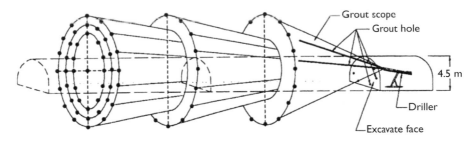

Figure 6.13 Schematic illustration of conical grouting in the Syueshan tunnel (Tseng *et al.*, 2001).

the uncertainty of the data, and the accuracy of analyses and assessment dictate the success or failure of tunnel design when dealing with engineering hazards.

6.3 ALEATORY UNCERTAINTY ANALYSIS OF DESIGN AND CONSTRUCTION FOR LONG DEEP TUNNELS

Moving now from the left-hand column to the right-hand column in Figure 6.3, the aleatory uncertainty will be discussed next, recalling that aleatory uncertainty comes from the Latin 'alea' for a game of chance. So this type of uncertainty is due to chance, intrinsic randomness, and is thus conceptually not fully resolvable. An example question is, "At exactly what chainage will water bearing fractures be encountered in the tunnel?"

6.3.1 Detailed geology variations

Both stiff structures and fractures control the shape of rockburst cavities (Figure 6.14) and can increase the intensity of rockbursts (Figure 6.15) but such stiff structures and fractures are difficult to locate before tunnelling. Also, sometimes advance exploration or drilling is not successful in obtaining information about these structures, especially fractures due to their closure properties and high cohesional strength before they are exposed by excavation. When exposed, these structures can of course be found at any location around the tunnel cross-section, such as the tunnel crown (Figure 6.14(a)), sidewall (Figure 6.14(b)), tunnel spandrel (or shoulder) (Figure 6.14(c)) and tunnel floor (Figure 6.14(d)). Different exposed locations may induce varying rockburst damage and rockburst mechanisms because of variations in the ground quality and the concentrated *in situ* stress.

Figure 6.15 shows severe rockburst damage which occurred in the drainage tunnel that was excavated by a TBM with a diameter of 7.2 m. This rockburst has a maximum depth of more than 8 m at the cross of the tunnel and nearly 30 m length of the rockburst cavity, and resulted in the destruction of the main TBM beam. It was established that this rockburst was related to a geological structure, as shown in Figure 6.15(d). Moreover, it is inferred that the mechanism of this rockburst relates to the fault-slip mechanism of the natural structure, which governs the boundary

Figure 6.14 Influence of stiff structural planes on the shape of rockburst cavities: (a) rockbursts affected by one stiff joint; (b) rockbursts affected by a set of joints; (c) rockbursts affected by two joints; (d) rockburst affected by two sets of joints.

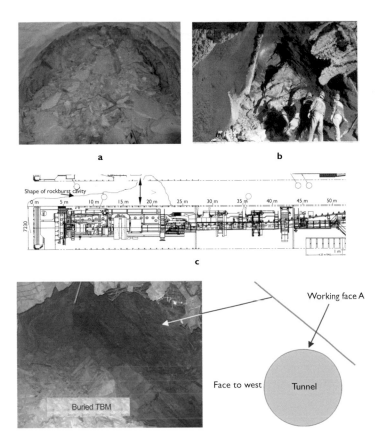

Figure 6.15 A rockburst effect exacerbated by a large stiff structural plane. The rockburst broke the main beam of the TBM. a) Rock blocks after the rockburst, direction towards the working face of the tunnel. (b) Maximum rockburst cavity, V-shape, after shotcreting. (c) Shape of the rockburst cavity in a longitudinal section along the tunnel axis. (d) Stiff structural plane at the cross-section of the tunnel in the rockburst cavity.

of the failure zone. Unfortunately, geometrical structures like the one shown in Figure 6.15(d) are often not exposed at the excavation surface and so are hidden in the surrounding rock mass and are not visually detectable. Hence, they involve high uncertainty and risk when they are not identified. After this rockburst, the drill and blast method was used for the remaining excavation of the tunnel.

6.3.2 Rock stress variations

The importance of the *in situ* stress on the stability of underground excavations is widely recognised. As discussed by Hoek and Marinos (2009), when anisotropic or high *in situ* stresses are present, failure and plastification phenomena such as rockbursts and squeezing may occur; so it is advantageous, if not essential, to establish the magnitudes and directions of the *in situ* principal stresses for any underground project, especially for deep tunnel conditions, As mentioned earlier, determining the *in situ* stress state is not an easy task, with the consequence that many tunnels are still designed without *in situ* stress measurements. Even when *in situ* tests are carried out, mean values can be assigned for the whole rock mass through which the tunnel will be driven, without considering the influence of geological variabilities on stress field and the whole tunnel profile (Hijazo *et al.*, 2012). The major reason for this is the complexity of factors causing *in situ* stress variation, as explained in Chapter 4, resulting in difficulties in accurately assessing the complete state of rock stress across the site.

According to Zang and Stephansson (2010), there are two scale factors that affect the state of rock stress: (i) continental scale factors, including the factors governing stress magnitude (e.g., the thickness, composition and age of the crust, crustal heterogeneities and the geothermal gradient) and the factors influencing stress direction (e.g., plate geometry and the distribution of the different plate boundaries); and (ii) engineering scale factors, such as geological and structural anisotropies, sedimentary loads, relief effects, glacial rebound, loads produced by submarine elevations or the convexity of the oceanic lithosphere and rock composition and geomechanical behaviour (Hijazo *et al.*, 2012). For a deep and long tunnel, the geological and structural anisotropies are among the most important factors.

i   the geological and structural factors governing *in situ* stress. Many cases analysed by Stephansson (1993) showed that faults are one of the main tectonic structures which can influence stress magnitude, and the increase or decrease of stress magnitude is due to the local influence of the discontinuities, faults, dykes, heterogeneities, intrusive bodies and folds, with variations of up to tens of MPa in faults or shear zones. Moreover, the lithological heterogeneities and structural anisotropies may also lead to stress concentrations (Hijazo *et al.*, 2012). For example, in contact zones between materials of different strengths, variations have been observed in horizontal stress magnitude with depth. Similarly, in rock masses with geological anisotropies and high Poisson's ratios, K values (the ratio between the principal stresses) higher than the regional K values have been recorded (Evans *et al.*, 1989; Cornet & Burlet, 1992; Gunzburger & Cornet, 2007).

ii The factors influencing gravity stress. High relief can exert considerable influence on the stress in rock volumes nearest to the surface. For most deep tunnels, their overburden range is usually from hundreds to thousands of metres, but they are rarely more than 3000 m in depth, so the surface relief of the engineering region still influences the state of rock stress along the tunnel. The *in situ* stress field is significantly influenced by the topography when a tunnel is located between steep-sided valleys, such as the Jinping II tunnels (Zhang *et al.*, 2012). Also, the geological history and evolution of rock structures, such as erosion and denudation processes, in relation to stress fields applied in the long term, may significantly influence the current *in situ* state of stress.

Although the important factors controlling the current *in situ* stress state have been identified, their impact on the *in situ* stress at deep tunnel conditions is only understood qualitatively at present because it is difficult to obtain the necessary information for quantitative analysis. Additionally, the induced stresses, i.e., the *in situ* stress perturbed by tunnelling, are modifications of the natural *in situ* stress in the region of the tunnel section. So, the complexity of the state of rock stress within the tunnel-peripheral rock mass is exacerbated.

To illustrate the variations of rock stress in deep and long tunnels, an important conclusion obtained by Hijazo *et al.* (2012) is reviewed and employed to demonstrate the relation between rock stress and geological anisotropies, and hence the variations of the *in situ* stress state along a tunnel. (These variations will also be discussed more deeply later in the major tunnel case example in this Chapter, i.e., the Jinping II tunnels.) Hijazo *et al.* (2012) studied *in situ* stress amplification due to geological factors along the Pajares tunnels in Spain, and the results of *in situ* stress amplification, are provided in Figure 6.16. The Stress Amplification Factor (SAF) estimates the increment of the principal horizontal stress in a tunnel rock mass due to geological and geomechanical local anisotropies. It can be seen from Figure 6.16 that the increments of the horizontal stress due to local anisotropies can range from 1.1 to 2.9 of the mean regional horizontal stress. This implies that the local geological and mechanical differences can lead to a remarkably altered distribution of the *in situ* stress state. If a tunnel is excavated through geological and mechanical anisotropies, such as faults, folds and/or variable lithology and rock mass properties, it is likely that the magnitudes, directions of the principal stresses and the ratios between the principal stresses will change along the tunnel and may be significantly different in different tunnel sections.

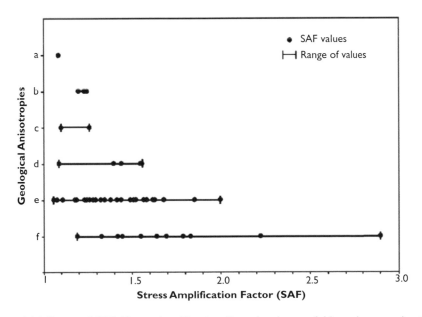

Figure 6.16 Range of SAF (Stress Amplification Factor) values. a: folds and geomechanical anisotropies; b: faults and geomechanical anisotropies; c: folds; d: faults and folds; e: faults; f: geomechanical anisotropies (Hijazo *et al.*, 2012).

6.3.3   Local water variations

Any tunnel constructed below the water table will be exposed to some level of risk associated with water inflows. The magnitude of this risk is highly variable and dependent upon the precise hydrogeological conditions and excavation circumstances, especially in aquifer systems. The uncertainty associated with groundwater inflow increases greatly with increasing hydrostatic head and complexity plus the geological conditions. As expected, in fractured rock aquifers, the uncertainties are greater than in porous media. When the karst aquifers through which tunnels will be excavated are the main aquifers, this uncertainty may be more complicated because of the extremely non-uniform distribution of groundwater, which cannot be easily assessed or probed. Such aquifers can lead to high local groundwater flow into the rock tunnels. Several disastrous events in tunnelling have been associated with large volumes of local groundwater inflow through geological features such as fault zones and open

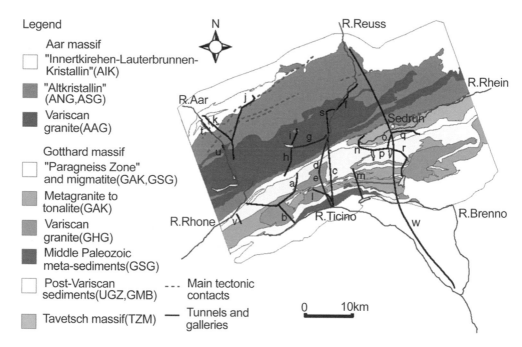

Figure 6.17a Map of the study area showing the main rivers, the main lithologies, the main tectonic units and tectonic contacts with black lines representing: Furka base tunnel Realp leg (a); Furka base tunnel Bedretto window gallery (b); Gotthard SBB railway tunnel (c); Gotthard A2 highway security gallery (d); Gotthard A2 highway main tunnel (e); KW Amsteg supply gallery (f); KW Goeschenen pressure gallery (g); KW Goeschenen Furkareuss supply gallery (h); KW Goeschenen Voralpreuss supply gallery (i); KW Oberhasli Gadmenwasser supply gallery (j); KW Oberhasli Handegg supply gallery (k); KW Lucendro pressure gallery (l); KW Ritom Garegna and Unteralpreuss supply galleries (m); KW Vorderrhein Val Val-Curnera gallery(n); KW Vorderrhein Tgom-Nalps gallery (o); KW Vorderrhein Curnera-Nalps gallery (p); KW Vorderrhein Sedrun-Medels gallery (q); KW Vorderrhein Nalps-St Maria gallery (r) KW Wassen supply gallery (s); Gas Transit Urweid gallery (t); Gas Transit Gstelli gallery (u); Gas Transit Obergesteln gallery (v); Gotthard Base Tunnel (under construction) (w).

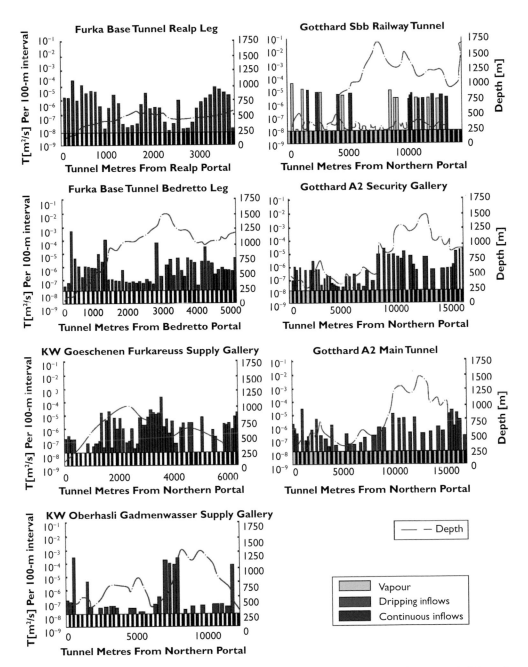

Figure 6.17b Transmissivity distribution along tunnel with dark bars, representing transmissivity values derived from continuous inflow rates stacked onto medium dark bars, representing transmissivity values derived from dripping inflow rates, finally stacked onto light bars, representing the minimum transmissivity derived from water vapour output measurements in the Gotthard Base Tunnel, for selected tunnels: a, b, h, j, c, d and e (Masset & Loew, 2010).

fractures. This indicates that the complexity of geological conditions, particularly the complicated distribution of regional geological units along a tunnel, can lead to high local groundwater flows being intersected by tunnelling.

However, for a deep and long tunnel, the hydrogeological issues are related to complicated geological structures, such as faults, folds, fracture systems, etc., along the tunnel line. Unfortunately, in this case little information can be made available regarding the geological structures or detailed hydrogeology of the aquifer systems, unless a large number of exploration drill holes are made. Even so, under deep and long tunnel conditions, there is always a lack of acceptable exploration information. Also, the collected hydrogeological data may have limitations in interpretation given a non-uniform distribution of the aquifer itself.

In his doctoral thesis, Olivier (2011) statistically analysed, by means of tunnel inflow data and other information, the transmissivity distribution along deep tunnels excavated in fractured crystalline rocks in the Central Alps (Switzerland). Figure 6.17(a & b) shows the locations of these tunnels, as well as their geological condition and transmissivity distribution. The Figure shows the variation of groundwater aquifers' properties, as well as the local characteristics of groundwater inflow, which depend on the geological units. This example variation of transmissivity distribution along tunnels illustrates the uncertainty of rock mass conditions and local distribution of groundwater.

6.3.4 Mechanical behaviour of the rock mass after excavation and in the long term

Under relatively high stress conditions, the main behaviour of hard rock is fracturing, including new cracking and crushing of the rock masses. This results in the reduction of rock strength and increase in rock deformation, which may cause instabilities. Importantly, this process of fracturing is a function of time; as time elapses, the depth and region of fracturing around the excavation surface may increase significantly, a process which has been validated by instrumentation. For example, a P-wave velocity test was conducted in the access tunnel in the Jinping II headrace project. The overburden at the testing location was up to 2300 m, the rock masses consisting of Baishan group marbles. In fact, fracturing induced by high stress was a major issue in the testing tunnel sections. The P-wave testing can be divided into two stages: (i) testing after the testing tunnel was excavated; and (ii) testing after one year. Based on the fracturing depth data obtained from the two types of test, numerical back analyses were used to assess the cohesional strengths of the rock masses. Figure 6.18 shows the results of two tests which indicated that the cohesional strength reduced to 10% of its original value. This reduction is directly attributed to the fracturing of the rock masses.

In addition, the same rock degradation has also been found in the Jinping II headrace tunnel #2, where one long-term testing hole was drilled in order to obtain the long-term behaviour of Jinping marbles under high stress. Figure 6.19 illustrates two wave velocity testing results at 36 days and 186 days after the tunnel section (chainage 13 + 085) was excavated. As shown in Figure 6.19, the fracturing zone or damage zone of the rock masses gradually increased, detected by the decrease of wave velocity and increase of the velocity-decreased hole depth (about 1.2 m).

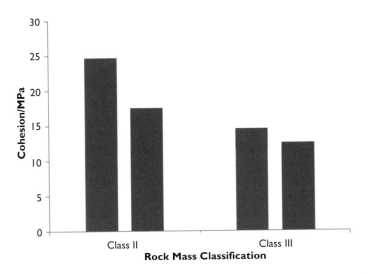

Figure 6.18 Reduction in cohesion strength with time in the Jinping access tunnels. (For both the Class II and Class III results, the first testing is the left of the two bars and the second testing is the right of the two bars.)

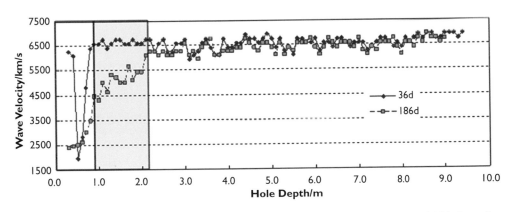

Figure 6.19 Wave velocities of rock masses obtained in Hole #6 at both 36 days and 186 days after the tunnel section (chainage 13 + 085) was excavated.

Such hard rock fracturing with time has a complicated dynamic evolution which is influenced by many factors, e.g., stress condition, rock mass structures, excavation size and shape, support system. Moreover, considering the groundwater conditions, such as the case where the groundwater recovers in the surrounding rock mass after the lining is installed, the long-term characteristics of such fractured rock masses near excavation surfaces may be somewhat different from those without groundwater.

6.4   METHODS TO ASSESS AND MITIGATE RISK FOR LONG DEEP TUNNELS

6.4.1   Rockbursts

6.4.1.1   Rockburst risk assessment

Rockbursts can occur at any location around deep tunnels, such as the crown, shoulder, sidewall, corner of the floor, floor, and even at the tunnel face, as shown in Figure 6.20, which shows rockburst field photos collected from the Jinping II headrace tunnels. The rockburst intensities are different under different excavation and geological conditions, so the risk assessment of rockbursts includes not only intensity and zone but also the location of the rockburst risk at the tunnel cross-section, depth of the rockburst cavity, and the consequence.

On the basis of different methods of rockburst hazard analysis, three different methods have been applied to assess the risks of rockburst in deep tunnels: (i) empirical criteria, including single index methods, as listed in Table 6.4, and multi-index methods, such as BPI (Mitri et al., 1993) and RVI (Qiu et al., 2011); (ii) the intelligent method or expert system, such as Support Vector Machines (SVM), Artificial Neural Networks (ANN), fuzzy mathematical comprehensive evaluation; and (iii) numerical methods or numerical indices, such as ERR (Cook et al., 1966), ESS (Ryder, 1988), ESR (Mitri et al., 1993), LERD & MGW (Wiles et al., 1998; Beck and Brady, 2002), FAI (Zhang et al., 2008), and LERR & ERE (Jiang et al., 2010).

The Rockburst Vulnerability Index (RVI) is a novel empirical multi-index system used to assess the rockburst risk and failure potential under deep tunnel conditions, which was proposed by Qiu et al. (2011) based on the rockburst cases collected from the Jinping II deep tunnels. By empirically quantifying the uncertainty, rockburst control factors, which include stress condition, rock strength, geological condition and engineering excavation environment, the maximum rockburst failure depth can be assessed by the empirical value, RVI (Qiu et al., 2011). The current output result of the RVI method is the potentially maximum depth of rockburst failure zones and potential maximum class of rockburst events.

'Intelligent methods' depend on the data mining of rockburst cases to identify the complex contributions to rockburst potential from the number of control factors (Feng & Wang, 1994). For example, a neural network model based on case studies of rockbursts is used to assess the failure depth and the intensity of the rockburst, with input parameters being the ratio of strength to stress, ratio of maximum principal stress to the minimum principal stress, a regional geological structure factor, intactness of the rock mass, brittleness index of the rock, and support strengths. Figure 6.21 shows a neural network model which may be used to estimate rockburst damage depth and rockburst intensity.

Numerical indices are efficient for assessing rockburst risk when considering the process of excavation and estimating the stress condition, energy evolution and released energy. In particular, they are helpful in optimising rockburst support parameters and an excavation design scheme. Several novel numerical indexes have been established, such as the FAI (Zhang et al., 2008) and the LERR (Jiang et al., 2010) (see Equation 6.1). The Local Energy Release Rate (LERR) and Elastic Release

Figure 6.20 Rockburst occurrence at different locations in the Jinping II tunnel cross-section. (a) crown, (b) shoulder, (c) sidewall, (d) corner of floor, (e) floor with cracking crossing the tunnel section, (f) upheaval of floor along the axis of the tunnel, (g) working face.

Table 6.4 Empirical rockburst criteria.

Type	Name (ref.)	Criterion equation	Thresholds	Rockburst classification	Remarks
Considering the influence of the tunnelling process and the characteristics of *in situ* stress and the induced stress around the tunnel	Hoek criterion (Hoek, 1990; 2010)	σ_{max}/σ_c	>0.7 =0.42~0.56 =0.34~0.42 <0.34	Severe rockburst Moderate damage Severe slabbing Little slabbing	σ_{max} Maximum tangential stress σ_v Vertical stress σ_c Uniaxial compressive strength
		σ_v/σ_c	>0.5 =0.2~0.5 <0.2	Rockbursting possible Spalling and slabbing Spalling with no supports	
	Russenes criterion (Jager, 1996)	$Is(50)/\sigma_c$	<0.083 =0.083~0.15 =0.15~0.20 >0.20	Severe rockburst Moderate rockburst Slight rockburst No rockburst	Is(50) Rock point load strength; σ_θ Maximum tangential stress
	Turchaninov criterion (Jager, 1996)	$(\sigma_\theta+\sigma_L)/\sigma_c$	<0.3 =0.3~0.5 =0.5~0.8 >0.8	No rockburst Rockbursting possible Rockbursting definite Severe rockburst	σ_θ Tangential stress around tunnel σ_L Axial stress around tunnel σ_c Uniaxial compressive strength
	The criterion from Erlang mountain tunnel (Xu L S, Wang L S, 1999)	σ_θ/σ_c	<0.3 =0.5~0.7 >0.7	No rockburst Must be rockbursting Severe rockburst	σ_θ Tangential stress around tunnels σ_c Uniaxial compressive strength

	Criterion	Index	Value	Rockburst description	Symbols
Not considering the influence of the process of tunnelling and the characteristics of the *in situ* stress and the induced stress around the tunnel	Barton criterion (Barton et al., 1974)	σ_c/σ_1	=5~2.5 <2.5	Moderate rockburst Severe rockburst	σ_1 Maximum principal stress σ_c Uniaxial compressive strength
	Tao criterion (Handbook of Geotechnical Engineering, 1994)	σ_c/σ_1	>14.5 =14.5~5.5 =5.5~2.5 <2.5	No rockburst Slight rockburst with slight AE Moderate rockburst with strong AE Severe rockburst with explosive sounds	σ_1 Maximum principal stress σ_c Uniaxial compressive strength
	China criterion (GB50218-94)	σ_c/σ_1	<4.0 =4.0~7.0	Rockbursting and rock ejection Rockbursting possible, spalling and collapses	σ_1 Maximum principal stress σ_c Uniaxial compressive strength
	Brittleness criterion (Xu L S, Wang L S, 1999)	σ_c/σ_t	<10 =10~14 =14~18 >18	No rockburst Slight rockburst Moderate rockburst Severe rockburst	σ_c Uniaxial compressive strength σ_t Uniaxial tensile strength
	Brittleness Index	U/U_1	<2.0 =2.0~6.0 =6.0~9.0 >9.0	No rockburst Slight rockburst Moderate rockburst Severe rockburst	U Deformation before peak strength U_1 Irreversible deformation before peak strength
	Elastic strain energy Index, Wet (Gu M C, 2001)	W_{sp}/W_{st}	<2.0 =2.0~3.5 =3.5~5.0 >5.0	No rockburst Slight rockburst Moderate rockburst Severe rockburst	W_{sp} Released strain energy W_{st} consumed strain energy due to irreversible deformation
	Impact Index (Gu M C, 2001)	$K_m/|K_s|$	<1.0	Rockbursting possible	K_m Loading stiffness on the stress-strain curve $|K_s|$ Unloading stiffness on the stress-strain curve after peak strength

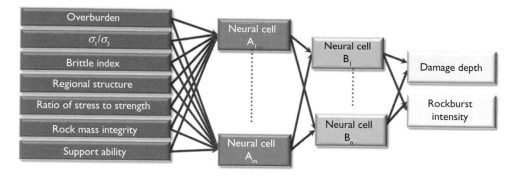

Figure 6.21 An ANN model to assess rockburst risk in the Jinping II tunnels.

Energy (ERE) are considered for assessing the location of the risk sections and the potential failure depths of rockburst occurrence.

$$LERR_i = U_{i\max} - U_{i\min}$$

(6.1)

where $LERR_i$ is the local energy release rate of an element and the peak strain energy and the minimum strain energy are given in Equations 6.2 and 6.3.

$$U_{i\max} = \left[\sigma_1^2 + \sigma_2^2 + \sigma_3^2 - 2v(\sigma_1\sigma_2 + \sigma_2\sigma_3 + \sigma_1\sigma_3)\right]/2E$$

(6.2)

$$U_{i\min} = \left[\sigma'^2_1 + \sigma'^2_2 + \sigma'^2_3 - 2v(\sigma'_1\sigma'_2 + \sigma'_2\sigma'_3 + \sigma'_1\sigma'_3)\right]/2E$$

(6.3)

Using the Local Energy Release Rate (LERR) index, the brittle failure intensity of the surrounding rock mass under high *in situ* stress can be estimated, and the position and scope of failure can be obtained. This overcomes the limitations of conventional indices in the stability analysis of underground projects. The index provides evidence for the scientific assessment of rockbursts in the excavation process of hard rock tunnels under high *in situ* stress conditions. This algorithm can be used to determine the cross-sectional size of tunnels in the excavation and the sequence and speed of excavation at which the excavation can be implemented. The method has been used to identify rockburst risk in the diversion tunnels at the Taipingyi Hydroelectric Plant, China (see Figure 6.22).

6.4.1.2   *Risk mitigation concepts in rockburst prone tunnels*

A novel three-step strategy was proposed by Feng *et al.* (2013) to reduce the hazards caused by rockbursts in the Jinping II tunnels and this method may be adopted for other tunnels which are considered to be subject to severe or extremely severe rockburst risks. The strategy consists of three steps or mitigation concepts, as shown in Figure 6.23.

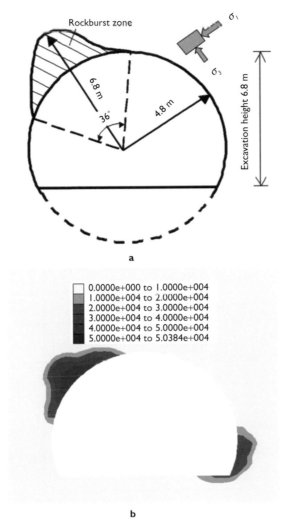

Figure 6.22 (a) Rockburst occurrence. (b) Numerical simulation of rockburst cavity and location in the diversion tunnel at Taipinyi hydroelectric plant (Jiang *et al.*, 2010).

i   *Reducing energy concentration.* This may be achieved by a series of optimisation processes, such as the optimisation of shape and size of the excavation section, optimisation of the number and height of excavation benches, optimisation of the excavation length and excavation rate, optimisation of the time at which excavation multiple tunnel faces change to a single tunnel face, and optimisation of location, shape, size and anterior distance of the pilot tunnel.

ii   *Destressing and transferring.* This is for extremely high stress conditions that can induce very severe rockbursts. The stress states may be improved by means of destressing holes or destressing blasts. From the viewpoint of destressing design,

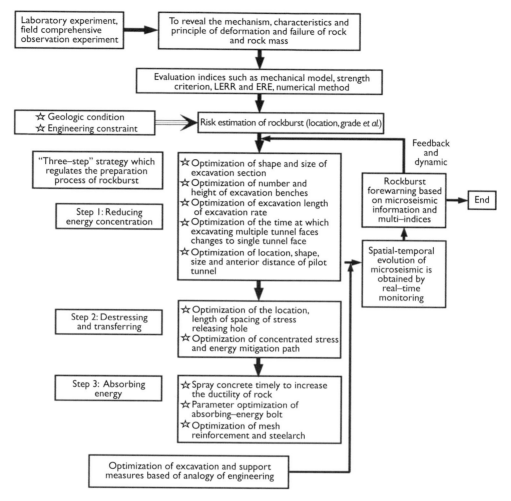

Figure 6.23 Risk mitigation concepts and dynamic control strategy in rockburst prone tunnels (Keynote ppt in Feng, 2011).

the location, length or spacing of stress releasing holes or blastholes, including blast parameters, must be optimised.

iii *Absorbing energy.* This is achieved through rock support systems, i.e., spraying concrete in a timely manner to increase the ductility of the rock mass, parameter optimisation of absorbing-energy rockbolts, optimisation of mesh reinforcement and steel arches, etc.

6.4.1.3 New approaches and optimisation of the risk-reduced construction procedures

Rockburst occurrences depend on many control factors such as stress fields, rock mass characteristics, geological setting, excavation actions, engineering layouts, etc.

From the viewpoint of reducing rockburst risks, except for the geological setting, the changes and adjustment of other factors may enable the decrease and control of severe and extremely severe rockbursts, and also to avoid the occurrences of minor or moderate rockbursts. Therefore, two methods for treating and controlling rockburst risks have been established: the strategic method and the tactical method. The strategic method involves the design of engineering layouts and excavations; while the tactical method refers to the support approaches, which may alter the effective rock mass properties and mechanical behaviour, and can also alter the states of the stress field.

6.4.1.3.1 Construction adaptations used to reduce rockburst risks

The optimisations of the excavation process and the construction adaptations are important strategic approaches in reducing rockburst risks in deep tunnelling or mining, their applications depending on the designs of the specific engineering layout and excavation. There are some key approaches that will be demonstrated in the following text.

To optimise excavation designs, numerous numerical methods and numerical indices have been proposed, as discussed in Section 6.4.1.1. In addition to the indices in Section 6.4.1.1, two energy indices (the Energy Release Rate (ERR) proposed by Cook et al. (1966) and the Elastic Release Energy (ERE) established by Feng et al. (2013)), are very helpful for optimising the excavation designs. Equation 6.4 provides the basis for the Energy Release Rate (ERR).

$$ERR = dW_r/dV \tag{6.4}$$

where dWr is the energy release in the surrounding rock at an excavation step and dV is the volume of the rock mass in this excavation step. The Elastic Release Energy (ERE) index is based on the Local Energy Release Rate (LERR), which can quantitatively describe elastic energy relief induced by the local instability of the surrounding rock mass in the construction process. This can serve as an evaluation index in construction schemes controlling rockbursts under high *in situ* stress conditions. By means of the ERE index, an optimisation algorithm has been developed (Feng et al., 2013), Equation 6.5. The key function is as follows

$$\text{Min } ERE = \sum_{i=1}^{n} LERR_i \times V_i \quad \text{s.t. excavation limitations} \tag{6.5}$$

where $LERR_i$ is the local energy release rate of an element, $LERR_i = U_{imax} - U_{imin}$ and V_i is the volume of the element. Also, the characteristics of plastic zones are important indices, and are also used to optimise the excavation designs.

1   Reduce cross-sectional size of the excavation

The optimisations of the tunnel cross-section refer to the tunnel shapes and tunnel sizes, which are limited by the functions of tunnel engineering and its demands, i.e., the size of the headrace tunnel, which depends on the design discharge area satisfying the design electric energy production. However, there may be different excavation shapes for the same design discharge area conditions. Figures 6.24(a) and (b)

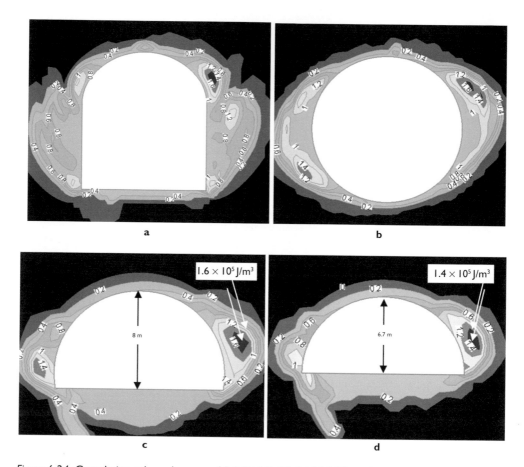

Figure 6.24 Cumulative released energy: (a) 2.88 MJ, (b) 2.14 MJ. The released energy per unit area around the tunnel: (c) 140.5 kJ/m², (d) 122.9 kJ/m² (Feng *et al.*, 2013).

demonstrate the differences of the plastic zone volume for circle and horseshoe excavation shapes which have the same discharge area. It should be noted that the circular excavation shape is a favourable shape because the released energy is lower than that in the horseshoe excavation shape.

However, during the optimisations of the tunnel cross-section, under the same excavation shape conditions, it is more important that the excavation height adopted in step tunnelling is adjusted to reduce rockburst risks. Figure 6.25 shows a case for optimising the height of the upper bench excavation in the Jinping II headrace tunnels excavated by the Drill and Blast (D&B) method. As illustrated in Figures 6.24(c) and (d) and Figure 6.25, when the excavation height is increased, the ERE index increases linearly, while the volume of the plastic zone decreases non-linearly. This indicates that the decrease of the excavation height may reduce the released strain energy and more plastic deformation may absorb the released strain energy, which then reduces the rockburst intensity and even avoids the occurrence of rockbursts.

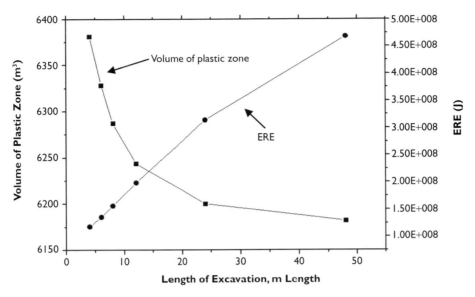

Figure 6.25 Comparison of plastic zone volume and ERE for different lengths of an excavation in deep tunnels (Feng *et al.*, 2013).

2   TBM changes to 'D&B and TBM'

As a particular case of optimising the construction size, the advance pilot tunnel ahead of the TBM full-section excavation is an effective excavation technology for dealing with extremely high rockburst risks faced by TBM tunnelling. This approach has been approved during the practices of the Jinping II tunnels (Zhang *et al.*, 2012) and is based on the method of tunnel cross-section optimisation. Different pilot tunnel shapes and locations should be evaluated by means of the numerical indices, and then chosen to reduce the released strain energy. This algorithm has been used for the optimisation of a 12.4 m diameter TBM excavation scheme in the Jinping II tunnels (see Figure 6.26(a)), where the seven schemes of an advance pilot tunnel were analysed to evaluate which is the best for controlling the released energy and reducing the rockburst risks. The algorithm answered the following questions: (i) Can the pilot tunnel excavated by D&B reduce the risk of rockburst during TBM excavation? (ii) Is the upper pilot tunnel better than the central pilot tunnel in this case? The real time monitoring of the microseismicity has verified this excavation scheme, and as seen from Figures. 6.26(b) and (c), there was a large reduction of microseismicity during the TBM-expanded excavation after the pilot tunnel excavation.

3   Excavation from opposite directions, then changing to one direction

During the construction of deep tunnels, and in order to shorten the excavation schedule, an excavation scheme in two opposing directions is commonly adopted. However, under rockburst prone tunnelling conditions, it has been found that this scheme may induce high rockburst risks. Figure 6.27(a) illustrates the strain energy conditions when the two tunnel faces approach each other. It can be seen that the stored elastic

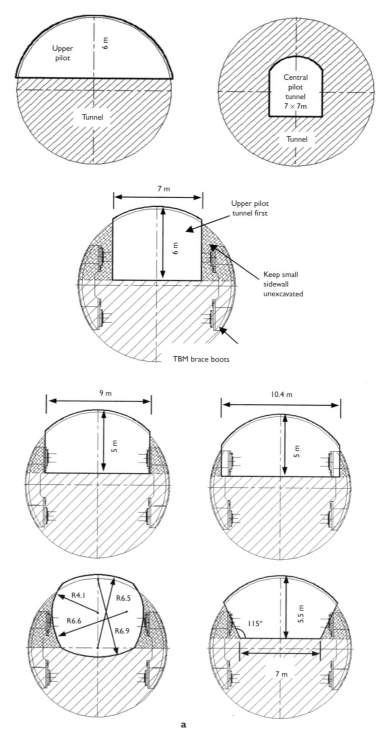

Figure 6.26 (Continued)

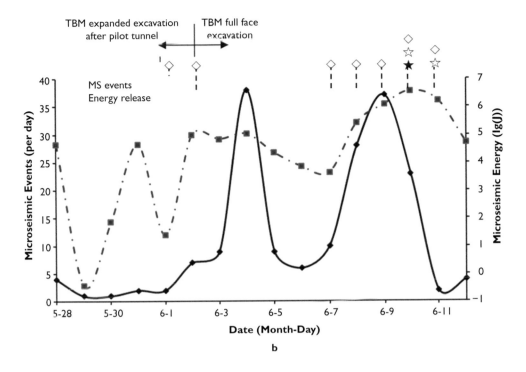

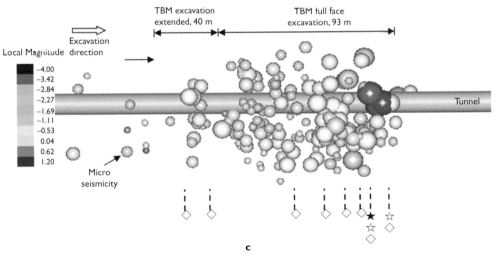

Figure 6.26 Change the TBM full surface excavation to D&B for the upper pilot tunnel excavation in advance and use TBM excavation for the rest of the cross-section. (a) Optimisation of excavation cross-section shape and size of D&B pilot tunnel. (b) Comparison of microseismic events, energy evolution and rockburst events during D&B pilot–TBM extension excavation and TBM full face excavation. (c) Comparison of microseismic events location and rockburst events during D&B pilot–TBM extension excavation and TBM full face excavation (◇: slight rockburst cases, ☆: moderate rockburst cases, ★: intense rockburst cases) (Feng et al., 2013). See colour plate section at the end of the book for the (c) component of this Figure.

energy gradually increases in the rock pillar between the two tunnel faces and reaches a peak value, when a large amount of energy can be released if the pillar fails. This was observed through the microseismic monitoring in the Jinping II tunnels, as shown in Figure 6.27(b), which was obtained from a case of face-to-face tunnelling conditions. In Figure 6.27(b), when the distance between the two tunnel faces so the pillar width decreased, then the microseismic events and released energy increased considerably. Based on this information, a moderate rockburst warning was implemented. Some mitigating treatments were employed, including ceasing the excavation at one tunnel face and the installation of systemic bolts around the different work faces. After these treatments, the released energy was decreased and the rockburst risk was successfully controlled, as shown in Figure 6.27(c).

4   Destressing and transfer of energy

The concentration of high induced stresses and energy is an important factor inducing the occurrence of rockburst hazards. To avoid or reduce this concentration, a destressing approach is used, such as stress release holes, destress blasting, etc. In fact, the small, advance pilot tunnel has the same function as the stress release holes. These approaches are one of the tactical treatments to reduce rockburst risks.

The design of the stress release holes and destressing blast holes is a process of tunnel design optimisation, and depends on the monitoring information indicating the characteristics of stress or energy concentration, and the evaluation by means

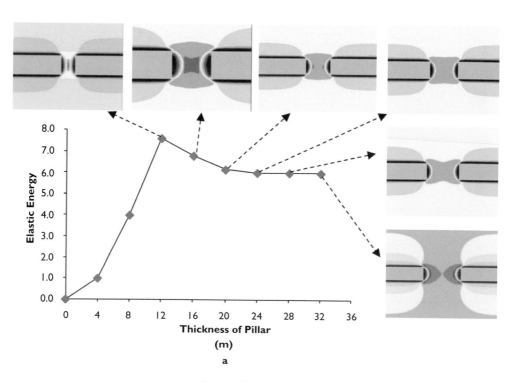

Figure 6.27 (Continued)

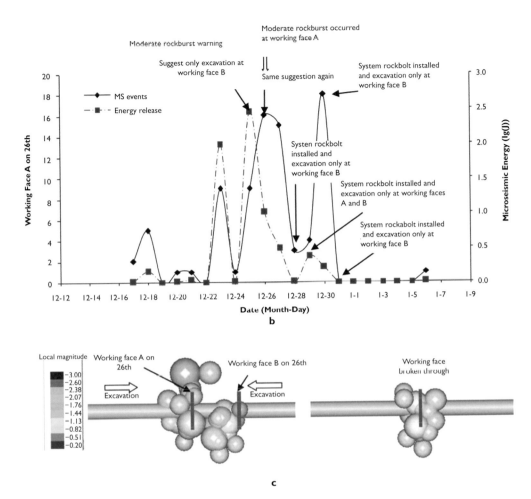

Figure 6.27 (a) Numerical simulation results showing the influence of the distance between two opposite working faces on the energy increase and release. (b) Warning and reduction of rockburst risk as the microseismicity evolution process continues due to the change in excavation from two opposite working faces to one working face with reinforcement. (c) Location of microseismicity during September 6–8 during which the rockburst risk was not controlled (left part of Figure) and during September 9–11 during which the rockburst risk was controlled successfully (right part of Figure) (Feng et al., 2013).

of numerical indices. Figure 6.28 illustrates the design process for the stress release holes based on the information obtained from AE monitoring during TBM excavation in the Jinping II headrace tunnel and numerical evaluation based on the stress and energy distributions. The effects of the stress release holes are realised: (i) to release strain energy prior to the excavation; and (ii) to transfer the stress concentration zone into the surroundings. As for the other outputs, the locations, length and spacing of these holes may be optimised.

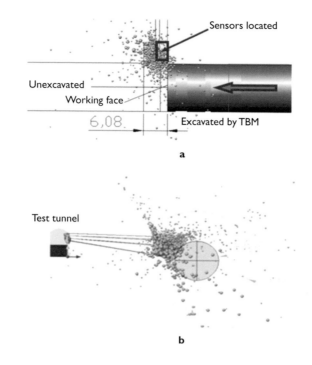

a

b

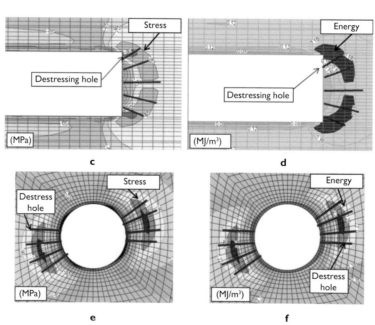

c

d

e

f

Figure 6.28 Determination of location and length of the destressing holes. (a) and (b) Acoustic emission monitoring around the working face and surrounding rocks during the TBM excavation at 1900 m overburden. (c) and (d) Numerical resolution for stress distribution and energy release distribution ahead of the working face. (e) and (f) Numerical resolution for stress distribution and energy release distribution ahead of the working face (Feng et al., 2013).

6.4.1.3.2   Design and optimal methods for the rockburst support system

The Canadian Rockburst Support Handbook (Kaiser *et al.*, 1996) presents an engineering approach to selecting rock support for burst-prone mines by systematically assessing both support demands and support capacities. Later, and intended for ground control engineers with a sound background in engineering design, rock mechanics, stress modelling and static ground control principles, "*The Guide to Rockburst Support Selection*" is aimed to be published in 2014. The guide presents detailed methods of analysis for support design for the following mechanisms.

i   Sudden volume expansion or bulking of the rock due to fracturing of the rock mass around an excavation.
ii   Rockfalls (or falls of ground), which have been triggered or been loaded by seismic shaking.
iii   Ejection of rock caused by: (a) momentum transfer from violently bulking (larger) blocks, or slabs of rock reduced to smaller blocks that are free to eject (e.g., between bolts), and (b) energy transfer from large remote seismic events to fractured rock near a stressed excavation.

The guide treats support design as a two-stage engineering process: determining the expected loading conditions or demand on support, and integrating the various elements into a support system to achieve a support capacity which exceeds demand. In addition to the conventional grouping of rockburst types into strain burst, pillar burst, and fault-slip burst, a distinction between static-load-induced or dynamically triggered and dynamically loaded strain bursts is essential for the support demand estimation. A software tool (BurstSupport™) has been developed in tandem to assist in applying the design methodology to a computer aided design package, which rapidly executes design calculations at all affected tunnel locations.

Based on the research and practices in the Jinping II tunnels, Feng *et al.* (2013) argued that the design of the support system under spalling and rockburst conditions shall satisfy the condition that support must control deformation:

i   use reinforcements to minimise bulking,
ii   use yielding bolts to follow deformations,
iii   use tough, deformable, retaining systems to prevent ravels.

The energy absorbing capability of rockbolts can be calculated using Equations 6.6, 6.7, and 6.8 as follows.

$$E_a^{ab} = \frac{1}{S_a^{ab} \times S_c^{ab}} \cdot E_u^{ab} \tag{6.6}$$

where E_a^{ab} is the energy absorbed by a rockbolt per unit area (kJ/m^2), E_u^{ab} is the critical energy absorbing capability of a rockbolt (kJ), S_a^{ab} is the axis spacing of rockbolts (m), and S_c^{ab} is the circumferential spacing of rockbolts (m). The spacing of rockbolts can be changed with the intensity of rockburst.

The length of rockbolts, L_{ab}, should be satisfied as:

$$L_{ab} = D_f + L_e \tag{6.7}$$

where L_e is the length of rockbolt reinforced effectively. For the water expansion rockbolt, $L_e > 1.2$ m (Soni, 2000), mostly, $L_e = 2.0 \sim 2.5$ m, D_f is the depth of the rockburst cavity.

The total energy absorbing capability of the support system can be calculated as

$$E_a = E_a^{ab} + E_a^{sh} + E_a^{st} + E_a^{gb} \tag{6.8}$$

where E_a is the critical energy absorbing capability of the support system per unit area (kJ/m^2), E_a^{ab}, E_a^{sh}, E_a^{st}, and E_a^{gb} are the critical energy absorbing capabilities of energy absorbing rockbolts, steel fibrous shotcrete, reinforced concrete spray layer, and permanent rockbolts per unit area, respectively (kJ/m^2).

If $E_a > E_v$, where E_v is the energy release of the rock mass per unit area during rockburst occurrence (kJ/m^2), then the support system can effectively control rockbursting. Otherwise, other measures must be adopted to reduce the concentration of energy, e.g., optimisation of the excavation cross-section size and advance rate and using destressing by drillholes or blasting.

6.4.1.3.3   Real time warning and dynamic control of rockburst

Microseismic activity represents the evolution of seismic sources and their potential trends so microseismic monitoring provides an approach for indicating the real time rock mass behaviour during tunnelling, which can also be used as a tool for the real time warning of rockburst risks. In the Jinping II tunnels, the microseismic monitoring provided new information, so one of the important results was that rockburst risks and damage potential could be assessed from the microseismic information or seismic parameters. Therefore, some treatment to reduce rockburst damage or some approaches to control rockburst intensities could be adopted and used. At the same time, microseismic monitoring can identify the effects and effectiveness of these treatments and control approaches. Also, some adjustment of the treatments and control approaches may be employed based on the evaluation of the effects and effectiveness. In this way, the process can be dynamic and be related to 'dynamic excavation and dynamic support'.

Figure 6.29 shows a representational case of this rockburst dynamic control process. On September 6th, after an advance length of about 12.02 m was completed, the released energy and microseismic events gradually increased. The following excavation of about 13.03 m led to the microseismic events increasing to nearly 40 per day, and the released energy also reached about 10^{10} J. This implied that the potential for a moderate rockburst was high. Therefore, a moderate rockburst warning was recommended on September 8th, which impelled the contractors to reduce the advances, from 9.55 m to 6.25 m, and then then to 6.64 m, and to locally increase the bolts to 6 m lengths. Adopting these dynamic control treatments, the released energy and microseismic events decreased remarkably after September 9th. Moreover, Figure 6.29(b)

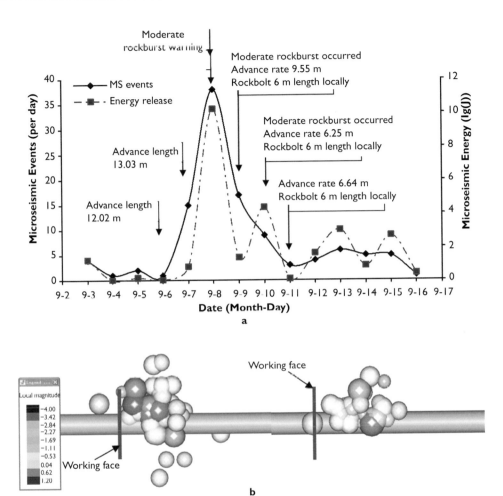

Figure 6.29 (a) Warning and reduction of rockburst risk as the microseismicity changes due to the change in TBM excavation advance rate and reinforcement. (b) Location of microseismicity during September 6th–8th during which time rockburst risk was not controlled (left part of (b)) and during September 9th–11th during which time the rockburst risk was controlled successfully (right part of (b)), (Feng et al., 2013).

shows the distribution difference of microseismic events and their energy before and after these treatments were implemented.

6.4.2   Water inrush

Research regarding the mechanism of water inrush during tunnel construction in mountainous regions, such as the water inrush mechanism and prediction of inrush water quantity is fruitful in providing advance information for tunnel construction

(Guan, 2003; Jiang, 2006; Zhang & Ren, 2010; Guan & Zhao, 2011). The characteristics of groundwater inrush in deep and long tunnels, i.e., in the context of this Chapter, can be considered to consist of four factors:

i   the large number of hydrogeological units that can be exposed during tunnelling;
ii the supply of groundwater around a deep tunnel may be large;
iii the magnitude and rates of water inrush can be large; and
iv the water head and groundwater pressures are high.

So, given the above-mentioned conditions, there are three main engineering issues which must be faced during the tunnelling. The first is the assessment of water inrush potential; the second is the prediction of the magnitude and rates of water inrush; and the last is the understanding and assessment of the stress-water pressure coupling effects which may induce hydro-fracturing.

6.4.2.1 Procedures for water inflow assessment

The following six steps are a useful guide for assessing the potential water inflow and hence decreasing the epistemic uncertainty before construction commences.

Step 1: Collect geological data, ground and groundwater information and permeability data from along and in the vicinity of the tunnel alignment, in order to prepare a geological longitudinal section together with the rock quality assessment along the tunnel.

Step 2: Prepare a rock mass permeability histogram chart of rock mass quality versus permeability. In this step, all available field and laboratory permeability testing data are reviewed in relation to the rock mass quality. In addition, groundwater monitoring records and seasonal effects should be assessed in order to determine the groundwater level for the tunnel. Once the generalised groundwater level has been determined, the tunnel is divided into a number of compartments for the estimation of water inflow.

Step 3: Estimate the water inflow to the tunnel using the above information together with the available information relating to geology, rock quality, rock mass permeability and water table. It is better if at least two methods can be used to compare the estimation results in the risk planning stage.

Step 4: Identify potential high water inflow zones (for the related criteria refer to Step 5) of the tunnel in order to propose additional ground investigation works and probe drilling during the tunnel construction. Based on the estimation results from Step 3, the number of high water inflow sections that may impact on the tunnel construction should be identified. These potential high water inflow sections should be reported together with the proposed appropriate defensive measures.

Step 5: Establish the grouting and ground treatment requirements in order to control groundwater inflow to the tunnel during tunnel construction. When the tunnel construction is in operation by means of an 'open' type of excavation (e.g., drill and blast method, open mode tunnel boring machine), the groundwater inflow may have an impact on the tunnel construction, and so grouting work is often necessary. The allowable water inflow limits and grouting requirements for the tunnel construction depend on the ultimate function of the tunnel, and the water-draw down limits. Also, they should be designed to ensure that the tunnel is being excavated in a safe manner and to reduce any impact on the environment above the tunnel to an acceptable level. The identified high water inflow zones and the grouting criteria play key roles in risk planning for tunnel projects, and should be documented in the construction contract of the Project Performance Requirements (or equivalent).

Step 6: Monitor and review the updated tunnelling records during construction. In addition to reducing the effects of inflow within the tunnel, the establishment of a comprehensive groundwater level and settlement monitoring programme during the construction phase is required. The monitoring programme should include continuous long-term monitoring of piezometers, settlement markers and tilt monitoring of sensitive structures (if relevant).

6.4.2.2   Assessment of water inrush potential

One major difficulty in predicting water inrush is understanding the relation between water inrushes and geological conditions. Studies have established correlations between the lithology, major and minor fracture distributions, thicknesses of relatively impermeable rock mass volumes and water inrush (Wang, 1998; Li, 1999; Zhang & Shen, 2004; Shao, 2006). Although most of these studies focus on water inrush in mines, that understanding of the water inrush control factors can be applied to the inrushes in civil engineering tunnels. The 'water inrush coefficient method' is a practical approach for predicting the possibility of water inrush and was established as a result of research into mine water inrush. However, although the method may also be used to address civil engineering tunnel water inrush, it should be noted that the water inrush coefficient method oversimplifies the geological, hydrogeological and other relevant conditions and, as a consequence, cannot accurately predict the possibility of a water inrush in our current context of long, deep tunnels (Meng et al., 2012).

In view of this, some researchers have proposed and introduced some novel approaches, such as grey system theory (Wang & Wang, 2006), fuzzy mathematics theory (Ge et al., 2009) and some other theories incorporating the GIS technique (Wu et al., 2006; Wu et al., 2011). However, although these ideas have advanced the subject, when applied, they are not fully satisfactory—for two reasons. The first reason is that the methods cannot thoroughly reveal the geological conditions; and the second is that the basic geological parameters used in these methods are often difficult to determine, with the result that the theoretical models are usually inconsistent with the actual circumstances. Therefore, it is important to be able to identify the geological conditions related to water inrushes, which can then lead to the establishment of a more acceptable correlation between the geological conditions and the risk of water inrush in hydrogeologically complex, deep tunnels.

The identification and study of complicated hydrogeological conditions requires the use of advanced prediction approaches, such as TSP (Tunnel Seismic Prediction), ground penetrating radar, the transient electromagnetic method, the infrared water detecting method, advanced borehole drilling, magnetic resonance differential, and so on. These approaches are necessary for providing early warnings prior to the occurrence of groundwater ingression (one case is illustrated in Figure 6.30).

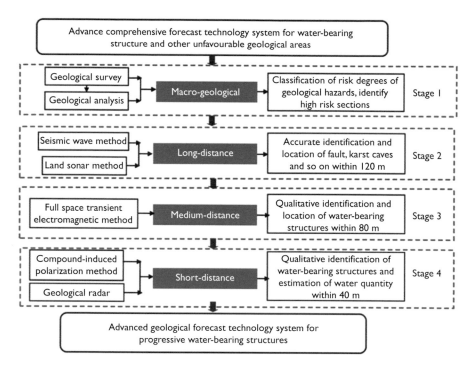

Figure 6.30 Comprehensive advanced geological forecast system (Zhao *et al.*, 2013).

6.4.2.3 *Assessment of tunnel water inflow*

Many methods have been proposed and applied in tunnel engineering to assess water inflow, such as the hydrogeology analogical method, simple water balance method, watershed water balance method, and 3D finite element or finite difference method. In addition, there are several analytical methods which may be used to assess water inflow, such as the Goodman method, Heuer and Raymer method, Heuer analytical method and IMS method, which have been reviewed in detail by Kong (2011).

6.4.2.4 *Treatment technologies for tunnel water inrush*

Depending on the site-specific groundwater environments, the treatment principles are draining-oriented, blocking-oriented or draining-and-blocking. The major principles have been summarized by Zhao *et al.* (2013), as follows.

i   **Draining-oriented:** such as energy relief and the systematic reduction method. This approach is mainly applied to large-scale, high-pressure, water-rich, karst and regional water-rich cavities.

ii **Blocking-oriented:** such as the advance curtain grouting method. This type is mainly applied to the cases in which the groundwater environment requirement is strict, the faults where the water pressure is less than 0.5 MPa, and the water leakage is from bedrock fissure(s).

iii **Draining-and-blocking**: such as pressure relief through water diversion tunnels, and grouting. This type of treatment is mainly applied to water-rich faults where the water pressure is over 0.5 MPa and there is water leakage from bedrock fissures, or karst after energy relief and pressure reduction.

The common treatment technologies have also been summarized by Zhao *et al.* (2013), as three method groups.

i The energy relief and pressure reduction method (Zhang, 2010), which means that, for a karst cavity with high pressure and significant water content encountered during tunnel construction, accurate blasting or advance drilling is applied to drain water, release mud, and reduce water pressure in the cavity, for the purpose of energy relief. During excavation, structural support and treatment after energy relief and pressure reduction may require a certain period of time. The energy relief and pressure reduction method should be considered for use during dry seasons; if it is used in rainy seasons, the constant supply of rainwater can pose a risk for tunnel construction.

ii Advance grouting technology (Mo & Zhou, 2008; Dai, 2009; Zhuang & Mu, 2009; Zhang *et al.*, 2010) which involves injecting appropriate grouting material into a karst cavity, fault (fractured) zone or large fractures, so as to achieve filling, reinforcement, and water-blocking in order to ensure the safe tunnel excavation and its long-term operation. Commonly used advance grouting technology includes full-face curtain grouting technology.

iii Advance jet grouting technology (Yang & Zhang, 2008; Zhao, 2012) which is employed for the purpose of the deformation control of surrounding rocks. If the states of the surrounding rocks, i.e., stable, temporarily stable, or unstable after excavation, can be predicted, then an information-supported design and construction method can be considered for control of the rock deformations.

Also, some novel treatment approaches have been applied in deep tunnels, such as the combination of the drainage pilot tunnel and sealing scheme, the distributary and decompression hole approach, the decompression and high pressure grouting method, the comprehensive drainage tunnel system, and so on. These technologies are illustrated in Section 6.5. with reference to engineering cases in the Jinping II tunnels in China. These technologies represent the some novel developments of common treatment technologies previously summarised by Zhao *et al.* (2013).

6.4.3   Large deformations of weak rock in deep tunnels

Large deformation hazards are particularly common in deep tunnels in weak rock, although significant convergences can also occur in shallower tunnels within very weak or over-stressed rock masses due to, for instance, tectonic or topographic effects (Shrestha & Broch, 2008; Hoek & Marinos, 2010; Hudson, 2010). When soft or weak rock mass is exposed by deep tunnelling, large deformation generally occurs, which may cause a decrease in the tunnel diameter, resulting in rock mass squeezing, difficulty of tunnel support, increased costs and delayed schedule of tunnelling.

The occurrence of such a large deformation depends on the *in situ* stress, rock mass properties, geological structures, and even the excavation process. Considering risk management, before tunnelling, work is required, such as the identification of the risk of large deformations, the risk assessment and classification, preliminary establishment of an action strategy. Once tunnels have been excavated, the detailed risk analyses continue to be necessary in order to manage the large deformation risks, reduction of such risks by suitable additional measures, transfer of this risk either partially or completely to third parties, and acceptance of this risk as a residual risk, as shown in Figure 6.31.

Large deformation risks include not only the hazards, damage and potential of large deformations, but also dangers faced by the workers and engineering machinery, the economic costs of tunnel construction, additional time consumption, etc. Note that the risk identification and the associated control approaches to large deformation are the most important procedures in the risk management for large deformation hazards in deep tunnelling.

For the identification process of the large deformation risk in deep and long tunnels, based on the engineering experiences from the Gotthard Base Tunnel in Switzerland, Ehrbar and Schoch (2006) stressed that it is essential to regard the geological conditions inducing large deformation as central risks. As discussed in Subsection 6.2.1.2, the geological setting along the length of deep tunnels is a key control

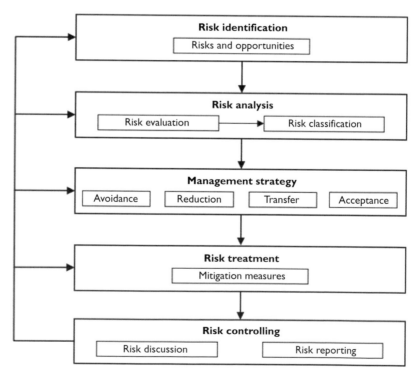

Figure 6.31 General process of risk management of large deformations in tunnels (Ehrbar & Schoch, 2006).

factor, as well as a major reason for inducing the potential of large deformation. In sub-Section 6.2.1.2, some suggestions and approaches, summarised from tunnel cases which underwent large deformation, are provided to highlight the geological identification processes of large deformation risk. The following sections will focus on the quantitative assessment of large deformation together with the control approaches and countermeasures.

6.4.3.1 Large deformation assessment

6.4.3.1.1 Reviews of large deformations in soft or weak rock masses in deep tunnels

Before discussing the assessment methods for large deformations occurring in soft or weak rock masses, simple reviews of the large deformation mechanism are necessary and helpful for supporting the risk assessment. Large numbers of investigations have indicated that squeezing and swelling are the two main modes relating to large deformations (Terzaghi, 1946; Nakano, 1974; Tan, 1982; Kidybinski & Dubinski, 1990; Aydan et al., 1993; Franciss, 1997; He et al., 2002; Li et al., 2010). Definitions of squeezing include the concepts of (i) non-elastic time dependent behaviour (e.g., Gioda & Cividini, 1996); (ii) failure of the rock mass due to concentration of stresses around the excavation; and (iii) large convergences, or large loads on the support, or both (Kovari & Staus, 1996; Panet, 1996). Gioda & Cividini (1996) reviewed in detail the characteristics and physical mechanisms. Aydan et al. (1993) investigated two tunnels (i.e., the Nabetachiyama and Shirasaka tunnels) in Japan, and indicated that the squeezing of soft rocks was responsible for the deformation.

The squeezing type of large deformation is generated by the concentration of shear stresses in the vicinity of the excavation, and is usually characterised by limited volume changes. Weak rock masses, such as shale, slate, phyllite, schist, and weakness/fault zones are incapable of sustaining high tangential stress components. When the induced stress reaches the rock mass strength, deformation (as time-dependent creep) occurs along the periphery of the tunnel. Immediately after each tunnel excavation increment has been completed, most of the rock stress is carried by the face itself. As excavation advances from a given point, there is a gradual decrease in this face effect, and tunnel deformation (inward movement) increases until it reaches its final value some distance from the face (Carranza-Torres & Fairhurst, 2000).

But a marked volume increment over time is observed in the case of swelling, caused by the increase in water content and a reduction of the volumetric effective stress. Note that while squeezing can be studied when considering the rock as a one-phase (solid) material, the analysis of swelling should consider the two-phase nature of the medium (Gioda & Cividini, 1996). Meng et al. (2013) illustrated several large deformation mechanisms, as shown in Figure 6.32, which were observed in the Zhegu mountain tunnel on the Sichuan-Tibet highway. These mechanisms consisted of plastic flow deformation of soft rock, shear sliding deformation of wedges, and bending deformation of thin-layered rock masses. The origins of these mechanisms are related to the lithologies exposed by tunnelling, including carbon phyllite and carbon slate, geological structures such as fault zones, and the in situ stress conditions. Moreover, the free swelling tests on rocks indicated that the swelling of soft rock is not a main

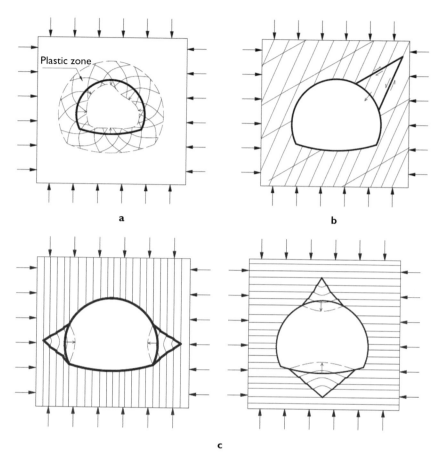

Figure 6.32 Large deformation mechanisms (Meng *et al.*, 2013): (a) Plastic flow deformation of soft rock; (b) Shear sliding deformation of a wedge; (c) Bending deformation in a thin-layered rock mass.

factor in the large deformations in the Zhegu mountain tunnel. However, Tan (1982) found that the swelling of surrounding rocks is the main reason for deformation in the Jinchuan mine in China. Therefore, due to the site-specific complexity of geological conditions and geostress, the mechanism of large deformations in a specific project requires tailored research. For example, Wang *et al.* (2009) studied the Wushaoling tunnel in China, and concluded that the stress-induced dilatancy of surrounding rocks led to the deformation.

6.4.3.1.2 *Prediction of large deformation (squeezing)*

In fact, large deformations discussed by most researchers really refer to rock squeezing because of its common occurrence during deep tunnelling, as well as its associations with difficulties during (and after) construction that normally require non-standard excavation and support methods (Jimenez & Recio, 2011). Therefore, the following discussion mainly focuses on the prediction of squeezing during deep tunnelling. Several methods are used to estimate large deformations during tunnelling, such as the empirical, numerical and experimental or monitoring methods.

1   Empirical Methods

The empirical method is the most important of all squeezing prediction methods, due to its simplicity and ease of use, as discussed by Shrestha (2005), noting that Jimenez and Recio (2011) have provided a review of empirical methods for squeezing prediction. Many methods for empirical squeezing prediction are based on the definition of competence factors which relate (or are indicators of) rock mass strength and stress at the tunnel depth. For instance, Jethwa *et al.* (1984) and Hoek and Marinos (2000) predicted tunnel squeezing based on the ratio between rock mass uniaxial strength, σ_{cm}, and lithostatic stress, $\sigma_v = \gamma H$. Table 6.5 lists the degree of squeezing via Jethwa's Method (Jethwa *et al.*, 1984).

On the other hand, Equation 6.9 was proposed by Hoek and Marinos (2000) to assess tunnel deformation. It is an approximate relation for the strain of the tunnel based on investigations via the axisymmetric finite-element model, considering a range of different rock masses, *in situ* stresses, and support pressures. Note that this equation is based on the conditions that the cross-sectional tunnel shape is circular, and that it is subjected to equal horizontal and vertical *in situ* stresses. As pointed out by Hoek (2001), the usual assumption for very weak rock masses is that they are incapable of sustaining significant differential stresses, and that failure occurs until the *in situ* horizontal and vertical stresses have been equalised. This assumption has been verified for squeezing rocks in a small number cases in which the tunnel deformations have been back analysed to estimate the ratio of horizontal to vertical *in situ* stresses.

$$\varepsilon_t = \left(0.2 - 0.25 \times \frac{p_i}{\sigma_v}\right) \times \left[\frac{\sigma_{cm}}{\sigma_v}\right]^{\left(2.4 \times \frac{p_i}{\sigma_v} - 2\right)} \tag{6.9}$$

where ε_t is the tunnel strain in percentage, σ_v is the overburden stress in MPa, σ_{cm} is the rock mass strength in MPa and p_i is the rock support pressure in MPa. Hoek (2001) proposed that values of $\sigma_{cm}/\sigma_v < 0.35$ are likely to produce squeezing (as defined by normalised convergences of more than 1% in unsupported tunnels). When the value of σ_{cm}/σ_v is less than 0.2, severe squeezing may occur. This indicates the onset of severe instability and, without adequate support, both the tunnel and face

Table 6.5 Degree of squeezing based on Jethwa's Method (Jethwa et al., 1984).

Degree of squeezing	Ranges
High	<0.4
Moderate	0.4–0.8
Slightly	0.8–2
Non-squeezing	>2

would collapse. For strain levels of less than 1%, experience suggests that there are few problems with tunnel stability. These strain levels generally occur in hard, strong rocks at relatively shallow depths, and the main stability problems are those caused by gravity falls of structurally defined blocks or wedges. Therefore, on the basis of tunnel closure assessment obtained using Equation 6.9, the squeezing ground conditions have been divided into the four classes of minor, severe, very severe and extreme squeezing ground conditions, as shown in Figure 6.33 (Hoek, 2001).

Panthi (2006) and Panthi and Nilsen (2007) suggested a methodology via a probabilistic approach for predicting squeezing. They argued that a probabilistic approach of uncertainty analysis which focuses on the effect of the variation in each input parameter is the most reliable way of predicting the extent of squeezing. Based on the researches performed by Panthi and Nilsen (2007), it can be concluded that the Hoek and Marinos approach (Hoek & Marinos, 2000) gives a fairly good estimation of tunnel squeezing. This indicates that the probabilistic approach of uncertainty analysis, considering the uncertainties of input parameters used in Hoek and Marinos' approach, has potential for predicting tunnel squeezing in future tunnel projects in weak rock mass conditions. It is emphasized that reliable data for the input variables are a key requirement for the successful use of such uncertainty analysis in predicting tunnel squeezing.

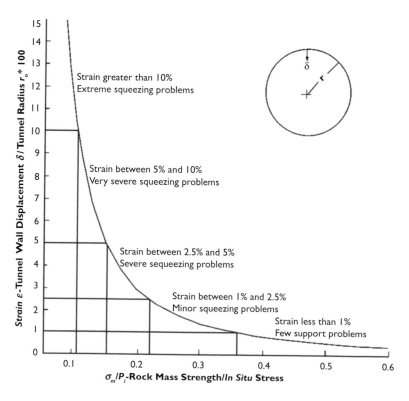

Figure 6.33 Squeezing classification associated with different levels of strain (Hoek, 2001).

Rock classifications have a long tradition of application in rock tunnelling, and it is typical to record RMR or Q values at the face as the tunnel advances; such records, in conjunction with observations of squeezing occurrence (or non-occurrence), can be used to develop empirical relations for squeezing prediction (Jimenez & Recio, 2011). As a result and based on the use of the geomechanical classifications, some empirical methods for squeezing prediction have been proposed. For instance, Singh *et al.* (2007) presented a well-known empirical correlation for anticipating squeezing conditions based on the Q value of the rock mass, in which tunnels deeper than $H = 350Q^{1/3}$ (with H in metres) could be expected to exhibit squeezing. Dwivedi *et al.* (2013) developed an empirical correlation between the squeezing deformation and the joint factor, $J_f = J_n/nr$, where J_n is the joint frequency, n is the inclination parameter depending upon the orientation of the joint plane with respect to loading direction, and r is the joint strength parameter dependent upon the joint condition (either clean and rough or filled-in joints), thickness of the joint, and joint alterations due to weathering, which is a measure of rock mass quality proposed by Ramamurthy and Arora (1994). This empirical correlation may be expressed by Equation 6.10:

$$\frac{u_p}{a} = \frac{5 \times 10^{-10}\, \sigma_v J_f^3}{K + 0.5} + 0.0052 \tag{6.10}$$

where u_p is the predicted radial deformation of the tunnel (m), a is the radius of tunnel (m), σ_v is the estimated vertical *in situ* stress component (0.027H), MPa, K is the support stiffness (MPa), and J_f is the joint factor.

Similarly, and given the practical difficulties for prediction of the Stress Reduction Factor (SRF) in the Q system, Goel *et al.* (1995) eliminated the influence of SRF on Q, and to this end they defined a Rock Mass Number as $N = (RQD/J_n)(J_r/J_a)J_w$. (Note that N is equal to Q when $SRF = 1$). Based on his research, the squeezing deformation can be assessed by Equation 6.11.

$$\frac{u_p}{a} = \frac{a^{0.12} H^{0.81}}{10.5 N^{0.27} K^{0.62}} \tag{6.11}$$

where H is the overburden thickness, and N is the rock mass number modified from the Q-system.

They also incorporated the influence of tunnel dimensions by considering the product $HB^{0.1}$, where H is the tunnel depth and B is the tunnel width (both in metres). For an in-depth review of these and other methods for empirical squeezing prediction, see Shrestha (2005) and Singh *et al.* (2007).

Other researchers have also proposed estimates of the degrees of squeezing intensity based on estimates of tunnel deformations. For instance, Aydan *et al.* (1993) transformed the competence factor concept into a strain concept (based on the analogy between the stress–strain response of rock in the laboratory and within the rock mass around tunnels), and proposed several levels of squeezing based on the ratio between the peak tangential strain at the tunnel boundary and the elastic strain (see Table 6.6). Similarly, Hoek (2001) proposed several levels of squeezing based on the strains produced by the excavation of an unsupported tunnel in a rock mass, see Table 6.6. Note

Table 6.6 Comparison between the Hoek method and the Aydan method for squeezing assessment.

	Hoek (2001)		Aydan et al. (1993)[a]		SI Index (Singh et al., 2007)	
Class number	Squeezing level	Tunnel strain ε_t	Squeezing level	Tunnel strain	Squeezing level	SI
1	Few support problems	$\varepsilon_t < 1\%$	No squeezing	$\varepsilon_\theta^a / \varepsilon_\theta^e \leq 1.0$	No Squeezing (NS)	$SI < 1.0$
2	Minor squeezing problem	$1\% < \varepsilon_t < 2.5\%$	Light squeezing	$1 < \varepsilon_\theta^a / \varepsilon_\theta^e \leq 2.0$	Light Squeezing (LS)	$1.0 < SI \leq 2.0$
3	Severe squeezing problem	$2.5\% < \varepsilon_t < 5\%$	Fair squeezing	$2.0 < \varepsilon_\theta^a / \varepsilon_\theta^e \leq 3.0$	Fair Squeezing (FS)	$2.0 < SI \leq 3.0$
4	Very Severe squeezing problem	$5\% < \varepsilon_t < 10\%$	Heavy squeezing	$3.0 < \varepsilon_\theta^a / \varepsilon_\theta^e \leq 5.0$	Heavy Squeezing (HS)	$3.0 < SI \leq 5.0$
5	Extreme squeezing problem	$\varepsilon_t > 10\%$	Very heavy squeezing	$\varepsilon_\theta^a / \varepsilon_\theta^e \geq 5.0$	Very Heavy Squeezing (VHS)	$SI \geq 5.0$

[a] The *UCS* of the rock mass was taken as 1 MPa.

that, the strain analogy for squeezing degree proposed by Hoek (2001) corresponds to the classification based on the stress ratio shown in Figure 6.34. Sakurai *et al.* (1993) defined warning levels for excessive deformation based on the concept of critical strain; and more recently Singh *et al.* (2007) proposed the use of their squeezing index (*SI*, defined as expected strain divided by critical strain) to predict levels of the squeezing potential in tunnels (see Table 6.6).

2   Numerical Methods
Numerical modelling is also used as a 'control method' in reducing the risk of tunnel construction failures (Shalabi, 2005; Ghiasi *et al.*, 2011). Due to the fact that some factors such as deformation are not completely predictable, using numerical modelling is an economical and capable method for predicting the behaviour of tunnel structures in various complicated loading conditions. Another benefit of using numerical simulation is in the graphical outputs which are used for both predicting the tunnel behaviour before, during and after construction and operation, and for communicating the information in an easily understandable manner.

The deformations and rock pressures can be estimated by means of numerical models, including plane strain models and three-dimensional models, which take into account the sequence of lining installation and excavation works. There have been significant efforts to develop tools for squeezing prediction and for estimation of convergences in tunnels using both rheological and time-dependent models. Most of the studies concerning these models were reviewed by Jimenez and Recio (2011). For instance, some researchers, such as Shalabi (2005), Debernardi and Barla (2009),

Sterpi and Gioda (2009), Barla (2010), Wang *et al.* (2014), established analytical solutions to compute creep deformations in squeezing rock on the basis of different constitutive laws, including elasto-visco-plastic models, a stress-hardening elastic-viscous-plastic constitutive law, hyperbolic and power creep laws, and a rheological damage model. Gioda and Cividini (1996) discussed some linear and non-linear viscous constitutive laws, based on which they included several applications of finite element analysis in the assessment and modelling of rock squeezing in tunnels.

From the viewpoint of engineering calculation for deep tunnelling, due to the high cost of three-dimensional analyses, in most cases in the past tunnel design calculations have been based on plane strain models which consider a tunnel cross-section. However, many researchers have found that pre-deformation (the ground displacement that occurs before supports are installed and which can only be estimated by three-dimensional models), governs the rock pressure and the ground responses of the system which consists of the rock mass and supports. Therefore, a large number of research efforts has focussed on the three-dimensional analyses of tunnel deformations, most of which are based on the axisymmetric problem of a cylindrical tunnel (Panet, 1979; Corbetta, 1990; Bernaud, 1991; Nguyen-Minh & Corbetta, 1992; Nguyen-Minh & Guo, 1993,1996; Bernaud & Rousset, 1996; AFTES, 2002; Shrestha, 2005). Recent papers have examined the influence of the tunnel shape and anisotropy or heterogeneity of the initial stress field (Carranza-Torres & Fairhurst, 2000; González-Nicieza *et al.*, 2008), as well as the effect of stress paths (Cantieni & Anagnostou, 2009). In summary, these numerical methods provide a suitable analysis tool for efficient use of the convergence-confinement method.

6.4.3.2   *Treatment technologies for large deformations*

Treatment technologies for dealing with large deformations in deep tunnels can be complicated, and are related to several factors such as, engineering geological conditions, excavation method, the large deformation mechanism, and the construction method and its requirements. Among these factors, the geological structures and deformation mechanism are more important than the others, as they govern the treatment strategies used to ensure the tunnel stability. Based on reviews of a number of studies, treatment technologies for large deformation may be divided into two types: (i) the excavation method, and (ii) the support method. In fact, in actual engineering practice, a combination of the two methods is generally employed. Sometimes during tunnelling, the modifications in excavation and support are implemented gradually and optimised in order to obtain better adaptability in controlling the large deformations. The excavation method and the support method are independently reviewed as follows.

6.4.3.2.1   *Excavation method*

One of the important control approaches for large deformation is an adaptable excavation strategy, including the optimisation of excavation sequence, excavation shapes and sizes, the over-excavation method for enabling deformation space, and the re-mining approach to deal with the occurrence of squeezing and reducing diameter after tunnelling. It must be noted that the excavation method is only a treatment strategy for large deformation by means of controlling and managing the deformation at each excavation step. However, the excavation method must be associated with support approaches in order to satisfy allowable deformation constraints and to ensure the stability of the rock mass. In other words, the practical approaches to treatment are mostly dependent on an excavation–support system.

1   Optimisation of excavation parameters

The selection of a particular construction procedure may have beneficial effects on large deformations. Steiner (1996) noted several typical squeezing cases which occurred during tunnelling, and stressed the importance of the construction procedure on the squeezing phenomenon. He found that the heading and benching method used in the Moffat Tunnel and Vereina Tunnel permitted a very short ring closure distance. However, for different excavation methods, i.e., drill and blast or TBM, there are different excavation parameters that must be optimised. In the drill and blast method, the optimisation of excavation parameters includes excavation sequence, excavation shape and its size. In the TBM method, the major parameters are TBM types, excavation rate, thrust forces and other characteristics of the machine and its operation. Sometimes, if the squeezing conditions are recognized in advance, the TBM can be designed to deal with large deformation problems. Mechanical excavation is generally used under weak or soft ground conditions (Moulton *et al.* 1995; Lombardi & Panciera, 1997; Kawatani *et al.*, 1999; Barla, 2001; Ramoni & Anagnostou, 2011), so we will discuss this further.

In fact, the excavation sequence, as well as the following support sequence, is an important design parameter, and relies on important factors such as rock types (lithology), strength and fragmentation of the rock mass, orientation of the rock structure, stress state, water pressure and the support systems. Most of these factors govern the intensity of large deformations and their distribution and mechanism. For some tunnels which pass through fault zones, their squeezing intensities are generally high, and the large convergence and heavy rock pressure phenomena are commonly the main difficulties. Kimura *et al.* (1987) has demonstrated an excavation method, namely the side drift method, for tunnelling through squeezing rock in two large

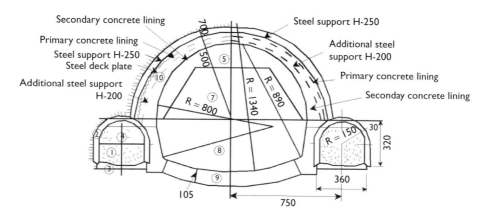

Figure 6.34 Side drift method of construction in the Enasan Tunnel (Kimura *et al.*, 1987).

fault zones of the Enasan Tunnel in Japan. The sequence of construction is shown in Figure 6.34, beginning with the side drifts of approximately 10 m² cross-sectional area, and followed by concrete filling which serves as the foundation of the primary and secondary concrete lining, with a total thickness of 1.2 m. The concept of the side drift method proposed by Kimura *et al.* (1987) is similar to the one relying on arch foot reinforcement by means of anchor piles, feet-lock bolts or grouting. Kovari and Staus (1996) have discussed in detail some advantages of the side drift method and the application of sequential excavation and full face excavation.

Additionally, both the benching method and pilot tunnel method are frequently used during deep tunnelling within heavily squeezing rock. Steiner (1996) illustrated the excavation sequences used in the heavy squeezing section in the Simplon tunnels, Switzerland (Figure 6.35). Central pilot tunnels were implemented and six or seven benches were excavated in sequence. Hoek (2001) suggested one excavation method for different squeezing intensities. This method indicates that, the higher the squeezing intensity, the more benches will be excavated, although this method depends more on experience-based judgment than on theoretical calculations, as argued by Hoek (2001). In addition to the bench excavation itself, the benching method was used together with variant support approaches and support types in Hoek's method.

2   Over-excavation method

In heavy squeezing conditions, the over-excavation method is a common approach used in order to increase the space and thus accommodate the deformation in deep tunnels. It is generally combined with deformable supports, which are discussed in sub-Section 6.4.3.2.2. For design, the key parameter which must be determined is the amount of over-excavation—which is decided according to changes in squeezing intensity through the use of advance probing, monitoring results and observations, as well as a risk assessment process made in tunnel regions previously excavated. This parameter also relies upon engineers' experience and geomechanical calculations. When estimating the required amount of over-excavation, the usual approach is to

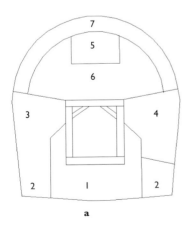
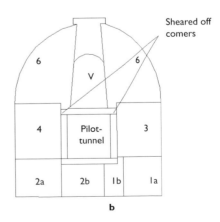

Figure 6.35 Excavation sequences in the Simplon Tunnel I (a) and II (b) in a strongly squeezing section (Steiner, 1996).

consider a tunnel cross-section far behind the tunnel face and assume plane strain conditions. However, on the basis of studies performed by Cantieni and Anagnostou (2009), the design estimation for the amount of over-excavation must consider the influence of the stress path and the characteristics of the support (Cantieni and Anagnostou, 2009).

3   Re-mining approach

Many tunnels, such as the Mucha highway tunnel in Taiwan (Hoek, 2001), Saint Martin La Porte access adit in Italy and France (Barla *et al.*, 2011), and the chlorite schist tunnel section in the Jinping II tunnels (Wu and Wang, 2011), encountered a cross-sectional reduction due to squeezing conditions. Once this phenomenon occurs, it signifies that re-mining or re-shaping to the original tunnel profile is necessary. During the re-mining of the failed rock masses, it is possible to involve heavy supports or secondary over-excavation. For example, in the process of the construction of the Mucha highway tunnel, as reported by Hoek (2001), the inward displacements of the roof and sidewalls of approximately 1.2 m (equivalent to a strain of about 15%) occurred when a fault zone was encountered. Heavy support using long, tensioned, grouted cables was introduced to support the failed rock mass while it was mined section-by-section. It was further stabilised by additional tensioned grouted cables, and the final concrete lining was placed as soon as possible after completion of the remedial work (see Figure 6.36).

Another example is the Saint Martin La Porte access adit (Barla *et al.*, 2011). After chainage 1267 m, the increased overburden caused the stresses in the lining to increase, and a yielding support was adopted. However, between chainages 1265 and 1340 m very large convergence took place with this support installed, as shown in Figure 6.37. To solve this problem, the advance was stopped at chainage 1384 m to allow for re-mining of the tunnel section beginning from chainage 1230 m. Moreover, a new yield-controlling support system, DSM, which consisted of sliding joints and highly deformable concrete elements, was used during re-mining.

Figure 6.36 Re-mining the perimeter of the 16 m span Mucha tunnel in Taiwan after severe squeezing in a fault zone. Photograph reproduced with permission from Sinotech Engineering Consultants Inc., Taipei (Hoek, 2001).

Figure 6.37 Saint Martin La Porte access adit at chainage 1325 m before remining the tunnel cross-section (Barla *et al.*, 2011).

6.4.3.2.2 Support method

The optimal support for large deformation conditions should be adjustable to the required deformability, which is a function of rock mass quality, support capacity, tunnel excavation and time. Many support technologies have been applied in deep tunnels in order to deal with the large deformation hazards. Significant efforts have been made by designers and engineers to deal with the large deformations through the improvement of support systems during deep tunnelling. Hoek (2001) reviewed some approaches commonly applied for tunnelling with large spans (i.e., 10–16 m) in severely squeezing ground. These approaches consist of advance supports, such as forepoles or grouting, and supports during and after excavation, such as steel arches, shotcrete, rockbolts or cables and the linings. Advance supports are commonly used for extremely squeezing ground, such as weak rock masses, fault zones or the tunnel sections influenced by exposed groundwater, and are also used to control the stability of tunnel faces. Note that Hoek (2001) also stressed the following:

> "*Instability of the face not only creates extremely dangerous conditions for the workmen in the tunnel, but it also has a major impact on the subsequent behaviour of the tunnel … this damage may require time-consuming and expensive treatment once the face has advanced through the fault or, if left untreated, it may cause problems later during the operating life of the tunnel.*"

This statement indicates that the support system for extreme squeezing ground should ensure the stability of the tunnel face. The same suggestion was also made by Lunardi (2000), who suggested that understanding and controlling the behaviour of the 'core' ahead of the advancing tunnel face are the secret to successful tunnelling under squeezing ground conditions. Figure 6.38 illustrates a case of the forepole umbrella support system shown by Hoek (2001) and used to control extreme squeezing of a 10 m span tunnel excavated using the full-face method. Note that, in Figure 6.38, grouted fibreglass dowels were installed in the tunnel face rock mass. The goal of this is to improve the stability of the rock mass ahead of the tunnel face and to control the deformation.

Most large deformation supports are installed during or after the tunnelling. In terms of their support effects, a shotcrete lining provides a surface support, the steel arches or ribs act as a linewise support, whereas rockbolts are effective in a pointwise manner, and thus are commonly used in poor rock conditions in order to increase the shearing resistance of the rock and to reduce and harmonise deformations. Note that, in large deformation ground, steel arches or steel ribs combined with a shotcrete lining may appear to be attractive and be commonly applied (Kovari & Staus, 1996), the erection of a steel rib requires only 15 to 20 minutes, and the space between the flange and the rock is wedged or shotcreted immediately. However, it should also be noted that, as argued by Hoek (2001), the above-mentioned heavy support system may lead to some problems: (i) Time consumption, i.e., the spraying of the shotcrete lining takes time and the shotcrete or concrete lining takes time to harden and achieve the required strength to provide adequate support; (ii) Technological difficulties such as the bending of the sets into the appropriate shape when using the steel arches with sliding joints as major support elements. Also, high stiffness shotcrete cannot accommodate rock convergence to obtain relief of rock pressure, and the shotcrete linings can fail under just a small eccentricity of the normal force (i.e., in the presence of bending moments), due to the lack of tensile strength (Kovari & Staus, 1996).

Therefore, the introduction of deformable supports improves the performance of traditional supports such as shotcrete and steel arches or ribs. These supports are able to deform without becoming damaged, in combination with a certain amount of over-excavation in order to accommodate the deformations. Based on the 'yielding principle', they can be structurally implemented in two main ways (Anagnostou & Cantieni, 2007): either by arranging a compressible layer between the excavation boundary and the extrados of a stiff lining, or through suitable detailing of the lining that will allow a reduction in its circumference. Figure 6.39 demonstrates two basic types of deformable supports to govern large deformations. As pointed out by Cantieni

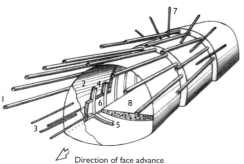

Direction of face advance

1   Forepoles–typically 75 or 114 mm diameter pipes, 12 m long installed every 8 m to create a 4 m overlap between successive forepole umbrellas.

2   Shotcrete–applied immediately behind the face and to the face, in cases where face stability is a problem. Typically, this initial coat is 25 to 50 mm thick.

3   Grouted fiberglass dowels–Installed midway between forepole umbrella installation steps to reinforce the rock immediately ahead of the face. These dowels are usually 6 to 12 m long and are spaced on a 1 m × 1 m grid.

4   Steel sets–installed as close to the face as possible and designed to support the forepole umbrella and the stresses acting on the tunnel.

5   Invert struts–installed to control floor heave and to provide a footing for the steel sets.

6   Shotcrete–typically steel fibre reinforced shotcrete applied as soon as possible to embed the steel sets to improve their lateral stability and also to create a structural lining.

7   Rockbolts as required. In very poor quality ground it may be necessary to use self-drilling rockbolts in which a disposable bit is used and is grouted into place with the bolt.

8   Invert lining–either shotcrete or concrete can be used, depending upon the end use of the tunnel.

Figure 6.38 A forepole umbrella support system used in a full face 10 m span tunnel excavation (Hoek 2001).

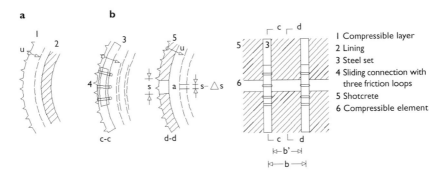

Figure 6.39 Basic types of deformable supports: (a) compressible layer between lining and excavation boundary; (b) yielding supports with steel sets, shotcrete and compressible insets (Cantieni & Anagnostou, 2009).

and Anagnostou (2009), in the case shown in Figure 6.39(a), the ground undergoes convergences while the clearance profile remains practically constant. This solution has been proposed particularly for shield tunnelling with very stiff segmental linings (Schneider *et al.*, 2005). In the case shown in Figure 6.39(b), the steel sets have sliding connections and are combined with shotcrete.

In fact, as early as nearly 80 years ago, deformable supports were used in deep mining, where excessive deformation occurred. At that time, heavy supports such as lining or steel arches were extensively used as major support elements controlling large deformations in rock masses. The design of the lining introduced deformable elements such as segments with intermediate timber to provide ductility, timber block elements, flexible shotcrete lining (Schubert, 1992), segmental shotcrete lining connected with deformable steel pipe groups or buckling inner tubes (Schubert, 1996, see Figure 6.40) and steel arches with sliding joints or joint gaps (Hoek, 2001; Hoek & Guevara, 2009). Some application cases can be found, including the Galgenber tunnel (Schubert, 1996), the Semmering pilot tunnel (Schubert *et al.*, 1996, 2000) and the Strenger tunnel (Budil *et al.*, 2004). Additionally, there are the relatively recently developed 'highly-deformable concrete' elements (Kovári, 2005; Thut *et al.*, 2006), which are composed of a mixture of cement, steel fibres and hollow glass particles. These collapse at a pre-defined compressive stress which is dependent on the composition of the concrete, thereby providing the desired deformability (Cantieni & Anagnostou, 2009). Also, recently steel fibre reinforced, high-strength shotcrete (SFRS) has been extensively applied into squeezing tunnels. Through comparisons of field measurement and analytical/numerical results, it is recognized that SFRS can be applied as a reasonable primary lining for tunnels excavated in ground with severe squeezing conditions (Hisatake, 2003).

In addition, Barla *et al.* (2011) suggested a new support system, namely the 'yield-control' DSM support system in which steel ribs with highly deformable concrete (HiDCon) elements are employed and which have been installed during the construction of the Saint Martin La Porte access adit. On the basis of this experience,

Figure 6.40 Support system with integrated, yielding, steel elements (Schubert, 1996).

Barla *et al.* (2011) summarised the main excavation/construction stages (see Figure 6.41) as:

i Face pre-reinforcement, including a ring of grouted fibre-glass dowels around the opening, designed to reinforce the rock mass ahead of the face and around the tunnel perimeter over 2–3 m thickness.

ii Mechanical excavation carried out in steps of 1 m lengths, with installation of a support system consisting of rock dowels (length 8 m) along the perimeter, yielding steel ribs with sliding joints (TH type), and a 10 cm thick shotcrete layer. The tunnel is opened in the upper cross-section to allow for a maximum convergence of 600 mm to take place.

iii The tunnel is opened to the full circular section at a distance of 20–30 m from the face, with the application of a 200 mm reinforced shotcrete lining, yielding steel ribs with sliding joints (TH type), and with longitudinal slots (one in the invert) fitted with Highly Deformable Concrete elements (in the following HiDCon elements). The tunnel is allowed to deform in a controlled manner to develop maximum convergence, which at this stage should not exceed 400 mm.

iv Installation of the final concrete lining is at a distance of 80 m from the face. Note that a clear difference exists between the DSM method suggested by Barla *et al.* (2011) and the method shown in Figure 6.38 proposed by Hoek (2001), namely the requirement of the controlled deformation rates. There is a rigorous limit for the deformation rates in the DSM method.

In the context of special engineering issues, such as groundwater problems in squeezing ground, intersection of deep tunnels, tunnel floor enhancements, several guideline squeezing support designs have been proposed. For example, Hoek (2001) offered suggestions concerning the treatments of groundwater in squeezing ground. Considering the support in the tunnel intersection area, which generally undergoes large deformation and significant failures, Hsiao *et al.* (2009) suggested some support guidelines based on the different squeezing conditions, as listed in Table 6.7.

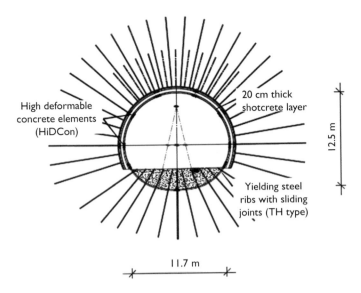

High deformable
concrete elements
(HiDCon)

20 cm thick
shotcrete layer

Yielding steel
ribs with sliding
joints (TH type)

12.5 m

11.7 m

Figure 6.41 Support system DSM (Barla *et al.*, 2011).

Table 6.7 Guidelines for tunnel support design in intersection areas (Hsiao *et al.*, 2009).

Geological condition	Support design suggestions
Slightly or non-squeezing rock ($\sigma_{cm}/P_0 \geq 0.5$)	Only strengthening support system to prevent potential wedge failure in the intersection area, such as increasing the thickness of shotcrete, and the density or length of rockbolt. Monitoring instruments should be installed to examine tunnel stability.
Moderately squeezing rock ($0.25 \leq \sigma_{cm}/P_0 \leq 0.5$)	Reducing rock mass rating in order to install heavier support or design a more conservative support system in the intersection area. Monitoring instruments should be installed to examine tunnel stability. Carry out detailed visual inspection of tunnel conditions and increase monitoring frequency during construction.
Severely squeezing rock ($\sigma_{cm}/P_0 \geq 0.25$)	Conservative excavation process and support system should be used in the intersection area. Auxiliary measures, such as ground improvement, should be made if necessary. 3D numerical analysis is suggested to assess the appropriateness of tunnel design. Monitoring instruments should be installed to examine tunnel stability. Carry out detailed inspection on tunnel conditions and increase monitoring frequency during construction.

The design and optimisation of support parameters, such as the length of rock-bolts, thickness of linings, spaces of steel ribs, and so on, should be implemented through an efficient numerical method. After simulating the observed performance during actual excavation, the analyses and assessment of the support systems with different support parameters can be obtained to guide the process of large deformation support design in the future tunnel sections.

6.4.4 Long term stability

Changes in the local stress state due to tunnelling activity and the effects of rock support perturb the stability of the rock mass surrounding the excavations—for better or worse. The readjustment of the rock to a new equilibrium state does not occur instantaneously, but rather as a gradual process over time. The readjustment process will continue until the engineering perturbations cease and the new stability state is reached—which may continue during the life of the deep tunnel, not only after construction but also during its operation. Consequently, the issues of long-term stability must be solved, especially the interactions between the rock mass and the support systems.

Much research and *in situ* observation and measurement have enriched our knowledge of this long-term behaviour (Ladanyi, 1974; Panet, 1979; Gioda, 1982; Windsor & Thompson, 1993; Tan, 1993; Hoek *et al.*, 1995; Malan & Spottiswoode, 1997; Oreste, 2003; Liu & Zhang, 2003; Corkum & Martin, 2007; He & Yang, 2007; Sandrone, 2008). In the context of the time-dependent behaviour of crystalline rocks, Hagros *et al.* (2007) have provided a detailed literature survey, noting that the importance of time-dependent behaviour in the form of rapid convergence of tunnels in weak rock is well known (Gioda, 1982; Panet, 1996). The time-dependency in both cases encompasses concepts such as the creep and stress relaxation of intact rock, creep of large scale discontinuities, delayed failure, and long-term strength.

To address these problems, several aspects must be solved, such as the long-term deformation or rheological behaviour of rock masses, deterioration of rock mass strength with time, and changes of the boundary conditions such as rock support implementation, evolution of damage or fractured zones, safety and loads on support elements. It has been noted that the ground pressure on the support system in a deep tunnel of a given shape is known to depend not only on the rock mass properties, ground stress conditions, geological structure conditions and damage or fractured zone characteristics near the excavation, but also on the type and the rigidity of the support system, such as the lining, and the time of its installation. After contact with the rock mass has been established, the stress on the support elements will vary with time, at first due to the advance of the tunnel face, then due to the gradual changes in the rock mass properties, which are caused by the combined action of processes such as fracturing, consolidation, swelling, creep, weathering and groundwater recovery (Ladanyi, 1974).

6.4.4.1 Long term stability assessment in deep tunnels

There are different mechanisms controlling the long-term behaviour of hard and of soft or weak rock masses. As presented by Malan (1998), the closure behaviour of hard rock is the result of the rheology of fracture zones around these excavations and the time-dependent extension of the zones following a mining increment. Malan (1998) presented an elasto-visco-plastic approach which was developed in order to simulate the time-dependent nature of a fracture zone. In contrast, large deformation occurring in soft ground is a function of plastic creep deformation, squeezing and swelling. Some closed-form solutions have been obtained. However, there are strong similarities linking the different theoretical or closed-form solutions suggested by the investigators. There are several common assumptions which apply generally to these analyses: (a) the virgin stress field is hydrostatic; (b) plane strain conditions exist and the distance from the tunnel face has no significance; (c) the rock mass surrounding the tunnel excavation is assumed to be homogeneous and isotropic; and (d) the excavation has a circular cross-section. However, despite the difference between realistic and simplified tunnel conditions, these theoretical methods are still used during analyses and design. Ladanyi (1974), in the development of a design criterion, considered a long-term strength concept of the rock surrounding the tunnel excavation, and examined the post-failure reduction of strength and associated volumetric dilution. This produces a formula which is applicable to high stress rock environments and various rock types, including creep-sensitive rock material.

Although there are some theoretical models and methods which may be used to analyse long-term behaviour (Lombardi, 1973; Ladanyi, 1974; Wilson, 1980; Detourney, 1986; Carranza-Torres & Fairhurst, 2000), long-term interactions are found to be complex and significant, and should be accounted for in numerical simulations. Some continuum mechanical models, which consider the time-dependent behaviour of deep rock masses, are used to assess the long-term stability of deep tunnel ground. These models commonly introduce rheology or creep elements: i.e., the visco-elasto-plastic model (Chen et al., 2007), which combines the Kelvin–Voigt element with the CWFS model; the CVISC rheological model (Jiang et al., 2008), which consists of the Maxwell, Kelvin and Mohr–Coulomb plastic models; the non-linear Drucker–Prager plastic coupling and creep constitutive model (Phienwej et al., 2007; Chen et al., 2013); the continuum visco-plastic approach, which was developed and implemented in a finite difference code (Malan, 1998); the VIPLA model (Lemaitre & Chaboche, 1996), based on Perzyna's overstress theory (Perzyna, 1966), which states that the strain rate tensor can be split into elastic and visco-plastic components; the SHELVIP model, the stress hardening, elastic, viscous, plastic model derived from Perzyna's overstress theory, (Debernardi & Barla, 2009), and so on.

A brief capability comparison among some of the above-mentioned models can be found in the paper by Barla et al. (2010). In these model analyses, several aspects, such as long-term deformation assessment around the excavation, excavation damage zone evolution, support pressure calculation and its safety evaluation and treatment scheme optimisation, require attention. Furthermore, considering the uncertainty of some of the data in geomechanical modelling, sensitivity analyses with respect to selected parameters are suggested in order to investigate the effects of their variability, within reasonable ranges, on performance. In addition to the consideration of the time-dependent behaviour of rock masses, many assessment methods attempt to achieve an integration of geotechnical and structural design.

Meanwhile, the detailed sequences of excavation, support, boundary transformation and their sensitivities to tunnel stability, throughout the entire construction period and tunnel life, are commonly concentrated on and analysed. This method regards the long-term stability of a deep tunnel as a ground-support interaction system, in which designs both to assess rock mass behaviour and to calculate support capacity are iteratively conducted (Boldini et al., 2003; Hoek et al., 2008). One of the most efficient methods is the convergence-confinement method which has been used to evaluate the long-term stability of the rock mass-lining system, as described by Sandrone et al. (2008). This method has the direct advantage that the rock mass-support interaction may be analysed during both tunnel construction and operational life.

The long-term analyses of hard rock and its support systems rely on the understanding of the long-term degradation of rock mass strength. It was observed that rocks loaded at certain stresses, lower than the short-term strength (typically measured in laboratories), but greater than the long-term or 'true' strength, can fail if the stress is maintained for a sufficiently long period of time. If the stress is less than the long-term strength (a fraction of the short-term strength), the rock will not fail, irrespective of the stress (load) duration. Damjanac and Fairhurst (2010) reviewed the issues and processes associated with such tests, and proposed that a driving-stress threshold must exist for massive rock masses below which no damage occurs, otherwise they would have failed already. Based on geological evidence and numerical modelling, they suggested that this lower limit occurs at a driving-stress of approximately 40 to 60% of the short-term laboratory strength, and is equivalent to the crack-initiation stress measured in unconfined laboratory samples. Two parameters control the predictions of time-dependent strength degradation (ITASCA, 2007): (i) time-to-failure, and (ii) damage evolution (rate) before time-to-failure. Time-to-failure is determined as a function of the stress state (i.e., the driving stress), as shown in Figure 6.42 and Figure 6.43. Note that the applied load in the axial direction and the confining pressure are respectively denoted by σ_1 and P_c. The axial load at failure during a short-term test is denoted by σ_f. The stress difference maintained during a static fatigue test conducted at a confining pressure of P_c is thus $\sigma = \sigma_1 - P_c$, then the stress difference at failure during a short-term test is $\sigma_s = \sigma_f - P_c$. Therefore, in Figure 6.42, the driving stress ratio is given by $\sigma/\sigma_s = (\sigma_1 - P_c)/(\sigma_f - P)$. The damage rates govern the degradation level of the rock mass strength, which may be denoted as follows in Equation 6.12:

$$M(t) = M_0 D(t) \tag{6.12}$$

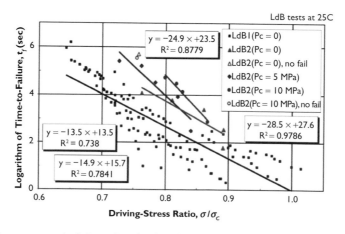

Figure 6.42 Laboratory static-fatigue data for Lac du Bonnet granite established by Schmidtke and Lajtai (1985) and Lajtai and Schmidtke (1986) for unconfined conditions (LdB1) and by Lau and Chandler (2004) for unconfined and confined conditions (LdB2) (ITASCA, 2007).

where M denotes the strength parameter, such as cohesional strength, internal friction angle or tensile strength, and $D(t)$ is the accumulated damage, which is generated using a micro-mechanical model, i.e., the PFC stress corrosion model.

The long-term stability of a deep tunnel focusses on the safety of lining structures, i.e., shotcrete lining, which may be divided into two types, temporary and final linings. The assessment aims at calculating the bending moments and axial thrusts in the linings, including their short- and long-term values, in order to ensure the lining design safety, as well as to optimise the design parameters. Rockbolt loads are also studied in order to adjust the support strategy and design parameters. These analyses are especially important when groundwater or other factors have a crucial influence on long-term stability. For example, considering a hydraulic tunnel like the Jinping II headrace tunnel, after the lining has been installed, the groundwater outside the lining will gradually recover. This indicates that the increase of groundwater pressure could result in a high lining load and lining failure, thus threatening the stability of the lining and the tunnel integrity. However, the analysis is complicated and is difficult to conduct using fully-coupled models. In practice, simplified models are used to solve the multi-physical field issues (Schweiger et al., 1991; Shin et al., 2002; Kishida et al., 2007; Pellet, 2009; Graziani & Boldini, 2012).

Also, it must be stressed that field measurement/monitoring is crucial, including the deformation and stresses in the rock mass and support elements, with emphasis on the long-term damage zone. Based on this approach and the monitored information obtained, the variables and parameters governing the long-term mechanisms can be identified and used for time-dependent models of the rock mass and rock structure long-term behaviour together with the support components. This is an essential component of the design concept of the New Austrian Tunnelling Method. Also, advanced technologies have been applied in deep tunnelling, i.e., the micro seismic approach, AE method, and digital drilling camera technology. These technologies will contribute to the goal of long-term monitoring and comprehensive assessment of deep tunnel long-term stability.

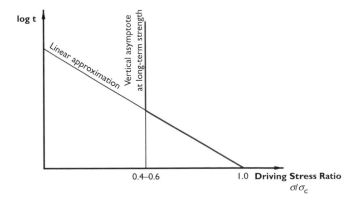

Figure 6.43 Static-fatigue curves (right-hand boundary lines) for intact rock, based on the approach by Damjanac and Fairhurst (2010).

6.4.4.2   Treatment technologies to ensure long term stability in deep and long tunnels

The treatment technologies are different when based on potential hazard types and are covered in Sections 6.4.1, 6.4.2 and 6.4.3. For hard rock tunnels, some approaches aim to govern the long-term degradation of the rock mass which locally can exhibit a gradual damage zone increase. To solve this long-term hazard, the initial support system must be installed as quickly as possible to provide the necessary support pressure. Meanwhile, energy-absorbing elements, such as cone bolts, water expansion anchor bars, shotcrete and mesh are used when the rockburst-prone ground is identified. Destressing approaches are also suggested to transfer the high stresses deeper into the rock mass around the excavation surfaces. For soft rock, methods may be used to control the load on the tunnel lining, such as over-excavation or the reaming method, a lining containing high deformation elements, and groundwater pressure control in terms of pressure-release holes. The designs based on the convergence-confinement method are commonly used to ensure the safety of the lining. One of the important reasons for this is that the method emphasises the effect of lining installation time and the capacity of the rock mass-support system.

Also, it is necessary to conduct field monitoring, including rock mass and support element monitoring, with treatments being adjusted on the basis of the monitoring results. However, this means that there are some long-term factors which cannot be identified, and thus many uncertainty issues can occur throughout a long engineering period. For these, the continued use of monitoring and a warning system are vital for long-term tunnel integrity.

6.5   ILLUSTRATIVE EXAMPLE: ASSESSMENT AND MITIGATION OF RISK FOR DEEP TUNNELS AT THE JINPING II HYDROPOWER STATION, CHINA

The Jinping II Hydropower Station is located on the Yalong River in Sichuan Province, China (see Figure 6.44), and has an installed capacity of 4.8 GW. This facility uses the water which flows along the 150 km long Yalong River and, as a result of the tunnels short-circuiting a loop in the river, there is a 310 m head for generating power. It has the highest water head of any station along the Yalong River, with an effective rating of 288 m, and the largest installed capacity. The main aspects of the hydropower station in the current context are the design and construction of the tunnel system, including four headrace tunnels, two access tunnels and one drainage tunnel, as shown in Figure 6.44. These tunnels are 16.7 km in length and cut through the Jinping Mountains, at a maximum depth of 2525 m, with over 75% of the tunnel depth being greater than 1700 m. The layout of the tunnel system is also shown in Figure 6.44. Headrace tunnels #1 and #3 have 12.4 m circular cross-sections and were excavated using tunnel boring machines (TBMs). Headrace tunnels #2 and #4 have 13 m horseshoe cross-sections and were excavated by the drill and blast (D&B) method. The centre lines between the four main tunnels are 60 m in length. Two access tunnels, A and B, were constructed parallel to the headrace tunnels. A drainage tunnel of 7.2 m diameter is located between access tunnel B and headrace tunnel #4. This has a length of 16.73 km and inter-axis distances of 35 m and 45 m from tunnels B and #4, respectively.

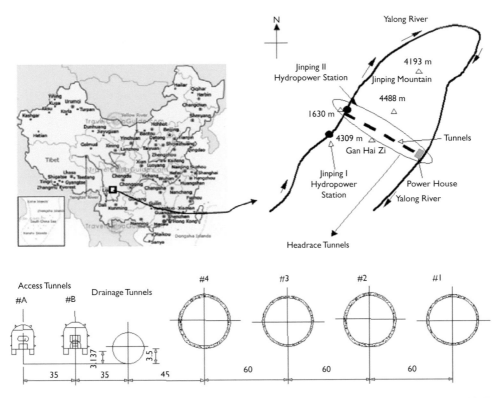

Figure 6.44 Location and tunnel layouts of the Jinping II Hydropower Station, China (Zhang *et al.*, 2012).

The hydrogeological condition is complicated in the area through which the Jinping II tunnels pass and a number of major technical difficulties met during construction exceeded the boundaries of existing knowledge. The technical issues encountered in the construction of the headrace tunnels included a large quantity of high-pressure water inrushes, ventilation issues during the tunnel excavation, and especially rockbursts and their treatment. During construction, the Jinping II tunnels experienced many geotechnical hazards and faced many engineering risks, such as rockbursts, ground squeezing with large deformations, water inrushes, karst caves, TBM jamming, roof collapses and support difficulties. Thus, the risks in the Jinping II tunnels represent some classic engineering hazards related to deep and long tunnels.

Figure 6.45 outlines the typical tunnel sections experiencing different engineering hazards. At the two ends of these tunnels, the major hazards were water inrushes and collapses, as well as the TBM becoming jammed in the poorer quality rock. With the increase of the overburden, failures caused by the high stresses occurred, which were manifested as cracking and fracturing, breakouts, spalling, slabbing and rockbursts, which damaged the rock masses significantly and led to support difficulties. The occurrence of these hazards related to the high stress conditions led to changes in the analysis and design approaches, along with some new applications of early warning, prediction and treatment methods in order to decrease the engineering uncertainties and risks. These approaches are described in the following sections. The studies on the prevention of rockbursts, treatment of sudden groundwater inrush, construction

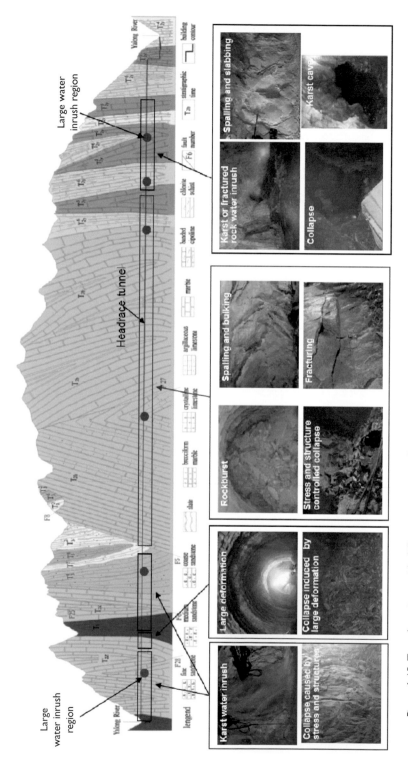

Figure 6.45 Tunnel sections with different engineering hazards, Zhang et al., (2012). See the colour plate section at the end of the book.

techniques for soft rock and the TBM method have enabled the development of new approaches to decreasing engineering risks and hence produced many improvements. These have laid a solid foundation for the construction of future hydro-electric power stations constructed in similar environments.

6.5.1   Epistemic uncertainty analysis of headrace long deep tunnels

Recalling that we have used the term 'epistemic uncertainty', as lack of knowledge, this sub-Section discusses the information available before tunnelling starts. Unexpected conditions subsequently encountered during tunnelling are discussed later in the context of 'aleatory uncertainty'.

6.5.1.1   Geological setting

1   Rock groups and lithologies
The geological-geotechnical data obtained for the project region were mainly based on surface charts, as well as the information from a 5 km long exploratory tunnel. The eastern ends of the tunnels are mainly buried in the T_{2y} marble of the Yantang Group, while the western ends are in the T_1 chlorite schist, T_3 sand slate, and T_{2z} marble of the Zagunao Group, and in the middle is the T_{2b} marble of the Baishan Group; 80% of the rock that the tunnels pass through is of marble lithology, as shown in Figure 6.46. The lithologies in the above-mentioned group are listed below.

- The Lower Triassic system T_1, which consists of biotite chlorite schist, metamorphic middle and fine sandstone, including thin bedded marble and psephitic or striped marble.
- The Yantang formation T_{2y}, including three groups: (i) Group T_{2y4}, which consists of grey or grey-green striped marble, local interlayered 0.3–1.5 m thicknesses of grey-white dolomitic marble; (ii) Group T_{2y5}, which consists of grey-black marble, pink medium coarse grained marble, and grey-white coarse-grain marble; the marble in the Yantang group has a uniaxial compressive strength of 70–110 MPa, Young's modulus of 20–35 GPa, and density 2670–2730 kg/m; and (iii) Group T_{2y6}, which is mainly grey or grey-black argillaceous limestone interlayered dark grey marble.

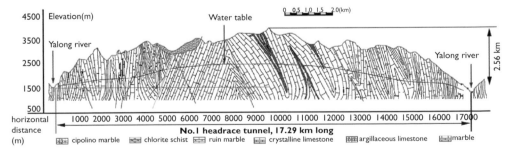

Figure 6.46 Geological section along the tunnels at the Jinping II hydropower station, China (HydroChina Huadong Engineering Corporation, 2005).

- The Baishan formation T_{2b}, which is a pink thick layer marble with a uniaxial compressive strength of 110–160 MPa, Young's modulus of 30–40 MPa, and density of 2780 kg/m³.
- The Zagunao formation T_{2z}, which consists of white or grey-white marble with a uniaxial compressive strength of 70–110 MPa, Young's modulus of 20–38 MPa, and density of 2370 kg/m³.
- Three formations of Permian Period T_3, which consists of sandstone with a uniaxial compressive strength of 90–130 MPa, Young's modulus of 18–25 MPa, density of 2720 kg/m³, and slate.

2 Geological structure and tectonic condition

Figure 6.47 shows the geological structures of the Jinping region. The project area contains a series of north–south directional, refolded, compact folds and high-angle compression or compression-torsion fractures. The faults run in the NS, NE, NEN, NWN, and NW directions with the N–S fault being consistent with the axis of the fold. The NE-directional fault is a branch of the NEN-directional fault, and is a sinistral strike-slip fault. The NWN fault is a dextral strike-slip branch of the NEN structure. The NW small-scale faults are a group of dextral strike-slip faults intersecting vertically with the NEN-structures. The fault structure in the eastern region is more highly developed than that in the western region. Most folds in the eastern region turn over to the west. Therefore, rumpled structures are more clearly developed in the western region. These geological structures were formed by both ancient and recent tectonic activities. The tectonic stress field has also been affected by these events.

Along the Jinping II headrace tunnels, there are two anticlinal structures and two synclinal structures between the west Yalong River and Fault F_6 (Jinping mountain fault), as shown in Figure 6.47. There is a compound syncline including sandstone and slate in the three formations of the Permian Period, as a core between Fault F_6

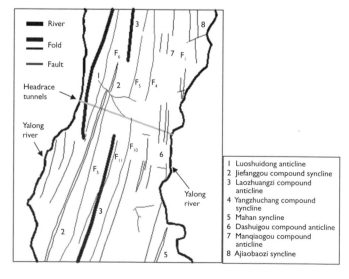

Figure 6.47 Structural geology in the Jinping II Hydropower station region, China (HydroChina Huadong Engineering Corporation, 2005).

and the Baishan formation, with very highly developed secondary folds. Two wings of the compound anticline to the east of Fault F_5 consist of the Baishan formation with interlayered but only slightly developed folds. The Yantang formation to the east consists of six small folds. There are four main faults along the tunnels:

i   Fault F_6 (Jinping mountain fault): N20°~50°E, NW or SE∠60°~87°. The width of the fault zone 1–4.2 m and the crush zone width is 6–37 m. It is developed with muddy zones, breccia zones and schistosity.

ii   Fault F_{28}: The orientation is N20°E with SE∠70°. The squeezed crush zone has a width of 1–2 m. The rock has been squeezed to a plate shape.

iii Fault F_5: The orientation is N10°–30°E with NW∠70°. It includes thick breccias in the main zone and the rock has schistosity and phyllitization.

iv   Fault F_{27}: The strike is N30°–40°W and dip NE and is located in the Baishan Formation T_{2b}.

It is known from two auxiliary tunnels that there are Faults F5 and F6, 14 structural planes in Class II, 22 structural planes in Class III-1, 76 structural planes in Class III-2, and 23 structural planes in Class IV, which are divided according to strike as nearly the South–North group, NE group, NW group and NWW–NEE group. There are bedding extrusion belts developed along the dolomite-marble zone. Moreover, it is also known from the exploration tunnels that there are several joint sets: (i) N5°~30°W, SW or NE∠30°~75°, mostly dense, smooth faced, and parallel to structure lines; (ii) N60°~80°W, SW∠10°~25° or ∠70°~85°, steep or low angle mostly open and with long elongation, which mainly act as elementary conduits in the headrace tunnels; (iii) N0°~30°E, SE or NW∠70°~90°, bedding joints, mostly closed but open locally and acting as elementary conduits in the headrace tunnels; (iv) N30°~60°E, SE∠10°~35°, low angle, mostly open, surface with undulations, long elongation; (v) N40°~50°E, SE or NW∠45°~80°; and (vi) N65°~80°E, NW or SE∠55°~80°, which act as conduits in the headrace tunnels.

6.5.1.2 Rock stress

Before the construction of the four headrace tunnels, there were some rock stress *in situ* measured data. However, these measuring points were all located in the shallow area of the headrace tunnels, and mainly in the two ends of the tunnels, as shown in Figure 6.48. The measured orientations of σ_1 at the east end of the tunnels are indicated in Figure 6.49. It can be seen that the directions of σ_1 average approximately 130°. Note that these data only represent the characteristics of the *in situ* stress field at the two entrance zones of the tunnels and cannot be applied to the sections that have large overburdens. Unfortunately and because of the depths involved, there were no data for the rock stress in these sections before the construction of the headrace tunnels. This was due to the phenomenon of rock discing, as shown in Figure 6.50. The fact that there were no stress measurement data for the deep buried tunnel sections caused epistemic uncertainty of the stress field. Moreover, it is known that the *in situ* stress field is strongly influenced by the geological conditions, especially the large fractures. So we note that: (i) The headrace tunnels are located in China's western region, a tectonically active area which has experienced significant tectonic movement

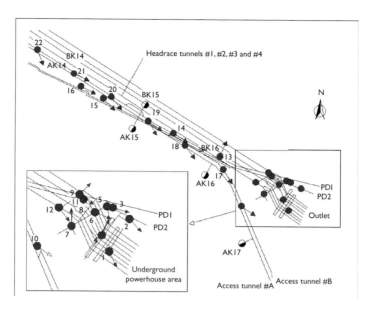

Figure 6.48 In situ stress measurement points near tunnels and the overall σ_1 orientations.

in the past and the role of the tectonic stress fields must not be ignored. (ii) Due to the development of the refolded compact folds along the tunnels, differences exist between the *in situ* stress values at the core of the syncline and the two limbs of the anticlines. (iii) The fault and fracture zones have an impact on the local *in situ* stress field.

To solve this difficulty, and to obtain an accurate assessment of the *in situ* stress, an integrated analysis method of *in situ* stress was proposed based on multi-source information (Zhang *et al.*, 2012), such as rock brittle failures, Excavation Disturbed/ Damaged Zones (EDZs), core discing, and large-scale deformation of soft rock. The method includes the analysis of the region's geological structures and topography integrated with analysis of the *in situ* stresses based on the multi-source information in the local tunnel sections. Several techniques were adopted, including multivariate regression, numerical simulation, and stereographic projection. The *in situ* stress regimes, orientations, and magnitudes in the tunnel sections at the different depths were obtained and verified. This enabled provision of the basic conditions for analysis of the stability of the surrounding rock mass during excavation. A flowchart of this method is shown in Figure 6.51. The basic principle and implementation process of this method include the following (Zhang *et al.*, 2012).

i　Analysis of the regional geological structures and topography, from which is obtained a preliminary estimation of the *in situ* stress regions along the tunnels, according to the inference about the orientation of the tectonic stress, through the analysis of the tectonic history and current tectonic activity within the project region, as well as an understanding of the influence of complex mountain and valley topography on the distribution characteristics of the *in situ* stress.

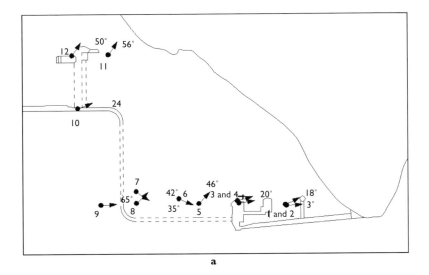

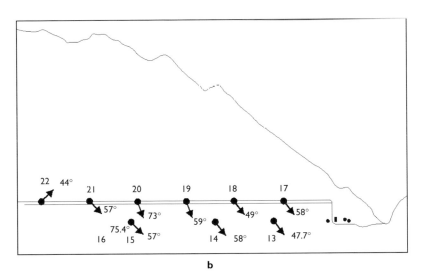

Figure 6.49 Inclination angles of σ_1 at all measuring points: (a) Inclination angle of σ_1 (N58°W) at measurement points in the underground powerhouse. (b) Inclination angle of σ_1 (N58°W) at measurement points in auxiliary tunnels and geological exploration tunnels at the east end of the tunnels.

ii   Integrated analysis of the *in situ* stress based on the multi-source information, through which the regimes, orientations, and magnitudes of the *in situ* stresses are obtained.

The major horizontal stress component along the tunnels is larger than the vertical stress component in the Jinping Mountains. However, the local assessment results

0 m

4.2 m

Figure 6.50 Rock discing in the rock surrounding access tunnel A at overburden 2430 m.

for the *in situ* stress also indicated that the vertical stress from the overburden is sometimes larger than the horizontal stress—caused by the structural setting in the local tunnel section. It can also be found that the *in situ* stress directions are easier to determine than their values. When large amounts of information related to *in situ* stress are collected and obtained in these sections, the integrated stress assessment method may be applied, and then the magnitudes of *in situ* stresses in local tunnel sections may be estimated. However, even if this method is effective and may be used to accurately assess the *in situ* stress, in some tunnel sections the local rock stress state is still difficult to determine due to the local variation of the geological conditions, which increases the aleatory uncertainty of rock stress along the tunnels. It is possible that in future this issue may be solved by the development of new rock stress measuring technologies, for example, the Borehole–Wall Stress Relief Method (BWSRM), developed by Ge and Hou (2011).

6.5.1.3 Hydrology

The Jinping Mountains at the project site are located near the Yalong River which is in a deep valley. The main peak has an altitude of 3900–4488 m and variations in height of 2560–3150 m (see Figure 6.46). There is a sequence in the river loop which is almost parallel to the strike of a main structure NNE, with a south–north length of 71 km, near the east–west width of 12–23 km having an area of 1126.7 km². The surface water in the karst in the block is not highly developed, and there are mainly dry valleys and seasonal dry valleys in the area.

The underground water in the block includes karst water, fracture water and porous water, all of which has originated from rainfall and is drained into the Yalong River. There is weak karst development throughout the project site which

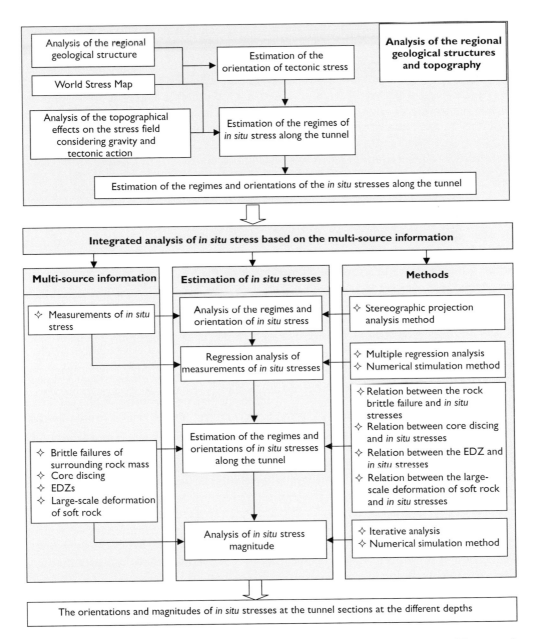

Figure 6.51 Workflow of the estimation method for *in situ* stress along deep tunnels (Zhang *et al.*, 2012).

is developed with water in the fractures and solution cracks in the marble. The network system consists of the NNE main structure line and NEE and NWW extension fractures or tensile-transtensional faults, tight folds, and steep strata control enrichment and migration of underground water. The large-scale uplift of the crust since the Cenozoic era and formation of the deep river valley has resulted in continuous changes of the hydrology. The karst is mainly developed vertically; there is no layered karst system.

Thus, the potential for sudden inundation of the tunnel during the excavation work must be examined. The underground water in the vicinity is conveyed by fissures and a network of channels with a continuous water source, resulting in the possibility of high pressure and large flow rates. During excavation of the adit and access tunnels, the maximum water inflow was about 4.9–7.3 m³/sec and high-pressure water inflow caused difficulties for excavation because it can lead to instability of the surrounding rock masses and poses a risk for the safety of the crews (Wu *et al.*, 2005).

6.5.1.4 *Properties of the rock mass*

Before tunnelling the Jinping II headrace tunnels, field exploration and laboratory assessment was conducted for the geology and rock mass conditions in two excavated access tunnels, as well as rock classification along the headrace tunnels, as shown in Figure 6.52, in which the GSI method established by Hoek (2001) was used in connection with the developmental conditions of the geological structures, such as folds and faults. Also, the mechanical parameters were estimated, as shown in Table 6.8. Based on these data, it can be inferred from Figure 6.52 that the chlorite schist tunnel section (GSI = 45; UCS = 20–40 MPa) and marble tunnel section (GSI = 50–70; UCS = 65–150 MPa) are two major hazard sources which may induce large convergence and brittle failure, respectively.

In order to reveal the mechanical features of hard rock under high geo-stress conditions, loading-unloading testing methods were employed, which can be used to estimate the evolution of damage and plastic deformation during the cracking and fracturing of hard rocks. Figure 6.53 illustrates the results of loading-unloading tests for Jinping T_{2b} marble specimens, in which four confining pressures were used. As shown in Figure 6.53, significant strain softening characteristics of the deep marble were observed. Moreover, as the confining pressure increases, the transformation from brittle to ductile behaviour is also demonstrated. In fact, for most deep rock testing work, which includes laboratory and field tests, it was found that characteristics such as strain-softening and the brittle-ductile transition were typical mechanical properties of these deep rocks. The properties demonstrate important evolutions of the mechanical parameters, as described in the following Section.

Figure 6.54 demonstrates that both of the strength parameters, cohesion and internal friction angle, vary with the plastic deformation. As the plastic parameter increases, the cohesional strength undergoes a process of softening, and the internal friction angle increases; physically, this implies that the micro-cracking and fracturing is induced by deviatoric stresses, so the increase of damage in the deep rocks gives rise to the variation of rock strength properties. The elastic parameters, namely shear modulus and bulk modulus, are also calculated, as shown in Figure 6.55. The shear modulus decreases with the plastic internal variable, showing that material damage is induced during the plastic deformation, and the bulk modulus increases slightly with the plastic internal variable, which is possibly due to the volumetric strain dilatancy.

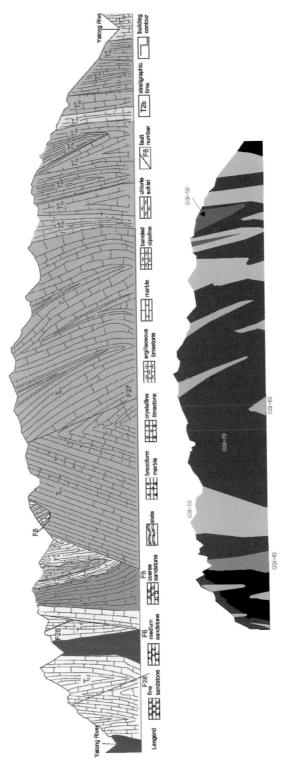

Figure 6.52 Some overall rock classification values along the Jinping II tunnel using the GSI method (HydroChina Huadong Engineering Corporation, 2005).

Table 6.8 Physico-mechanical parameters of rocks in the project area (Wu & Wang, 2011).

Class	Lithology	Unit weight (kN/m³)	Uniaxial compressive strength (MPa)		Modulus of deformation (GPa/m)		Modulus of elasticity (GPa)		Poisson's ratio
			Dry	Wet	Horizontal	Vertical	Horizontal	Vertical	
II	Mid thick bedded fine grained sandstone (T_3)	27.4	104–152	71–114	10–12	11–15	18–25	25–35	0.23–0.27
	Marble of the Zagunao group (T_{2z})	27.2	70–90	55–78	8–10	12–14	20–25	30–38	0.22
	Mid thick bedded marble (T_{2b})	27.7	90–100	75–85	16–20	15–18	30–40	30–40	0.18
	Striped mica marble (T_{2y4})	28.0	85–90	55–62	13–15	10–12	20–25	15–20	0.21
	Mid thick bedded marble (T_{2y5})	27.1	70–95	65–85	10–16	9–13	15–35	20–30	0.21–0.22
	Argillaceous limestone (T_{2y6})	27.0	70–75	60–70	9–11	8–10	16–17	13–15	0.27
III	Mid thick bedded fine grained sandstone (T_3)	27.1	98–139	71–110	7–9	8–10	15–21	20–25	0.27
	Layered sandstone slate (T_3)	27.6	70–95	42–53	6–9	8–10	10–18	16–21	0.26–0.3
	Marble of the Zagunao group (T_{2z})	27.2	65–72	55–65	7–9	9–11	16–20	14–25	0.25
	Chlorite schist (T_1)	26.5	40–50	30–40	6–7	5–6	9–13	8–10	0.28
	Mid thick bedded marble (T_{2b})	27.6	75–85	60–70	10–12	9–10	20–25	18–20	0.20
	Striped mica marble (T_{2y4})	27.5	70–85	50–60	8–11	7–10	9–16	8–15	0.23–0.26
	Mid thick bedded marble (T_{2y5})	26.6	65–90	55–80	6–11	5–10	11–17	7–15	0.23–0.27
	Argillaceous limestone (T_{2y6})	26.5	60–70	50–65	6–9	5–8	9–15	6–12	0.28–0.3
IV	Slate (T_3)	26.2	30–40	22–26	2–4	3–5	15–18	8–16	0.31
	Chlorite schist (T_1)	26.1	30–40	20–25	2–4	3–5	10–15	8–10	0.32
	Fault and fracture	–	45–55	40–45	0.6–1.5	0.4–1.0	1.0–1.5	1–2	0.35

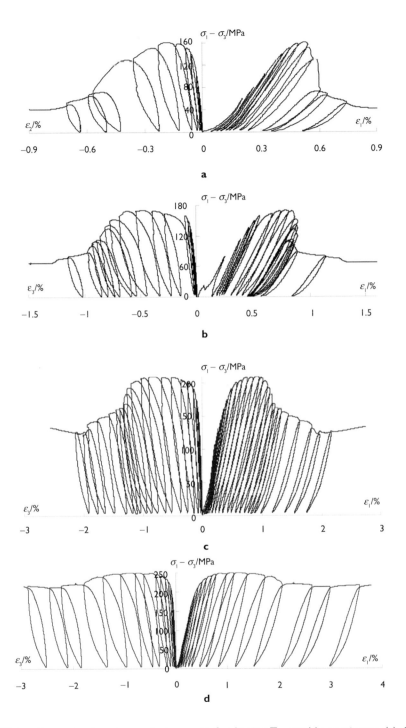

Figure 6.53 Loading–unloading stress–strain curves for Jinping T$_{2b}$ marble specimens: (a) Confining pressure 5 MPa; (b) Confining pressure 10 MPa; (c) Confining pressure 20 MPa; (d) Confining pressure 40 MPa (Feng et al., 2013). Note the significant energy content of the hysteresis loops—which is typical for marble.

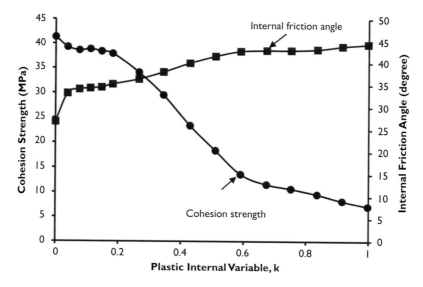

Figure 6.54 Cohesional strength and friction angle parameters at different values of the plastic internal variable (Feng *et al.*, 2013).

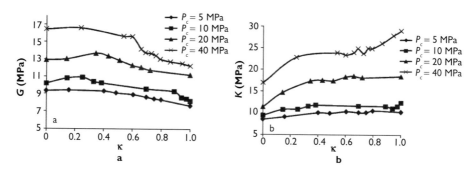

Figure 6.55 Elastic parameters at different values of the plastic internal variable under various confining stresses. (a) Shear modulus; (b) bulk modulus (Zhou *et al.* 2010).

6.5.1.5 Specific project location

From the perspective of rock mechanics, determining the location of the Jinping II headrace tunnel line relies to a great extent on the assessment of rockbursts and the tunnel stability. In order to reduce the risk of failure of the rock mass after excavation, the direction of the tunnel line should be parallel or sub-parallel to the maximum horizontal principal stress, all other factors being equal. The second issue is the determination of spacing between the tunnels which is based on the fact that the Jinping II tunnels were designed as a seven-tunnel system, containing two access tunnels, one drainage tunnel and four headrace tunnels. The design and decision-making process regarding the tunnel locations and spacing is discussed in the following text.

The tunnel route in the project zone for the Jinping II hydropower station is at a height of 1,600 m ASL and is 16 km long, with overburdens ranging from 1,270 to 2,525 m. As already mentioned, the geology along the tunnel through the Jinping Mountains, with heights of up to 4,300 m, mainly consists of medium to thick beds of marble with compressive strengths extending up to 150 MPa. Due to the fact that there are or will be several hydropower stations on the upstream portion of the Yalong River, two access tunnels (A and B) have been designed for the transportation needs of these power stations. Additionally, another purpose of the two access tunnels is geological exploration ahead of the four headrace tunnels (see Figures 6.44 and 6.45 above).

The initial design was that there would be only the two access tunnels A and B, and four headrace tunnels, and there was no drainage tunnel. The spacings between the two access tunnels and between access tunnel B and headrace tunnel No. 4 were 35 m and 80 m, respectively. The spacing between each pair of headrace tunnels was 60 m. The overburden of the headrace tunnels varies from 1900 to 2525 m. Access tunnels A and B, which have respective sizes of 6.5 × 7 m and 7.5 × 8 m, were excavated completely in 2007 before the construction of the four headrace tunnels. There were significant water inrushes during the excavation of access tunnels A and B. In order to collect water during the excavation of the four headrace tunnels, an additional drainage tunnel was added between access tunnel B and headrace tunnel No. 4. The spacing between access tunnel B and the water drainage tunnel is 35 m, and that between the drainage tunnel and headrace tunnel No. 4 is 45 m.

6.5.1.6   Excavation and support method

The construction of the Jinping II headrace tunnels underwent an adjustment of the excavation and support design as a result of the occurring engineering issues. Wu and Wang (2011) reviewed the detailed construction of the Jinping II headrace tunnels and, based on their reviews, the generalised construction process is summarised as follows.

Considering the excavation methods used in the tunnelling, three stages were used during the construction of the Jinping II headrace tunnels, namely the D&B method at first, followed by a combined method of D&B and TBM, and finally D&B once again. In the first stage of construction, the D&B method was adopted to excavate the portals of the headrace tunnels and assembly tunnels for TBM equipment, then the TBM was used in combination with D&B in order to obtain high excavation rates and shorten the engineering schedule. All of the diversion tunnels were excavated by the D&B method, which entailed high risks of rockburst hazards, water inrush and large deformations in the chloritic schist tunnel section as well as the karst caves encountered. In fact, the TBM and D&B methods can co-ordinate well with each other: when some locations are not suitable for TBM construction, reverse construction at the back of the TBM can be conducted by D&B through the creation of an additional adit. When the TBM successfully passes through the working face of the D&B within a certain distance, sufficient working space for D&B can be provided through the branch holes, and thus the construction progress can be accelerated.

For tunnel excavation, several engineering stages can be classified with respect to the design modifications.

i Preparatory stage. This began in 2004 and ended in 2008 with the excavation of two access tunnels. During this stage, the two parallel access tunnels, which are both 17.5 km in length, were excavated by the D&B method. These tunnels not only served as transportation tunnels for the following construction of the headrace tunnels, they could also be utilised as exploration tunnels to reveal the potential engineering hazards and risks, as well as to identify the complicated geological conditions.

ii Preliminary design. A large number of design decisions were formulated, including the tunnel location layout, tunnel shapes and sizes, spaces between the headrace tunnels, excavation methods and planned support schemes. At this preliminary design stage, according to the hydropower requirements, five tunnels were planned to be excavated, including the four headrace tunnels and one drainage tunnel. Due to the topographical constraints, it was impossible to arrange adits and inclined shafts in the middle of the tunnels to shorten the construction period: excavation could only begin from either end of a tunnel. The excavation direction of the drainage tunnel and headrace tunnels was designed, as shown in Figure 6.56(a), in terms of two excavation directions. From east to west, headrace tunnels Nos. 1 and 3 (12.4 m in diameter) and the drainage tunnel (7.2 m in diameter) were excavated by TBM, and the others were excavated by D&B. The headrace tunnels excavated by D&B are 13.0 m in diameter. From west to east, the water drainage tunnel and headrace tunnels were all excavated by the D&B method with diameters of 7.2 m and 13.0 m respectively. The D&B excavation scheme of headrace tunnels included two benches with 8.5–9 m height for the upper bench, as shown in Figure 6.57.

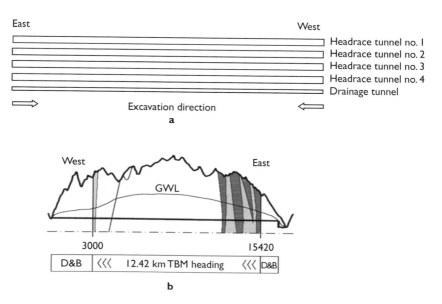

Figure 6.56 Excavation of the drainage tunnel and four headrace tunnels: (a) Excavation direction; (b) Planned excavation methods in longitudinal section (HydroChina Huadong Engineering Corporation, 2005).

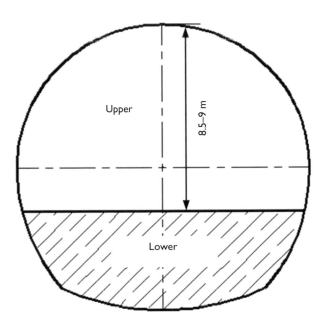

Figure 6.57 D&B excavation scheme for the headrace tunnels (HydroChina Huadong Engineering Corporation, 2005).

iii Excavation stage. Due to the fact that their construction is performed in advance, two access tunnels are helpful to widen the working face and excavate the long tunnel step by step. Therefore, during the practical excavation process, the headrace tunnels were divided into three segments, namely the east segment, middle segment and west segment. In each segment, the excavations were implemented from two opposite tunnel faces exposed by the access–headrace tunnel adits or drainage–headrace adits. Figure 6.58 shows the excavation processes in the middle segment, in which both the D&B and TBM methods were used. Note that, in order to ensure excavation safety under severe or extremely severe rockburst conditions, the TBM operation was ceased and the machine disassembled in favour of the D&B method. In particular, and with reference to the different geological conditions, the construction scheme for the headrace tunnels was 'dynamically modified', including an increase in the number of adits and a combination of D&B and TBM methods. Importantly, the adits, such as the access–headrace, access–drainage and drainage–headrace adits, can create enough working faces so as to accelerate the construction progress and hence enable power generation to be expected in 2012.

Through unremitting hard work, headrace tunnel #1 was successfully completed on June 6, 2011. Due to the fact that 15% of the length of headrace tunnel #1 was in a strong rockburst area, different support methods were applied according to the different rockburst intensities, and detailed improvements were proposed. During

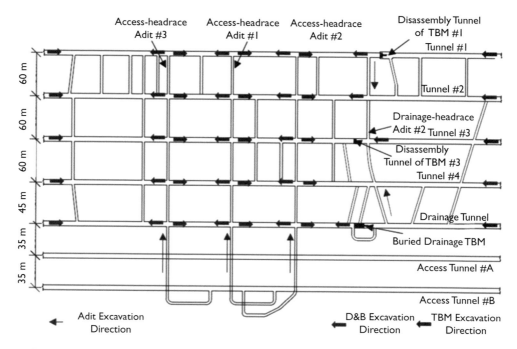

Figure 6.58 Excavation designs in the middle segment during the construction of the Jinping II headrace tunnels.

excavation, many new materials, techniques and methods were used to ensure the safety and speed of construction, such as advance blasting for stress relieving, water swelling anchors and nano-material injection. Moreover, microseismic monitoring technology was used for the tunnel construction, and a new microseismic monitoring system was built. With this system, it is possible to monitor and analyse the microseismic activity of rocks continuously, which contributes significantly to rockburst forecasting.

The initial design for the support systems for the four headrace tunnels, the support parameters, such as bolt types and their length and spacing, shotcrete types and their thicknesses, suspended net, steel arch, etc., was optimised in terms of the field conditions, risk assessments and the changes of excavation methods. For particular engineering hazards, such as rockbursts, large deformations, groundwater inrushes and karst caverns, effective support systems were used to reduce the rock mass damage and risks of rock instabilities. These parameters are illustrated in the following sections in the context of hazard treatments.

6.5.1.7 Water inrush

As illustrated in Subsection 6.5.1.1 titled 'Geological settings', the conditions for the construction of the headrace tunnels for the Jinping Hydropower Station II are complicated, such as large depth, long tunnel length, and large tunnel diameter, thus

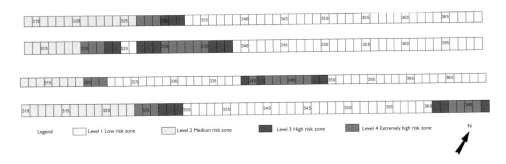

Figure 6.59 Risk assessment of water inrush at the west end of Jinping II headrace tunnels (Li & Li, 2014).

the prevention and treatment of the high pressure and concentrated water inrushes became one of the most important aspects. Therefore, two important tasks, namely groundwater assessment and engineering treatment, were carried out.

To assess the groundwater status and predict the groundwater parameters, such as infiltration intensities, cone of depression, inflow rates and inrush magnitudes, the three-dimensional finite element method was applied and the transport of groundwater, throughout the entire construction process of the Jinping II tunnels, was studied and analysed (Zhang *et al.*, 2010). A pre-excavation primary assessment of the water inflow was obtained as a result of the tunnelling and without any supporting measures to deal with the inflow. Moreover, the coupled Hydro-Mechanical (HM) behaviour of the fractured rock during tunnelling was analysed with a three-dimensional stochastic continuum model, which could account for the heterogeneity of the permeability (Chen & Ruan, 2007).

Later, Li and Li (2014) proposed a GIS method to assess the risk of groundwater inrush in the Jinping II tunnels. They identified the factors influencing karst water inrush and established an index system to assess the potential of water inrush. Various influencing factors exist relating to karst development, which include primarily lithology, geological structure, groundwater hydrodynamic conditions, and geographic and geomorphic conditions, as well as neo-tectonic movement characteristics. According to their assessment, the extremely high-risk area of water inrush accounts for 8.99% of the total, that with high risk accounts for 22.98%, that with medium risk accounts for 47.02%, and that with low-risk accounts for 21.01%. Figure 6.59 shows the risk assessment of water inrush at the west end of the Jinping II headrace tunnels (Li & Li, 2014).

6.5.1.8 Rockbursts

Rockbursts were observed in the massive marble in the reference projects as from approximately 1,700 m overburden. This is the same as the assessment given by Gong and Li (2007), Zhang and Fu (2008), and Feng *et al.* (2013), based on empirical rockburst criteria. Table 6.9 shows the results of rockburst risk assessment in the Jinping II headrace tunnels. It can be seen that there is a risk potential of moderate and severe

Table 6.9 Assessment results of the Jinping II headrace tunnels based on empirical rockburst criteria (Feng et al., 2013).

Criterion name	σ_θ/σ_c		σ_c/σ_1		$(\sigma_\theta+\sigma_L)/\sigma_c$		σ_c/σ_t		Wet		Assessment results
	Maximum	Minimum	Maximum	Minimum	Maximum	Minimum	Maximum	Minimum	Maximum	Minimum	
Hoek criterion	2.21	0.86									Severe rockburst
Turchaninov criterion					2.63	1.15					Severe rockburst
Russenes criterion	2.21	0.86									Severe rockburst
The criterion from Erlang mountain	2.21	0.86									Severe rockburst
Tao criterion			3.24	1.62							Moderate and severe rockburst
Barton criterion			3.24	1.62							Moderate and severe rockburst
China criterion (GB50218-94)			3.24	1.62							Rockbursting and rock ejection
Brittleness criterion							40.24	16.29			Moderate and severe rockburst
Elastic strain energy index, Wet									2.52	3.52	Moderate rockburst

rockbursts. This preliminary estimate of the rockburst potential in the Jinping II tunnels before their excavation leads to the consideration of support and excavation designs to deal with the potential rockburst hazards. However, there are uncertainties which cannot be assessed by the empirical rockburst criteria, such as the potential rockburst locations, failure parameters, e.g., rockburst intensity and range, hazards to workers and excavation machinery, support difficulties, and the configuration of the monitoring scheme.

Advanced approaches are required for the assessment of rockburst risk in the Jinping II tunnels. These approaches consist of several numerical models, numerical indices, comprehensive empirical indices and intelligent analysis methods. A number of numerical and material models were proposed by Jiang *et al.* (2008) (rock mass deterioration model, RDM), Huang *et al.* (2008) (GPSEdshs model,), Zhou *et al.* (2010) and Zhang *et al.* (2010) (elastoplastic coupling mechanical model), to accurately describe the mechanical behaviour of the Jinping deep buried marbles. Several rockburst numerical indices also were established by Zhang *et al.* (2011, FAI index), Jiang *et al.* (2010, LERR index) and Qiu (2011, RERI index).

The failure approaching index (FAI) is an index of the damage degree for the surrounding rock masses and is based on the geometrical analysis of the relation between the stress state at a point and the yield surface as defined in principal stress space, and the equivalent plastic shear. Zhang *et al.* (2011) demonstrated the applications of this index in the Jinping II tunnels. Figure 6.60(a) shows a FAI result which was calculated to analyse the spalling failure event at a depth of 2,500 m in access tunnel A. The FAI distribution accurately identified the spalling zone shown in Figure 6.60(b).

Excavation for underground engineering works leads to a change in the stress boundary conditions because the unsupported tunnel periphery becomes a principal stress plane with zero stress acting perpendicular to it. This concentrates the stress in the adjacent rock mass which can then cause a sudden release of high intensity energy. This released energy not only induces failure of the rock mass, but also generates dynamic energy and high speed displacements of the crushed rocks. In view of this, Jiang *et al.* (2010) applied an energy index, i.e., the Local Energy Release Rate (LERR), to assess the potential of rockbursts. Figure 6.61 illustrates a successful application of the LERR index to study one rockburst event which occurred at chainage

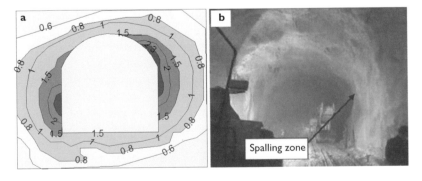

Figure 6.60 Tunnel sections at a depth of 2,500 m in access tunnel A: (a) distribution of FAI (Failure Approaching Index) and (b) actual spalling zone (Zhang *et al.* 2011).

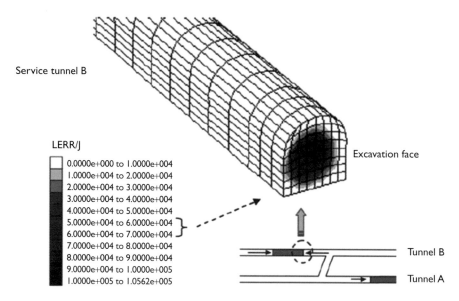

Service tunnel B

LERR/J

0.0000e+000 to 1.0000e+004
1.0000e+004 to 2.0000e+004
2.0000e+004 to 3.0000e+004
3.0000e+004 to 4.0000e+004
4.0000e+004 to 5.0000e+004
5.0000e+004 to 6.0000e+004
6.0000e+004 to 7.0000e+004
7.0000e+004 to 8.0000e+004
8.0000e+004 to 9.0000e+004
9.0000e+004 to 1.0000e+005
1.0000e+005 to 1.0562e+005

Excavation face

Tunnel B

Tunnel A

Figure 6.61 Simulated rockburst at excavation face using local energy release rate (LERR) at chainage K9 + 512 in the access tunnel #B (Jiang *et al.* 2010).

K9 + 512 in access tunnel B. It can be seen that the heavy concentration occurred at the face of this tunnel.

In addition to these numerical indices, a novel empirical index, the Rockburst Vulnerability Index (RVI), was also proposed to assess the failure associated with rockbursts (Qiu *et al.* 2011). The index is based on the fact that before the four headrace tunnels were excavated, the excavations of two access tunnels, A and B, had been completed in 2007. During the tunnelling of these two tunnels, many severe and extremely severe rockburst events had occurred. These rockburst cases could thus provide knowledge about the rockburst control factors for the conditions of the Jinping II tunnels, such as *in situ* and induced stresses, geology conditions, rock lithological influence, excavation effects and mechanical properties. The RVI index was established to quantify the control effects of these rockburst factors. Qiu *et al.* (2011) successfully adopted the RVI index, to assess the potential and failure depths of rockbursts along the headrace tunnels before their excavation.

In fact, due to the use of the different excavation methods (i.e., TBM in headrace tunnels #1 and #3 and the drainage tunnel, and D&B in headrace tunnels #2 and #4), the excavation progress among these tunnels differed. So, the excavated tunnels could be seen as effectively the pilot tunnels for the yet to be excavated tunnels, with the rockburst risk in the non-excavated tunnels being assessed by the rockburst conditions in the excavated tunnels. This is known as "the rockburst engineering analogy method", which is illustrated in Figure 6.62. As shown in the Figure, the rockburst risks in the non-excavated tunnels #2 and #4 could be evaluated by the rockburst events occurring in the two access tunnels, the drainage tunnels, and even headrace tunnels #1 and #3.

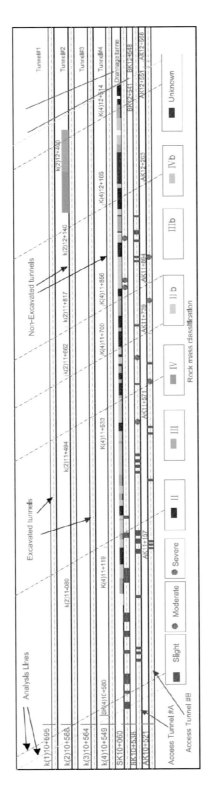

Figure 6.62 A rockburst assessment case based on the 'engineering analogy' method considering the rockburst information obtained from two access tunnels (see colour plate section at the end of the book).

6.5.1.9   Large deformations

In the Jinping II tunnels, the chlorite schist is one of the major strata and one which may exhibit large deformation and squeezing behaviour. The maximum deformation of many sections after the primary support was installed reached 0.5–0.7 m, which is 7.6–10.6% of the radius of the tunnel. According to assessment methods for the squeezing deformation of a rock mass as proposed by Hoek and Brown (1980) and Barla (1995) (see Table 6.10), and based on empirical statistical data for the large deformation of soft rocks, the chlorite schist in the Jinping tunnels can be classified into the levels of 'Severe' and 'Extremely severe'. However, due to the different mechanical properties of the chlorite schist and the variability of the *in situ* stress conditions along the Jinping tunnels, the squeezing behaviour varies in different chloritic schist tunnel sections. Figure 6.63 shows the percentages of large deformations, accounting for the total length of the chlorite schist tunnel section in two headrace tunnels, i.e., #1 and #2. For these two tunnels, the sectional percentages of 'Severe' and 'Extremely severe' compressive deformations are as high as 74% and 59%, respectively, whilst those of moderate compressive deformation are 23% and 37%. Insufficient pre-support and delay in the support structure led to inefficient control of the development of the plastic zone, and thus resulted in large deformation and even collapse.

However, the above assessment results were obtained after the excavation of the tunnels, as well as being based on the information from deformation monitoring. This is unsuitable for the risk assessment before the excavation, due to the

Table 6.10 Assessment of squeezing behaviour of the rock mass.

Method	Deformation percent (%)			
	Slight	Moderate	Severe	Extremely severe
Hoek and Brown (1980)	1–2.5	2.5–5	5–10	>10
Barla (1995)	1–3	3–5	>5	–

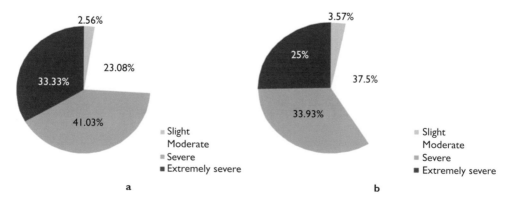

Figure 6.63 Percentage of different levels of squeezing behaviour: (a) headrace tunnel #1; and (b) headrace tunnel #2 (Zhou et al., 2014).

lack of deformation data from the rock masses. In fact, to evaluate the risk of large deformation in the chlorite schist of the Jinping tunnels before excavation, several methods may be adopted. The first is the numerical method to assess the magnitude of the chlorite schist. By applying this method, a cautious approach is to determine the mechanical model for the chlorite schist and its parameters, which must be obtained from mechanical tests, empirical methods proposed by Hoek and Brown (1980), or back-analyses of field deformation. For example, due to the adoption of the upper and lower bench excavation method, the deformation behaviour can be identified after the excavation of the upper bench, through which the mechanical parameters of the rock masses may be back calculated. These results can be used to calculate the deformation behaviour during the excavation of the lower bench, as well as to analyse the long-term stability of the rock masses.

The second method is the analogy method, which is based on the fact that the Jinping tunnels have a parallel and reasonably adjacent layout. So, according to the deformations in the excavated tunnel sections which have the same or similar geological and excavation conditions, the deformation behaviour in the yet-to-be-excavated tunnel sections may be assessed.

6.5.1.10 Long term stability

According to the excavation information feedback and monitoring data, the stability of the surrounding rocks in the construction period is relatively high. The deformation is not large and the tunnelling conditions are good, except for the chloritic schist area at the west end. However, due to the site-specific engineering conditions, such as large overburden depth and brittle marble conditions, fracturing and spalling often occur around the tunnels. The long-term relaxation cannot be ignored, as it will influence the long-term stability of the structure during the operating period. Some rockbolt systems may fail under high *in situ* stress and high water pressure conditions. The strength of the surrounding rocks in wet conditions is lower than that in dry conditions. Furthermore, constructing a bolt system at this time may not be appropriate, as there could be risks during the operating period if the shotcrete support is only used in the construction period. Thus, it is necessary to carry out secondary reinforcement concrete support to ensure the long-term stability of the surrounding rocks and improve the safety level.

6.5.2 Aleatory uncertainty analysis of the headrace tunnels

Recalling that 'aleatory uncertainty' refers to those hazards that are not easily predicted and must be dealt with essentially when they occur during tunnelling, we now outline the problems encountered during construction of the headrace tunnels.

6.5.2.1 Geological variations at different chainage intervals

In Section 6.5.1.1, the detailed geological setting was reviewed and information collected based on the exploration and investigation of field engineering geology. This work only provides a relatively rugged geological framework. In order to conduct deeper analyses

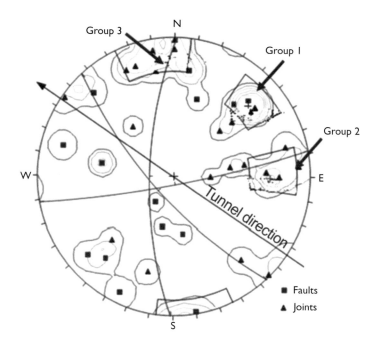

Figure 6.64 Geological structures representing rockburst risks in the Jinping II tunnels (Qiu *et al.* 2011).

and assessment of tunnel hazards and the required design excavation and support, we need detailed and local geological information, such as the rock mass structures, i.e., faults, folds and fractures, groundwater and karst conditions, local stress field conditions, etc. For example, to assess the rockburst risk, Qiu *et al.* (2011) proposed an empirical method, namely the RVI method, in which major geological structures representing potential rockburst risks were identified, as shown in Figure 6.64. Therefore, in the practical application of the RVI method, the potential for specific structures must first be assessed. This assessment relies on the information either obtained by advance geological prediction or as exposed in the excavated tunnels.

In fact, the latter may also be used as an approach to reveal the geological 'anomalies', such as groundwater bearing faults or other hydrogeological structures. One of the advanced methods for detecting such geological anomalies is the Tunnel Seismic Prediction (TSP) method. Also, the Ground Penetrating Radar (GPR) signal response to water-bearing structures was used for theoretical derivations, and 3D tomography of the Transient Electromagnetic Method (TEM) was used to develop an equivalent conductance method. Based on this, a technical system for the reliable prediction of adverse geological features was developed by analysing the advantages and disadvantages of all of the prediction methods. The Analytic Hierarchy Process (AHP) was developed for prediction optimisation.

Figure 6.65(a) and (b) illustrates a realistic application of the GPR method for revealing the geological conditions near test tunnel #2-1 between headrace tunnels #2 and #4. As shown in Figure 6.65(b), four fractures, namely S1, S2, S3 and S4,

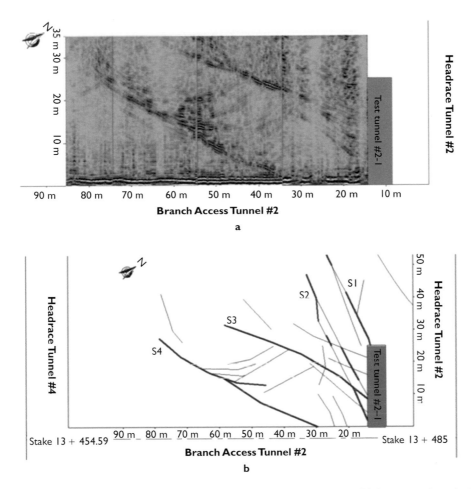

Figure 6.65 Geological assessment in Jinping II tunnel: (a) GPR result; and (b) Interpreted geological condition (HydroChina Huadong Engineering Corporation, 2005).

were identified. This indicates a complicated local geological condition. However, these four geological structures were not exposed during the tunnelling of headrace tunnels #2 and #4 and branch access tunnel #2, where relatively intact rock masses were found. Clearly, local geological variations had an important role in the actual rock mass conditions around test tunnel #2-1. In view of this variation, in the regions 30 m ahead of the tunnel face and in two sidewalls, the GPR and TSP methods were employed to obtain detailed local tunnel rock mass conditions during the process of construction of the Jinping II tunnels. In addition, during the several months after excavation and lining installation, sound wave testing and the GPR method were also used several times in the local tunnel sections in order to observe the changes of the rock mass structures and tunnel lining. Furthermore, permeability testing was also implemented to observe the effects of grouting and the hydrogeological properties of the local geological structures.

Table 6.11 Principal *in situ* 3D rock stress components (Ge & Hou, 2011).

Principal stresses				Elastic modulus (GPa)	Poisson's ratio
	Magnitude (MPa)	Trend (°)	Plunge (°)		
σ_1	63.1	125.6	0.5		
σ_2	57.7	215.5	15.2	68.0	0.11
σ_3	38.6	37.4	74.8		

6.5.2.2 Rock stress variations affecting the three-dimensional stress field

In order to improve the reliability of rock stress measurement, Ge and Hou (2011) developed the Borehole–Wall Stress Relief Method (BWSRM) and used it in 2012 to obtain the values of the three-dimensional rock stress components around the testing tunnel, as shown in Table 6.11 (Ge & Hou, 2011). This was done after all of the headrace tunnels and the water drainage tunnel had been excavated. This stress result provides the values in the middle sections of the Jinping II headrace tunnels, where the overburden is over 2300 m. However, to assess the 3D stress field along the Jingping II tunnels, a comprehensive stress determination method is necessary, which has already been discussed in Section 6.3.2.

6.5.2.3 Local water variations based on prediction in advance

The distribution of fracture and karst water is complex, due to the local variations in the geology and hydrogeological conditions. At the east and west ends of the Jinping II tunnels, karst structures were exposed by excavation, and explored by drilling and TSP testing. Therefore, karst water represents a major groundwater type at both tunnel ends. In the middle of the tunnels, where there is a large overburden, the groundwater distribution was inhomogeneous and concentred in local tunnel regions where water inrush could occur. The groundwater types consisted of both fracture and karst water.

The advance drilling and Tunnel Seismic Prediction (TSP) methods were extensively applied in the prediction of groundwater and hydrogeological structures during the construction of the Jinping II tunnels. Additionally, the distribution of the hydrogeological conditions and groundwater inrush regions exposed by the two access tunnels, which were excavated prior to the excavation of the headrace tunnels, also directed the understanding and assessment of groundwater risk in the headrace tunnels.

6.5.2.4 Mechanical behaviour of the rock mass after excavation and in the long term

In the Jinping II headrace tunnels, several advanced *in situ* testing and monitoring technologies have been employed in order to understand the mechanical behaviour of the rock mass after excavation and in the long term, such as the digital panoramic

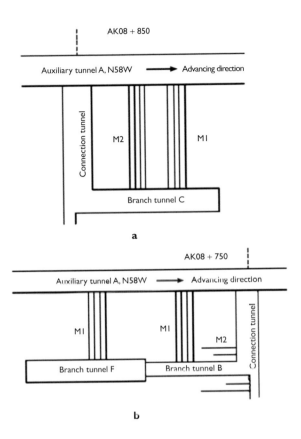

Figure 6.66 Configuration of the access-tunnel testing tunnels: (a) Test zone of branch tunnel C; (b) Test zone of branch tunnels B and F (Li et al., 2012).

borehole camera technique (Li *et al.*, 2010, 2012), acoustic emission method (Chen *et al.*, 2010), real-time microseismic monitoring (Chen *et al.*, 2010, 2011, 2012), sliding micrometer and cross-hole acoustic wave testing (Li *et al.*, 2011).

6.5.2.4.1   Field testing and monitoring

Controlling and establishing the extent and evolution characteristics of the Excavation Damaged Zone (EDZ) are key problems in the design of tunnel excavation and supports. In order to study this problem, several specific test tunnels with different overburden and geometries were excavated from the Jinping II access tunnels and between headrace tunnels #2 and #4. These test tunnels may be divided into two groups: (i) access-tunnel testing tunnels, as shown in Figure 6.66; and (ii) branch-tunnel testing tunnels, named as testing tunnel #2-1, see Figure 6.67. Table 6.12 lists the testing tunnel parameters. Two testing goals were achieved: the excavation of the access-tunnel testing tunnel served as a tool to reveal the evolution of marble fracturing and damage zone using the D&B method; whereas, testing tunnel #2-1 was designed to determine the mechanical behaviour of the rock mass after TBM excavation.

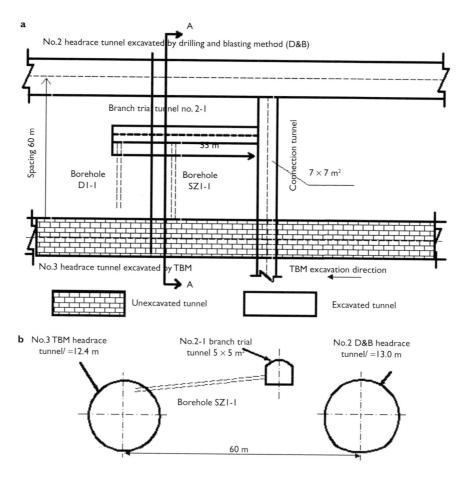

Figure 6.67 Profile seen through two of the four large tunnels showing the location of testing tunnel No. 2-1 and testing boreholes: (a) Configuration of the test scheme, and (b) Profile of the A–A section (Li et al., 2012).

Table 6.12 Testing tunnel parameters shown in Figures 6.68 and 6.69.

Tunnel no.	Cross-sectional size (m)	Length (m)	Overburden (m)	Excavation method	Lithology
Branch tunnel B	5.0 × 5.0	30	2370	D&B (Full-face)	T_{2b} marble
Branch tunnel C	3.0 × 2.0	30	2430	D&B (Full-face)	
Branch tunnel F	7.5 × 8.0	40	2370	D&B (Benches)	
Testing tunnel #2-1	5.0 × 5.0	25	1900	D&B (Full-face)	T_{2y5} marble
Access tunnel A	6.7 × 6.3	17500	0–2375	D&B (Full-face)	Marble, sandstone
Headrace tunnel #2	φ13.0	16660	0–2525	D&B (Benches)	and slate, chlorite schist
Headrace tunnel #3	φ12.4	16670	0–2525	TBM (Full-face)	

In the two testing zones, a digital borehole camera, sliding micrometer, cross-hole acoustic wave equipment and acoustic emission apparatus were adopted, installed and tested in pre-installed facilities and pre-drilled boreholes. The typical properties of the surrounding rock mass, including cracking, deformation, elastic wave velocities and micro-fractures, were measured throughout the entire process of tunnel excavation. The extent and characteristics of the formation and evolution of the EDZs were analysed for the different construction methods involving TBM and D&B, with the test tunnels being excavated by full-face or benches, respectively.

6.5.2.4.2 *Rock mass damage and fracturing evolution*

As an illustration, the results of the damage zone evolution as obtained by a digital borehole camera in testing tunnel #2-1 are shown in Figures 6.68 and 6.69. A comparison of the images taken at different times during the period of TBM driving permitted the study of changes in the EDZ surrounding headrace tunnel #3. These testing results indicated a complicated cracking evolution of the rock mass after tunnel #3 was excavated by the TBM method.

A series of new cracks formed during the excavation ahead and behind the borehole monitoring section. The new crack locations, as identified by the digital borehole image analyses, are shown in Figure 6.69. A small part of the monitoring borehole beyond 32.3 m was removed by TBM excavation. The properties of the new cracks are described as follows.

1   New crack formation

Two new cracks appeared with widths ranging from 1–3 mm at the borehole depth of 29.5–30.0 m, at a distance of 2.2–2.7 m from the sidewall of headrace tunnel #3. The trace lines and timing of crack generation at this location are shown in Figure 6.69.

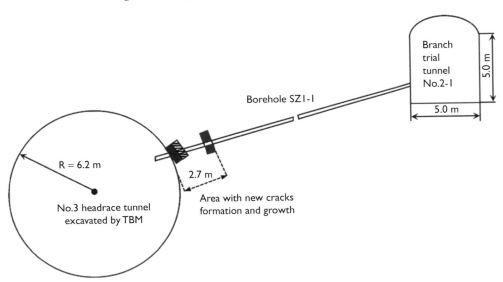

Figure 6.68 Schematic drawing of the location of new cracks based on digital image analysis of the test borehole SZ1-1 (Li *et al.*, 2010).

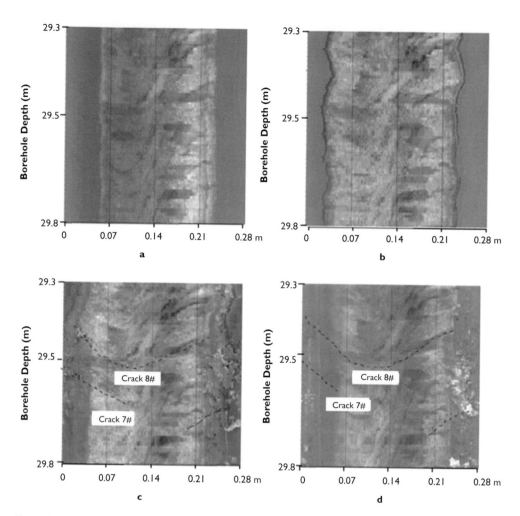

Figure 6.69 Borehole images of fractures in the excavation damaged zone at borehole depth 29.5–29.8 m measured on: (a) seven days before excavation (5th December, 2009), no crack; (b) two days before excavation (10th December, 2009), no crack; (c) three days after excavation (15th December, 2009), cracks observed; (d) 15 days after excavation (27th December, 2009), width of crack reduced and crack closure observed due to disappearance of the left part of crack No.8. (Dashed lines are traces of the cracks) (after Li *et al.*, 2010).

At the borehole depth of 31.1–31.5 m, where the distance from the sidewall of headrace tunnel #3 is 0.7–1.1 m, a new crack appeared with a width of 2.6 mm (Figure 6.70). At borehole depth 31.8–32.3 m, where the distance from the sidewall of headrace tunnel #3 is 0.1–0.4 m, five new cracks appeared with widths of 2–8 mm (Figure 6.71).

2   Changes to pre-existing joint cracks

The monitoring results from the exploratory boreholes indicate that pre-existing intersecting joints changed their width, length and shape during TBM excavation. For

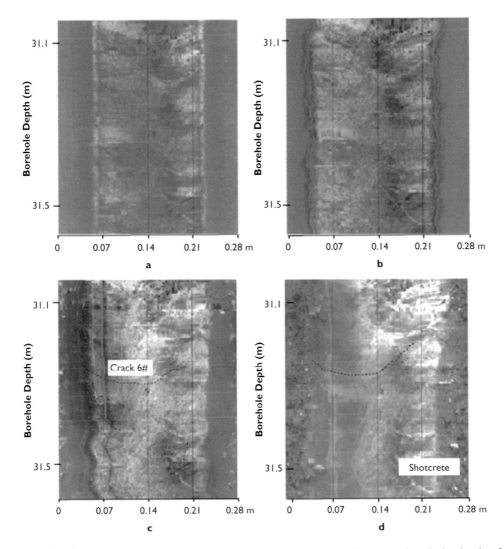

Figure 6.70 Borehole images of fractures in the excavation damaged zone at borehole depth of 31.1–31.5 m measured on: (a) seven days before excavation (5th December, 2009), no crack; (b) two days before excavation (10th December, 2009), no crack; (c) three days after excavation (15th December, 2009), cracks observed; (d) 15 days after excavation and tunnel supported (27th December, 2009), crack closure observed. (The dashed lines are traces of the cracks) (Li *et al.,* 2010).

example, crack No. 5 at a depth of 31.82 m first exhibited increased width, as shown in Figure 6.71(b) and Figure 6.71(c), during the excavation of headrace tunnel #3; however, its width decreased after application of reinforcement, as shown in Figure 6.71d.

3 Crack closure

Two days after the TBM advanced, the tunnel was reinforced by mesh shotcrete and rockbolts. Naturally, supporting the tunnel sidewalls had a considerable effect on

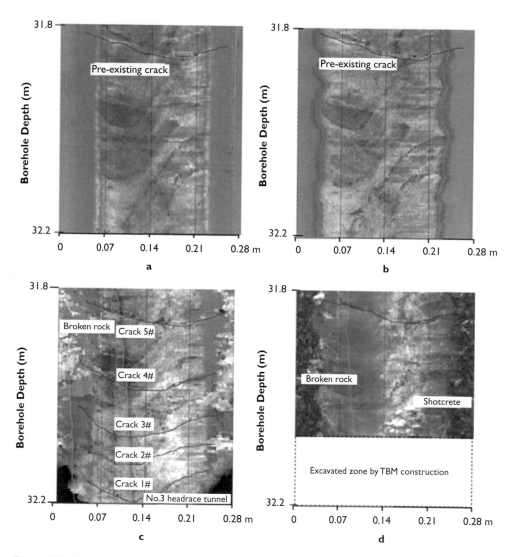

Figure 6.71 Borehole images of fractures in the excavation damaged zone at the borehole depth of 31.8–32.2 m measured on: (a) seven days before excavation (5th December, 2009), no new crack; (b) two days before excavation (10th December, 2009), no new crack; (c) three days after excavation (15th December, 2009), new cracks observed; (d) 15 days after excavation and tunnel supported (27th December, 2009), width of crack reduced or crack closure. (Dark lines are the traces of the cracks) (Li *et al.*, 2010).

the fracturing pattern in the EDZ. After grouting and application of tunnel support, some new cracks gradually closed, such as crack No.3 and crack No.4, shown in Figure 6.71, and crack No.8 shown in Figure 6.69.

Another testing example is the Acoustic Emission (AE) monitoring implemented in testing tunnel #2-1 by Chen *et al.* (2010; 2011). This testing can enable the damage zone evolution during tunnelling to be determined. Before AE testing, several

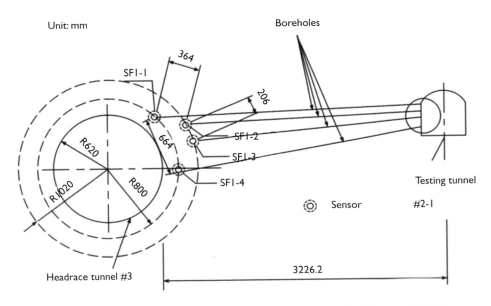

Unit: mm

Boreholes

364

SFI-1

206

R620

664

SFI-2

SFI-3

SFI-4

R1020

R800

Testing tunnel

Sensor

#2-1

Headrace tunnel #3

3226.2

Figure 6.72 Location of AE sensors through pre-drilled boreholes (Chen *et al.* 2010).

pre-drilled boreholes were drilled, in which AE sensors were installed into designated locations, as shown in Figure 6.72. Figure 6.73 shows the AE testing results, which indicate that a large number of AE events were concentrated in the region having a constant distance from the tunnel surface. For the damage zone induced by the excavation of headrace tunnel #3 (12.4 m in diameter), the damage depth is about 9 m, where the severely fractured zone is about 3 m. Similar testing results were also reported by Cheng *et al.*, (2013).

In addition, under the rockburst prone ground conditions in the Jinping II tunnels, having a large overburden over 2000 m, microseismic technologies have also been employed as an early warning system, which will be discussed in Section 6.5.3.2.

6.5.3   Assessment and mitigation of local risk during the construction of the headrace tunnels

6.5.3.1   Water inrush

There are four approaches for handling the potential, high-pressure, large inflows, namely probing, draining, controlling and sealing (Wu *et al.*, 2011). Some overview details of these approaches are as follows.

i   In addition to the theoretical assessment of groundwater inrush, most applicable approaches to evaluate the inflow potential are drilling, probing and advanced geological prediction, which were employed in the process of tunnelling. Probing involves investigating the geological conditions in front of the tunnel face through various forecast measures before excavation. This not only probes for potential

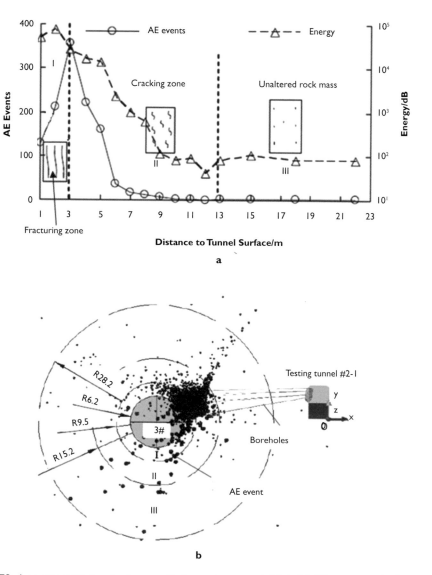

Figure 6.73 Acoustic emission testing results in testing tunnel #2-1: (a) AE events with distance to tunnel surface; and (b) Spatial distribution of AE events (Chen *et al.*, 2010).

water inflow, but also helps to ensure safe excavation. Excavation must occur through the underground water bearing zone for faster tunnelling progress. Figure 6.74 shows a water inrush event, which was successfully explored by advanced drilling from one D&B tunnel face. In tunnels excavated by TBM, rock drills are employed to drill ahead of the TBM, probing for changing geological and hydrological conditions. Information collected in this manner is used to specify pre-excavation rock consolidation and water cut-off grouting programmes, as well as to anticipate near-future rock support measures. It is imperative that any incoming water flow be limited, to allow continued excavation by the TBM.

Figure 6.74 A water inrush case explored by drilling and probing (HydroChina Huadong Engineering Corporation, 2005).

During tunnelling, several geological prediction technologies were adopted to identify the hydrologic conditions, such as Bore-Tunnelling Electrical Ahead Monitoring (BEAM) applied in TBM tunnel sections of headrace tunnel #1 (Gao, 2009), GPR in the D&B tunnels (Su & Fu, 2011; Li *et al.*, 2008), and TSP approaches (Wu *et al.* 2006). These prediction technologies have provided significant amounts of advanced geological information during tunnelling to assess the potential of geological hazards. For example, Li *et al.* (2008) proposed a four-colour warning method of tunnel geological problems, for which corresponding contingency plans are developed based on comprehensive geological prediction and optimisation. This warning method has been applied successfully in many projects, including the Jinping II tunnels, to avoid casualties and serious economic losses.

ii   Draining involves diverting a large water inflow from the working site through a special drainage channel, such as the drainage tunnel. This plan specifies dewatering the mountain by draining water into the 7.2 m drainage tunnel, which was excavated by TBM before excavation of the headrace tunnels. The construction design, including the TBM design, allows for large volumes of water to be drained through the bored headrace tunnels as they are excavated, minimising the impact on excavation logistics and TBM operations. Figure 6.75 shows the draining system at the east end of the Jinping II tunnels. Several cross tunnels were excavated after the excavation of the headrace tunnels, through which groundwater flowed into the drainage tunnel and thence into the Yalong River.

iii The control approach makes it easier to either stop inflow or allow the water in. This is a necessary preparation for handling the inflow. If control is lost, it will be time consuming to seal a large inflow, thus resulting in construction delays. This system should be organised as soon as possible after bypassing a large water inflow. The concept of this control step is to give the constructor the ability to

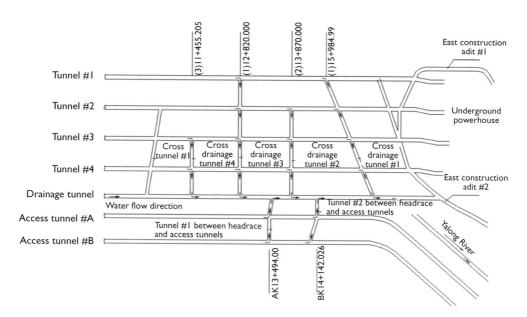

Figure 6.75 East end layout of drainage treatment in Jinping II tunnels (HydroChina Huadong Engineering Corporation, 2005).

control the rate at which the groundwater is drained into the tunnel, from every point in the excavated tunnel. In this way, it is hoped that the water can be allowed to flow into the bored tunnel at the maximum allowable volume rate which will allow continued TBM operations. Ideally, if successful, the system should permit the constructor to drain where and when necessary to maintain operations. Of course, this will require high quality water cut-off grouting, drain pipes and valves.

iv   It is also important to seal the water inflow when it is not needed, in order to allow drainage of the underground water at the inflow locations.

There are several successful control cases of very large groundwater inflow, to which different control approaches have been applied.

Case 1: High pressure inflow at lower sidewall near AK14+760.
On 30th March 2005, a very large inflow was encountered in the lower left side-wall near chainage AK14+760, and its instantaneous flow rate was about 5–7 m³/s, which decreased to 2.68 m³/s after one week. The inflow point located in T_{2y5} was grey and white marble. There are two group structures which were observed in the rock masses near the point: (a) one group with an orientation of N85°E, NW∠55°; and (b) the other group with an orientation of N70°~85°E, NW∠65°~75°. A water outlet was found, which was a relatively large fracture, about 1.3 m in width, 1.8 m in height and 4.2 m in depth. Figure 6.76 shows a photo of this inflow case and its groundwater inrush condition. To control the inflow, one method consisting of the

Figure 6.76 High pressure flow point at lower sidewall near AK14+760 in Access Tunnel #A (HydroChina Huadong Engineering Corporation, 2005).

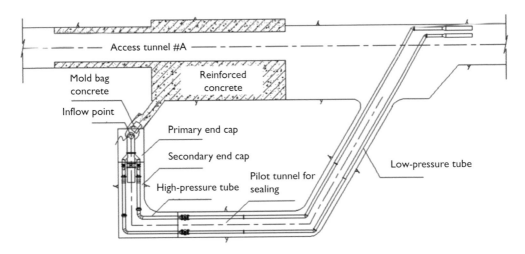

Figure 6.77 Layout of drainage tunnel and sealing scheme for the high pressure flow point in the lower sidewall near AK14+760 in Access Tunnel #A (HydroChina Huadong Engineering Corporation, 2005).

excavation of pilot tunnel and inflow sealing was applied, as shown in Figure 6.77. This method utilised the high-pressure steel tubes set up in an excavated pilot tunnel to drain the water into the drainage channel. Figure 6.78 is one of the field photos, and shows the successful control effect based on the method illustrated in Figure 6.77.

Figure 6.78 Treatment for the high pressure flow point in the lower sidewall near AK14+760 in Access Tunnel #A (HydroChina Huadong Engineering Corporation, 2005).

Case 2: High pressure inflow in the lower sidewall near AK10+612.
The tunnel section near chainage AK10+612 was the second water inrush zone, and its inflow condition is shown in Figure 6.79(a). The flow gushed into access tunnel #A from the hydrogeological structures and karst channels exposed on the north tunnel crown, and the water inflow was about 1.5 m³/s. An approach, namely the distributary and decompression hole approach, was applied to this case, and it worked efficiently. As shown in Figures 6.79(b) and (c), many decompression holes were drilled in the crown of the tunnel, these holes crossing through the hydrogeological structures. The water pressure in the hydrogeological structures was decreased, then high-pressure grouting was completed, which sealed the groundwater channels.

Case 3: Heavy inflow exposed on the tunnel floor.
Under the condition that a large water inflow was exposed on the tunnel floor, an approach called the 'caisson and sealing approach' was proposed and applied by designers from the HydroChina Huadong Engineering Corporation. This approach can solve the problem of sealing high-pressure, highly concentrated and large inflows. Figure 6.80 illustrates a design case of this caisson and sealing approach. The cover weight concrete can truncate the inflow into tunnels and limit the water so that it can only be drained into high pressure tubes. This approach has been successfully employed at the tunnel section chainage K13+785–13+836 m in headrace tunnel #3.

a   b

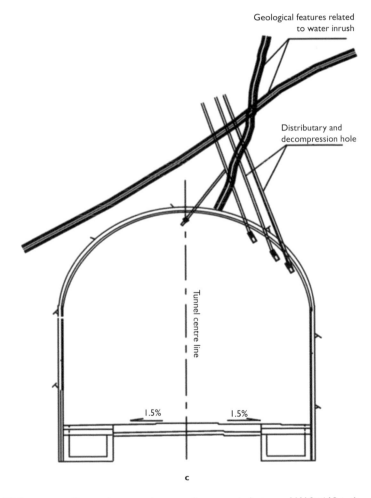

Geological features related
to water inrush

Distributary and
decompression hole

Tunnel centre line

1.5%   1.5%

c

Figure 6.79 High pressure flow point near the tunnel crown at chainage AK10+612 in Access Tunnel #A:
(a) before treatment; (b) after treatment; and (c) schematic of the approach termed "distributary and decompression hole" (HydroChina Huadong Engineering Corporation, 2005).

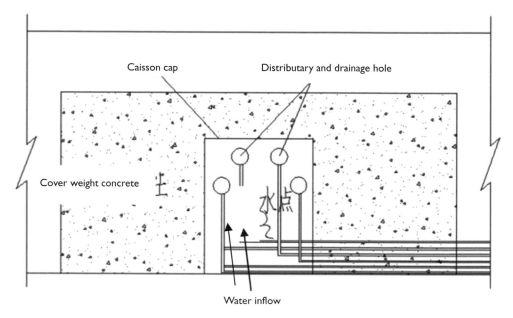

Figure 6.80 Schematic of the caisson and sealing approach (HydroChina Huadong Engineering Corporation, 2005).

6.5.3.2   Rockburst: monitoring, in situ tests, warning and mitigation

6.5.3.2.1   Field Monitoring

Deep rock excavation under the Jinping II tunnel conditions will induce rockbursts and key factors associated with the excavation methods (TBM or D&B), cross-section size and shape, advance rate, etc., have a considerable influence on the rockburst evolution process. Therefore, *in situ* tests and monitoring during the tunnelling processes are helpful in investigating and in understanding the mechanisms of the rockburst evolution processes, as well as the associated assessment of rockburst risk. Such *in situ* tests and monitoring may consist of microseismic or AE monitoring, extensometer deformation testing, the use of optical fibre and sliding micrometer instruments, sonic wave measuring, borehole ground penetrating radar, and the use of a digital panoramic borehole camera.

Table 6.13 gives the monitoring/measuring objective of these *in situ* tests and monitoring items. In practice, for the Jinping II tunnels, a series of *in situ* test results obtained by digital borehole cameras, including deformation and wave velocity monitoring (Li *et al.*, 2011, 2012a, 2012b), indicated the following: (i) the existing joints opened, closed and propagated during the evolution of rockbursts; (ii) new cracks were generated, then opened, closed, propagated and connected during the evolution of the rockbursts; (iii) the deformation of the rock mass was small before rockburst occurrence; and (iv) the wave velocity in the rock mass in the Excavation Damaged Zone (EDZ)

Table 6.13 *In situ* tests and monitoring items adopted for the rockburst prone tunnel sections in the Jinping II tunnels.

Testing/monitoring approaches	Monitoring or measuring objective
Microseismic; AE	Locations of cracking events; Released Energy; Microseismic evolution
Extensometer; Optical fibre; Sliding micrometer	Deformation behaviour
Sonic wave measuring; Borehole GPR	Geological structure identification; Rock mass integrity; Damage zone evolution
Digital panoramic borehole camera	Evolution of micro-cracking and damage zone; Identification of rock mass structure altered by high stresses

between two boreholes was reduced during the evolution of a rockburst event. These observations are helpful for understanding the mechanism of rockbursts occurring in the Jinping II tunnels. Moreover, these testing results led to a novel support concept and new design methods, namely the 'cracking-restraining' method and 'cracking, restraining plus energy-absorbing' method. The results emphasise the importance of controlling the process of rock mass cracking and fracturing by means of a combination of systematic bolts and shotcrete plus the application of excavation design.

In view of the objective of assessing the rockburst risk, a real-time warning system is also needed to predict the intensity of rockbursts and potential rockburst failure depth. The real-time microseismic monitoring system can be used as this warning system, which has been applied successfully to warning and preventing rockbursts in the Jinping II headrace tunnels. The implementation of this approach and its necessity depend on the preliminary assessment of rockburst risks and the rockburst knowledge obtained from the two access tunnels and the drainage tunnel, which show that the occurrences of severe or extremely severe rockbursts were related to the uncertainty of local tunnel engineering conditions, such as local geological structures, local *in situ* stress characteristics, special excavation actions and local support conditions. The microseismic monitoring approach can provide a helpful tool to identify and observe the responses of rock masses under the above-mentioned local conditions, and also can be used as an early-warning tool, as it is more accurate than the empirical and numerical assessment of rockburst hazards. Also, the determination of the temporal fractal, energy fractal and spatial fractal dimensions of microseismicity (see Figure 6.81) indicate that microseismicity measured during the evolution process of rockbursts can be used to predict the final intensity of the rockburst.

The microseismic events were analysed using the moment tensor method and the wave analysis method, which revealed that there were differences in terms of mechanism between the rockbursts of different types (Feng *et al.*, 2012). For example, (i) the 'immediate strain rockburst' consists mostly of tensile cracking with a few shearing cracks its evolution, Figure 6.82(a); (ii) the 'immediate strain–structure slip' rockburst mainly consists of tensile cracking with several shear and mixed cracks during its evolution, Figure 6.82(b); (iii) the 'time-delayed' rockburst also consists of tensile cracks, shear cracks and mixed cracks during its evolution, Figure 6.82(c).

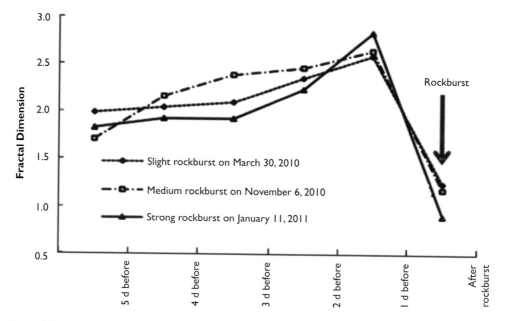

Figure 6.81 Temporal fractal behaviour of the rockburst evolution process having different intensities (Feng et al., 2012).

6.5.3.2.2   Rockburst risk assessment based on microseismic monitoring

The real-time monitoring results for microseismicity in deep tunnels indicate the different evolution processes of the experienced rockbursts.

i   From the initiation to the full occurrence of the 'immediate strain' rockburst, the microseismicity is active. The microseismic energy index increases at the beginning, decreases, and then increases again before the occurrence of the rockburst. The cumulative apparent volume continues to increase during the short period of temporary stability (Figure 6.83(a)) (Chen et al., 2011, 2012).

ii   From the initiation to the full occurrence of the 'time-delayed' rockburst, the initial microseismicity is the same as that in the immediate rockburst, but there is a quiet period of several days to several tens of days before the main occurrence of the rockburst (Figure 6.83(b)).

Based on the microseismic information, as shown in Figure 6.83, the intensity potential of different rockbursts can be evaluated by an assessment formula, i.e., Equation 6.13,

$$P_i = \sum_{j=1}^{6} w_j P_{ji} \qquad (6.13)$$

where i is the rockburst intensity (extremely intensive, intensive, moderate, weak, none); j is the microseismic information (number of events, energy, apparent volume, events rate, energy rate, apparent volume rate); P_{ji} is the probability distribution

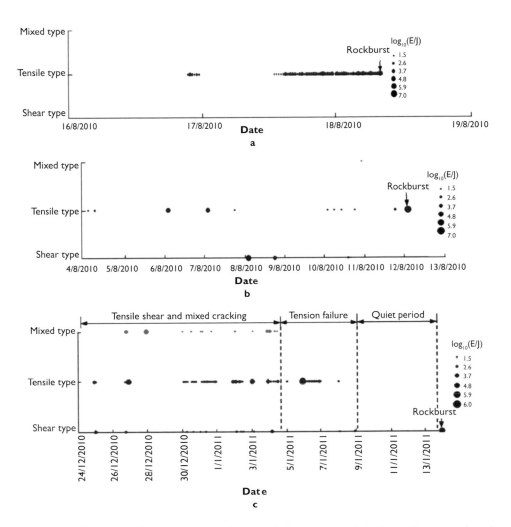

Figure 6.82 Difference in the mechanisms of the evolution process of (a) 'immediate strain' rock-burst, (b) 'immediate strain–structure slip' rockburst, and (c) 'time-delayed' rockburst (Feng *et al.,* 2012).

function, based on microseismic information *j*; and w_j is the weight coefficient of the probability distribution function, which is obtained through the search by Particle Swarm Optimisation (PSO). Another method is to establish a neural network model to describe the real-time monitored microseismicity with the intensity and failure depth of the rockbursts. The microseismicity evolution-based neural network model can be used to predict the intensity and failure depth of the rockburst in the vicinity of a given zone, as described by Chen (2012).

Figure 6.84 shows a case where the rockburst risk assessment was conducted. A comparison of the prediction with actual occurrence of rockbursts in the given zone of tunnel #3 (TBM) indicated that the prediction was highly accurate. This signifies that the warning system for the approach of microseismicity was effective in practice.

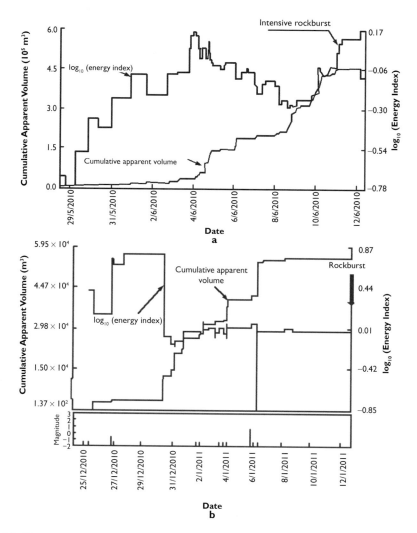

Figure 6.83 Change of microseismicity during the rockburst evolution processes of (a) 'immediate' rockburst and (b) 'time-delayed' rockburst (Feng et al., 2012).

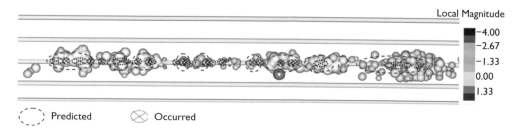

Figure 6.84 Comparison of the predicted rockburst zones and the actual rockburst occurrence zones (Feng et al., 2012). The upper four horizontal bands are the four headrace tunnels at the Jinping II site. See the colour plate section at the end of the book.

Therefore, this method, as a major warning tool, was applied at the sections in the Jinping II tunnels which had large overburdens.

6.5.3.2.3   Rockburst warning in the headrace tunnels based on microseismic information

The real-time microseismic monitoring and rockburst warning were implemented at the four headrace tunnels and the drainage tunnel. During the continuous microseismic monitoring, the accuracy of rockburst warning was about 88%, as shown in Figure 6.85. A large number of microseismic events was identified and located during tunnelling. The rockburst potential and its intensity was also predicted, which provided an early-warning and a basis for the assessment of rockburst risk.

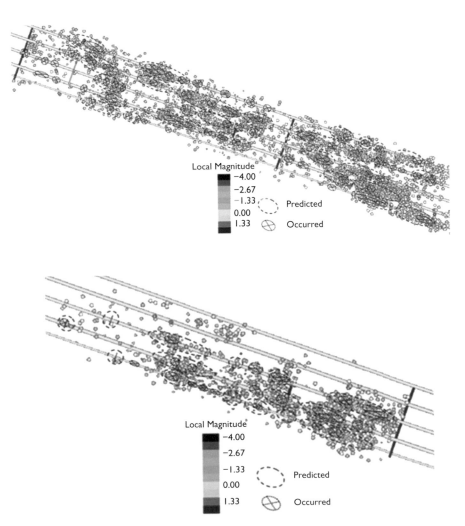

Figure 6.85 Comparison of the predicted zones and actual occurrence zones of rockbursts in the Jinping II tunnels (Feng *et al.,* 2012).

6.5.3.2.4 Mitigation and dynamic control of rockburst risks

During excavation, the monitored microseismic information can be used as guidance to mitigate the effects of rockbursts. In particular, the energies of microseismicity for different intensities of rockbursts can be identified (see Table 6.14). The actual rockburst risk is evaluated and, if necessary, the excavation and support must be checked to determine whether or not they meet the rockburst resistant design. The revealed geological conditions can be updated as inputs to the rockburst risk assessment methods listed in the Section above, in order to calibrate the intensity of the rockburst. If the intensity of the predicted rockburst is different, the excavation and support design are modified accordingly.

In fact, the mitigation of rockburst risks in the Jinping II tunnels included two methods: (i) the strategic method, and (ii) the tactial method. Table 6.15 gives the

Table 6.14 Energy levels of micro-seismicity for different intensities of rockbursts at the Jinping II project site (Feng et al., 2012).

Intensity of rockburst	Common logarithm of energy monitored by the microseismic technique E (J)
No rockburst	$(-\infty, 0)$
Slight rockburst	$(0, 2)$
Moderate rockburst	$(2, 4)$
Intensive rockburst	$(4, 7)$
Extremely intensive rockburst	$(7, +\infty)$

Table 6.15 Strategic and tactical treatment methods for different intensities of rockbursts at the Jinping II project site (Feng et al., 2013).

Class	Geological survey countermeasures	Excavation countermeasures	Support countermeasures

Extremely intensive rock-burst	1 To survey the occurrence of faults and structural planes, and to study the space relations between the occurrence and the excavation
	2 To investigate local geological anomalies (e.g., anticline, including its core and flanks)
		1 Size optimisation:
		  a Optimise section shape and size
		  b Change the bidirectional tunnelling layout to unidirectional before this tunnel excavation system is fixed
		2 Special methods:
		  a Stress releasing hole
		  b Control stress release on structures
		  c Optimise pilot tunnel layout
		  d Real-time microseismic monitoring and pre-warning
			1 Support type and its parameter (The design value of absorbed energy is about 50 kJ/m^2):
			  a Conduct timely shotcrete, absorbing energy 10.9 kJ/m^2
			  b Set up systematic energy-absorbing bolts with steel plate, absorbing energy about 39.1 kJ/m^2. Rockbolts should intersect the plane of controlling structure at a large-angle
			  c Suspend nets, set up steel arches, conduct second shotcreting
			2 Support process of D&B: Conduct timely shotcrete; set up systematic energy-absorbing bolts; suspend nets; set up steel arches, conduct second shotcreting

(Continued)

Table 6.15 (Continued)

Class	Geological survey countermeasures	Excavation countermeasures	Support countermeasures
			3 Support technique process of TBM: In range L1, conduct timely shotcrete; suspend nets; set up systematic energy-absorbing bolts; set up steel arches; in range L2, conduct second shotcreting
Major rock-burst			1 Support type and its parameter (The design value of absorbed energy is about 22–50 kJ/m²) 2 Conduct timely shotcrete, absorbing energy 10.9 kJ/m²; Set up systemic energy-absorbing bolts to absorb energy about 11.1–39.1 kJ/m²; Other parameters are the same as those of the extremely intensive rockburst
Moderate rock-burst		1 Size optimisation Optimise the excavation plan 2 Excavation methods (D&B or TBM), section shape, section size are all consistent with the corresponding parameters under the condition of non-rockburst sections 3 Real-time microseismic monitoring and pre-warning	1 The design values for bearing shock energy is in range of 13–22 kJ/m² 2 Conduct timely shotcrete, absorbing energy 4.7–10.9 kJ/m²; Set up systematic energy—absorbing bolts with steel plate, absorbing energy about 8.3–17.3 kJ/m²
Minor rock-burst		Excavation rate, excavation methods (D&B or TBM), section shape, section size are all consistent with the corresponding parameters under the condition of the non-rockburst sections	1 The design value for bearing shock energy is less than 13 kJ/m² 2 Conduct timely shotcrete 3 Set up 'random' bolts with steel plate in the concentration zone of stress or in the section of high rockburst risk

detailed countermeasures used in both the strategic and tactical methods, and also lists the workflow for the support and its major function. To minimise rockburst risk, the reasonableness of a 60 m distance between each pair of headrace tunnels was validated. For the D&B tunnels, the favourable upper-bench excavation heights were continuously adjusted to take account of the geological conditions and rockburst risks. For the TBM tunnels, when the excavation was conducted in tunnel sections with a high overburden and which were highly prone to rockbursts, the pilot tunnels were excavated before the TBM excavation of the main tunnel. The pilot tunnel excavation strategy to significantly reduce the risk of rockburst by TBM excavation was adopted in headrace tunnels #1 and #3 (see Feng *et al.*, 2012b). Figure 6.24 (in Section 6.4.1.3) shows the advantages and effects of reducing the rockburst risk after this excavation strategy was adopted. As illustrated in Figures 6.24(b) and (c), the micro-seismic events and their seismic energies were remarkably controlled. The high intensity rockbursts, were also decreased. In addition, if two tunnel faces are excavated face to face, adjustment of the tunnel strategy may be necessary to pause the excavation of one tunnel face according to the released energy analyses (Feng *et al.*, 2012b).

The tactical method mainly refers to the use of rockburst supports, i.e., the system of bolt-shotcrete support in the Jinping II tunnels. The principles for the support conditions include three items (Feng *et al.*, 2013): (i) the cracking-restraining philosophy; (ii) the control of the energy released; and (iii) the absorption of released energy. Most importantly, these principles are combined with the excavation progress by means of the dynamic support design concept, the support parameters being adjusted according to the assessment of rockburst risks. As listed in Table 6.16, on the basis of rockburst classifications in the Jinping II tunnels, variable schemes for different support elements may be adopted so that different magnitudes of absorbed energy may be obtained. The workflows for these support elements follow those listed in Table 6.15.

Thus, an optimisation concept has been applied to the support designs of the Jinping II tunnels, with a large number of numerical assessments for the support programmes, many comprehensive evaluations of the rockburst support capabilities, and relatively accurate characterisations of the damage zones. Table 6.17 gives the optimised bolt parameters, which were calculated using Equation 6.14 (Feng *et al.*, 2013):

$$L_{ab} = D_f + L_e \tag{6.14}$$

where L_{ab} is the designed length of the bolt (m); D_f is the effective anchoring length of the bolt (m), the part in the range of FAI > 1; and L_e is the predicted maximal rockburst damage depth (m) calculated by the RVI method. Figure 6.86 shows a field support case for headrace tunnel #3.

According to a large number of support practices, it can be established that, if the shock energy of a rockburst is greater than 50 kJ/m², then the excavation optimisation strategy or other strategic methods should be adopted to reduce the released energy. This is due to the fact that the energy level, i.e., 50 kJ/m², is the upper limit of shock energy controlled by the system of rockbolt-shotcrete support.

In summary, based on the assessment and mitigation of the rockbursts mentioned above, in the Jinping II tunnels, the rockburst warning and predictions were conducted in 241 tunnel sections, which are about 7605 m in length. The avoidance of major rockburst events was achieved in 135 tunnel sections which were assessed as severe or extremely severe rockburst risks, and about 4082 m in length. The intensities and the associated risks were decreased in another 13 tunnel sections, which are about 418 m in length.

6.5.3.3 Large deformation: monitoring and treatment

To control the deformation of chloritic schist at the west end of the Jinping II headrace tunnels, an excavation sequence with two benches, upper and lower benches, was implemented. At the design stage, the support system consisted of steel ribs,

Table 6.16 Support parameters for different intensities of rockbursts at the Jinping II project site (Feng et al., 2013).

Rockburst classification		Minor (I_b)	Moderate (II_b)	Major (III_b)	Extremely intensive (IV_b)
First shotcrete	Concrete type	Steel fibre	Steel fibre	Steel fibre	Steel fibre
	Thickness (mm)	10	10–15	15	15
Rockbolt	Type	Absorbing energy	Absorbing energy	Absorbing energy	Absorbing energy
	Length (m)	3.0	3.5	4.5	>4.5
	Setting	Local, random	Local, random	Systematic	Systematic
	Spacing (m × m)	1.0–2.0 × 1.0–2.0	1.0–2.0 × 1.0–2.0	0.5–1.0 × 0.5–1.0	0.5 × 0.5
Absorbed energy (kJ/m²)		8–22	13–22	22–50	50
Steel suspended nets	Diameter (mm)	8	8	8	8
	Mesh size (mm × mm)	200 × 200	200 × 200	150 × 150	150 × 150
Rockbolt	Type	Permanent	Permanent	Permanent	Permanent
	Length (m)	Determined by numerical analysis			
	Setting	Systematic	Systematic	Systematic	Systematic
	Space (m × m)	1.0–2.0 × 1.0–2.0	1.0–2.0 × 1.0–2.0	1.0–2.0 × 1.0–2.0	1.0–2.0 × 1.0–2.0
Second shotcrete	Concrete type	Steel fibre	Steel fibre	Steel fibre	Steel fibre
	Thickness (mm)	20	10–15	15	15
Absorbed energy (kJ/m²)		19–35	24–35	33–50	50

Note: Total of absorbed energy cannot be greater than 50 kJ/m².

Table 6.17 Rockburst support parameter optimisation at Jinping II project site (Feng et al., 2013).

Classification of rock mass	Rockbolt type	Suggested length (m)	Designed length (m)
IIIb (Intensive rockburst)	Grout rockbolt (permanent)	6.0	6.0
	Expansion shell anchored rockbolt (permanent)	6.0	6.0
	Swelling dowel bolt (temporary, absorbing-energy)	4.5	5.0
IVb (Extremely intensive rockburst)	Grout rockbolt (permanent)	6.0/9.0	6.0/9.0
	Expansion shell anchored rockbolt (permanent)	6.0/9.0	6.0/9.0
	Swelling dowel bolt (temporary, absorbing-energy)	6.0	5.0

Figure 6.86 Example of support in headrace tunnel #3.

shotcrete and rockbolts, which was insufficient to control the deformation of the chloritic schist. This support system is shown in Figure 6.87, in which large deformation led to the failure of the steel ribs at the location of the foot arch, as shown in Figure 6.87(b). The distributions of compressive deformation based on empirical statistical data show that excavation in the chloritic schist formation can generate large compressive deformations. The maximum deformation in many sections (after the primary support) reached 0.5–0.7 m, which is 7.6–10.6% of the tunnel radius. For example, tunnels #1 and #2 were excavated by the D&B method and supported by reinforced concrete lining. After the excavation and introduction of the support for the tunnel upper sections in the chloritic schist stratum, the stresses in the anchors were high, with some of the anchors at their design strength. Large convergence occurred with consequential encroachment into the excavation space. This indicated that a new support system was required because there was insufficient support in the original design. Insufficient pre-support and/or delay of support structure may not efficiently control the development of the plastic zone, resulting in large deformation and even collapse.

Accordingly, two design schemes were proposed to control the deformation of the upper bench section before excavating the lower bench, as shown by Figure 6.88. New designs for controlling the deformation of the chloritic schist are based on numerical analyses. The two schemes adopted were an anchor bar pile in advance at the arch foot (ϕ 32 mm, L = 9 m and @ 1.0 m) and pre-stressed anchors at the spandrel (ϕ 32 mm, L = 9 m, T = 150 kN and @ 1.5 m), where ϕ, L, T and @ denote the diameter, length, pre-stress and spacing of anchors, respectively. Scheme 1 used a

Figure 6.87 Large deformation and re-mining process after large convergence of rock masses in the chloritic schist stratum: (a) re-mining process; (b) failure of foot steel ribs due to the large deformation; (c) pre-stressed bolts to reinforce arch feet.

downward pre-stressed anchor cable at the side wall of the central section (L = 15 m, T = 1000 kN and @ = 3.0 m), while Scheme 2 used a horizontal pre-stressed anchor cable at the side wall of the central section (L = 15 m, T = 1000 kN and @ = 3.0 m). Note that this support system consisted mainly of pre-stressed anchors or bolts, anchor bar pile, steel ribs with deformable elements, and fibre shotcrete, which is a heavy support system for large deformations. This enabled moderate convergence in the over-excavation spaces.

The prediction of deformation for the two support schemes was obtained using FLAC3D models. The predicted deformations of the tunnel section after excavating the invert section with two reinforcement schemes are shown in Figure 6.89 and Table 6.18. The deformation of the surrounding rocks was 50–150 mm after the excavation of the tunnel invert, with the distribution of deformation being asymmetrical but following a clear shape. The largest displacement occurs at the lower section of the sidewall, and significant upheaval is observed at the sidewall at the right invert and floor, so that immediate support is necessary to form a closed-loop system. The deformations of the middle and top sections are less significant than that of the invert

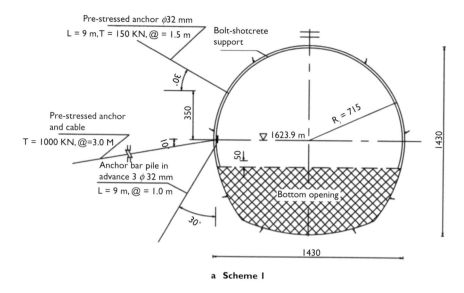

a   Scheme I

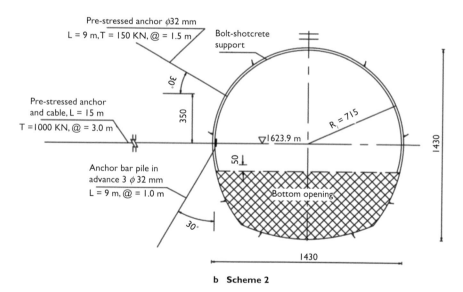

b   Scheme 2

Figure 6.88 Layout of reinforcement schemes for the tunnel invert (Zhou et al., 2014).

section. The advanced anchor bar pile at the arch foot is helpful for the stability of the surrounding rocks. In this case, the lower section of the sidewall can be reinforced, and it plays an important role in the connection and support of the upper section of the arch. A suitable anchor bar pile spacing is about 1.0 m. Having grouting pressure at the zone of the cracked surrounding rock mass, the pre-support strip at the arch

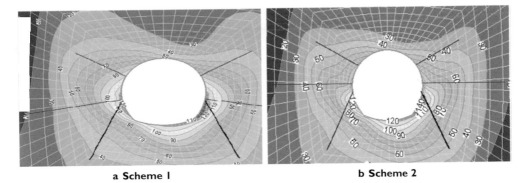

a Scheme 1   b Scheme 2

Figure 6.89 Deformation distributions for two reinforcement schemes (units: mm, Zhou *et al.*, 2014).

Table 6.18 Predicted deformations of the tunnel section after excavating the invert section with two reinforcement schemes (Zhou *et al.*, 2014).

| | Maximum deformation and position | | | |
| | Zone with extremely severe compressive deformation less than 500 mm | | Zone with extremely severe compressive deformation more than 500 mm | |
Scheme	Above mid part of sidewall	Below mid part of sidewall	Above mid part of sidewall	Below mid part of sidewall
1	100 mm, mid part of the two sidewalls	120 mm, mid part of the right arch foot	110 mm, mid part of the two sidewalls	140 mm, mid part of the right arch foot
2	50 mm, mid part of the two sidewalls	100 mm, mid part of the right sidewall	90 mm, mid part of the two sidewalls	140 mm, mid part of the right sidewall

foot ensures the stability of the temporary sidewall at the invert. The pre-stressed anchor and cable decrease the upward displacement induced by the excavation of the tunnel invert section. The predicted results show that scheme 2 with a horizontal arch cable is better than scheme 1. Therefore, scheme 2 was used to reinforce the top section at the construction site (see Figure 6.90). The observed results after excavating the tunnel invert section reveal that the deformation of the surrounding rock mass can be controlled within an acceptable level when using scheme 2.

In addition to the application of the new support system described above, the over-excavation and re-mining methods were also used to integrate with the support system. For non-excavated chloritic schist in Class III, a diameter of 14.6 m was excavated and associated with a lining of 0.6–0.8 m in thickness, while the 15 m diameter with lining of 0.8 m thickness was used for Class IV. Note that the original design diameter was 13 m. One re-mining task was implemented in the excavated and deformed rock masses, as shown in Figure 6.87(b).

a   b

Figure 6.90 Scheme 2 at the construction site. (a) Anchor steel-stake in advance at arch foot. (b) Pre-stressed anchor and cable at mid part of the sidewall (Zhou *et al.*, 2014).

6.5.3.4   Long term stability

6.5.3.4.1   Long term stability assessment in the Jingping II headrace tunnels

The #1 headrace tunnel was broken through in June 2011. The # 2, 3 and 4 headrace tunnels and the water drainage tunnel were broken through before the end of November 2011. The #1 headrace tunnel began to transfer water to generate electricity in December 2012, and the # 2 headrace tunnel in 2013. It is expected that #3 and #4 will begin to transfer water in 2014.

For the marble tunnel sections, the complex geological conditions, including high stress induced by the great depth and high seepage pressure, led to many engineering problems, especially the long-term stability problem related to the brittle marble. In order to reflect the character of the visco-plastic deformation, Chen *et al.* (2007) proposed a combined model of the cohesion weakening-friction strengthening model and the visco-elastic model. Based on the *in situ* displacement monitoring data, a visco-elasto-plastic model and its parameters can be obtained by the intelligent method based on the GP and PSO methods. The rheological mechanical characters of the surrounding rock mass in the headrace tunnels are simulated using FLAC3D. Through analysis of the calculated results, the long-term stability of the tunnels is estimated by using the integrating multi-index method including displacement, displacement velocity, stress, failure approach index, and internal force/stress of the support structures. Wu *et al.* (2005) analysed the deformation, stress distribution and plastic zone of the surrounding rock mass during the processes of excavation and support using an elasto-plastic FEM. The influence of the different seepage control schemes on the working behaviour of the liner and surrounding rock were studied.

Similar analyses of the marble's long term stability around the Jinping II tunnels were completed by Wu *et al.* (2013). Liu *et al.* (2011) simulated the time-dependent failure of the Jinping marble by the Parallel-bonded Stress Corrosion (PSC) model

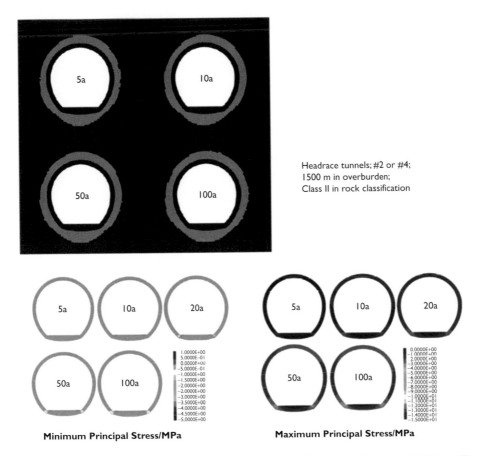

Figure 6.91 Minimum and maximum principal stresses in the lining at a depth of 1,500 m (Zhang et al., 2013).

in PFC. Using the same method suggested by Liu *et al.* (2011), Zhang *et al.* (2013) assessed the long-term fracturing characteristic and liner pressures after 100 years, as shown in Figure 6.91. The microscopic behaviour in PFC can help us to understand the brittle rock strength over time. Further studies have revealed the time-dependent behaviour of the deep buried marbles in the Jinping II tunnels, i.e., creep tests conducted by Zhao *et al.* (2012), Zhang *et al.* (2011; 2013) and Liu *et al.* (2013), and the field AE tests completed by Chen *et al.* (2010) and Cheng *et al.* (2013).

For the chloritic schist tunnel sections, and in order to evaluate their long-term stability, laboratory and field tests were conducted to evaluate the deformation behaviour, as shown in Figure 6.92. The creep behaviour is clear when the stress is over 30 MPa, where accelerating creep processes are observed when close to the long-term strength. The rheology phenomenon was also found in the results obtained from convergence monitoring. In Figure 6.92(b), the monitoring was conducted at a tunnel cross-section at chainage 1+715 in tunnel #1. As shown in Figure 6.92(b), at 100 days after the excavation of the lower bench, the maximum displacement, which occurred

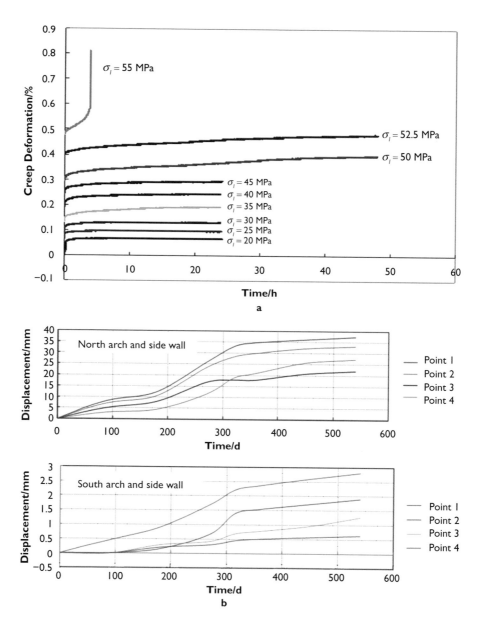

Figure 6.92 Laboratory and field testing results for the Jinping chloritic schist: (a) Creep testing; (b) Field displacement monitoring at Chainage 1+715 in headrace tunnel #1.

at the location of the north arch, reached up to 9 mm, then 12.3 mm at 180 days. At 320 days after the second excavation of the lower bench, the displacement was nearly 32.9 mm, then 37.4 mm at 540 days. Afterwards, the displacement value tended to remain unchanged. Similar phenomena were also observed at the south arch and side wall, but the displacement values were very small, only up to 2.8 mm.

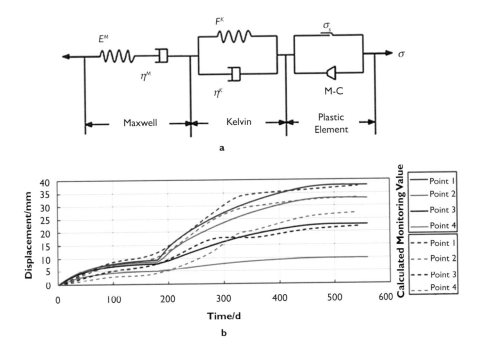

Figure 6.93 The model used to analyse the rheological behaviour of chlorite schist: (a) CVISC model; and (b) comparison between the monitoring and calculated displacements (Zhou et al., 2014).

Based on the testing and monitoring results, the CVISC model (Hooke–Kelvin rheology model) in FLAC3D, shown in Figure 6.93(a), was employed to quantify the chlorite schist behaviour with time, as well as to assess the design schemes for the concrete lining and its load-bearing capability. Figure 6.93(b) provides a comparison between the monitoring and calculated displacements by means of the CVISC model, in which back-analysed rock mass parameters were used on the basis of the monitoring displacement at chainage 1+715 in tunnel #1. Following the identification of the model parameters, the loads on the lining structure were analysed, as shown in Figures 6.94(a) and (b). This method was used to direct the design of the lining parameters. When considering the groundwater, the method used to assess the long-term stabilities of rock masses and the lining will be discussed in detail in the following section.

6.5.3.4.2   Treatment strategies to ensure long term stability of the Jingping II headrace tunnels

Considering the long-term issues of the Jinping II tunnels, regardless of the hard or soft rock type, the rock support designs give attention to two main aspects, namely the short- and long-term stability. Therefore, the concept of combining temporary and permanent support elements was implemented during tunnel construction, i.e., the combination of three rockbolt types (swelling dowel bolt, expansion shell

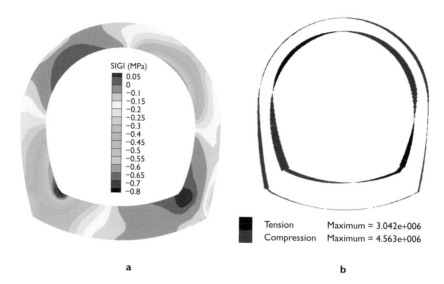

SIGI (MPa)
0.05
0
−0.1
−0.15
−0.2
−0.25
−0.3
−0.4
−0.45
−0.5
−0.55
−0.6
−0.65
−0.7
−0.8

Tension   Maximum = 3.042e+006
Compression   Maximum = 4.563e+006

a

b

Figure 6.94 Loads on the concrete lining: (a) Stress distribution; and (b) Stress on steel bar in lining.

anchored rockbolt, and grouted rockbolt) were used in rockburst-prone tunnel sections. The application of the swelling dowel bolt was aimed at ensuring temporary safety near the tunnel faces, and also to absorb the released energy from the rockburst events. The expansion shell anchored rockbolt, and grouted rockbolt had the purpose of long-term stability after tunnel excavation. From the viewpoint of the operational life of the Jinping headrace tunnels, they will operate under high internal and external water pressures. This means that the design of the concrete lining may play an important role in the tunnel safety during the tunnel operation life.

In other words, the support design for the Jinping II tunnels relies on the concept of mobilising the load bearing capacities of the rock mass itself, and also of controlling groundwater by means of grouting. Therefore, a united load-bearing system is provided by means of rock supports of shotcrete and bolts, seepage-proofing grouted zones and concrete lining structures. The shotcrete and bolts support system of not only ensured stability during the tunnelling, but also prevented any gradual increase in the fractured zones. The seepage-proofing grouting formed a low permeability zone, so that it could address the disadvantage of high permeation pressure. Moreover, it also provided an improvement of the rock mechanics properties, such as deformation modulus and rock mass strength. More importantly, the united load-bearing system provides confining pressure and at the rock boundary transforms one or two dimensional stresses into three dimensional stress systems. All these strategies play potentially crucial roles in the long-term stability of the Jinping headrace tunnels.

Figure 6.95 shows the united load-bearing system design and the numerical model which was used to assess the load on the lining and evolution of the damage zone for the long-term operational life of the tunnels. In addition, the effects of water pressure-release holes on the rock mass and lining stabilities can be evaluated.

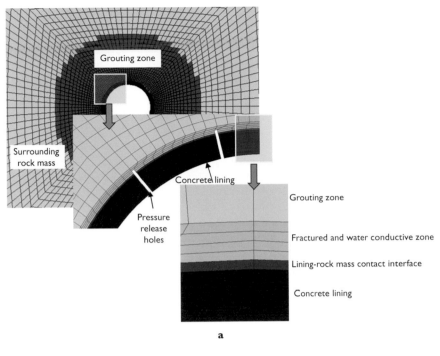

Grouting zone

Surrounding
rock mass

Concrete lining

Pressure
release
holes

Grouting zone

Fractured and water conductive zone

Lining-rock mass contact interface

Concrete lining

a

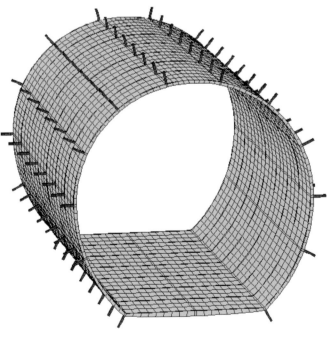

b

Figure 6.95 United load-bearing system design and its numerical model: (a) Simulation of the grouting zone and concrete lining; and (b) Layout of the water pressure-release holes.

6.6 CHAPTER SUMMARY

In this Chapter, we have presented a major case example of methods to reduce risks when excavating long, deep tunnels—with specific reference to the Jinping II hydropower headrace tunnels in China, which are 17 km long and have a maximum overburden of 2,500 m. We began by providing statistics on the world's longest tunnels currently under construction and being planned, and we reviewed rockburst risk in Chinese tunnels. Then, in line with the Frontispiece governing flowchart, we provided a follow-on flowchart for epistemic and aleatory uncertainty analyses as, respectively, initial risk management and dynamic and final risk management. The geological setting at the Jinping II site was described, together with the *in situ* stress and hydrogeology, plus the proposed construction procedures. The analysis of rockbursts was presented and the flowchart for risk mitigation and the dynamic control strategy was given in Figure 6.23. Examples of microseismicity related to rockbursts were included, as were the procedures for mitigating water inrushes and large deformations.

Thus, the illustrative major case example described in this Chapter described a real case of hazard predictions and countermeasures used—the risks' severity, potential failure locations, risk warnings, mitigation of rockburst occurrence, large deformation analyses of weak surrounding rock in different excavation sectional forms and the optimisation of the support system in large deformation-prone rock masses, the prediction and treatment of groundwater inrushes, and so on. The Chapter summarises a large amount of work completed by numerous engineers and scientists, especially by the research team led by this book's second author, whose analysis results have guided the design and construction and ensured rapid and safe excavation of the headrace tunnels.

This illustrative example indicates that risk assessment in deep and long tunnels is a complicated and challenging task, due to the occurrence of the complex failure modes, including rockbursts, spalling, large deformations and water inrushes, as well as major collapses which arise from the complicated engineering and geological conditions. These conditions involve a number of epistemic and aleatory uncertainties, which means that a great deal of information must be collected and the control factors and their control mechanisms also need to be analysed in order to identify the separate rock mass failure types and hazards. However, the successful implementation of risk assessment and treatment in the Jinping II headrace tunnels demonstrates that this risk assessment methodology and its approach to risk control, treatment and mitigation are able to successfully address the complicated risk analysis tasks. Hopefully the content of this Chapter will serve as a useful reference for similar projects.

* * * * *

The upper half-section of the #1 headrace tunnel was bored through in June 2011, as shown in Figure 6.96, and the tunnel began to generate hydroelectricity on schedule in December 2012. The #2 headrace tunnel began to generate electricity in 2013. It is expected that #3 and #4 will begin to generate electricity in 2014.

Figure 6.96 The upper half-section of #1 headrace tunnel was bored through in June 2011.

Chapter 7

Risks associated with hydropower cavern groups

7.1 INTRODUCTION

In Chapter 6, we discussed the range of risks and their mitigation associated with the design and construction of deep and long tunnels. In this Chapter, we present a similar overview for the design and construction risks associated with hydropower cavern groups—as a further illustrative case example. The book's thematic flowchart in the Frontispiece and included in the previous Chapter as Figure 6.3 is also the structural guide to the content of this Chapter.

7.1.1 Development of large hydropower cavern groups

The underground caverns of hydropower stations which were built in China from the 1950s to 70s are relatively small in size. Their widths are usually less than 20 m, and heights less than 30 m. In the 1970s, the size of caverns increased: for example, the main cavern at the Dinorwig pumped storage hydroelectric scheme in North Wales, UK, which was constructed in the 1970s has dimensions 23 m wide, 51 m high, and 180 m long. Since the 1980s, the underground powerhouses of the Lubuge and Ertan hydropower stations in China were successfully built using modern design and construction technology, and the Ertan hydropower station, with its large span and high side-wall, is China's largest hydropower station built in the 20th century (25–31 m wide, 65 m high, 280 m long). In fact, the construction technology of large underground caverns in China has undergone vigorous development over the past several decades, and this has been covered in the books, by Pan *et al.* (2000) and Chen *et al.* (2000).

Since the turn of the century, many large-scale hydropower projects have been built or will be built in western China, such as (1) the Xiaolangdi and Laxiwa hydropower projects along the Yellow River, (2) Ludila, Wudongde, Baihetan, Xiluodu and Xiangjiaba hydropower projects along the Jinsha River, (3) Lianghekou, Yangfanggou, Jinping I, Jinping II, Guandi and Daduhe hydropower projects along the Yalong River, (4) Shuangjiangkou, Changheba, Huangjinping, Dagangshan and Pubugou hydropower projects along the Dadu River, and (5) Xiaowan, Nuozhadu and Dachaoshan hydropower projects along the Lancang River. Their locations are shown in Figure 7.1.

Large-scale underground caverns are some of the main structures of hydropower projects; the spans of these caverns are usually more than 30 m, their heights more than 50 m, the highest being close to 90 m. The main and auxiliary powerhouse, transformer

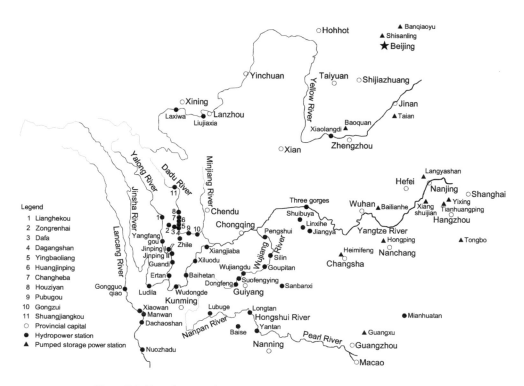

Figure 7.1 Distribution of many of China's large hydropower projects.

chamber and draft tube gate chamber are usually arranged in parallel, and are cross-linked with the headrace tunnels, tailrace tunnels and bus tunnels. For example, the underground powerhouse dimensions of the Xiluodu hydropower station are 436 m × 33.8 m × 78.2 m (Fan *et al.* 2011); Jinping I hydropower station, 284.8 m × 25.5 m × 66.7 m (Li *et al.* 2009); Baihetan hydropower station, 439 m × 32.2 m × 78.5 m (Xu *et al.* 2012); Laxiwa hydropower station, 309.7 m × 30 m × 73.8 m (Jiang *et al.* 2010); and Dagangshan hydropower station, 226.5 m × 30.8 m × 30.8 m (Zhu *et al.* 2013).

The large underground powerhouses lie in western China, where the geological conditions are complex and the *in situ* stress is high. The maximum component of *in situ* stress at the Laxiwa hydropower station underground powerhouse is 22–29 MPa; that of the Jinping I hydropower station underground powerhouse is 20–37.5 MPa; and that of the Jinping II hydropower station headrace tunnel is 42 MPa, while the maximum *in situ* stress component of some sections for the planned west route project of the South-to-North water diversion may be 50 MPa, estimated by Wang and Ma (2002). High *in situ* stress becomes the main factor affecting the stability of the rock surrounding these large underground caverns, especially through the generation of rockbursts. During the excavation of underground caverns, such large *in situ* stress values can make the hard, brittle rock mass fracture and be ejected suddenly—as we discussed in Chapter 6. This is dangerous for the safety of both workers and equipment and decreases the stability of the surrounding rock. This situation has occurred during the excavation of the caverns, such as the Tianshengqiao, Ertan, Pubugou and Laxiwa hydropower stations. Peng (1998) pointed out that during the excavation of the Ertan caverns, dozens of rockburst instances occurred, which caused the pre-stressed anchor cables to lose their integrity. The rockbursts caused failure in the cavern-peripheral rock and the construction progress was seriously affected. The safety of construction personnel was also repeatedly threatened by rockbursts during the excavation of the Laxiwa underground powerhouse and the Jinping II subsidiary tunnel.

The world's largest underground powerhouse (main powerhouse, 434.0 m × 34.0 m × 86.7 m; transformer chamber, 378.4 m × 21 m × 39.8 m; wire outlet shaft, ϕ50 m × 134.28 m) of the Baihetan hydropower station is planned to be built in China. There are multiple interlayer shearing belts in the region of the underground powerhouse, which intersect with the vault and sidewall. The maximum *in situ* stress component in the left and right bank are respectively 19 MPa and 25 MPa, and their orientations are similar to those of the shearing belts. Such geology conditions will have a great influence on the stability of the caverns, which we have discussed in Chapter 4 and which have been analysed by Xu *et al.* (2012). The Wudongde hydropower station is another large-scale hydropower station to be constructed in China. There are two caverns, one each on the banks of the Jinsha River. Each one includes a main powerhouse (321 m × 31.8 m × 86.9 m), transformer chamber (255.6 m × 18.5 m × 34.5 m), draft tube gate chamber, six headrace tunnels, six bus tunnels and six tailrace tunnels. The strata of this region are composed of limestone, marble and dolomite, their dip angle being in the range 50–80°. The intersection angle between the rock layers and cavity axis is small, which will lead to rock block failure during the excavation of the caverns. This geology situation has been covered by Wang (2013) in his PhD thesis.

7.1.2 Current status of design and risk management for large rock caverns

For these large caverns, the problems are caused by their size, the geological setting may be complex and contain many different geological units, and the rock instability mechanisms and their propagation can also be complex. In addition, historical data from previous cavern construction is often incomplete with the mechanical parameters being uncertain, and there are many site and project factors affecting the construction of each cavern, so it is difficult to draw overall conclusions from past experience. Because of these factors, different design methods have been presented by different scholars, such as the engineering analogy method, numerical analysis method, probability analysis method, etc. Among these design methods, today the engineering analogy method, which is based on engineering experience, is still used in most underground engineering design and it is the main method for the design specifications. When using this method, the design parameters are determined according to the engineering geology conditions, rock mass classification results, relevant specifications and similar projects. It is widely used to determine the layout of caverns, sectional form, size parameters and support type, not only for the small single tunnels but also for complex large caverns. The shortcoming of this method, however, is that it depends greatly on engineering experience which has been affected by local conditions. So, when the designer does not have local site experience on which to base the design, the scheme may be either over-designed or unsafe and hence be inefficient. In particular, the risks may be increased.

At this time, the theoretical design method may be much more suitable. This method is gradually being improved along with the development of rock and soil mechanics theory and computing research. It has experienced many periods, such as the rigid structure design period, elastic structure design period, elastic-plastic numerical analysis design period and structure reliability design period, as well as the load–structure model, rock–structure model and convergence–constraint model, and so on. Following all these developments, the theoretical design method makes the design results more credible, and enables quantitative design parameters to be obtained. Of course, much depends on the local geological conditions, but we have now reached the stage where these are included as key design supporting information.

However, due to complex engineering geological conditions, incomplete information, uncertainty during the design and construction process and limitations in understanding of the rock mass mechanical characteristics, safe and economic design schemes and the associated construction processes may still be difficult to obtain fully via the theoretical design method. When significant local changes occur during the construction, such as adverse geological phenomena, excessive construction disturbance and serious damage in the surrounding rock, the collapse of caverns may occur unless the original design scheme is changed. Therefore, with the development of shotcrete and rockbolt support technology, the dynamic design method, having predicting, monitoring, evaluating and correcting capabilities, gradually developed into the most popular design concept and method. Müller and Blazevic proposed the New Austrian Method (NATM) theory in 1934. This approach makes the supporting structures safe and economical through shotcrete, rockbolt support and field monitoring technology, which makes good use of the supporting capability of the surrounding rock.

Feng and Hudson (2003, 2011) created a modern rock engineering design methodology, and discussed the eight basic types of methods for rock engineering modelling and design supported by sets of Protocol Sheets to provide the necessary information by recording the data and decisions leading to the rock engineering design, including the during- and post-construction feedback required for back analysis. These seven sets of Protocol Sheets cover the following subjects: objectives and overall approach to the project, geological setting and site investigation, modelling, design, technical auditing, feedback and back analysis. These Protocol Sheets can be found in Feng and Hudson (2011). Currently, the dynamic feedback optimal design method is widely used to analyse and design large caverns.

For the uncertainties existing during the design and construction process, research on risk assessment has been developed since the 1870s. Risk assessment standards or specifications were proposed by the insurance associations and engineering management groups, particularly in Australia, Austria, Canada, and United States, plus the International Tunnelling Association (Eskesen *et al.* 2004) and the International Road Federation. The risk assessment norms and guidelines were also formulated by the departments of, for example, railways and highways in China. In the context of this book, risk assessment research has focussed on the underground traffic engineering field, including subway tunnels, highway tunnels, railway tunnels and underwater tunnels. However, the risk assessment research results relating to large hydropower caverns with complex geological conditions are insufficient.

Due to the complex and uncertain geological conditions, high *in situ* stress and large excavation dimensions, there are more risks in the process of design and construction of caverns compared to tunnels. Therefore, it is necessary to develop further work on the large-scale underground caverns based on current research results. Risk assessment standards, including frequency classification, consequence classification, risk classification, risk acceptance criteria and other risk management tools are an important part of risk assessment but at present and for large caverns with complex geological conditions and high *in situ* stress, the methodology has not yet been established. There are many risk assessment methods available from other subject applications: those currently used mainly include multi-risk, event tree analysis, decision tree analysis, Monte Carlo simulation, Analytic Hierarchy Process (AHP), neural networks, Bayesian networks, fuzzy comprehensive evaluation, and so on. These methods have been used as described in papers or research theses, such as Sturk *et al.* (1996), Kampmann *et al.* (1998), Feng (2000), Clark and Borst (2002), Jafar *et al.* (2010), Sousa, (2010), Sousa and Einstein, (2012), and Brown (2012). In addition, the Tunnel Cost Model (TCM), Decision Aids in Tunnelling (DAT) and Cost Estimate Validation Process (CEVP) have also been used to evaluate tunnel construction risk by Einstein *et al.* (1992), Einstein (1996), Reilly, (2003), Reilly and Brown, (2004), Goricki *et al.* (2004), and Isaksson and Stille (2005). In order to use these existing methods as appropriate for planning large underground caverns, further development work is required as we discuss in the next Section.

7.1.3   Why is a new method of risk management required?

For large underground caverns, a 'dynamic optimal design method' based on the geological conditions and monitoring data obtained during construction is usually used to analyse stability. An ideal design method should optimise the design and construction scheme throughout the project life cycle process, which means accommodating the uncertainty associated with the geological conditions and the construction events—and which should lead to the lowest construction cost. Therefore, a dynamic risk evaluation and management method should be utilised. The reasons are further amplified as follows.

1   Due to the large size of the caverns, complex engineering geological conditions, intersecting chambers and long construction periods, the existing underground engineering construction risk assessment method and evaluation standards are not suitable for analysis of such underground caverns.

2   Tunnel engineering can be regarded as a type of 'linear engineering', the length being much more than the sectional size, whereas for caverns the dimensional ratios are not so extreme. Also, the high sidewalls and chamber/tunnel intersections cause the surrounding rock to unload over a greater area which is more severely affected by natural structural planes in the rock mass. Also, the deformation of the surrounding rock may be large and affected by any soft strata, faults, fracture intersections and high *in situ* stress, which can be amplified as the caverns are created by the downward extraction of successive rock benches. For these reasons, the risk assessment process for caverns should be a dynamic operation.

3   The heights of the main powerhouse, wire outlet shaft and draft tube gate chamber are significant. The rock support is installed as the excavation proceeds in layers/benches. If the supporting scheme is not sufficient, the deformation of the surrounding rock will be excessive when the bottom layer is excavated. Then, any reinforcing treatment will be difficult to implement because the sidewalls and

vaults are so high, and their long-term stability put in jeopardy. Therefore, the construction risk and long-term stability risk should be reduced through a risk evaluation incorporating the field monitoring and geological information.

4   The excavation sections of large underground caverns are substantial, and the engineering geological information of these regions is usually not complete and obtained from much smaller sampling areas, i.e., boreholes and small tunnels. Faults, joints and fissures exposed during the process of construction taken together with a high *in situ* stress will cause the surrounding rock to have a large relaxation region, large deformations and local instability potential. The local instability may be due to spalling and slabbing, unloading of fractures, collapse, instability of rock blocks, buckling, structural plane sliding or rockbursts, which are difficult to analyse or judge with the numerical methods, because there are insufficient detailed geological data before the excavation. Although the geological information can be obtained during the excavation in layers, the uncertainty and complexity of the geological conditions and parameters cannot be incorporated into the engineering analogy method and numerical analysis method. Therefore, a new risk assessment method which can incorporate the uncertain information should be proposed for large underground cavern groups.

7.1.4   Outline flowchart for risk management for large hydropower cavern groups

The outline flowchart for the overall risk management method for a large hydropower cavern group is shown in Figure 7.2. At various stages, it may be necessary to go back to check each of the steps to update the models and methods.

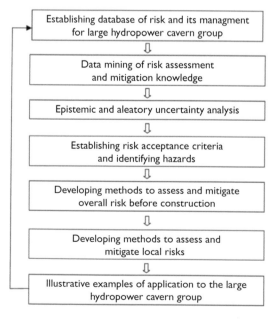

Figure 7.2 Flowchart to establish risk management methods for large hydropower cavern group.

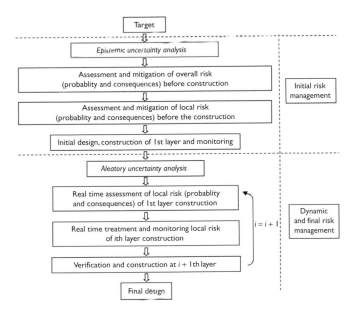

Figure 7.3 Flowchart of risk management for a large hydropower cavern group before and during construction.

7.1.5 Initial and final risk management for assessing and mitigating the risks for a large hydropower cavern group

The risk management of large hydropower cavern groups involves two main stages: initial risk management and dynamic/final risk management. At the initial risk management stage, it includes epistemic uncertainty analysis, assessment and mitigation of the overall risk (probability and consequences) and local risk (probability and consequences) of the large hydropower cavern group before the construction. This is followed by initial design and construction for the first layer (top bench) of the cavern. At the dynamic and final risk management stage, the risk is assessed and mitigated during the construction of each cavern excavation layer according to the revealed geological conditions and actual behaviour of the surrounding rock. The flowchart for assessing and mitigating risk in a large hydropower cavern group before and during the construction is shown in Figure 7.3.

7.2 DATABASE OF 60 LARGE HYDROPOWER CAVERN GROUPS IN CHINA

7.2.1 Principles for establishing a database

The risk assessment of large-scale hydropower underground caverns, which are either under construction or planned, should be informed by similar construction experience. Therefore, the first types of information available should be systematically analysed, such as engineering geological investigation data, direct geological information exposed during excavation, excavation schemes, supporting parameters, failure types, reinforcement measures, field monitoring information, and so on. The corresponding qualitative and/or quantitative relations, such as the relation between the geological conditions and excavation scheme, or surrounding rock failure types and reinforcement measures/support parameters, plus field monitoring information, should also be analysed. Then, the overall risk and local risk of the caverns in the process of construction can be evaluated based on the aforementioned geological, monitoring and testing data from the existing projects. The required section or local areas which should be observed or monitored can then be determined. The excavation scheme and supporting parameters of the local area may be adjusted according to the risk evaluation results. Finally, based on the overall and local risk evaluation results, the dynamic risk assessment can be carried out using the geological information and monitoring data obtained during construction in order to evaluate the stability of the excavated areas and to contemporaneously predict the risk level of unexcavated areas and suggest the risk control measures. Thus, a database of large hydropower underground cavern groups should be established in order to obtain the qualitative and quantitative relations that can be established from the various types of information available. Generic information is useful but specific information from excavations in similar geological settings is the most useful.

When collecting the information to form the database, the following conditions should be satisfied, if possible.

1   The cavern group should be made up of a main powerhouse, transformer chamber, bus tunnel and tailrace tunnel. The draft tube gate chamber and wire outlet shaft can be considered as one part of the cavern group.
2   The main powerhouse and transformer chamber should be connected by the bus tunnel. If the transformer chamber is one part of the main powerhouse or built on the ground, this project is not so useful for the database purpose.
3   The width of the main powerhouse should be more than 20 m. Its width near the vault, which is greater, is regarded as the appropriate parameter for the cavern width for the database purpose.
4   The height of the main powerhouse is from roof to the floor. The distance between the main powerhouse and the transformer chamber is the shortest distance between their sidewalls.

Based on the above principles, 60 projects were selected to form the database by studying construction information, research reports and scientific papers. In the process of the data collection to form the project database, firstly the database documents were separately evaluated according to the geological information, excavation schemes, supporting parameters, failure types of the surrounding rock, reinforcement measures and field monitoring information. Then, the collected data were put into the EXCEL document for convenience of access. The hydropower stations in the established database are shown as Figure 7.4 via the EXCEL page. Finally, the different types of information were analysed as described earlier.

7.2.2 Content of the database

The information relating to 60 projects forms the database. The contents mainly include geological information, excavation schemes, supporting parameters, failure

Design Parameters Database of Underground Cavern Groups

	Hydropower station	Main powerhouse design parameters			Transformer chamber design			Distence between height and transformer chamber
		Length	Width between side wall	Width near roof	Height	Length	Width	Height	
4	Ertan	280.3	25.5	30.7	65.4	214.9	18.3	25.0	40.0
5	Longtan	388.5	28.9	30.7	75.1	400.0	19.8	34.2	43.0
6	Xiaowan	298.1	28.3	30.6	82.0	230.6	19.0	22.0	50.0
7	Pubugou	294.1	26.8	30.7	70.2	249.1	18.3	26.6	42.0
8	Laxiwa	311.8	27.8	30.0	74.8	233.6	29.0	51.5	51.0
9	Jingping I	277.0	25.6	28.9	68.8	201.6	19.3	32.7	45.0
10	Xiangjiaba	245.0	31.0	33.0	85.5	175.0	26.0	23.0	40.0
11	Xiluodu	439.7	28.4	31.9	75.6	349.3	19.8	33.2	50.0
12	Jingping II	352.4	25.8	28.3	72.2	374.6	19.8	34.1	45.0
13	Nuozhadu	418.0	25.8	29.0	81.6	348.0	19.0	22.6	46.0
14	Dagangshan	226.6	27.3	30.8	74.6	144.0	19.3	25.8	47.0
15	Baihetan	438.0	31.0	34.0	86.7	368.0	21.0	39.5	61.0
16	Dachaoshan	234.0	24.9	26.4	67.3	137.4	16.2	30.8	64.0
17	Taian	180.0	24.5	25.9	53.7	164.0	17.5	18.2	35.0

Figure 7.4 Some of the projects in the large-scale hycropower underground cavern groups database.

types and degrees of surrounding rock, reinforcement measures and field monitoring information, etc., with the main content of each aspect being shown as follows.

1   Geological survey data

According to the geological survey reports for different design stages of the cavern groups together with previous information, the geological survey data have been collected. They include the lithology, rock uniaxial strength, fracture distributions, fault character and distribution, *in situ* stress test results, rock mass quality classification results and their distribution, rock mass integrity evaluation results and distribution, groundwater distribution, and field monitoring information. The uniaxial strengths of the different rocks are included as their interval values and mean values. The information on the fracture distributions includes their classification, occurrence and distribution along the powerhouse length direction. The location, occurrence and scale of faults near the cavern groups are used to form the fault character and distribution information. The maximum and minimum values and their directions of the *in situ* stress have been obtained.

2   Geological information obtained during excavation

Based on the early geological survey reports, the geological information should be revised and improved according to the geological sketches made during the cavern bench excavation. Attention should be paid to the character and distribution of new joints, fractures or faults exposed by excavation and any changes in the fault properties.

3   Excavation schemes

According to the construction scheme in each case, the excavation sequence of the main powerhouse roof and cavern groups, thickness of each excavation layer, the blasting method for the layers, excavation method for the rock anchor beam, monitoring information relating to blasting vibration and deformation and stress, adjusted excavation parameters and excavation effect should be collected.

4   Size and support parameters

The designed size parameters of the cavern group, supporting parameters of bolts and cables, spray layer thickness, support time and support parameters adjusted to the local area should be gathered. The designed size parameters include the powerhouse roof span, sidewall span, powerhouse height, transformer chamber span and height, span between powerhouse and transformer chamber, vertical and horizontal dimensions of the powerhouse. The support parameters relating to bolts and cables include their length, spacing, diameter, load bearing capacity and location distribution. The adjusted support parameters and corresponding geological conditions and monitoring results should also be gathered.

5   Failure types and degree of the surrounding rock

Three main failure types are those caused by the rock stress, rock mass structure and a combination of both. The specific data should be collected according to the surrounding rock failure mode classification standards for the large-scale cavern groups, as proposed by Xiang *et al* (2012) and shown in Figure 7.5. Attention should be paid to the failure location and engineering geological conditions, degree of failure, main failure reasons, reinforcement measures and monitoring information before and after failure.

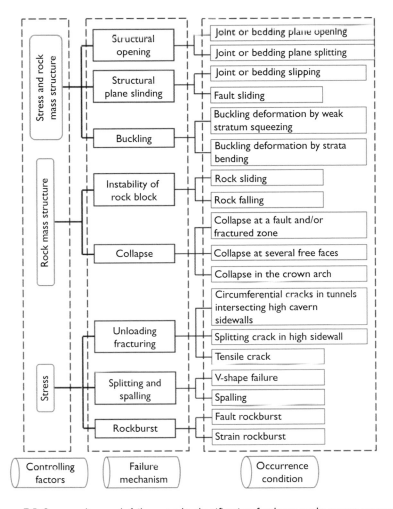

Figure 7.5 Surrounding rock failure mode classification for large-scale cavern groups.

6   Reinforcement measures

The data concerning the rock reinforcement measures should be related to the surrounding rock failure types. The differences between original support measures and modified reinforcement schemes should be observed.

7   Field monitoring information

Field monitoring information includes the surrounding rock deformation in different sections, bolt stress, anchor cable stress and acoustic test results of the rock mass relaxation depth during excavation. The *in situ* stress supplementary survey results, geological CT test results and overloaded anchor cables should also be focussed on, especially their relations with the geological conditions, surrounding rock failure types and reinforcement measures.

7.2.3 Statistical analysis of key issues

7.2.3.1 Lithological character and rock mass quality

The overall quality of the host rock where the hydropower cavern groups are constructed is usually quite high. However, there may be different kinds of rock near to the caverns where the cavern size is significantly large; then, the rock strengths may be different, with some high and some low. In this case, the rock mass quality, which affects the stability of the surrounding rock after excavation, is variable across the site. The statistics and analysis results of the lithology and rock mass quality of the cavern group area can be used to evaluate the rock mass stability. They can simplify the relations between the lithology, rock mass quality and failure modes of the surrounding rock and reinforcement measures, and may also be used to guide hydropower cavern group development in the future.

The lithology and rock mass quality statistics results for 60 hydropower projects in the database are shown in Table 7.1. See the note below the Table for an explanation of the rock mass quality grades.

7.2.3.2 Structure and strength of the rock mass

Rock mass strength parameters and rock structural characteristics are the key factors that affect the stability of the cavern groups. The rock mass strength parameters can be obtained by considering the influence of rock mass integrity and structural surfaces; these values are the basis of studying the relations between them with design size, failure model and degree of surrounding rock involved. Structural characteristics include fracture intensity and groupings, spatial distribution and especially the occurrence of fault or bedding plane shears, as well as the relation between these and the cavern group strike values. Based on these, an 'influence rule' among the rock structural characteristics, failure model and degree of surrounding rock can be obtained. The statistics relating to the structural characteristics and rock strength parameters and of the projects in the database are shown in Table 7.2.

7.2.3.3 Stress conditions

The *in situ* stress conditions near cavern groups are the main factors influencing the failure modes of surrounding rock during and after cavern excavation: failure modes such as rockbursts, splitting and spalling, unloading fracturing, joint or bedding plane opening, and buckling deformation by strata bending, are all associated with the *in situ* stress conditions. The *in situ* stress magnitude, rock strength–stress ratio, principal stress directions and intersection angle with the cavern group axis direction are the chief factors affecting the rock mass failure mode, location and degree. Through analysing these, the relations between the rock strength–stress ratio and surrounding rock deformation, failure modes, support parameters and local reinforcement measures can be obtained. The key stress parameters are the maximum principal *in situ* stress magnitude, direction and intersection angle with the cavern group axis. The stress condition results for the projects in the database are shown in Table 7.3.

Table 7.1 Lithology and rock mass quality statistics for large hydropower cavern groups.

No.	Project name	Lithology	Rock mass quality
1	Baihetan	Basalt	Right bank: II (52%), III1 (34%), III2 (15%); Left bank: II (16%), III1 (37%), III2 (47%).
2	Baishan	Granite	I and II more than 80%
3	Baise	Diabase	II and III, III more than 50%
4	Dachaoshan	Basalt	II (51%), III (46%), IV (3%)
5	Dagangshan	Granite	II and III (85%), IV (15%)
6	Ertan	Syenite and basalt	II (80%), III (10%), IV (10%)
7	Goupitan	Limestone	II (69%), III (17%), V (14%)
8	Guandi	Basalt	II more than 80%
9	Guangxuyiqi	Granite	II and III more than 80%
10	Heimifeng	Granite	II (80%), III (10%), V (10%)
11	Houziyan	Limestone	III1 more than 80%
12	Bailianhe	Granite	II and III more than 80%
13	Jiangya	Limestone	II more than 80%
14	Jinping II	Marble	III (83%), II (14%), IV (2%)
15	Jinping I	Marble	III1 (85%), IV1 (15%)
16	Laxiwa	Granite	II (90%), III (10%)
17	Longtan	Sandstone	II–III1 (87%), IV (15%)
18	Lubuge	Limestone and dolomite	II and III more than 80%
19	Ludila	Metamorphic sandstone	II and III more than 80%
20	Nuozhadu	Granite	III (47%) II (37%),
21	Pubugou	Granite	II and III (85%), IV, V (15%)
22	Shisanling	Conglomerate	II and III (71%), IV, V (29%)
23	Shuangjiangkou	Granite	II (66%), III (23.4%)
24	Silin	Limestone and dolomite	II and III more than 80%
25	Yangfanggou	Granite	II and III more than 80%
26	Taian	Granite	II and III (85%)
27	Tianhuangping	Tuff	II and III more than 80%
28	Tongbai	Granite	II (85%) and III less than 15%
29	Xiluodu	Basalt	I and II (> 50%), III (50%)
30	Xiangshuijian	Granite	II more than 80%
31	Xiangjiaba	Sandstone	I and II (85%), IV, V class (15%)
32	Xiaolangdi	Sandstone	II (85%)
33	Xiaowan	Gneiss	I and II more than 80%
34	Yixing	Sandstone	IIIb (51%), IV, V (49%)
35	Changheba	Granite	II, III (85%) IV (15%)
36	Renzonghai	Granite	II and III (85%)
37	Dafa	Limestone	III (85%), IV (15%)
38	Manwanerqi	Rhyolite	III and II more than 80%
39	Pengshui	Limestone and Shale	II more than 95%
40	Sanxia	Granite	I and II (75%), III (10%), IV (15%)
41	Shuibya	Limestone	IV and V more than 80%
42	Yele	Diorite	III (85%), IV (15%)

Note: The Roman numerals I, II, III, IV and V stand for the rock mass quality grade according to the Standard for Engineering Classification of Rock Masses in China.

Table 7.2 Rock mass strength parameters and structure characteristics statistics results of large underground cavern groups.

| No. | Project name | Uniaxial compressive strength of rock (MPa) | | Elasticity modulus (MPa) | | Structural characteristics of the surrounding rock |
		Minimum value	Maximum value	Minimum value	Maximum value	
1	Baihetan	70	90		10*	Caverns on both sides affected by bedding fault zone, left bank affected by C2, right bank affected by C4 and C5; maximum in situ stress field of the right bank was 20–30 MPa higher than that of the left bank.
2	Baishan		125			A dyke crosses the rock mass, 0.3 m width, broken rock mass, large inclined angle between its strike direction and powerhouse axis direction, large dip angle, no effect on stability of surrounding rock.
3	Baise					Hard and brittle rock, developing joints and fissures. Structural alteration belt J163 close to the upstream sidewall; strike direction of joint T1 parallel with sidewall, N60–75°W.
4	Baoquan	85	100	10	27	Wet uniaxial compressive strength 150 MPa.
5	Dachaoshan	60	85	8*	10*	Rock mass: blocky structure, monoclinal structure, leaning to the upstream at a small angle, including a tuff soft interlayer. Large fault F217, dip angle 75–88°, between the transformer chamber and the main powerhouse; F168 passing through the main powerhouse at a small oblique angle, unstable blocks formed by the faults, other structural planes and tuff interlayers.
6	Dagangshan		80		8(III)	Simple geological structure, no regional fracture structures, development of large numbers of extrusion fracture zones, small faults, joints and fissures along with rocky matrix. Faults F56–60 pass through the underground cavern group.
7	Dongfeng					Developed rock layers are filled. Their direction is NE45–75°, and dip direction is NE12–16°.
8	Ertan	100	173	5	35	Rock mass structure is simple. Small faults and compressed fragmented belts are developed in local areas. Their width is 50–200 mm, and length is short.
9	Goupitan	60	80			Strike direction of rock layer is 30–35°, dip direction is NW, dip angle is 45–48°. The intersection angle between the rock layer strike direction and powerhouse axis direction is 40–45°. The strike direction of the faults is NW. Their dip angle is large. Their scale is small. Their width is 50–200 mm. Their density is high, average spacing is 4 m, some fault width is 500–600 mm, crack group is 4. There are large karst caves near the intersections of faults.

No.	Name					Description
10	Guandi	100	190			There are no faults or large soft structural surfaces in the cavern group area. Rock mass quality is good. Rock mass structure is blocky. There is confined water found in the drilling.
11	Jinpingerji	65	80			The vertical in situ stress is 12–17 MPa. The horizontal in situ stress is 9–15 MPa. The dry uniaxial compressive strength is 80–90 MPa. The wet uniaxial compressive strength is 65–80 MPa.
12	Jinpingyiji	60	75	13/16*	22*	Blocky structure, high rock uniaxial compressive strength, rock mass compact and hard, with high integrity, and only a small number of faults.
13	Laxiwa	100	120		26.4*	
14	Longtan		130	16	22	Rock layer strike direction is N5–15°W, its dip direction is NE57–60°. The inter-formational disturbed belts are developed.
15	Lubuge	40	55	15	40	
16	Ludila	45	164			
17	Nuozhadu	45	80		20*	
18	Pubugou	80	100	35*	40*	'Gravel' structure, thick layered structure. Dry uniaxial compressive strength is 70–80 MPa, and wet uniaxial compression strength is 50 MPa.
19	Shisanling	50	98		40	
20	Shuang-jiangkou	53	115	21.9	53	The intersection angle between the powerhouse axis and rock layer strike direction is small. There are thin and very thin limestone layers present near the bottom of the side wall. The rock mass deformation modulus is low. The quality grade is poor.
21	Silin		90			

(Continued)

Table 7.2 (Continued)

No.	Project name	Uniaxial compressive strength of rock (MPa)		Elasticity modulus (MPa)		Structural characteristics of surrounding rock
		Minimum value	Maximum value	Minimum value	Maximum value	
22	Tianhuang-ping	100		10	30	Rock mass is fresh, hard, homogeneous, complete and without large structural surfaces. Small scale and less developed level III structural surface; small scale and developed level IV structural surface. Saturated uniaxial compressive strength is 120.5 MPa.
23	Xiluodu	60	170	17	36	Rock mass is hard, fresh and intact. Stability is good. Fissures in the powerhouse area are not developed.
24	Xiangjiaba	75	80	12	20	Rock layer occurrence is NE60–80°/SE∠15–20°. There are shear zones along with the distribution of muddy soft rock or soft rock. The rock mass wet uniaxial compressive strength is 130 MPa, and dry uniaxial compressive strength is 75–80 MPa.
25	Xiaolangdi					Rock mass is hard, with little effect on the faults and fracture zones.
26	Xiaowan	90	140	22	28	There are three level III faults crossing the underground powerhouse. Their intersection angle is 25–50°.
27	Yixing	22	56	1	7	There are large faults F220 and F204 located in the north and south ends. The exposed width of F204 is 5–15 m. There are cataclasite, breccia, mylonite and fault gouge fillings in the fault zone.
28	Pengshui	60	80			Rock mass strike direction is 22–25°, dip direction is 110–115°, and dip angle is 60–70°.
29	Sanxia					
30	Shuibuya	30	70			The underground cavern group is arranged between tensional faults F2 and F3. Its surrounding rock mass is composed of hard and soft layered rock.
31	Yele	104	127			Developed small scale faults, shear bedding fault zones and structural fractures, poor rock mass integrity, large differences in rock mass properties in different sections.

Note: The numbers with * are the elasticity modulus; others are the deformation modulus.

Table 7.3 In situ stress conditions for large cavern groups.

No.	Project name	Principal stress values (MPa)		Orientation		Dip angle (°)		Angle with cavern axis (°)	Strike direction of cavern axis
		Minimum value	Maximum value	Minimum value	Maximum value	Minimum value	Maximum value		
1	Baihetanzuo	19	23	N30W	N50W		25	10	N20E
	Baihetanyou	22	26	N30W	N50W		13		N10W
2	Baishan		9.58		N49E	5		70	
3	Baoquan	23.4	27.6						N75W
4	Dachaoshan	13	16.9	N18E/44	N61E				N55E
5	Dagangshan	11.37	22.2						
6	Dongfeng		12.2	N10E	N30E			77	N6W
7	Ertan	30	40		N43E		22		N75E
8	Goupitan	11	14	N28.7W	N53W			32	N5E
9	Guandi	25	35				30	40	
10	Guangxuyiqi		14						
11	Hemifeng	4	8						
12	Houziyan	21.5	36.4	N41W	N75W	21	47		NE35
13	Jinping II	10	22.9	S43E	S47.4E	20	50	19.3	N65W
14	Jinping I	21.7	35.7	N28.5W	N71W		10		N25E
15	Laxiwa	14.6	29.7		N				N3E
16	Lianghekou		23.6		N46E				N50W
17	Longtan	12	13	N20W	N80W			30	
18	Lubuge	13.5	18						N50E
19	Ludila	10	15						N76E
20	Nuozhadu	6.55	9		NE			20–40	
21	Pubugou	21.1	27.3	N45E/54	N84E			26.7–36.7	N42E

(Continued)

Table 7.3 (Continued)

No.	Project name	Principal stress values (MPa)		Orientation		Dip angle (°)		Angle with cavern axis (°)	Strike direction of cavern axis
		Minimum value	Maximum value	Minimum value	Maximum value	Minimum value	Maximum value		
22	Shisanling	9.5	11.7	N45W	N60W				
23	Shuangjiang-kou	16	38	N3W	N50W			10	N18W
24	Silin	8.5	10.96		N20E				N5W
25	Yangfanggou	12.6	13						
26	Taian		12						N40W
27	Tianhuang-ping	17	27						
28	Tongbai	10.9	12.2						
29	Xiluoduzuo	16	18	N60W	N70W	15	35	36–46	N24W
	Xiluoduyou	16	21	N60W	N70W		25		N70W
30	Xiangshui-jian		9.8						
31	Xiangjiaba	8.9	12.2	N25E	N35E				N30E
32	Xiaolangdi		5						
33	Xiaowan	16.4	26.7	N49W	N64W	49	53	9–24	N40W
34	Yixing	8.26	16						
35	Changheba	20	36	N45W	EW				
36	Huangjin-ping	9.2	24.6	N80W	EW				N70W
37	Gongguoqiao	10	13	N28E	N30E				N50E
38	Manwanerqi								N40E
39	Pengshui		10		N20E				N24E
40	Sanbanxi		14.5					34–56	
41	Sanxia	7	11.73		S32W			78.5	
42	Shuibuya		5.62	N76E	N80E		58.9		NW64
43	Yele		8.8	N2E				24	N22W

Using the results already presented, plus the background information, the maximum principal stress, rock maximum stress:minimum strength, rock maximum strength:maximum stress ratio, rock mean strength:maximum stress ratio, rock mean strength:mean stress ratio can been seen in Figures 7.6–7.8. From Table 7.3, the maximum principal *in situ* stress of the Changheba, Guandi, Houziyan, Jinping I, and Shuangjiangkou cavern groups are greater than 30 MPa, so we would expect that the surrounding rock stability of the cavern groups would be affected by the stress after excavation and unloading. The high *in situ* stress condition is the main influence factor for failure modes such as spalling, splitting cracks in high sidewalls, circumferential cracks in the tunnels intersecting the high cavern sidewalls and buckling deformation by strata bending, which occurred in the Guandi and Jinping I cavern groups.

Taking the statistical results of rock strength shown in Table 7.2 and the *in situ* stress conditions in Table 7.3, the statistical results for the strength–stress ratios of the rock at the different sites can be obtained and are shown in Table 7.4. The results for the maximum principal stress, minimum strength to maximum stress ratio, maximum strength to maximum stress ratio, mean strength to maximum stress ratio, and mean strength to mean stress ratio can been seen in Figures 7.6–7.8. From Figure 7.6, the maximum principal *in situ* stress for the Jinping I, Shuangjiangkou, Houziyan, Changheba and Guandi hydropower stations are greater than 30 MPa. The surrounding rock stability is affected by the stress after excavation and unloading. This condition of high *in situ* stress is the main influencing factor causing the failure modes, such as spalling, splitting cracks in a high sidewall, circumferential cracks in tunnels intersecting the high cavern sidewalls, and buckling deformation through strata bending, which occurred in the Jinping I and Guandi hydropower station underground cavern groups.

Some general rules can be developed. If possible, the hydropower cavern groups should not be located in areas where large faults, developed joints and fissures, fracture zones or high *in situ* stress exist. The intersection angle between the main powerhouse axis direction and the main fault strike direction should be large. Also, attention should be paid to the negative impact of fracture sets on cavern stability. The intersection angle between the powerhouse longitudinal axis direction and the maximum principal stress strike direction should be small when the *in situ* stress is high. The depth of the powerhouse should be more than twice its excavation span. The thickness between powerhouse and transformer chamber should not be less than 1–1.5 times their spans. The thickness between up and down cavity should not be less than 1–2 times the smaller cavern span.

7.2.3.4 *Arrangement of cavern group by size*

The common arrangements of the cavern groups are shown in Table 7.5, which are explicitly covered in the Company Standard of Hydrochina Corporation (Q/HYDROCHINA 009–2012). The sizes of the large hydropower cavern groups in China can be seen in Figures 7.9–7.13.

7.2.3.5 *Excavation scheme and parameters*

We expect the unloading degree and relaxation depth of the caverns and intersection areas to be more severe when the stress field is high, and particular attention should be paid to the intersection areas. Firstly, the tunnels intersecting with the sidewall

Table 7.4 Strength-stress ratio statistical results for the large cavern groups.

Project name	Principal stress magnitudes (MPa)		Saturated uniaxial compressive strength of rock (MPa)		Mean strength/ maximum stress in rock
	Minimum	Maximum	Minimum	Maximum	
Ertan	30	40	100	173	3.41
Shuangjiangkou	16	38	53	115	2.21
Houziyan	21.5	36.4	60	80	1.92
Changheba	20	36	60	80	1.94
Jinping I	21.7	35.7	60	75	1.89
Guandi	25	35	100	190	4.14
Laxiwa	14.6	29.7	100	120	3.70
Baoquan	23.4	27.6	85	100	3.35
Pubugou	21.1	27.3	80	100	3.30
Tianhuangping	17	27	100	–	3.70
Xiaowan	16.4	26.7	90	140	4.31
Baihetanyou	22	26	70	90	3.08
Huangjinping	9.2	24.6	–	121.8	4.95
Lianghekou		23.6	34	88	2.58
Baihetanzuo	19	23	70	90	3.48
Jinping II	10	22.9	65	80	3.17
Dagangshan	11.37	22.2	–	80	3.60
Xiluoduyou	16	21	60	170	5.48
Lubuge	13.5	18	40	55	2.64
Xiluoduzuo	16	18		108	6.00
Dachaoshan	13	16.9	60	85	4.29
Yixing	8.26	16	–	56	3.50
Ludila	10	15	45	164	6.97
Sanbanxi	–	14.5	100	250	12.07
Goupitan	11	14	60	80	5.00
Guangxuyiqi	–	14	–	81	5.79
Yangfanggou	12.62	13.04	–	90	6.90
Longtan	12	13	–	130	10.00
Tongbai	10.9	12.2		120	9.84
Xiangjiaba	8.9	12.2	75	80	6.35
Taian	–	12	116	160	11.50
Sanxia	7	11.73	–	79.74	6.80
Shisanling	9.5	11.7	50	98	6.32
Silin	8.5	10.96	–	90	8.21
Pengshui	–	10	60	80	7.00
Xiangshuijian	–	9.8	–	100	10.20
Nuozhadu	6.55	9	45	80	6.94
Yele	–	8.8	104	127	13.13
Shuibuya	–	5.62	30	70	8.90
Xiaolangdi	–	5	60	150	21.00

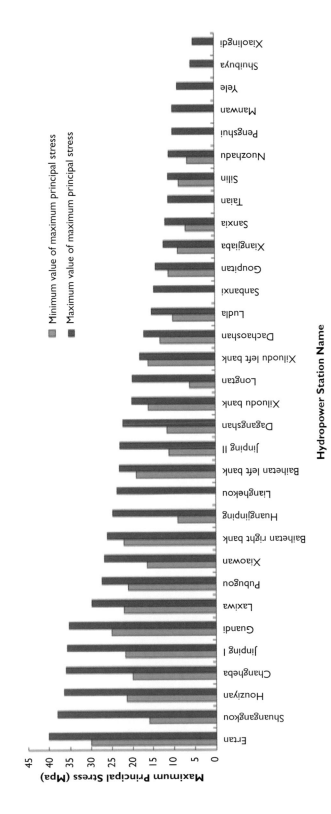

Figure 7.6 Statistical results for the maximum principal stress magnitude in the large cavern groups.

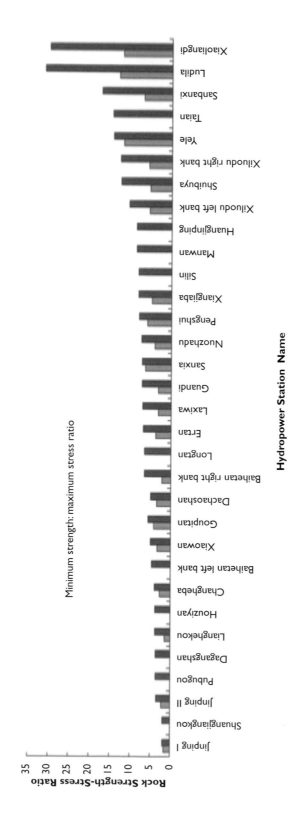

Figure 7.7 Statistical results for the strength-stress ratio for the large cavern groups.

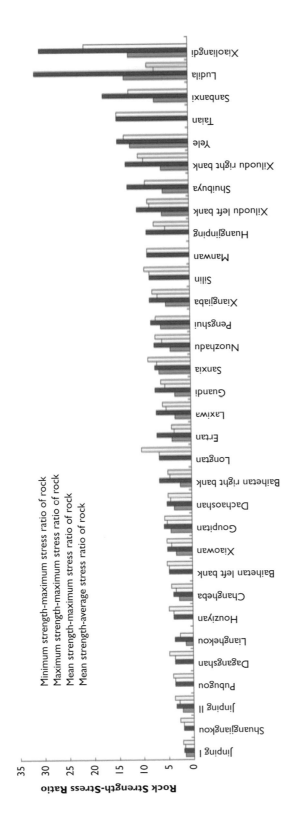

Figure 7.8 Rock strength–stress ratios for the large underground cavern groups.

Table 7.5 Arrangement and characteristics of the cavern groups.

No.	Location of transformer chamber	Diagram	Advantage	Disadvantage
I	Locating between main powerhouse and draft tube gate chamber		Layout compact, easy maintenance; separated arrangement can reduce degree of accidental failure.	The distance between the main powerhouse and draft tube gate chamber cannot be too small, or it is detrimental to the cavern stability.
2	Locating upstream of the main powerhouse		Layout compact, easy maintenance; accident failure degree can be reduced; tailrace tunnel can be shortened.	It is difficult to arrange the transformer chamber and design the drainage system. Surrounding rock stability is relatively poor.
3	Locating in the main powerhouse		Bus tunnel shorter, small power loss; convenient management; tailrace tunnel can be shortened.	Transformer close to generator unit, large failure degree of fire and explosion; surrounding rock mass stability may be poor due to the increased main powerhouse size.
4	Locating above the main powerhouse and draft tube gate chamber		The distance between main powerhouse and draft tube gate chamber cannot be shortened; accidental failure degree can be reduced.	Inconvenient operation and maintenance; longer bus tunnel; increased investment, large power loss, more equipment for ventilation and transportation.

should be excavated in order to reduce the blasting effect on the high sidewall. For example, the high pressure shaft, bus tunnel and tailrace tunnel should be excavated first, then the three large caverns (main powerhouse, transformer chamber and draft tube gate chamber).

Through comprehensive analyses, the existing hydropower cavern group excavation methods, analysed by Lu *et al.* (2011), can be determined as follows.

1 Arch excavation scheme

Construction scheme one (AEP-I): Both side drifts are excavated first; the rock pillar in the middle is excavated later.

Construction scheme two (AEP-II): Middle drift is excavated first, then the expanding excavation to both sides is completed.

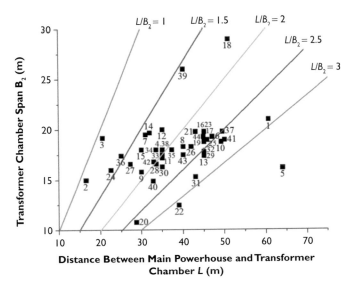

Figure 7.9 Distribution and boundary values of B_2 and L (see captions on graph axes).

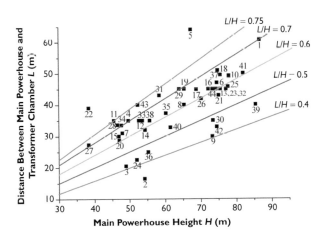

Figure 7.10 Distribution and boundary values of L and H (see captions on graph axes).

The contour of the arch should be excavated with the smooth wall blasting method.

2 Rock anchor beam excavation scheme

Beam Excavation Plan I (BEP-I): First, presplit blasting is implemented to the outside portion of the rock mass protective layer. Then, cut blasting is used to excavate the middle rock mass. Finally, the middle segment of the protective layer can be excavated by presplit blasting and smooth wall blasting.

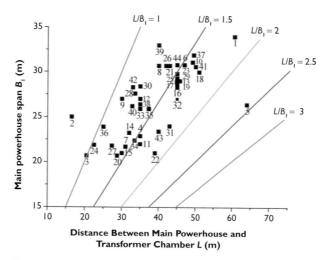

Figure 7.11 Distribution and boundary values of B_1 and L (see captions on graph axes).

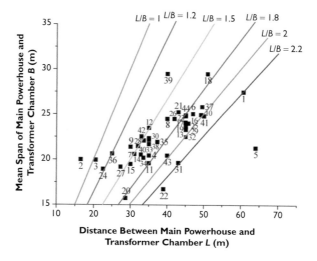

Figure 7.12 Distribution and boundary values of B and L (see captions on graph axes).

Beam Excavation Plan II (BEP-II): First, cut blasting is used to excavate the middle rock mass. Then, smooth wall blasting is implemented on the outside part of rock mass protective layer, and the middle segment of the protective layer can be excavated by smooth wall blasting.

3   High sidewall excavation scheme

The rock mass excavation scheme used to form the sidewall in layers can be seen in Figure 7.14. The excavation process is shown as follows.

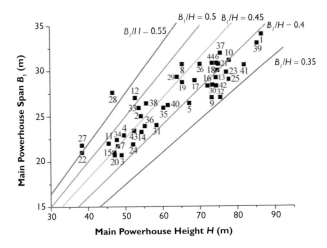

Figure 7.13 Distribution and boundary values of main powerhouse height (*H*) and span (*B*) (see captions on graph axes).

Wall Excavation Plan I (WEP-I): AA (presplit blasting of the contour) → IV1 (cut blasting in the middle section) → IV2 (expanding excavation blasting) → IV3 (cushion blasting).

Wall Excavation Plan II (WEP-II): IV1 (cut blasting to the middle section) → IV2 (expanding excavation blasting) → IV3 (loose blasting of the protective layer) → AA (smooth blasting of the contour).

Wall Excavation Plan III (WEP-III): BB (presplit blasting of the protective layer) → IV1 (cut blasting of the middle section) → AA (presplit blasting of the contour) → IV2 (expanding excavation blasting) → IV3 (cushion blasting).

Wall Excavation Plan IV (WEP-IV): IV1 (cut blasting to the middle section) → IV2 (expanding excavation) → BB (smooth blasting of the protective layer) → IV3 (loose blasting of the protective layer) → AA (smooth blasting of the contour).

Wall Excavation Plan V (WEP-V): BB (presplit blasting of the protective layer) → IV1 (cut blasting of the middle section) → IV2 (expanding excavation) → IV3 (cushion blasting) → AA (smooth blasting of the contour).

Wall Excavation Plan VI (WEP-VI): IV1 (cut blasting of the middle section) → BB (presplit blasting of the protective layer) → IV2 (expanding excavation) → IV3 (cushion blasting) → AA (smooth blasting of the contour).

Wall Excavation Plan VII (WEP-VII): IV1 (cut blasting of the middle section) → BB (presplit blasting of the protective layer) → IV2 (expanding excavation) → AA (presplit blasting of the contour) → IV3 (cushion blasting).

When the *in situ* principal stress component perpendicular to the powerhouse axis is less than 10 MPa, presplit blasting or smooth wall blasting can be selected to excavate the contour. The excavation sequence may be WEP-I, WEP-II, WEP-III, WEP-IV or WEP-V. If the blasting vibration must be strictly controlled, then presplit

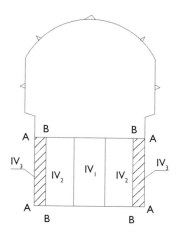

Figure 7.14 High sidewall excavation sequence for an underground powerhouse.

blasting should be used first, the sequence may be WEP-I, WEP-III or WEP-V. When the perpendicular stress value is more than 10 MPa, the excavation sequence with cut blasting of the middle section then presplit blasting or smooth wall blasting should first be used. The sequence may then be may be WEP-II, WEP-IV, WEP-VI or WEP-VII. If the blasting vibration must be strictly controlled, then WEP-VI or WEP-VII can be implemented first.

A smooth excavation face can be obtained through presplit blasting or smooth wall blasting. In the case of presplit blasting, a smaller amount of simultaneously detonated explosive is used in the closely spaced, parallel blast holes, so that a fracture plane is created through all the blastholes, thus forming a relatively planar surface. This surface then acts as a defensive surface for the subsequent blasting, reflecting stress waves and preventing fractures produced in the later blasting area from extending to the surrounding rock. However, during the powerhouse excavation of Ertan and Pubugou with a high *in situ* stress field, the blasting test results showed that the presplitting crack quality was not good without cut blasting of the middle section in advance. When the cut blasting of the middle section was performed first, followed by expanding the excavation and smooth blasting, then the excavation was improved.

For the Jinping I and Jinping II powerhouses, the presplitting crack quality was not optimal when the blasting was performed outside the protective layer before cut blasting of the middle section. Therefore, implementing the cut blasting of the middle section first is important regardless of whether smooth wall blasting or presplit blasting is used. During the excavation of the Guandi hydropower station, the downwardly excavated 2nd to 4th layers of the main powerhouse were first excavated by cut blasting of the middle section, followed by expanding the excavation on two sides with protective layers, then presplit blasting along the contour of the protective layers, and finally smooth wall blasting. Parts of the main powerhouse excavated after the 4th layer, the transformer chamber and pressure adjustment shaft, were first excavated by cut blasting of the middle section, followed by expanding the excavation on two sides with protective layers, and finally smooth wall blasting. The blasting excavation height was less than 6 m, and the actual blasting excavation effect was good. The blasting vibration monitoring results showed that the vibration velocity was less than the recommended standard value.

Large powerhouse excavation sequences and the contour blasting method for different hydropower projects performed in China can be seen in Table 7.6. Details of the support data for different schemes are given in the following sub-Section.

7.2.3.6 Support parameters

Zhang *et al.* (2006) notes that the surrounding rock of the Pengshui hydropower station cavern group is supported by systematic bolts and shotcrete. The shotcrete is sprayed steel fibre concrete with thickness 150 mm. The systematic bolts for the vault are hollow grouted anchor bolts, with row spacing 1.5 m × 1.5 m and length 8 m or 10 m arranged alternately. The bolt diameter is 32 mm and the installed load is 150 kN. The upstream and downstream sidewalls are supported with systematic mortar bolts and pre-stressed anchor cables. The intersections of the caverns are reinforced with longer and larger diameter bolts. The mortar bolt row spacing is 1.5 m × 1.5 m, and lengths 6 m or 9 m. The anchor cable row spacing is 4.5 m × 4.5 m, with length 25 m, and installed load 1500 kN.

Zhang *et al.* (2012) points out that the surrounding rock of the Guandi hydropower station cavern group is supported with steel mesh (φ8 mm @ 200 mm × 200 mm), shotcrete (concrete strength C25 and thickness 150–200 mm) and systematic bolts. In addition, the anchor cables are used to support the high sidewall. For the main powerhouse, the bolt diameter is 32 mm, length is 6 m or 9 m, arranged alternately, and

Table 7.6 Large powerhouse excavation sequence and contour blasting method (Lu *et al.* 2011).

Project	Powerhouse excavation size (length × width × height) (m)	Lithology	Maximum principal in situ stress (MPa)	Excavation sequence and contour blasting method		
				Vault	High sidewall	Rock anchor beam
Ertan	280.3 × 25.5 × 63.9	Syenite, gabbro and basalt	15.1	AEP-II	WEP-II	BEP-II
Longtan	388.5 × 30.7 × 77.3	Sandstone and mudstone	5.2	AEP-I	WEP-III	BEP-I
Shuibuya	150.0 × 23.0 × 68.3	Limestone	2.5	AEP-II	WEP-V	BEP-I
Xiaowan	298.1 × 30.6 × 82.0	Granitic gneiss	14.9	AEP-II+ WEP-VI	WEP-III	BEP-I
Pubugou	294.1 × 30.7 × 70.2	Granite	13.5	AEP-II	WEP-II+ WEP-VI	BEP-II
Xiluodu	443.3 × 31.9 × 75.6	Basalt	10.0	AEP-II	WEP-III	BEP-I
Laxiwa	311.7 × 30.0 × 75.0	Granite	9.7	AEP-II	WEP-III	BEP-I
Sanxia	329.5 × 32.6 × 86.2	Granite and diorite	5.0	AEP-II	WEP-II+ WEP-V	BEP-I
Xiangjiaba	255.4 × 33.4 × 88.2	Sandstone	6.7	AEP-II	WEP-I	BEP-I
Jinping I	276.9 × 25.9 × 68.8	Marble	20.0–35.7	AEP-II	WEP-V	BEP-I
Jinping II	352.4 × 28.3 × 72.2	Marble	12.0–24.0	AEP-II	WEP-V	BEP-I
Nuozhadu	418.0 × 31.0 × 81.7	Granite	3.5	AEP-II	WEP-II	BEP-I

row spacing is 1.5 m × 1.5 m. The anchor cable length is 15 m or 20 m, row spacing is 4.5 m × 4.5 m, and locking force is 1500 kN or 1750 kN. For the transformer chamber arch, the bolt diameter is 28 mm, length is 6 m, and row spacing is 1.5 m × 1.5 m. For its sidewall, the diameter is 32 mm, and length is 6 m or 9 m, arranged alternately. Three rows of pre-stressed anchor cables with a 2000 kN locking load are designed to support the rock mass between the transformer chamber and powerhouse, and another three rows of cables for the rock mass between the transformer chamber and pressure adjustment shaft, and their row spacing is 4.5 m × 4.5 m. For the pressure adjustment shaft arch, the bolt diameter is 32 mm, length is 6 m, and row spacing is 1.5 m × 1.5 m. For its sidewall, the diameter is 32 mm, and length is 6 m or 9 m, arranged alternately. Its anchor cable length is 15 m or 20 m, the row spacing is 4.5 m × 4.5 m, and locking force is 1500 kN or 1750 kN.

Zhang (2011) describes the supporting parameters of the Jinping I hydropower station underground main powerhouse as follows. Arch area: (1) Steel mesh with $\phi 8$ mm @ 200×200 mm, shotcrete with C25 concrete and 150–200 mm thickness. (2) Mortar bolt diameter 32 mm and length 7 m, pre-stressed bolt diameter 32 mm and length 9 m, pre-stress 120 kN, average row spacing 1.2 m × 1.4 m, alternative arrangement. Sidewall area: (1) Steel fibre reinforced shotcrete with C30 concrete and 50 mm thickness, steel mesh with $\phi 8$ mm @ 200×200 mm, shotcrete with C25 concrete and 100 mm thickness. (2) Mortar bolt and pre-stressed anchor diameter 32 mm, row spacing 1.0 m × 1.5 m, mortar bolt length 6 m or 9 m, pre-stressed bolt length 12 m, anchor cable length 15 m or 20 m, row spacing 3.0 m × 3.0 m, locking load 1750 kN. Arch area of transformer chamber: (1) Steel mesh with $\phi 8$ mm @ 200×200 mm, shotcrete with C25 concrete and 150–200 mm thickness. (2) Mortar bolt diameter 32 mm and length 7 m, pre-stressed bolt diameter 32 mm and length 9 m, pre-stress 120 kN, row spacing 1.2 m × 1.5 m. (3) F_{14} fault reinforced with shotcrete and arch ribs. Sidewall area: (1) Steel fibre reinforced shotcrete with C30 concrete and 50 mm thickness, steel mesh with $\phi 8$ mm @ 200×200 mm, shotcrete with C25 concrete and 100 mm thickness. (2) Mortar bolt diameter 28 mm, row spacing 1.5 m × 1.5 m, length 6 m or 9 m, anchor cable length 15 m or 20 m, row spacing 3.0 m × 3.0 m, locking force 1750 kN.

Xiang (2010) described the supporting parameters of the Jinping II hydropower station main powerhouse as follows. Arch area: (1) Steel mesh with $\phi 8$ mm @ 200×200 mm, shotcrete with C25 concrete and 150 mm thickness. (2) Pre-stressed bolt diameter 28 mm or 32 mm, row spacing 1.5 m × 1.5 m, length 6 m or 8 m, pre-stress 120 kN. Sidewall area: (1) Shotcrete with C25 concrete and 150 mm thickness, steel mesh with $\phi 8$ mm @ 200×200 mm. (2) Mortar bolt diameter 28 mm, length 6 m, row spacing 1.5 m × 1.5 m; pre-stressed bolt diameter 32 mm, row spacing 1.5 m × 1.5 m, length 9 m, pre-stress 120 kN; anchor cable length 20 m or 45 m, row spacing 4.5 m × 4.5 m, locking force 1500 kN or 1750 kN. Arch area of transformer chamber: (1) Steel mesh with $\phi 8$ mm @ 200×200 mm, shotcrete with C25 concrete and 150 mm thickness. (2) Pre-stressed bolt diameter 28 mm, length 7 m, pre-stress 120 kN, row spacing 1.5 m × 1.5 m. Sidewall area: (1) Steel mesh with $\phi 8$ mm @ 20×20 mm, shotcrete with C25 concrete and 120 mm thickness. (2) Mortar bolt diameter 28 mm, row spacing 1.5 m × 1.5 m, length 5 m; pre-stressed bolt diameter 32 mm, length 7 m, pre-stress 120 kN; anchor cable length 45 m, row spacing 4.5 m × 4.5 m, locking load 1750 kN.

Further experience of the support parameters for various hydropower station large cavern groups located in China, developed by the Company Standard of Hydrochina Corporation (Q/HYDROCHINA 009-2012), can be seen in Table 7.7.

Table 7.7 Support types and associated parameters for large-scale underground cavern groups (see notes below the Table).

Rock type	Location	Cavern span (m) 20 < B ≤ 25	25 < B ≤ 30	30 < B ≤ 35
I	Arch	Mortar bolt ϕ = 220–250 mm, L = 4.0–6.0 m @ 1.5–2.5 m; shotcrete δ = 1000–1500 mm; local steel mesh	Mortar bolt ϕ = 250–280 mm, L = 5.0–8.0 m @ 1.5–2.0 m; shotcrete δ = 120–150 mm; local steel mesh	Mortar bolt ϕ = 28–32 m, L = 6.0–9.0 m @ 1.5–2.0 m; shotcrete δ = 120–200 mm; local steel mesh
	Sidewall	Mortar bolt ϕ = 220–250 mm, L = 4.0–6.0 m @ 1.5–2.5 m; shotcrete δ = 100–120 mm; local steel mesh	Mortar bolt ϕ = 250–280 mm, L = 5.0–8.0 m @ 1.5–2.0 m; shotcrete δ = 100–150 mm; local steel mesh	Mortar bolt ϕ = 28–32 m, L = 6.0–8.0 m @ 1.5–2.0 m; shotcrete δ = 120–200 mm; local steel mesh
II	Arch	Mortar bolt ϕ = 250–280 mm, L = 5.0–8.0 m @ 1.5 m, local longer or anchor bar; shotcrete or steel mesh and shotcrete δ = 150–200 mm;	Mortar bolt ϕ = 280–320 mm, L = 6.0–9.0 m @ 1.5 m, local longer or anchor bar; steel fibre reinforced shotcrete or shotcrete with steel mesh $\delta \geq$ 20 mm	Mortar bolt ϕ = 280–320 mm, L = 8.0–9.0 m @ 1.2–1.5 m, local longer or anchor bar; steel fibre reinforced shotcrete or shotcrete with steel mesh δ = 200–250 mm
	Sidewall	Mortar bolt ϕ = 250–280 mm, L = 5.0–8.0 m @ 1.5 m, local longer or anchor bar; shotcrete δ = 150–200 mm; anchor cable L = 15.0–20.0 m @ 4.0–6.0 m, T = 1500–2000 kN	Mortar bolt ϕ = 280–320 mm, L = 6.0–9.0 m @ 1.5 m, local longer or anchor bar; steel fibre reinforced shotcrete or shotcrete with steel mesh $\delta \geq$ 200 mm; anchor cable L = 15.0–20.0 m @ 4.0–4.5 m, T = 1500–2000 kN	Mortar bolt ϕ = 280–320 mm, L = 8.0–9.0 m @ 1.2–1.5 m, local longer or anchor bar; steel fibre reinforced shotcrete or shotcrete with steel mesh δ = 200–250 mm; anchor cable L = 15.0–20.0 m @ 3.0–4.5 m, T = 1500–2000 kN
III	Arch	Mortar bolt ϕ = 250–320 mm, L = 6.0–8.0 m @ 1.2–1.5 m, local pre-stressed anchor; steel fibre reinforced shotcrete or shotcrete with steel mesh δ = 150–200 mm	Mortar bolt ϕ = 280–320 mm, L = 6.0–9.0 m @ 1.2–1.5 m, mortar bolt and pre-stressed anchor arranged alternately; steel fibre reinforced shotcrete or shotcrete with steel mesh $\delta \geq$ 200 mm	Mortar bolt ϕ = 280–320 mm, L = 8.0–9.0 m @ 1.2–1.5 m, mortar bolt and pre-stressed anchor arranged alternately; local anchor cable; steel fibre reinforced shotcrete or shotcrete with steel mesh δ = 200–250 mm, local shotcrete arch ribs

(Continued)

Table 7.7 (Continued)

Rock type	Location	Cavern span (m)		
		$20 < B \leq 25$	$25 < B \leq 30$	$30 < B \leq 35$
	Sidewall	Mortar bolt ϕ = 250–320 mm, L = 6.0–8.0 m @ 1.2–1.5 m, local longer; shotcrete or shotcrete with steel mesh δ = 150–180 mm; sometime anchor cable L = 15.0–20.0 m @ 4.0–4.5 m, T = 1500–2000 kN	Mortar bolt ϕ = 280–320 mm, L = 6.0–9.0 m @ 1.2–1.5 m, local anchor bar; steel fibre reinforced shotcrete or shotcrete with steel mesh δ = 150–200 mm; anchor cable L = 15.0–20.0 m @ 3.0–4.5 m, T = 1500–2000 kN	Mortar bolt ϕ = 300–320 mm, L = 8.0–9.0 m @ 1.2–1.5 m, local anchor bar; steel fibre reinforced shotcrete or shotcrete with steel mesh $\delta \geq$ 200 mm; anchor cable L = 15.0–20.0 m @ 3.0–4.5 m, T = 2000–2500 kN
IV	Arch	Mortar bolt ϕ = 250–320 mm, L = 6.0–8.0 m @ 1.0–1.2 m, mortar bolt and pre-stressed anchor arranged alternately, local longer pre-stressed anchor; steel fibre reinforced shotcrete or shotcrete with steel mesh $\delta \geq$ 200 mm, systematic steel arch or shotcrete grid arch @ 0.8–1.2 m; reinforcing concrete lining if necessary	Mortar bolt ϕ = 280–320 mm, L = 6.0–9.0 m @ 1.0–1.2 m, mortar bolt and pre-stressed anchor arranged alternately, some area with longer pre-stressed anchor or anchor cable; steel fibre reinforced shotcrete or shotcrete with steel mesh $\delta \geq$ 200 mm, systematic steel arch or shotcrete grid arch @ 0.8–1.2 m; reinforcing concrete lining arch @ 0.8–1.0 m if necessary	Mortar bolt ϕ = 300–320 mm, L = 8.0–9.0 m @ 0.8–1.0 m, mortar bolt and pre-stressed anchor arranged alternately, local anchor cable; steel fibre reinforced shotcrete or shotcrete with steel mesh $\delta \geq$ 200–250 mm, systematic steel arch or shotcrete grid arch @ 0.8–1.0 m; reinforcing concrete lining thickness 1.2–1.6 m
	Sidewall	Mortar bolt ϕ = 250–320 mm, L = 6.0–8.0 m @ 1.0–1.2 m, mortar bolt and pre-stressed anchor arranged alternately, local longer pre-stressed anchor or anchor cable; steel fibre reinforced shotcrete or shotcrete with steel mesh δ = 180–200 mm, systematic steel arch or shotcrete grid arch @ 0.8–1.2 m; anchor cable L = 15.0–20.0 m @ 4.0–4.5 m, T = 1500–2000 kN; reinforcing concrete lining if necessary	Mortar bolt ϕ = 280–320 mm, L = 6.0–9.0 m @ 1.0–1.2 m, mortar bolt and pre-stressed anchor arranged alternately, local anchor bar; steel fibre reinforced shotcrete or shotcrete with steel mesh $\delta \geq$ 200 mm, systematic steel arch or shotcrete grid arch @ 0.8–1.2 m; anchor cable L = 20.0–25.0 m @ 3.0–4.5 m, T = 1500–2000 kN; reinforcing concrete lining if necessary	Mortar bolt ϕ = 300–320 mm, L = 8.0–9.0 m @ 1.0–1.2 m, mortar bolt and pre-stressed anchor arranged alternately, local anchor bar; steel fibre reinforced shotcrete or shotcrete with steel mesh $\delta \geq$ 200–250 mm, systematic steel arch or shotcrete grid arch @ 0.8–1.0 m; anchor cable L = 20.0–25.0 m @ 3.0–4.5 m, T = 2000–2500 kN; reinforcing concrete lining thickness 1.2–1.6 m

V		Span columns		
V	Arch	Mortar bolt ϕ = 250–320 mm, L = 6.0–9.0 m @ 0.8–1.0 m, mortar bolt and pre-stressed anchor arranged alternately, local longer pre-stressed anchor; steel fibre reinforced shotcrete or shotcrete with steel mesh $\delta \geq$ 200 mm, systematic steel arch or shotcrete grid arch @ 0.8–1.2 m; reinforcing concrete lining thickness 1.0–1.5 m	Mortar bolt ϕ = 280–320 mm, L = 8.0–9.0 m @ 0.5–0.8 m, mortar bolt and pre-stressed anchor arranged alternately, systematic anchor cable; steel fibre reinforced shotcrete or shotcrete with steel mesh $\delta \geq$ 200–250 mm, systematic steel arch or shotcrete grid arch @ 0.8–1.0 m; reinforcing concrete lining thickness 1.2–1.6 m	Mortar bolt ϕ = 300–320 mm, L = 9.0–12.0 m @ 0.5–0.8 m, mortar bolt and pre-stressed anchor arranged alternately, systematic anchor cable; steel fibre reinforced shotcrete or shotcrete with steel mesh $\delta \geq$ 250 mm, systematic steel arch or shotcrete grid arch @ 0.5–0.8 m; reinforcing concrete lining thickness 1.5–2.0 m
	Sidewall	Mortar bolt ϕ = 250–320 mm, L = 6.0–9.0 m @ 0.8–1.0 m, mortar bolt and pre-stressed anchor arranged alternately, local longer anchor or anchor bar; steel fibre reinforced shotcrete or shotcrete with steel mesh $\delta \geq$ 200 mm, systematic steel arch or shotcrete grid arch @ 0.8–1.0 m; anchor cable L = 15.0–20.0 m @ 4.0–4.5 m, T = 1500–2000 kN; reinforcing concrete lining thickness 1.0–1.5 m	Mortar bolt ϕ = 280–320 mm, L = 8.0–9.0 m @ 0.8–1.0 m, mortar bolt and pre-stressed anchor arranged alternately, local anchor bar; steel fibre reinforced shotcrete or shotcrete with steel mesh $\delta \geq$ 200–250 mm, systematic steel arch or shotcrete grid arch @ 0.8–1.0 m; anchor cable L = 20.0–25.0 m @ 3.0–4.5 m, T = 2000–2500 kN; reinforcing concrete lining thickness 1.2–1.6 m	Mortar bolt ϕ = 300–320 mm, L = 9.0–12.0 m @ 0.8–1.0 m, mortar bolt and pre-stressed anchor arranged alternately, systematic anchor bar; steel fibre reinforced shotcrete or shotcrete with steel mesh $\delta \geq$ 250 mm, systematic steel arch or shotcrete grid arch @ 0.5–0.8 m; anchor cable L = 20.0–25.0 m @ 3.0–4.0 m, T = 2500–3000 kN; reinforcing concrete lining thickness 1.5–2.0 m

Notes:

1 B—cavern width (m); L—anchor bolt or anchor cable length (m); @—anchor bolt, anchor cable, steel arch or shotcrete grid arch span (m); ϕ—bolt diameter (mm); δ—steel fibre reinforced shotcrete or shotcrete thickness (mm); T—bolt or cable designed tension (kN).

2 The underground caverns with span exceeding 20 m or 25 m are not suitable when the rock mass quality is IV or V. If they are to be built, special demonstrations are required. The supporting parameters in the Table are only for reference.

3 The supporting design parameters for the caverns should be adjusted according to their location in the cavern group.

7.2.3.7 Monitoring

The deformation monitoring data from the cavern group after excavation are collected, including the deformation of the main powerhouse arch, transformer chamber arch, upstream and downstream sidewall and rock anchor beam. The relations between the deformations at different locations, cavern sizes and rock strength–stress ratios are shown. The statistical results can be seen in Table 7.8. The relation between rock anchor beam displacement and strength–stress ratio is shown in Table 7.9. The statistical relations between monitored deformation of the rock anchor beam and cavern span, rock strength–stress ratio can be seen in Figures 7.15 and 7.16.

7.2.3.8 Rockbolt stresses

The anchor stress monitoring results for some of the hydropower station main powerhouses have been collected in order to analyse the bolt stresses under different conditions. The results can be seen in Table 7.10. According to these statistics, the anchor stress monitored at the Pubugou hydropower station is greater than the other projects, and at some times being greater than the bolt tensile strength.

The rockbolt stress data results for the Jinping II powerhouse and transformer chamber after excavation are shown in Figure 7.17.

7.2.3.9 Stress in cable anchors

Anchor cable force monitoring results for large cavern groups have been collected and are listed in Table 7.11. The anchor cable force data results for the Jinping II powerhouse and transformer chamber after excavation are shown in Figure 7.18. It can be observed that the number of anchor cables containing greater force than the design value is 30%. These anchor cables are most numerous in the high stress areas, which are at the powerhouse downstream skewback area and the transformer chamber upstream bottom area.

7.2.3.10 Relaxation depth of the surrounding rock

The acoustic wave test results for the underground cavern group show that most rock failure depths are less than 5 m, and the unloading relaxation depth is less than the cavern span. The relaxation depth test results for typical sections of different cavern groups, such as the Ertan, Dagangshan, Pubugou, Xiluodu and Jinping I hydropower stations, can be seen in Figures 7.19–7.23. The data are shown in Table 7.12.

The field test results for the Jinping I hydropower station powerhouse showed that the acoustic wave velocity of the pre-construction surrounding rock was around 6000 m/s; while for the unloading rock mass as a result of construction it was less than 4500 m/s. According to the acoustic wave and deformation test results for the main powerhouse peripheral rock, the changes in the relaxation zone can be seen in Figure 7.24.

The distribution and changes of the main powerhouse peripheral rock relaxation zone are described as follows.

Table 7.8 Displacement data results for different cavern groups.

Project name	Lithology	Maximum in situ stress component (MPa)	Cavern	Cavern size/m	Monitoring displacement at different sites		
					Vault (mm)	Upstream sidewall (mm)	Downstream sidewall (mm)
Manwan	Rhyolite		Main powerhouse	73 × 22 × 69	9.13	32.04	17.50
Longtan	Sandstone	12	Main powerhouse	398.9 × 29 × 75.1	4.6	80.3	40.1
Xiaolangdi	Sandstone	5	Main powerhouse	251.5 × 26.2 × 61.4	17	23.15	10.5
			Transformer chamber	174.7 × 14.4 × 17.9	1.5	8.5	5.5
Suofengying	Limestone	14	Main powerhouse	133 × 24 × 7	0.12	4.87	4.14
			Transformer chamber	75.0 × 15.3 × 25.4	0	0.14	3.12
Ertan	Syenite	38	Main powerhouse	280.3 × 30.7 × 65.4	7.7	117.4	124.17
			Transformer chamber	214.9 × 18.3 × 25	2.53	185.2	44.47
Dachaoshan	Basalt	17	Main powerhouse	233.9 × 26.4 × 62.9	1.95	14.37	29.44
Sanbanxi	Sandstone	14.5	Main powerhouse	147.5 × 22.7 × 60.1	3.91	8.3	9.53
Tongbai	Granite	–	Main powerhouse	172 × 24 × 54.8	–1.2	6.67	9.76
Lubuge	Dolomite	–	Main powerhouse	125 × 18 × 38.4	–	30.5	–
Shisanling	Conglomerate	10–15	Main powerhouse	145 × 23 × 46.6	15.23	29.95	40.50
			Transformer chamber	141.7 × 16.5 × 25.6	9.54	30.98	26.24
Jiangya	Granite	–	Main powerhouse	108 × 20 × 47	2.19	5.7	4.5
Laxiwa	Granite	22–29	Main powerhouse	129.5 × 21.9 × 52.08	4.2	54.07	31.67
			Transformer chamber	107.5 × 16.0 × 20.1	2.01	45.39	11.74
Xiaowan	Granite	27	Main powerhouse	298.4 × 30.6 × 79.38	10.2	112.7	65.5

Table 7.9 Rock anchor beam displacement and strength-stress ratio.

Project name	Powerhouse upstream sidewall			Powerhouse downstream sidewall		Mean of sidewall maximum displacement (mm)	Mean strength: maximum stress ratio
	Anchor beam largest displacement (mm)	Anchor beam mean displacement (mm)	Sidewall mean displacement (mm)	Anchor beam largest displacement (mm)	Anchor beam mean displacement (mm)		
Ludila	34.49			65.67		50.08	6.97
Jinpingerji	90.09	65.50	38.28	43.63	11.05	66.86	2.84
Jinpingyiji	31.81	23.20	39.80	75.87	56.40	53.84	1.89
Laxiwa	35.60	18.60	22.15	28.40	25.70	32.00	3.70
Pubugou	65.12	39.76	37.46	77.39	35.16	71.26	3.30
Dagangshan	34.41	19.69	16.29	21.92	12.89	28.17	3.68
Guandi	36.05	18.72	15.53	23.57	12.33	29.81	4.14
Xiluoduzuoan	15.20	8.90	14.55	41.10	20.20	28.15	6.00
Xiluoduyouan	33.40	13.90	12.60	22.80	11.30	28.10	5.48

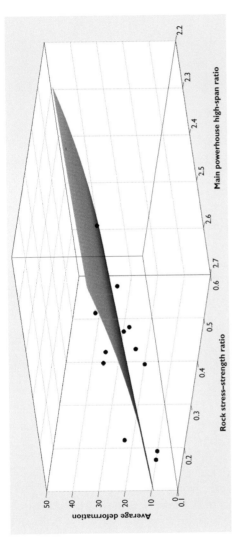

Figure 7.15 Statistical result for the rock anchor beam mean deformation, $z = 49.42x^2 + 16.43xy + 2.817y^2 + x + y + 22.06$, $R^2 = 0.7477$, where, z is the deformation of the rock anchor beam (mm); x is the rock strength stress ratio; y is the main powerhouse high-span ratio.

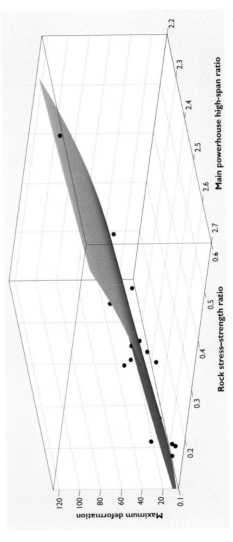

Figure 7.16 Statistical result for the rock anchor beam maximum deformation, $z = 278.9x^2 + 19.4xy - 6.583y^2 + x + y + 46.86$ $R^2 = 0.8959$, where z is the rock strength stress ratio; y is the main powerhouse high-span ratio; x is the maximum deformation of the rock anchor beam (mm).

Table 7.10 Monitoring results for main powerhouse anchor stress (MPa).

Project name	Vault			Upstream sidewall			Downstream sidewall		
	Maximum	Minimum	Mean	Maximum	Minimum	Mean	Maximum	Minimum	Mean
Shuibuya	160		100						
Dachaoshan	112			238	44		135	78	
Manwan	126	30		124	60		177	87	
Goupitan							137		
Jinpingyiji	44	15					232	60	
Nuozhadu	164			120			126		
Pubugou	320		80	426		160	476		220
Dagangshan	167		139	247		217	413		119
Xiluodu				293		<100			
Xiangjiaba	50			50					

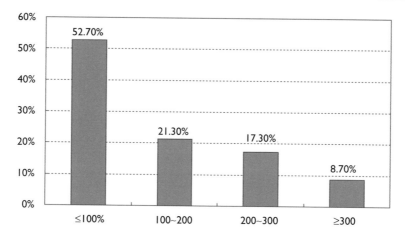

Figure 7.17 Rockbolt stress distribution for the Jinping II powerhouse and transformer chamber.

1 The surrounding rock relaxation depths around the main powerhouse, upstream and downstream, are large, and their relaxation depths are 6.5 and 8 m, respectively. The relaxation depth of the vault is small, generally less than 2 m.

2 The relaxation depth of the surrounding rock is mainly attributed to blasting damage depth and unloading relaxation depth, being between 6.5 and 8 m.

3 The relaxation depths at the top and bottom of the upstream sidewall are similar. However, for the downstream sidewall, the relaxation depth at the top is greater than that at the bottom.

4 The relaxation zone of the upstream sidewall expands downward during the excavation, but the downstream sidewall relaxation zone develops more in the horizontal direction than in the vertical direction.

7.2.3.11 Fractures in the surrounding rock mass

The shotcrete cracks in the Jinping I underground cavern group first appeared in the 0+101–0+145 m section and at an elevation of 1670–1672 m in the main powerhouse

Table 7.11 Anchor cable force monitoring results for the main powerhouse in different hydropower projects.

Project name	Elevation	Locking load	Current or failure load	Mean load	Load increment	Proportionate increase	Failure date
Dachaoshan	Upstream 815.25	1724.2	2454.8		730.6	42.4	2001.11.13
	Upstream 825.75	1707.2	2250.8		543.6	31.8	2001.10.12
	Vault	820	997		177	20.9	
	Vault	886.4	528.9		−357.5	40.3	
	Downstream 825.75	1005.8	1891.7		885.6	88	2002.11.12
	Downstream 815.25	867.4	1133.5		266.1	30.7	2001.03.11
	Downstream 808	1397.4	1684.8		287.4	20.6	2001.03.11
	Upstream 815.25	1931.6	2472.7		541.1	28	2002.10.12
Goupitan		70%–85%				2.9%–21.34	
Jinpingyiji							
Xiaowan	EL1019.5–1007	1000	946.3–1443 3	1100		0.94–1.44	
	EL1003.5980	1800	1617–1899	1750		0.9–1.05	
	EL1005.2–982.7	1800	1602–2441	2000		0.89–1.34	
Dagangshan							
Guandi		80%			15%–18%		
Xiluodu					10%–24%		
Xiangjiaba		1700–1890	1156–2390	1630.2		−5.76–7.36	

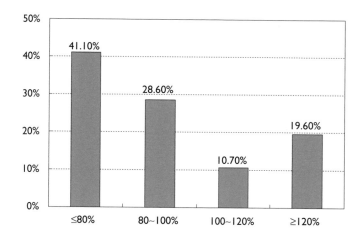

Figure 7.18 Anchor cable force distribution in the Jinping II powerhouse and transformer chamber, as percentages of the design values.

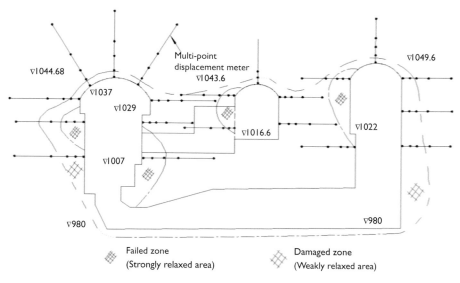

Figure 7.19 Rock mass relaxation depths at Ertan underground cavern group, typical section (Li, 1997).

downstream in late April 2008; then later the cracks continued developing. By August 2008, there were many shotcrete layer cracks near the arch crown of the powerhouse downstream side, which could be seen clearly in the 0–005–0+170 m section, especially in the section 0+115–0+145 m. The cracks which appeared first were discontinuous, extending in a sawtooth pattern, with the overall extension direction being horizontal. The fracture opening width was about 20–60 mm. Some shotcrete layers had collapsed and the cracks were formed by the expanding and shearing of the shotcrete layers. In fact, the shotcrete layers were found to be separated from the rock mass surface when the cracked shotcrete layers were stripped in the 0+132–0+185 m section at elevation 1670–1671 m of the powerhouse downstream vault in March 2009.

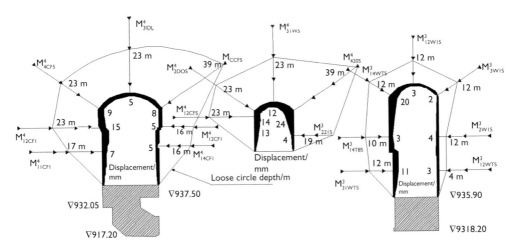

Figure 7.20 Rock mass relaxation depths at Dagangshan cavern group, typical section (Li & Wu, 2011).

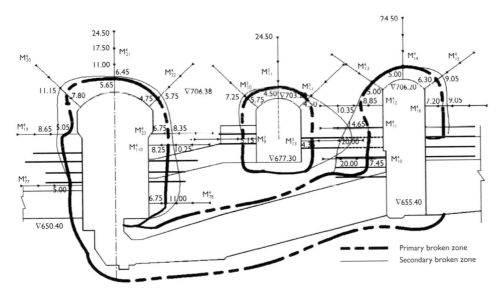

Figure 7.21 Rock mass relaxation depths at Pubugou underground cavern group, typical section (Zou & Xiao, 2010).

Brittle damage occurred in the fresh marble rock mass in the vault. The types of damage were mainly splitting and bending, as well as local crushing. The damaged rock mass shape was irregular slab like or in the form of rock fragments having 100–200 mm thickness. The fracture surfaces were fresh and relatively flat. Their occurrence was commonly orientated at N50–60°W/NE<30–40°. Rock beams were formed by splitting at the arch parallel with the excavation surface and the steel arch rib bent into the powerhouse under the load. These features can be seen in Figure 7.25.

In December 2008, some shotcrete layer cracks were found in the upstream sidewall of the transformer chamber, located between the bus tunnels and at the bottom

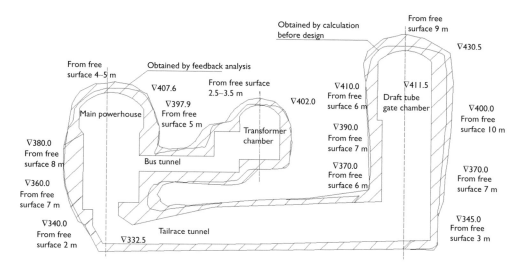

Figure 7.22 Rock mass relaxation depths at Xiluodu underground cavern group, typical section (Li *et al.*, 2013).

of the sidewall. Their strike direction mainly tended towards the outside of the mountain, with a dip angle of 50–70°. They were tensile cracks, with flat sides and no shear movement, as shown in Figure 7.26 with their opening width being 1–3 mm at first; then the cracks continued expanding, and the maximum opening width reached 10 mm by March 2009. The cracks in the sidewall near the bus tunnels changed the most.

When layer IV of the transformer chamber was excavated, the cracks near the downstream arch foot occurred first, and then expanded rapidly. By December 2008, the shotcrete layer cracks appeared almost everywhere throughout the downstream sidewall, especially near the arch foot, as can be seen in Figure 7.27. There were shotcrete layer cracks near the arch foot of the 0+000–050 m section and skewback of the 0+130–142 m section (the 'skewback' is the sloping face of the abutment on which the extremity of an arch rests). The character of the cracks near the transformer chamber downstream were similar to those of the main powerhouse. The lengthwise cracks extended in a sawtooth pattern and their opening widths were 5–100 mm. The damage near arch foot and skewback was mainly shear failure, which caused the shotcrete layer to crack and collapse.

Ring cracks were mainly distributed near the #6 bus tunnel, contacting the hole and cable tunnel. The distance between the cracks and powerhouse or transformer chamber was usually 6–10 m, and the maximum distance was about 15 m, located near the #2 bus tunnel arch crown. The cracks were generally flat and nearly vertical, as shown in Figure 7.28.

According to the crack monitoring results of the cavern groups during excavation, among the smaller hydropower stations, such as Shisanling, Guangxu, Jiangya and Dachaoshan, there may be 1–2 bus tunnel ring cracks within 2 m of the powerhouse downstream sidewall. However, among the larger hydropower stations, such as Longtan

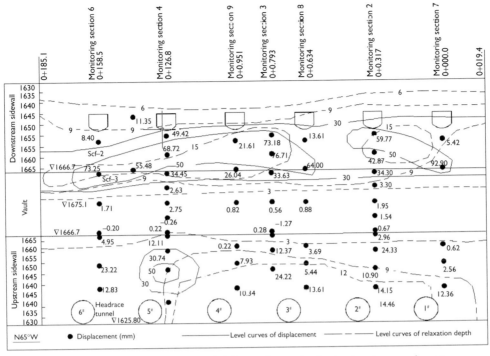

a Rock mass relaxation depths and deformation at the Jinping I main powerhouse.

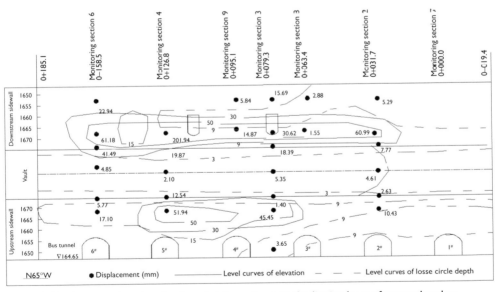

b Rock mass relaxation depth and deformations at the Jinping I transformer chamber.

Figure 7.23 Rock mass relaxation depth test results at the Jinping I main powerhouse and transformer chamber (Wei et al., 2010).

Table 7.12 Relaxation depth and failure depth test result of different hydropower underground cavern group.

Project name	Main powerhouse					Transformer chamber				
	Location or type	Vault (m)	Upstream sidewall (m)	Downstream sidewall (m)	Mean value (m)	Location or type	Vault (m)	Upstream sidewall (m)	Downstream sidewall (m)	Mean value (m)
Jinping II	Failure depth	2.8/1.4	3.4/1.2	3.6/1.2			3	3.2/2.0	3.0/1.6	6.5–8
Jinping I	Relaxation depth				6.5–8					
Nuozhadu	Failure depth				0.6					
Laxiwa	Failure depth		2.2/1.4	3.0/2.0				1.6	2.4/1.4	
Ertan	Failure depth				5/2.5					
Pubugou	#3–#4 relaxation depth			20				12		
	#4–#5 relaxation depth			15				15		
Dagangshan	0+135–158	39/23	11	11		0+102–140	39/23	23	19	
	0+116–130		23	9		0+64–89	23/11	23	10	
	0+74–88	39/23	23/17	16		0+1–30	39			
	0+17–23	23/11	23/11	27/16						
	0–28–38	11	11	11						
Guandu	Failure depth (II)	1.1/0.6				II	1.0			
	Failure depth (III)	3.3/0.9				III	1.7/0.5			
Xiluodu left	General relaxation	3/1	local	12						
Xiluodu right	Local relaxation depth	5/3			6					
Pengshui	Strong unloading	4.5/3.5								
	Weak unloading	7.1/6.8								
Silin	Strong unloading	4.35/3.8								
	Medium unloading	7.1/6.8								

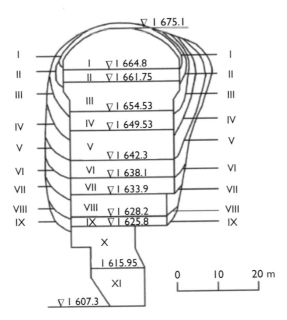

Figure 7.24 Disturbed rock zones around the Jinping I main powerhouse as a function of the progressive excavated layers (Wei & Deng, 2010).

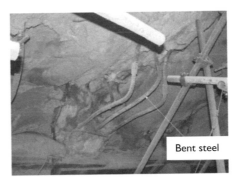

Bent steel

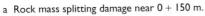

a Rock mass splitting damage near 0 + 150 m.   b Steel rib archbending damage near 0 + 120 m.

Figure 7.25 Surrounding rock mass failure phenomena in Jinping I main powerhouse vault (Lu et al., 2010).

and Ertan, there may be multiple ring cracks within 8 m of the powerhouse downstream sidewall, and their widths may be greater than 10 mm. There were multiple ring cracks in the bus tunnel 15 m near the powerhouse downstream sidewall in the Laxiwa hydropower station, as can be seen in Figure 7.29, the maximum crack width being 15 mm.

7.2.3.12 Typical failure modes

Typical failure modes of the large hydropower cavern groups were analysed, and the data can be seen in Table 7.13.

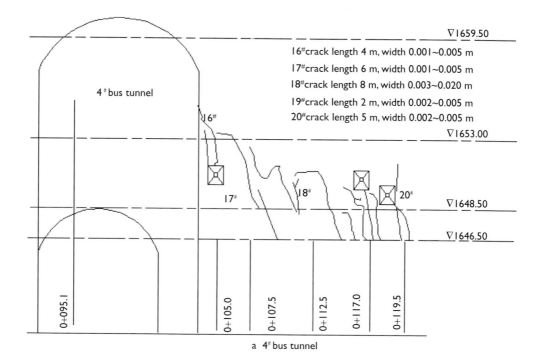

∇1659.50

16#crack length 4 m, width 0.001~0.005 m
17#crack length 6 m, width 0.001~0.005 m
18#crack length 8 m, width 0.003~0.020 m
19#crack length 2 m, width 0.002~0.005 m
20#crack length 5 m, width 0.002~0.005 m

∇1653.00

∇1648.50

∇1646.50

4# bus tunnel

16#

17#   18#   20#

0+095.1   0+105.0   0+107.5   0+112.5   0+117.0   0+119.5

a 4# bus tunnel

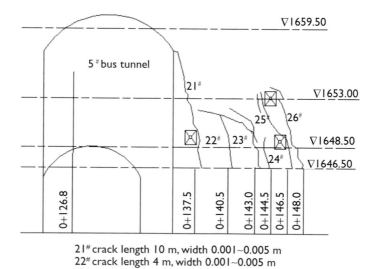

∇1659.50

5# bus tunnel

21#

∇1653.00

25#   26#

22#   23#   ∇1648.50

24#   ∇1646.50

0+126.8   0+137.5   0+140.5   0+143.0   0+144.5   0+146.5   0+148.0

21# crack length 10 m, width 0.001~0.005 m
22# crack length 4 m, width 0.001~0.005 m
23# crack length 5 m, width 0.001~0.003 m
24# crack length 2 m, width 0.001~0.003 m
25# crack length 6 m, width 0.003~0.007 m
26# crack length 7 m, width 0.002~0.003 m

b 5# bus tunnel

Figure 7.26 Crack distribution in the Jinping I transformer chamber, upstream sidewall (Lu *et al.*, 2010).

Figure 7.27 Shotcrete layer cracks near the Jinping I transformer chamber downstream skewback (Lu *et al.*, 2010). The 'skewback' is the sloping face of the abutment on which the extremity of an arch rests.

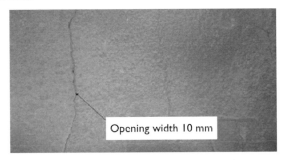

Figure 7.28 Ring crack in Jinping I 2# bus tunnel (Lu *et al.*, 2010).

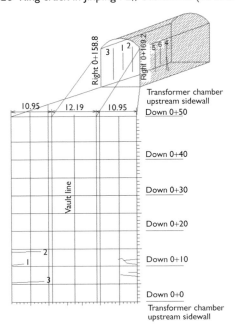

Figure 7.29 Crack distribution at Laxiwa hydropower station bus tunnel (Li *et al.*, 2008).

Table 7.13 Typical failure modes in the large hydropower cavern groups.

Project name	Typical failure modes
Baise	The unstable blocks were formed by the intersection and combination of several groups of fractures. The cohesion between the structures was weak. The excavation blasting caused the joints and structures to open, or the opening width to increase in size.
Longtan	The dip angle of the layered rock mass is high. There were many weak and steep structural surfaces, which were bedding surfaces, and bedding fault zones and joints. The stability of the surrounding rock was poor. Some local rock blocks were unstable. The force in some bolts underwent large and sudden changes.
Dagangshan	The caverns were excavated at shallow depths, with a thin rock mass overlying the main powerhouse arch crown and the cavern width is large. There are many fractures in the rock mass.
Ertan	Rockbursts occurred in some local areas, the surrounding rock spalled and ring cracks appeared along the bus tunnel.
Jinping I	During the excavation process, large deformations in the surrounding rock mass were clear, the shotcrete layer near the main powerhouse downstream spalled severely, fractures in the high sidewall moved, and the rock mass exfoliated. At the same time, the disturbed zone depth became larger and the bolt/cable loads were greater than their design values.
Guandi	The bedding fault zones and fractures were the main factors affecting the surrounding rock mass stability. Rock block instability problems were severe.
Jinping II	The most unfavourable geological conditions for the upstream side were the steep bedding surfaces, fault F65 and the associated joints. The rock mass was cut into blocks by the faults and joints, and some blocks collapsed.
Baishan	The distance between the transformer chamber and powerhouse is only 16.5 m, which is 0.66 times that of the main powerhouse span. This is smaller than the value given by the standard. The deformation of the sidewall was large. The bolt and shotcrete support could not satisfy the stability of the rock mass.
Manwan	A rock block near the arch crown was formed by the faults. The geological condition near the 23–80 m section was poor.
Shisanling	There were some faults and weak fracture zones.
Yixing	There were unstable blocks near the powerhouse arch crown which were caused by the faults and joints.
Sanbanxi	The local stability of the arch crown and sidewall was poor, which was affected by the weak structures. Rock blocks collapsed easily during the construction.
Pubugou	Rockburst damage occurred.
Dagangshan	There were unstable blocks near the arch crown. The rock mass collapsed along with the dyke strike direction. The fracture zone width of the hanging side was large and the heading side width was small.
Pengshui	Local surrounding rock was loose and fractured. The rock mass in the upstream sidewall toppled over. The rock mass in the downstream sidewall slipped along with the rock layer.
Silin	There were unstable blocks near the arch crown and downstream sidewall, which were affected by the bedding and fault zone and fractures. The rock mass in the downstream sidewall mainly slipped along the rock layer. The rock mass in the arch crown mainly collapsed along with the interlayers and fractures.
Suofengying	The surrounding rock mass stability was affected by the fault in the 0–0+11.7 m section.
Wujiangdu	The demand for seepage control was high because of the karst.
Goupitan	Developed karst caused rock outbursts and mud outbursts to occur readily. The karst scale was large, and many caverns were affected.
Xiaowan	The main failure mode was block collapse. There were relaxation and deformation phenomena occurring in some local sections, and there were cracks at the end of the powerhouse. The cracks near fault F10 expanded continuously.

(Continued)

Table 7.13 (Continued)

Project name	Typical failure models
Shuibuya	The soft rock thickness was 3.75 m in total, which accounted for 56% of the rock mass in the 6–9 m thickness in the upper main powerhouse. There were also four bedding fault zones, their quality was poor, and shear strength low. The stability problem was serious during the construction and operational periods.
Dachaoshan	There were soft tuff interlayers, their thickness being 0.3–1.0 m, and maximum thickness ~ 1.70 m. There were low-angle dip interlayers crossing through the main powerhouse. The tuff interlayer T4 crossed through the arch crown, which was extremely dangerous for the arch crown stability and construction safety.
Nuozhadu	The local surrounding rock mass was affected by the groundwater, especially to the level III structural surface.
Sanxia	The main rock mass failure modes were tensile rupture, slipping and shear failure, which were affected by excavation unloading and blasting.
Tianhuangping	The shotcrete layer cracks near the downstream arch foot were mainly affected by large displacement of the surrounding rock mass and uneven rock surfaces.
Xiaolangdi	There were three continuous mud layers within the 23 m thickness of the upper arch crown.
Laxiwa	There were unstable blocks near the arch crown, and ring cracks along the bus tunnels.
Yele	The extent of a planar structural surface was large, which was the main factor affecting the stability of the powerhouse arch crown. The blocks collapsed easily after excavation.
Yutan	The rock layer was thin. Some rock blocks fell down along the surface of the joints.
Jiangya	There were many ring cracks. Fault F15 was filled with mud and stones. Its thickness was 100–500 mm. The corrosion influence depth was greater than 3 m. The intersection angle between the fault and powerhouse axis was about 10°. The arch crown, skewback, rock-anchored beam and upstream sidewall were all affected by the faults.
Xiluodu	The main problem affecting the cavern stability was a weak zone. Sliding blocks could be formed by the bedding fault zones and fractures, which affected the stability of the arch crown and high sidewall. There were large plastic deformations near the fracture zone.
Xiangjiaba	The rock mass failure modes were mainly controlled by rock structures. The modes controlled by stress were mainly located in the areas where the stress concentrates and the rock was soft. The rock mass stability was affected by the soft interlayers.
Jiangkou	Rockbursts occurred during the engineering geological exploration, which in turn caused spalling of the surrounding rock and affected its stability.
Liujiaxia	Local collapse of the upstream sidewall occurred near the elevation of 1635 m during the excavation, which was affected by the fractures and excavation face.
Maerdang	The in situ stress was high. There were disc-shaped rock cores present during the drilling. Rockbursts may occur.
Shiziping	The unloading caused by excavation and structural surfaces combined with a high slope was disadvantageous to the stability of the powerhouse.
Dafa	One large fault which crosses through the arch crown and was revealed during the II layer excavation was extremely adverse to the long-term stability of the surrounding rock.
Xiaotiandou	The thickness of the fractures and fault zone was 20 mm. The effect on rock mass stability was slight.

(Continued)

Table 7.13 (*Continued*)

Project name	Typical failure models
Taian	The stability of the surrounding rock mass at the powerhouse end was poor. The strike direction of dyke β 24 was parallel with the axis, which was adverse to the sidewall stability.
Yantan	The stability of the caverns was affected by faults F48 and F211, a quartz dyke, and cavern intersections. Large displacements, stress concentrations and plastic zones usually appeared near the intersections of caverns and faults.
Foziling	There is a large fault fracture zone crossing through the main powerhouse. Its width is 400–600 mm, strike direction is N40W, dip direction is NE, and dip angle is 70–75°.
Langyashan	There was one alteration zone between fracture sets #1 and #2, which affected the stability of the surrounding rock mass.

7.2.3.13   Effect of loss of cable anchors and rockbolts

After the excavation of the Jinping II hydropower station underground powerhouse and transformer chamber, some anchor cables between the two excavations were shown to fail during the field inspection in September 2010. The failure modes are described below and can be seen in Figure 7.30.

1   **Steel cage loosening or collapsing:** The steel cage and steel shim of the anchorage pier loosened or fell off completely due to the impact produced by the sudden rupturing of the steel strands. This can be seen in Figure 7.30(a).
2   **Steel cage punctured by the steel strand:** The steel cage was punctured by the steel strands when they ruptured suddenly and moved with high speed and had large elastic strain energy. This can be seen in Figure 7.30(b).
3   **Steel strands catapulting outwards:** The steel strands catapulted outwards when they ruptured at high speed as a result of the released the elastic strain energy. This can be seen in Figure 7.30(c).
4   **Steel strand shrinking into the anchor hole:** The steel strands shrank into the anchor hole slowly if the clamp locking resistance was not sufficient during the increasing anchor cable load. This can be seen in Figure 7.30(d).
5   **Imperfect fracture:** The steel strands of the failed pre-stressed anchor cable are not all loose or damaged. Most of the time, only several strands ruptured.

Regarding the steel cage loosening or falling off as a failure evaluation criterion, 38 anchor cables in five rows between the powerhouse and transformer chamber failed: 16 failed cables in the first row, 18 failed cables in the second row, and 4 failed cables in the third row. These are indicated in Figure 7.31.

There were also some overloaded anchor cables in the Jinping I underground powerhouse, which were detected via the monitoring procedure. By August 30th, 2011, the anchor cable loads in 22% of the tested cables on the upstream sidewall were larger than their design values (where the number of tested cables was 55), while 22% of the tested cables in the downstream sidewall were overloaded. The maximum overload was 40%. The locations of these can be seen in Figure 7.32.

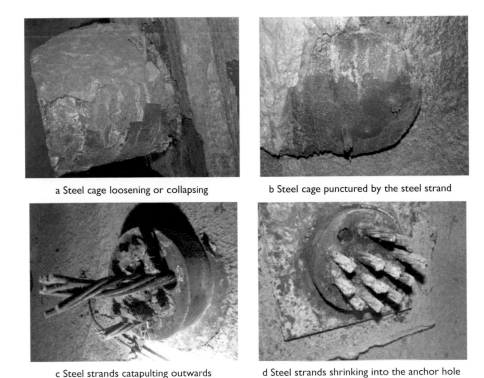

a Steel cage loosening or collapsing   b Steel cage punctured by the steel strand

c Steel strands catapulting outwards   d Steel strands shrinking into the anchor hole

Figure 7.30 Jinping II pre-stressed anchor cable failure types (Jiang *et al.*, 2013).

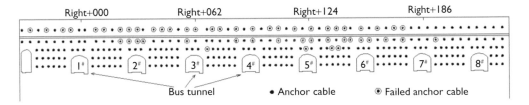

Figure 7.31 Distribution of failed cables at the Jinping II downstream sidewall (Jiang *et al.*, 2013).

7.2.3.14   Measures used to reduce local risks

1   Anchor cable overloading treatment

Anchor cable overloading is related to large-scale rock deformation. The overloading can be overcome by reducing the initial locked-in load of the anchor cables. For installing the anchor cables, the installation should be later than the originally planned time (commensurate with safety), or the locked-in load should be reduced according to the monitoring results and feedback analysis. For example, the locked-in load of the Jinping I hydropower station underground powerhouse cables was reduced to 50–60% of the design value from 80%. For the already installed anchor cables, it is difficult to release the locked-in load. One method is to add additional anchor cables near the failed ones. Another method is to weaken the local rock mass under the anchor head which can reduce the cable tension. This is illustrated in Figure 7.33.

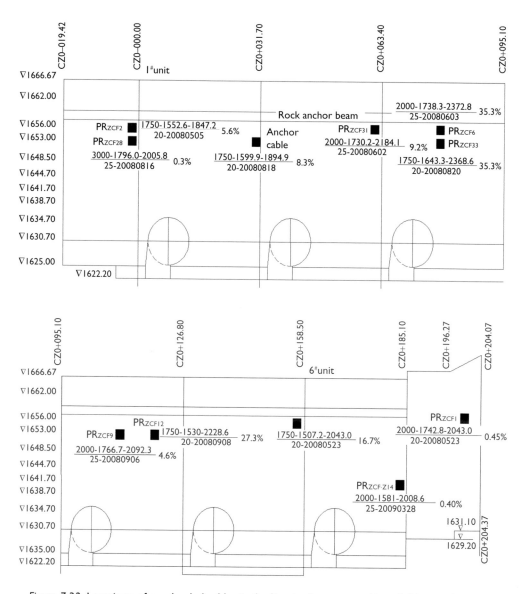

Figure 7.32 Locations of overloaded cables in the Jingping I upstream sidewall (Hou *et al.*, 2012).

2 Method to reinforce a loosened rock mass zone

Consolidation grouting was used to reinforce the loosened rock mass affected by the excavation of the Penshui hydropower station, with acoustic testing and crack permeability rate data being collected. The test results showed that the wave velocity of the surrounding rock mass within a 6 m depth was improved, being initially lower than 4000 m/s and later greater than 4000 m/s. The mean improvement was 10%.

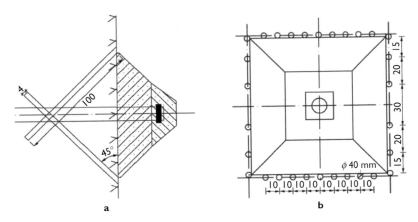

Figure 7.33 Method to reduce anchor cable tension load (units: cm) (Li et al., 2009).

7.3 EPISTEMIC UNCERTAINTY ANALYSIS

Following Section 7.1 which described aspects of hydropower cavern risk and the Section 7.2 overview of the hydropower caverns in China and experiences with excavation and support, we now discuss epistemic uncertainty (concerning basic lack of knowledge) in this Section and aleatory uncertainty (concerning unpredictable or unexpected events) in this Section.

7.3.1 Geological setting

The engineering geological conditions of a hydropower cavern group are mainly determined through the data obtained at each site investigation stage. The strata and lithology, characteristics of the strata, rock mass structure, together with location, scale and properties of fault zones should be studied carefully, especially any loose, soft, soluble and karst strata distributions. The hydrogeological conditions should also be inspected, such as the distribution and characteristics of any strongly permeable zones, aquifers and aquitards. The maximum water emission and influence on the surrounding rock stability should be estimated and the probability of water bursts or mud bursts during excavation should be predicted. The *in situ* stress should be measured, or at least estimated if measurement is not possible because of the project depth, in order to predict its influence on the stability of the surrounding rock and to predict the likelihood and intensity of rockbursts. The underground powerhouse location and axis direction should then be proposed according to the engineering geological conditions and rock types. Then, the stability of the cavern group should be predicted.

Although the geological data obtained before excavation is likely to enable anticipation of the main engineering problems, the stability of the main powerhouse vault, upstream and downstream high side walls and cavern intersections are difficult to evaluate fully before the excavation process. The engineering geological information obtained through boreholes and exploration tunnels is limited in terms of that required for a full prediction capability. In particular, the precise influence of fractures on the rock mass stability at the specific site may be difficult to determine.

These uncertainties may lead to significant differences between the excavation and support schemes proposed before construction and those actually required.

Case Example: The distribution of the main faults at Section #4 of the Jinping II hydropower station cavern group, obtained through the Feasibility Study Report of Jinping II Hydropower Station along the Yalong River (Huadong Engineering Corporation Ltd., Engineering Geology, Vol. 2 (2005)), can be seen in Figure 7.34. Only fault F_{37} was revealed in exploration tunnel #3. The locations and orientations of faults F_{65}, F_{16} and F_{36} and their strata boundaries could not be determined through the exploration tunnel and boreholes, but they could be obtained to some extent through the ground outcrop and geological information available in other exploration tunnels. Thus, there was uncertainty about the fault distribution in Section #4.

For example, the range of the fault F_{16} strike direction is from N30°E to N50°E, and the range of its dip angle is from SE∠50° to SE∠61°. This uncertainty may not appear to be very large, but the difference can have a great effect on the surrounding cavern rock stability: for example, if the dip angle of fault F_{65} increases by just one or two degrees, then the influence on the stability of the rock beam may be large; or, if the dip angle of fault F_{16} decreases by one or two degrees, then the influence on the stability of the cavern intersection may be large. The numerical simulation model for the excavations can enable stability predictions for these different geological conditions and hence provide the basis for useful strategic guidance, but it is the actual conditions that are required as input.

In addition to the faults *per se*, there are other types of discontinuities in the rock mass, e.g., joints, bedding plane separations and fracture zones, around the cavern group volumes which affect the local stability of the surrounding rock after excavation. If their characteristics can be determined before excavation, then the local

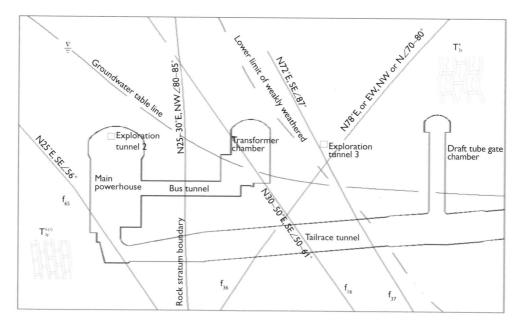

Figure 7.34 Geological section for the #4 unit of the Jinping II hydropower station cavern group, China.

risk can be controlled. However, there are many uncertainties concerning the geological setting; thus, sometimes it is difficult to reduce the overall risk and localised risks. The uncertainty of the geological setting can be reduced to some extent by more investigation before excavation and engineering geological surveys during excavation, but the uncertainty cannot be avoided completely.

7.3.2   *In situ* rock stress

The most widely used stress measurement methods are the overcoring and hydraulic fracturing methods. It must be remembered that stress is a tensor quantity having six independent components—which are usually presented as the magnitudes and directions of the three principal components of the stress tensor. But this means that a stress measurement method must enable six measurements to be made in each measurement case: this is possible with some overcoring methods, but not with the basic hydraulic fracturing method. However, the HTPF method (Hydraulic Testing of Pre-existing Fractures) does enable the normal opening pressure for six differently orientated fractures to be measured, and hence the full stress tensor to be determined.

The most direct and reliable method to provide an understanding of the *in situ* rock mass stress field is measurement, but there are two main problems: firstly, such measurements only provide 'point' properties; and secondly, it is difficult to conduct measurements at significant borehole depths, i.e., more than a few hundreds of metres. Also, the discontinuities in the rock mass structure affect the local stress field, so the measurements can only reflect the local stress field at the measurement location. However, in recent years, powerful numerical modelling, in which the discontinuities can be explicitly included, has been developed enabling both direct simulation and the ability to establish the influences of uncertainties about the rock mass structure. This type of modelling and direct simulation enables an improved approach to the planning of cavern locations, excavation schemes and surrounding rock stability.

Case Example: 16 locations around the underground powerhouse area at the Jinping I hydropower station were tested to obtain the *in situ* stress tensor. The results can be seen in Figure 7.35.

The principal stress magnitudes and orientations at an elevation near 1660 m were estimated as follows. (Note that, whereas the orientation of a fracture plane is characterised by the dip direction and dip values, the equivalent orientations of stress components, which are lines, are characterised by the 'trend' and 'plunge' values.)

- The range of σ_1 was between 16.1 and 35.7 MPa, with mean value 23.2 MPa.
- The range of σ_2 was between 7.4 and 25.6 MPa, with mean value 14.9 MPa.
- The range of σ_3 was between 4.1 and 22.2 MPa, with mean value 9.2 MPa.
- The trend of σ_1 was between N28.5°W and N71°W, and its mean value was N48.7°W.
- The plunge of σ_1 was between 20° and 50°, and its mean angle was 34.2°.

Thus the area is a high *in situ* stress area. The *in situ* stress measuring points around the underground cavern group and the distribution of *in situ* stress values, with trends and plunges changing along with the elevation are shown in Figures 7.36–7.38.

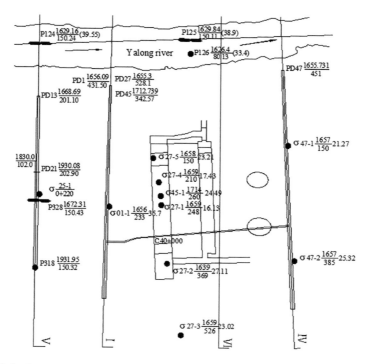

Figure 7.35 In situ stress measuring locations at the Jinping I hydropower station (Zhang, 2011).

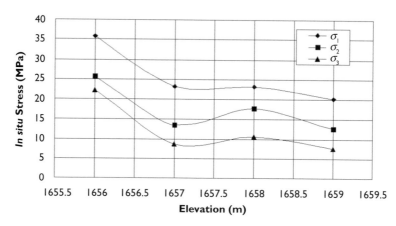

Figure 7.36 Principal stress variation near elevation 1660 m of the Jinping I hydropower station (Zhang, 2011).

From these Figures, it can be seen that three principal stress values decrease in line with the increase of elevation. Near the elevation of 1656 m, the stress values were between 20 and 35 MPa; when the elevation was 1657–1659 m, the stress varied from 10 to 25 MPa.

The 150° direction of the principal stress σ_1 changed little with the height, the intersection angle with the powerhouse axis being 30° and the plunge angle was not

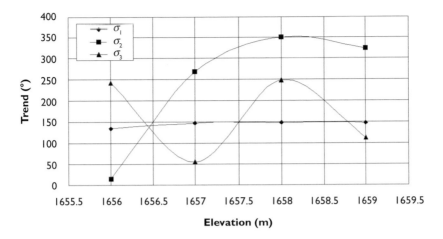

Figure 7.37 Variation in principal stress trends near elevation 1660 m of the Jinping I hydropower station (Zhang, 2011).

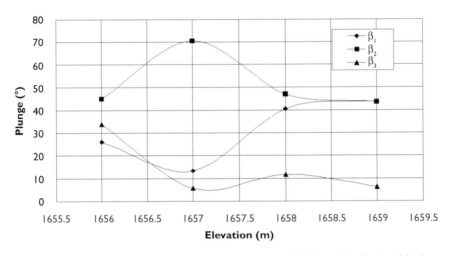

Figure 7.38 Principal stress plunge angles near elevation 1660 m of the Jinping I hydropower station (Zhang, 2011).

large. The direction of principal stress σ_2 varied between 250° and 350° along with the height, and the intersection angle with the powerhouse axis was high, together with the plunge. These values were disadvantageous in terms of the surrounding rock stability. There were no obvious regularities to the direction of principal stress σ_3. So, because of the above analysis, it is known that there is significant uncertainty in the *in situ* stress testing results within three metres. These types of results are by no means uncommon but they do indicate that there are great uncertainties within the main powerhouse excavation height of 68.8 m.

The changes of different principal stresses along with the width near the elevation of 1658 m are shown in Figure 7.39. The stress concentration region appears for the depths of 230 to 360 m. The depth of the Jinping I underground cavern group is

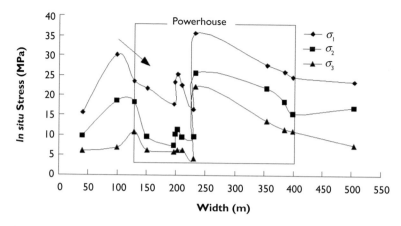

Figure 7.39 Variation of principal stress magnitudes with width near elevation 1658 m at the Jinping I hydropower station (Huang *et al.*, 2011).

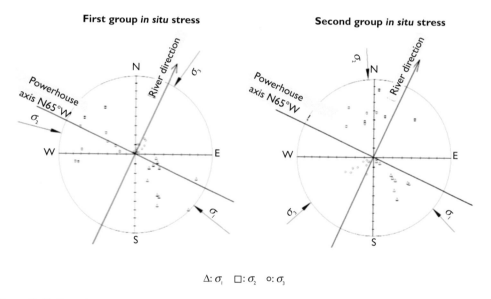

Figure 7.40 Distribution of *in situ* principal stress trends at Jinping I hydropower station (left: first group; right: second group) (Zhang, 2011).

between 112 and 388 m, which is in the higher stressed zone—and which is adverse in terms of the surrounding rock stability.

The principal stress trends for the *in situ* stress testing results can be divided into two groups, as shown in Figure 7.40. It should be understood that these directions are not exact values.

From Figures 7.36–7.40 and the associated text, it is evident that there is great variability in the *in situ* stress values, trends and plunges for the different 'horizontal depths', elevations and locations—which is caused by the spatial variability of the

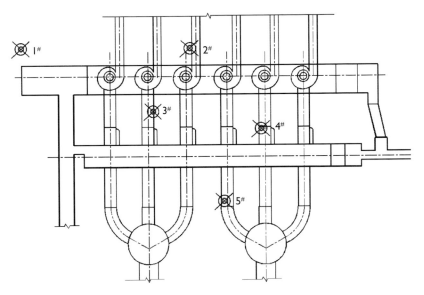

Figure 7.41 Supplementary *in situ* stress testing locations at the Jinping I hydropower station (Zhang, 2011).

rock mass characteristics. Thus, it is inappropriate to characterise the *in situ* stress as a single set of values. During the third layer excavation, five locations were further tested to obtain the *in situ* stress, Figure 7.41. These showed that the *in situ* stress field is high and the principal stress directions are greatly affected by the rock mass unloading as a result of excavation.

The uncertainty about the rock stress can be reduced by comprehensive analyses of field testing results, back analyses, and numerical modelling calibrated by the test results. Most importantly, the field rock stress testing results obtained during the initial excavation of cavern groups can be used to estimate the types of damage in the surrounding rock. In this way, the risk introduced by the rock stress uncertainty may be contained.

7.3.3   Hydrogeology

Hydrogeology is important in relation to cavern stability because a cavern group is affected by the spatial distribution and seepage characteristics of the groundwater, so these should be determined via testing data and a numerical simulation method. The distribution and characteristics of the rock permeability, aquifers, aquicludes, catchment structures and highly permeable zones should be established during the regional hydrogeological investigation for the cavern groups, with special emphasis if the location is in a karst area. The possibility of sudden water inrush should be predicted. The maximum water discharging volume and its influence on rock mass stability should also be estimated, together with the water pressure reduction coefficient. In addition, the underground water level, water temperature, water pressure, recharge, runoff and discharge conditions should be determined. However, and similar to the *in situ* stress state discussed in the previous sub-Section, there is uncertainty during the hydrogeological analysis caused by the uncertainty of the engineering geological data, parameters and analysis models.

The uncertainty can be divided into two types: objective uncertainty and subjective uncertainty. The objective uncertainty is caused by the intrinsic stochastic characteristics of the groundwater, which are related to the temporal and spatial changes in the hydrogeological variables, such as precipitation, river flow and water levels. The subjective uncertainty is caused by incomplete understanding, which is carried through to the uncertainty of the mathematical models and parameters.

The distribution of drilling holes and exploration tunnels for the Jinping II hydropower station can be seen in Figure 7.42. According to the test results, the groundwater type for Exploration Tunnel 2 was shown to be fissure water. There was considerable groundwater in section T2y5-(1) of the powerhouse, the water pressure was low, as evidenced from the test results of DK32 and DK33. In addition, water bursting in an exploratory tunnel was related to the NEE and NWW structural fractures. The opening width of fault F_{28} in PD2 was about 50–300 mm, with outflowing groundwater at 4.25 l/s; whereas, the opening width of fault F_{30} in PD1-2 was about 100–200 mm, and the initial flowing quantity was 80 l/s. The difference is the fact that fracture characteristics and size of the rock mass are different, so the groundwater character is different.

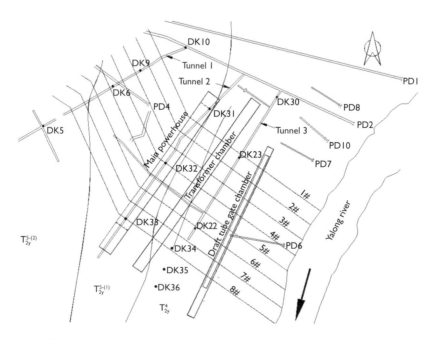

Figure 7.42 Distribution of drilling holes and exploration tunnels at the Jinping II hydropower station (Feasibility Study Report of Jinping II Hydropower Station along the Yalong River: Huadong Engineering Corporation Ltd., Engineering Geology, Vol. 2 (2005)).

The field test results indicate that the weakly permeable rock accounts for 52% of the rock mass, the medium permeable rock for 21%, and low permeable rock for 27%. The underground water level depth at PD1 was between 212 m and 242 m, whereas the depth of PD2 was between 211 and 274 m. The depth in the upstream pressure adjustment shaft was between 212 and 228 m. A local depth of 351 m was affected by fault F_7. So, the groundwater flowpaths are uncertain, and are strongly influenced by the rock mass structural characteristics, especially the faults. From the Feasibility Study Report of Jinping II Hydropower Station along Yalong River (Huadong Engineering Corporation Ltd., Engineering Geology, Vol. 2 (2005)), the hydrogeological differences for PD1 and PD2 can be seen in Table 7.14 and so it could be anticipated that there would be significant inhomogeneity and hence in the overall hydrogeological characteristics of the powerhouse region. The methods to reduce the uncertainty can be field monitoring and numerical feedback analysis based on the survey results.

7.3.4   Properties of the rock mass

The rock mass consists of many types of rock and structural surfaces, which are affected by *in situ* stress and underground water. There are the different 'DIANE' features of discontinuity, inhomogeneity, anisotropy and non-elasticity in the rock mass. Also, the rock mass quality in different areas is not the same, due to differences of the rock structure, rock weathering, mechanical properties and degree of unloading.

Example:
The rock mass quality grade is typically used to evaluate the rock mass characteristics. The methods include the RMR method developed by Bieniawski, Q system developed by Barton, GSI system developed by Hoek, and the BQ (Basic Quality) system which is the Chinese standard (See Appendix B of this book for an explanation of the BQ system). The contributory factors to the different evaluation indices of these rock mass classification systems and their algebraic formulations are different, and so the rock mass quality grades obtained by the different methods are not the same. The RMR and Q method results for exploration tunnel 2 at the Jinping II hydropower station can be seen in Figure 7.43, and the results for the exploration tunnel at the

Table 7.14 Hydrological characteristics of PD1 and PD2.

	PD1	PD2
Hydrological characteristics	T_{2y}^5 is a strong water-rich stratum, with the groundwater being fracture flowing water, accounting for 75% of the water—which is abundant. The water was flowing at up to 80–230 L/s. The total flow quantity in this region can reach more than 0.5 m³/s, and the maximum may reach 0.7 m³/s in the flood season.	T_{2y}^4 is a weak, water rich stratum. Most sections of the tunnel were dry. The groundwater was fissure water. T_{2y}^{5-1} is a strong water-rich stratum. The water flowing speed was not large. The total flow can reach >0.15 m³/s, and the maximum water flowing speed may reach up to 0.2 m³/s in the flood season.

Xiluodu hydropower station in Figure 7.44. It can be seen from the curves that the relations for the different projects obtained by the same method are different, as are the relations of the same project obtained by different methods. These uncertainties are mainly caused by the spatial variability of the rock mass characteristics.

7.3.5 Specific project location

The selection of the location for the hydropower cavern group is of course important during the design period. Factors such as general layout, total investment, construction time and operational management should all be considered. If possible, the hydropower cavern group should be constructed in the area where the rock mass is strong, intact, weakly weathered and without special geological problems. The inter-

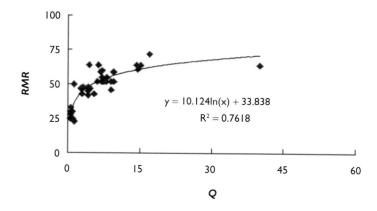

Figure 7.43 RMR and Q system classification results for Jinping II exploration tunnel rock mass.

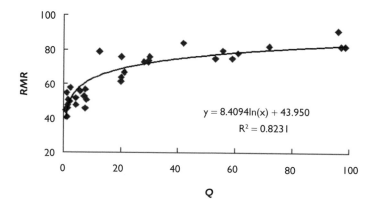

Figure 7.44 RMR and Q system classification results for Xiluodu exploration tunnel rock mass.

section angle between the underground powerhouse axis direction and the maximum principal stress direction should be small, and the intersection angle between the axis direction and fracture strike direction should be large. For rock masses with massive structure and large extent, the angle should be greater than 30°; for those with steep strata, the angle should be greater than 45°.

Example:

The underground powerhouse at Guandi hydropower station was initially located near the right bank of the dam axis, according to the needs of the hydraulic structural layout during the feasibility study stage. At this location, the *in situ* stress is high, the buried depth is great, the scale of bedding fault zones is small, there are no large faults or soft structural surfaces, and the rock mass integrity is good. Therefore, N67°E was chosen as the powerhouse axis direction. The intersection angle between it and the bedding fault zone strike direction is 40–90°, and the intersection angle with the fracture strike direction is also large. Having a large intersection angle between the axis direction and structural surfaces strike directions is beneficial in terms of the stability of the surrounding rock. However, the intersection angle between the axis direction and the maximum horizontal principal stress direction is also large, about 70°, but this is detrimental to the stability of the cavern-peripheral rock. These orientations can be seen in Figure 7.45. Finally, the powerhouse axis direction was changed to N5°E. The intersection angle between the axis direction and principal stress direction was then 40° and the intersection angle with the structural surface direction was then 80°. The altered powerhouse axis is nearly parallel to the dam axis, which

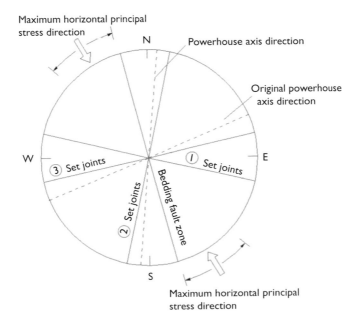

Figure 7.45 Angular relations between the Guandi powerhouse axis, *in situ* stress and main rock mass fracture direction (Xiao *et al.*, 2006).

is beneficial for the cavern group arrangement and reduces costs. This uncertainty related to the specific project location can thus be reduced by studies of the detailed rock stress and geology setting: if the magnitude of the major principal rock stress is high, its effect should be considered first; otherwise, the greatest attention should be given to the effect of the geology setting. But, ideally, their total effects with respect to the cavern surrounding rock mass should be the least.

7.3.6 Excavation and support method

The safety of the underground cavern group is affected by the excavation and supporting methods, which should be determined according to, *inter alia*, the rock mass category, rock mass strata, depth, *in situ* stress, cavern group spatial distribution. Due to the fact that the geological conditions are usually complex, the geological information is incomplete, and the calculating parameters are difficult to determine, the 'engineering analogy' method has often been used for underground engineering design, i.e., a method based on engineering experience. However, if there have been no similar projects, then it is difficult to recommend the excavation and supporting methods in new circumstances. So, in this case, once an overall scheme has been established, a dynamic feedback optimisation design method is suggested—which allows changes in the excavation and supporting parameters according to real-time monitored data.

In line with this, a modern rock engineering design method has been proposed by the authors in 2003 and is described in detail in our previous and companion book "Rock Engineering Design" published in 2011, also by CRC Press, Taylor & Francis. Initially, it consists of eight basic methods related to rock engineering modelling and design. The design process is divided into seven steps: namely, project goal, characteristics of rock mass, engineering and site, selecting the design method, selecting the modelling method, suggesting the preliminary design scheme, back analysis, final design parameters, and verification. This method has been applied to several large underground cavern groups.

Example:
The main powerhouse dimensions at the Laxiwa hydropower station are 311.75 m × 30.00 m × 30.00 m (length × width × height). Because of the large height and span, the excavation method proposed involved preparing the vault first then excavating down layer by layer. There were nine excavation layers, each having depths of between 6 and 9 m, as illustrated in Figure 7.46. The support used was shotcrete, bolts and anchor cables. According to the excavation monitoring data, the depth of the loose rock zone was ~5 m and the depth of the transition zone between 5 and 12 m. The mechanical parameters of the three zones in sections A2 and A5 were back-analysed according to the deformation increase through the C3 and C7 layer excavations. The results were then used to predict the surrounding rock stability and to recommend the support parameters. Meanwhile, the reinforcement scheme proposed for the rock between the main powerhouse and transformer chamber was analysed. Anchor cables with design parameters tension T = 2000 kN @ 4 m × 6 m, and length 50 m were proposed. The numerical analysis results showed that the rock surrounding the cavern group is stable after excavation. The arch deformation is less

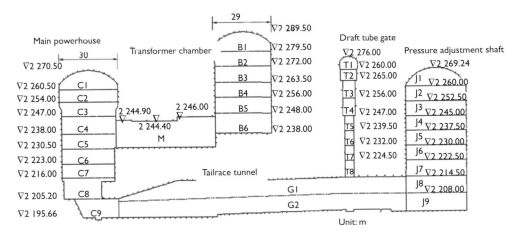

Figure 7.46 Excavation layers at Laxiwa underground powerhouse (Yao *et al.*, 2011).

than 30 mm, and the maximum convergence displacement at the middle of the side-walls does not exceed 120 mm.

The uncertainty introduced by the excavation and support method can be reduced by the dynamic feedback analysis method. Firstly, the excavation and support method can be proposed based on similar projects; then the optimised excavation and support method can be determined according to the field monitoring results, such as the surrounding rock displacement, relaxation depth, anchor cable loads, rockbolt loads and so on. Finally, the surrounding rock can be further reinforced.

7.4 ALEATORY UNCERTAINTY ANALYSIS

Following the previous discussion about reducing the epistemic uncertainty at the design stage, we now discuss aleatory uncertainty, i.e., dealing with unforeseeable uncertainties/occurrences during construction. Of course, considering the whole construction project from original conception to a final working hydropower station, it is not possible to make a hard and fast distinction between epistemic and aleatory uncertainties, but we find the concepts useful, especially in the contexts of 'before' and 'during' construction and in presenting the information in this book.

7.4.1 Detailed geology variations

The general engineering geology location of large hydropower cavern groups is determined through the survey results of different potential locations and the conditions exposed on the mountain surfaces. There is uncertainty due to the fact that some faults and/or fracture zones may not be revealed at the surface. Even for geological information obtained via exploration tunnels, only the tunnel proximate rock can be studied and the exact geological conditions around the cavern group cannot be obtained. So the engineering construction will be affected by this uncertainty.

Example:
The lithology of the Dagangshan hydropower station main powerhouse is granite with diabase dykes β_6, β_{80} and β_{81} crossing it. The granite is fresh, strong and has a high integrity. The *in situ* stress σ_1 is 11–22 MPa and the rock mass quality is grade II and III. The surrounding rock is basically stable but the local area stability can be adversely affected by faults and dykes. During the first layer excavation of the underground powerhouse, some rock blocks near dyke β_{80} collapsed when the rock mass in the upstream area I- ③ (0+133.0–0+135.0) was blasted. A collapse occurred near the 4th unit of the powerhouse vault, and the 3rd unit vault was also affected. The cavity created by the collapse was located near the vault on the upstream side of the fourth unit and vice powerhouse (see Figure 7.47); its dimensions were 19 m long, 15 m wide and 31 m high. The lower size of the collapsed body was 39 m long and 15 m wide; its top size was 15 m long and 4 m wide; the volume was about 3000 m³. Four diabase dykes, β_{132}, β_{80}, β_{81} and β_{c2}, traversed the collapsed mass. The dyke characteristics are given in Table 7.15.

Short rockbolts, shotcrete with steel netting, anchor cables and consolidation grouting were used to reinforce and contain the surrounding rock. Meanwhile, 10 sets of multi-displacement monitoring instruments, bolt load detectors and five sets of anchor cable dynamometers were added to monitor the deformations and loads. The results showed that the rock mass deformation values changed little, and the deformation rates were small.

The causes of the collapse were as follows: (1) Diabase dyke β_{80} was broken up and blasting caused the surrounding rock to become loose and collapse;

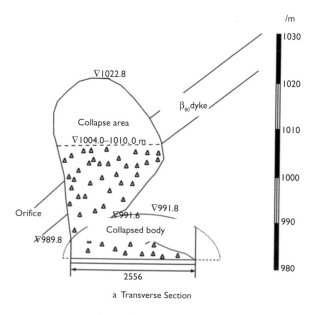

a Transverse Section

Figure 7.47 (Continued)

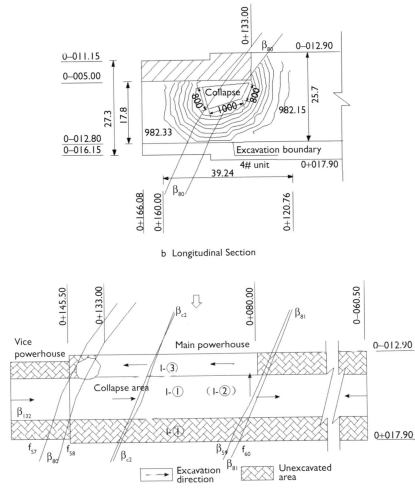

Figure 7.47 Features of the collapse at the Dagangshan main powerhouse (Zhang 2010, Zhu *et al.*, 2013).

Table 7.15 Dyke characteristics (Cai *et al.*, 2012).

Dyke number	Occurrence	Width (m)	Character description
β_{80}	N15°E/NW∠50°–N25°W/SW∠65°	3–4	Cataclastic structure, fault type contact
β_{132}	N45°E/NW∠84°	0.3–0.5	Mosaic structure, fault type contact
β_{c2}	N20°W/SW∠80°	0.1	Block structure, fracture type contact
β_{81}	N15–20°W/SW∠70–80°	1.5–2	Block–Mosaic structure, fault type contact

(2) The excavation process and excavation area partition were not suitable, the blasting was performed without support to the diabase dyke, plus the support was installed too late.

In terms of the aleatory uncertainty, this was caused by detailed geology variations but this can be reduced by detailed geological surveys during excavation. The occurrence and location of the exposed joints and faults should be tested and compared with previous survey results, so that their effect on the surrounding rock stability can be evaluated. When the joints and faults are exposed for the first time, they should be given appropriate attention.

7.4.2 Rock stress variations

There can be great uncertainty concerning the *in situ* stress in engineering construction volumes, which is caused by inhomogeneous rock mass structures. The variation in the stress can be obtained to some extent through measurement results, but the complete variation is difficult to obtain. Moreover the stress field is of course affected by the excavation unloading effects.

Example:
At the Three Gorges underground powerhouse in the right bank, five locations at three test sections were selected to measure the *in situ* stress to establish the induced stress field in the surrounding rock near blocks #18 and #19, which are located on the downstream side of units #3 and #4. The location of blocks #18 and #19 can be seen in Figure 7.48, and the test point layout can be seen in Figure 7.49.

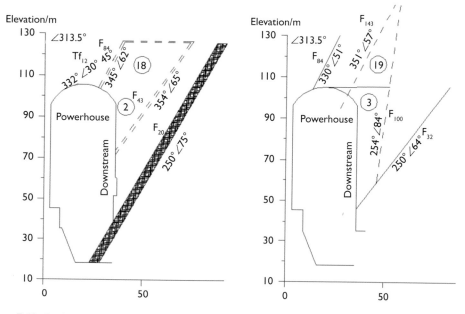

Figure 7.48 Outline geological profile of blocks #18 and #19 at the Three Gorges underground powerhouse (Huang, 2007).

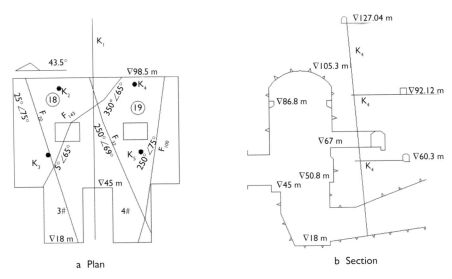

Figure 7.49 Induced stress field test points layout at the Three Gorges underground powerhouse (Huang, 2007).

The depth of the hole K_1 (Figure 7.49(a)) was 108 m, and the lengths of the holes K_2, K_3, K_4 and K_5 were 30 m. The hydraulic fracturing technique and the stress relief method were used to measure the stress in the rock mass. This was performed after layer V was excavated. By comparing the original stress field and the stress test results, the differences can be stated as follows.

1   Change of stress direction

The mean azimuth direction of the maximum horizontal principal *in situ* stress component near the powerhouse area was 302°: its direction changed from NW to NE after the powerhouse excavation.

2   Change of stress magnitude

The maximum stress component in the test area was 5.7 MPa. The surrounding rock, 8–10 m in from the excavation boundary, exhibited significant relaxation and unloading causing the stress magnitude to be 10–12 MPa in the maximum principal stress direction to 1–2 MPa in the minimum principal stress direction.

The original rock stress and its variability may be difficult to be determine before excavation, but the rock stress variation during excavation can be reduced by suitable excavation and supporting methods.

7.4.3   Local water variations

The groundwater distribution and flow characteristics are related to the rock mass structural characteristics. If the rock mass integrity is high, with few joints and cracks, then the underground water permeability will be low. If there are faults and fracture zones and karst channels throughout the rock mass, then the groundwater will flow more easily. When these geological structures are altered during construction, underground water can leak out and affect the rock mass stability. The method which may be used to avoid this situation is to understand the detailed groundwater distribution in the rock mass, but this is difficult to establish. Therefore, uncertainty exists about the groundwater characteristics.

Example:

According to the site investigation information for the Goupitan hydropower station, the underground water would flow out from the #8 karst system and W24 karst system during the underground powerhouse excavation. The potential quantity of water flowing out was estimated to be 432–7776 m³/d. For the transformer chamber, only a small amount of water seepage may occur in the local area, while the actual water quantity from the W24 karst system increased significantly 6 to 12 hours after rainfall. The maximum water quantity was 7000 m³/d, and the maximum hourly water quantity was about 6000 m³/h, continuing for 70 minutes. The maximum gushing mud quantity was 1000 m³ on one occasion. Then it was found that the W24 karst system connects with the surface, and the gushing of water or mud occurred several dozen times. The direction of the underground water flow and the points where water gushed out can be seen in Figure 7.50.

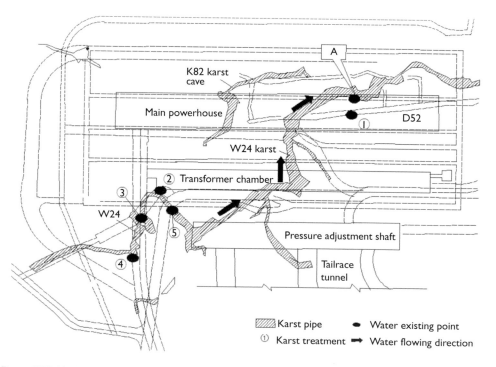

Figure 7.50 Underground water flow directions and water outflow points at Pigoutan (Zi *et al.*, 2009).

The local water variations should be surveyed carefully during excavation and, if there is any discharged underground water on the rock surfaces, it should be noted and treatment methods, such as draining the water off through pipes or plugging water with chemical materials, should be implemented early according to the monitoring results.

7.4.4 Mechanical behaviour of the rock mass after excavation and in the long term

There can be uncertainty regarding changes in the mechanical properties of the excavation peripheral rock—which are influenced by the geological conditions, degree of unloading, excavation disturbance, excavation to support installation time, and other factors, especially in high crustal stress areas.

Example:
The monitoring and geophysical data indicated that the depth affected by deformation in the downstream side of the Jinping I main powerhouse and transformer chamber was greater than in the upstream side. By October 2009, the maximum affected depth in the upstream sidewall region was 8.3 m and the maximum depth in the downstream sidewall was greater than 10 m, with individual points exceeding 15 m, which exceeded typical powerhouse values. The borehole monitoring results for the P-wave velocity and borehole camera information at elevation 1649 m near the 0+093 m upstream side indicated that the wave velocity did not reduce before the new cracks appeared, but it reduced after rockbursts, as shown in Figure 7.51. The dip angle of the new fracture surface, shown via the borehole camera results in the Figure, was large and the image has clear crack opening characteristics.

The mechanical behaviour of the rock mass after excavation can be evaluated based on the displacement, borehole wave velocity and borehole camera monitoring results. It can be controlled through the number and lengths of anchor cables or bolts.

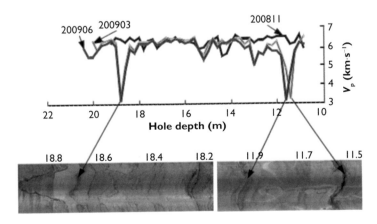

Figure 7.51 Test results of borehole wave velocity and associated borehole camera images from Jinping I example (Huang *et al.*, 2011).

Through continued monitoring the large displacement risk can be reduced, so the surrounding rock stability will be acceptable in the long term.

7.5   RISK ASSESSMENT METHOD FOR A LARGE HYDROPOWER CAVERN GROUP

7.5.1   Principles

The construction risk assessment for a large hydropower cavern group is performed based on the location, layout, design size, supporting parameters, and so on. The risk can be divided into three phases: namely, the overall risk assessment, the local risk assessment before construction, and the dynamic risk assessment during construction.

- The overall risk assessment is conducted for large areas based on engineering geological exploration data, although, at the time of the assessment, the geological information is not comprehensive and detailed. Moreover, the total evaluation area is large compared to specific areas such as the powerhouse cross-section, each bus tunnel, transformer chamber cross-section, etc. being regarded as independent evaluation units.
- For overall risk assessment areas with a potentially high risk, the local risk occurrence probability of typical excavation cross-sections will be assessed according to the numerical analysis results. The local areas with high risk will then be determined, and local risk controlling measures will be adopted.
- According to the geological information revealed by early cavern excavation and dynamic feedback analysis results based on monitoring information, the risk occurrence probability of the surrounding rock mass support is evaluated. The effectiveness of local risk management measures will also be evaluated. Meanwhile, the risk associated with subsequent excavating layers will also be predicted, the assessment being performed for each layer during cavern group excavation.
- Finally, the overall risk assessment will be performed after excavation and after all local defensive risk arrangements have been completed, and further advice will be suggested to ensure that the cavern group is acceptably safe.

7.5.2   Method for assessment and mitigation of overall risk for a large hydropower cavern group before construction

The overall risk assessment for large underground cavern groups before construction is implemented based on the engineering geological investigation data obtained by drillholes, tunnels and geophysical methods. The evaluation results provide the decision-making basis for determining the construction plan and optimising the design parameters. The overall risk assessment areas are usually large. For the powerhouse and transformer chamber, each evaluation area should contain one machine unit, as illustrated in Figure 7.52. Each bus tunnel and part of the tailrace tunnel is regarded as a separate evaluation unit. A flowchart of the overall risk assessment process is presented in Figure 7.53.

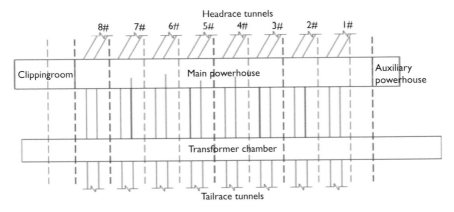

Figure 7.52 Overall risk assessment section divisions for hydropower cavern groups.

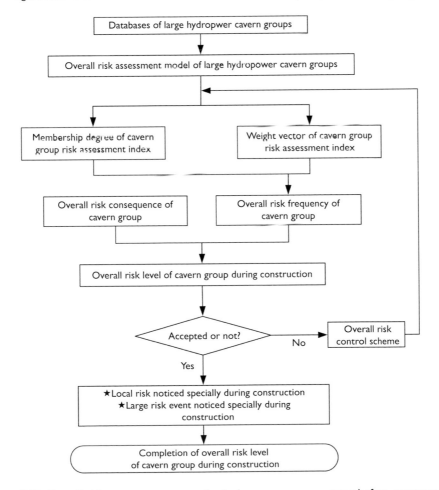

Figure 7.53 Overall risk assessment process for hydropower cavern groups before construction.

According to the engineering geological information and excavation supporting parameters of the cavern groups before construction, the rock mechanics properties, structural characteristics of the host rock, cavern layout features, groundwater characteristics, special geological conditions and the construction excavation features are used to evaluate the overall risk. The analysis model is included here in Figure 7.54.

The overall risk occurrence probability has five classification levels: namely, I, II, III, IV and V. I represents the minimum risk grade, and V the maximum risk grade. The risk level can be determined by the fuzzy comprehensive evaluation method. The membership degrees of the assessment indices and the method to determine the weight vector and analysis method should be confirmed before calculating the risk level.

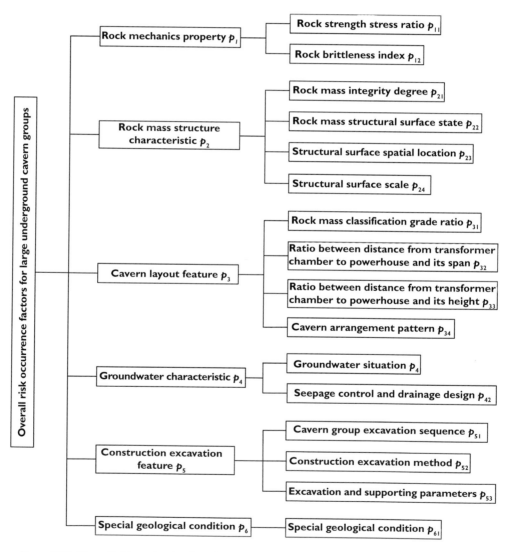

Figure 7.54 Overall risk evaluation flowchart for hydropower cavern groups before construction.

7.5.2.1 Method to determine the membership degree of the assessment index

The membership function of each index used to evaluate the overall risk can be presented through the database, which includes 60 large hydropower underground cavern groups, described as follows.

7.5.2.1.1 Rock mechanics property, p_1

The rock mechanics characteristics and relation with the *in situ* stress are determined by the strength–stress ratio for the rock and the brittleness index, as described below.

1 Rock strength stress ratio, p_{11}
The strength–stress ratio of the rock is obtained using the rock's uniaxial compressive strength and the *in situ* stress state. When there are many rock types in the analysis area, their mean value will be used. The standard values for the different levels are given in Table 7.16. Their membership functions can be seen in Figure 7.55.

2 Rock brittleness index, p_{12}
The rock brittleness index is obtained using the rock's uniaxial compressive strength and tensile strength. The standard values for the different levels are given in Table 7.17 and their membership functions in Figure 7.56.

Table 7.16 Rock strength–stress ratio values for the different evaluation grades.

Evaluation grade	I	II	III	IV	V
Rock strength–stress ratio, R_c/σ_1	>7	7–4	4–2	2–1	<1

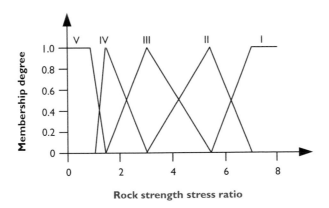

Figure 7.55 Membership functions for the rock strength–stress ratio.

Table 7.17 Rock brittleness index values for the different evaluation grade.

Evaluation grade	I	II	III	IV	V
Rock brittleness index, R_c/σ_1	<10	10–14	14–18	18–22	>22

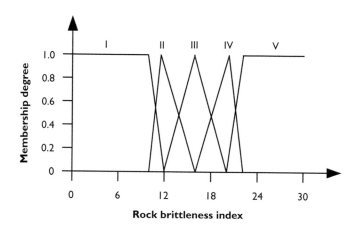

Figure 7.56 Membership function for the rock brittleness index.

7.5.2.1.2   Rock mass fracture characteristic, p_2

The rock mass fracture characteristic of caverns is determined through the rock mass integrity degree, fracture surface state, fracture surface spatial location and fracture surface scale.

1   Rock mass integrity degree, p_{21}
The rock mass integrity degree in the analysis area is determined by the volume-weighted mean value and the engineering geology investigation report. The standard values of the different levels are given in Table 7.18 and their membership functions in Figure 7.57.

2   Rock mass fracture surface state, p_{22}
The rock mass fracture surface state can be determined by the fracture surface opening width, filling and roughness conditions. It is divided into five types. If the structural surface has multiple sets, the value can be obtained through their weighted values. The standard values for the different levels are given in Table 7.19, and their membership functions in Figure 7.58.

3   Fracture spatial location, p_{23}
The effect of a steep fracture on the stability of the high sidewall and that of a horizontal or sub-horizontal fracture on the stability of the arch crown should be carefully considered. If the fracture is a complex set of fractures with multiple sets, the evaluation value should be obtained with their weights. The standard values of the different levels can be seen in Table 7.20, and their membership functions in Figure 7.59.

4   Structural surface scale p_{24}
The fracture plane width and length are suggested for evaluating the structural surface scale. The standard values of the different levels can be seen in Table 7.21, and their

Table 7.18 Rock mass integrity degree values for different evaluation grades.

Evaluation grade	I	II	III	IV	V
Rock mass integrity coefficient, K_v	$K_v > 0.75$	$0.75 \geq K_v > 0.55$	$0.55 \geq K_v > 0.35$	$0.35 \geq K_v > 0.15$	$K_v \leq 0.15$

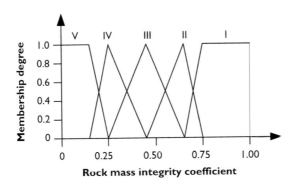

Figure 7.57 Membership function for the rock mass integrity coefficient.

membership functions in Figure 7.60. The exponential values for the fracture plane width and length are used to calculate the membership degree, and the total value can be obtained by their weighted values, which are 0.4 and 0.6, respectively.

7.5.2.1.3   Cavern layout feature, p_3

The cavern layout feature index can be evaluated through the rock mass classification grade, the ratio between the distance from the transformer chamber to the powerhouse and its span (L/B), the ratio between the distance from the transformer chamber to the powerhouse and its height (L/H), and the cavern arrangement pattern.

1   Rock mass classification grade ratio p_{31}
The ratio of each rock mass classification grade can be estimated from the engineering geological investigation data in the tunnel. Rock masses graded I, II or III are those mainly considered. The standard values of the different levels can be seen in Table 7.22, and their membership functions in Figure 7.61.

2   Ratio between distance from transformer chamber to powerhouse
    and its span (L/B), p_{32}
The stabilities of the powerhouse downstream sidewall and transformer chamber upstream sidewall are affected by the distance between them. The unloading depth is affected by the powerhouse span. There is a relation between the two, thus their ratio (L/B) is used to evaluate the cavern group stability. According to statistical results

Table 7.19 Rock mass fracture surface state values for different evaluation grades.

Structural surface state

	Closed W < 0.5		Micro opening 0.5 ≤ W < 5.0									Opening W ≥ 50	
Opening width, W (mm)													
Filling	—		No filling			Rock debris			Muddy			Rock debris	Muddy
Rough conditions	Rough	Flat and smooth	Rough	Rolling smooth or straight rough	Flat and smooth	Rough	Rolling smooth or straight rough	Flat and smooth	Rough	Rolling smooth or straight rough	Flat and smooth		
Values													
Hard rock	27	21	24	21	15	21	17	12	15	12	9	12	6
Slightly hard rock	27	21	24	21	15	21	17	12	15	12	9	12	6
Soft rock	18	14	17	14	8	14	11	8	10	8	6	8	4

Notes:
1. If the length of the structural surface is less than 3 m, three should be added to the value for hard rock and slightly hard rock, and two should be added for soft rock. If its length is more than 10 m, two or three should be subtracted from the value.
2. If the opening width of the structural surface is more than 10 mm and without filling, the value should be zero.

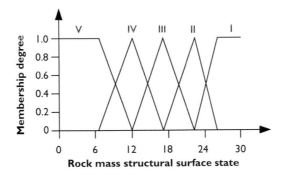

Figure 7.58 Membership function for the rock mass structural surface state.

Table 7.20 Fracture surface spatial location values and different evaluation grades.

Intersection angle between fracture surface and cavern axis	60–90°				30–60°				<30°			
Fracture surface dip angle	>70°	45–70°	20–45°	<20°	>70°	4–70°	20–45°	20°	>70°	4–70°	20–45°	<20°
Arch	0	−2	−5	−10	−2	−5	−10	−12	−5	−10	−12	−12
Sidewall	−2	−5	−2	0	−5	10	−2	0	−10	−12	−5	0

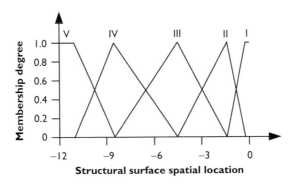

Figure 7.59 Membership function of fracture surface spatial locations.

Table 7.21 Fracture surface scale values and the different evaluation grades.

Evaluation grade	I	II	III	IV	V
Structural surface width (m)	>10	1.0–10	0.1–1.0	0.01–0.1	Joint or fracture (<0.01)
Structural surface extension length (m)	Regional fault	>1000	100–1000	10–100	<10

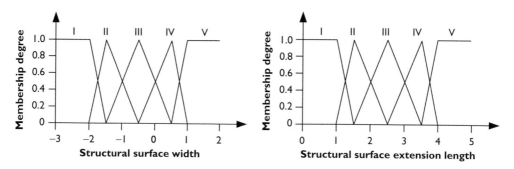

Figure 7.60 Membership function of fracture surface scale.

Table 7.22 Rock mass classification grade ratio values for different evaluation grades.

Evaluation grade	I	II	III	IV	V
Ratio of rock mass classification grade better than grade III (%)	>85	80–85	75–80	70–75	<70

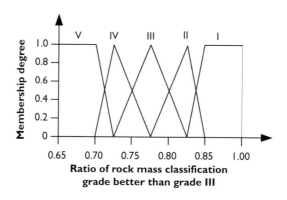

Figure 7.61 Membership function for rock mass classification grade ratios.

from the existing large cavern groups (Figure 7.62), the standard values of the different levels can be defined as shown in Table 7.23, and their membership functions in Figure 7.63.

3   Ratio between the distance from transformer chamber to powerhouse
     and its height (L/H), p_{33}

The stability of the powerhouse upstream and downstream sidewalls and the transformer chamber upstream sidewall is affected by the distance between them and the powerhouse height. Their ratio, L/H, is used to evaluate the cavern group stability.

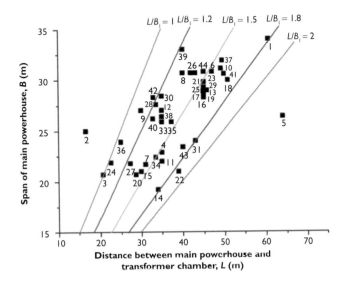

Figure 7.62 Values of *L* and *B* for existing hydropower cavern groups.

Table 7.23 L/B values to different evaluation grades.

Evaluation grade	I	II	III	IV	V
L/B	>1.8	1.5–1.8	1.2–1.5	1.0–1.2	<1.0

Note: If the *in situ* stress is high and the rock is hard, brittle or fragile, the grade should be reduced by one grade.

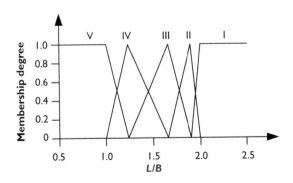

Figure 7.63 Membership function for *L/B* ratios.

According to the data from existing large cavern groups (Figure 7.64), standard values of different ratios can be defined as shown in Table 7.24, and their membership functions in Figure 7.65.

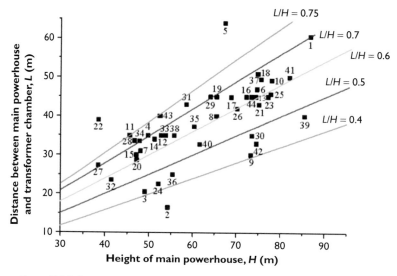

Figure 7.64 Data for *L/H* ratios from existing hydropower cavern groups.

Table 7.24 *L/H* evaluation grades.

Evaluation grade	I	II	III	IV	V
L/H	>0.8	0.7–0.8	0.6–0.7	0.5–0.6	<0.5

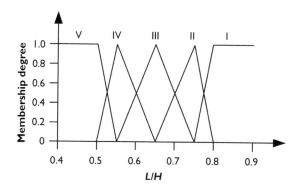

Figure 7.65 Membership function for *L/H* ratios.

4   Cavern arrangement pattern, p_{34}

Other factors, such as equipment layout requirements, overall rock mass stability, project investment, construction and maintenance, should be considered. According to the arrangement of the cavern groups, they can be divided into four types, as indicated by the sketches in Table 7.25. The standard values of the different levels are given in this Table, and their membership functions in Figure 7.66.

Table 7.25 Cavern arrangement patterns and different evaluation grades.

Evaluation grade	I	II	III	IV	V
Layout	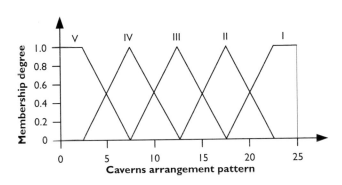				
Value	20–25	15–20	10–15	5–10	0–5

Note: The value should be determined according to the relative distance between caverns, intersection angle between strike direction of powerhouse axis and maximum principal stress direction, intersection angle between strike direction of powerhouse axis and main fracture surfaces (which is better when it is larger than 60°), the vertical and horizontal sizes of the cavern group.

Figure 7.66 Membership function for cavern arrangement pattern.

7.5.2.1.4   Groundwater characteristic, p_4

The rock mass stability is affected by the efficacy of the groundwater seepage control scheme. It can be analysed through the groundwater circumstances, seepage control and drainage design.

1   Groundwater situation, p_{41}

Although seepage control and drainage measures are used to reduce the influence of groundwater on cavern stability, they may not be effective if the water head pressure is high. Therefore, the groundwater situation is regarded as a single factor. The standard values of the different levels are defined as shown in Table 7.26, and their membership functions can be seen in Figure 7.67.

2   Seepage control and drainage design, p_{42}

According to the rationality and validity of the seepage control and drainage scheme, the standard values for the different levels are defined as shown in Table 7.27. Their membership functions can be seen in Figure 7.68.

Table 7.26 Groundwater situations and evaluation grades.

State			Dry or water dripping	Flowing water	Waterburst	
Water quantity q (L/min • 10 m)			q ≤ 25	25 < q ≤ 125	q >125	
Water head pressure H (m)			or H ≤ 10	or 10 H ≤ 100	or H >100	Grade
Rock mass quality	I	Values	0	0–2	2–6	I
classification	II		0–2	2–6	6–10	II
	III		2–6	6–10	10–14	III
	IV		6–10	10–14	14–18	IV
	V		10–14	14–18	18–20	V

Note: The rock mass quality classified grade is a value that has not been corrected according to the overall groundwater condition and fracture surface occurrence.

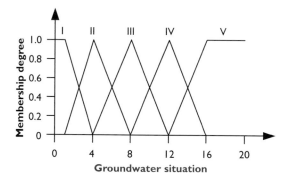

Figure 7.67 Membership function for groundwater situation.

Table 7.27 Seepage control and drainage design values and the different evaluation grades.

Evaluation grade	I	II	III	IV	V
Effectiveness of groundwater treatment	Very reasonable and effective	Reasonable and effective	Normal	Poor	Poorer
Values	20–25	15–20	10–15	5–10	0–5

Note: These factors should be considered in evaluating the cavern waterproof effectiveness, such as anti-seepage curtain (usually grouting holes 1–2 row, row spacing 1.50 m–3.00 m, grouting pressure 2–5 MPa), distance to the cavern surface (1.5–2.0 times cavern span), arrangement form (fully closed or semi-closed), lap length and construction quality. Also, the arrangement of drainage curtain (fully closed or semi-closed), distance to the cavern surface (1.0–1.5 times cavern span) and rationality of the drainage tunnel should be considered to determine the cavern drainage effectiveness.

7.5.2.1.5 Construction excavation scheme, p_5

The construction excavation scheme can be evaluated through the excavation sequence, excavation method and supporting parameters.

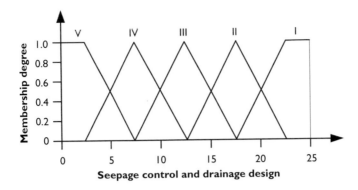

Figure 7.68 Membership function for seepage control and drainage design.

Table 7.28 Cavern group excavation sequence values and different evaluation grades.

Evaluation grade	I	II	III	IV	V
Reasonability of excavation sequence	Very reasonable	Reasonable	General	Poor	Very bad
Values	20–25	15–20	10–15	5–10	0–5

Note: The excavation sequence should be from arch to sidewall, from outside to core, from top to bottom, excavation and support step by step.

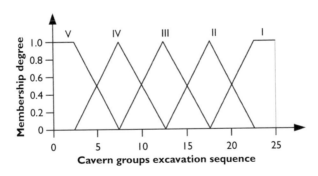

Figure 7.69 Membership function for cavern group excavation sequence.

1   Cavern group excavation sequence, p_{51}

The standard values for the different levels are defined as shown in Table 7.28 and their membership functions can be seen in Figure 7.69. The values are determined by the construction organisation design.

2   Construction excavation method, p_{52}

The excavation blasting method and excavation height are mainly considered in evaluating the reasonability of the excavation method. The standard values for the

Table 7.29 Construction excavation method values and associated evaluation grade.

Evaluation grade	I	II	III	IV	V
Reasonability of excavation method	Very reasonable	Reasonable	General	Poor	Very bad
Values	20–25	15–20	10–15	5–10	0–5

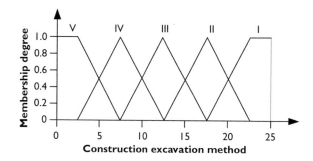

Figure 7.70 Membership function for construction excavation method.

Table 7.30 Excavation and support parameters values and the evaluation grades.

Evaluation grade	I	II	III	IV	V
Reasonability of excavation and supporting parameters	Very reasonable	Reasonable	General	Poor	Very bad
Values	20–25	15–20	10–15	5–10	0–5

different levels are defined as shown in Table 7.29 and their membership functions can be seen in Figure 7.70.

3   Excavation and supporting parameters, p_{53}

The excavation and supporting parameters should be first established by empirical analogy, then the parameters can be optimised through the results of the monitoring, model testing and numerical analyses. 'Soft' supporting and system supporting should be used first, and 'stiff' support and local support used as supplements. The standard values for the different levels are defined as shown in Table 7.30, and their membership functions can be found in Figure 7.71.

7.5.2.1.6   Special geological conditions, p_6

Special geological conditions which cannot be avoided may have a strong influence on cavern stability, such as the karst system at the Goupitan hydropower station, bedding fault zones at the Baihetan hydropower station, and diabase dykes at the Dagang-shan hydropower station. Their effects on the cavern stability should be considered.

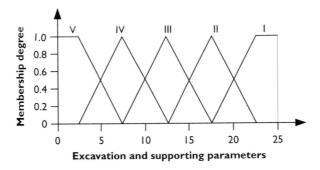

Figure 7.71 Membership function for excavation and supporting parameters.

Table 7.31 Special geological condition values to different evaluation grade.

Evaluation grade	I	II	III	IV	V
Reflection of special geological conditions	Very little	Little	General	Large	Very large
Values	20–25	15–20	10–15	5–10	0–5

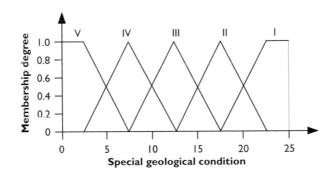

Figure 7.72 Membership function for special geological conditions.

The standard values of the different levels are defined as shown in Table 7.31, and their membership functions can be found in Figure 7.72.

According to the determining principles described above, the membership degree can be established by the evaluation index values for the actual project in hand. The membership degree of evaluation indices, p_{11} and p_{12}, for the different grades can be expressed as follows (Equation 7.1).

$$[N_1] = \begin{bmatrix} \overset{I}{u_{1I}} & \overset{II}{u_{1II}} & \overset{III}{u_{1III}} & \overset{IV}{u_{1IV}} & \overset{V}{u_{1V}} \\ u_{2I} & u_{2II} & u_{2III} & u_{2IV} & u_{2V} \end{bmatrix} \qquad (7.1)$$

level   I   II   III   IV   V

7.5.2.2 Weight vector determining method

The weight vectors of the different evaluation indices can be calculated by the fuzzy hierarchy method; the process is shown in Figure 7.73. The importance contrast value for the evaluation index can be seen in Table 7.32, and the corresponding membership functions are shown in Figure 7.74.

The first grade evaluation indices, p_1–p_6, of the overall risk evaluation model were compared one by one according to Table 7.32 and Figure 7.74. Eleven experts were

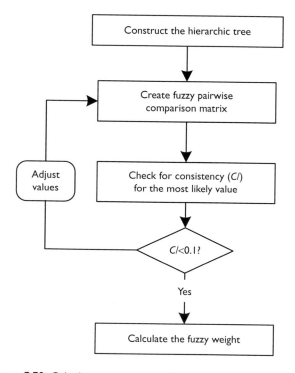

Figure 7.73 Calculation process for the fuzzy hierarchy method.

Table 7.32 Terms for the evaluation criteria.

Terms	Fuzzy number	Triangular fuzzy scale	Triangular fuzzy reciprocal scale
Equal Importance (EI)	$\tilde{1}$	$(1, 1, 1)$	$(1, 1, 1)$
Intermediate (IMI)	$\tilde{2}$	$(1, 2, 3)$	$(1/3, 1/2, 1)$
Moderate Importance (MI)	$\tilde{3}$	$(2, 3, 4)$	$(1/4, 1/3, 1/2)$
Intermediate (ISI)	$\tilde{4}$	$(3, 4, 5)$	$(1/5, 1/4, 1/3)$
Strong Importance (SI)	$\tilde{5}$	$(4, 5, 6)$	$(1/6, 1/5, 1/4)$
Intermediate (IVSI)	$\tilde{6}$	$(5, 6, 7)$	$(1/7, 1/6, 1/5)$
Very Strong Importance (VSI)	$\tilde{7}$	$(6, 7, 8)$	$(1/8, 1/7, 1/6)$
Intermediate (IEXI)	$\tilde{8}$	$(7, 8, 9)$	$(1/9, 1/8, 1/7)$
Extreme Importance (EXI)	$\tilde{9}$	$(8, 9, 9)$	$(1/9, 1/9, 1/8)$

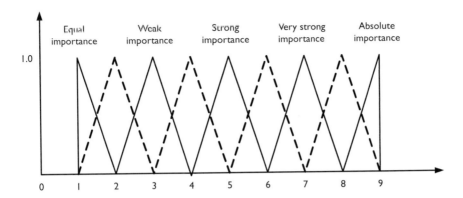

Figure 7.74 Membership functions for qualitative descriptors in the fuzzy-AHP estimation scheme.

Table 7.33 Weight and most probable value assigned by one expert.

	p_1	p_2	p_3	p_4	p_5	p_6	w
p_1	1	1/2	1/2	5	3	7	0.1957
p_2		1	1/2	6	4	8	0.2726
p_3			1	7	5	9	0.3730
p_4				1	1/3	2	0.0435
p_5					1	4	0.0868
p_6						1	0.0284

invited to suggest the comparison matrix content. The most likely value of the triangular fuzzy number comparison matrix given by expert one can be seen in Table 7.33.

The comparison matrix $\tilde{A}_k$ determined by the kth expert is shown as follows (Equation 7.2),

$$\tilde{A}_k = \begin{bmatrix} \tilde{a}_{k11} & \cdots & \tilde{a}_{k1j} & \cdots & \tilde{a}_{k1n} \\ \vdots & \vdots & \vdots & \vdots & \vdots \\ \tilde{a}_{ki1} & \cdots & \tilde{a}_{kij} & \cdots & \tilde{a}_{kin} \\ \vdots & \vdots & \vdots & \vdots & \vdots \\ \tilde{a}_{kn1} & \cdots & \tilde{a}_{knj} & \cdots & \tilde{a}_{knn} \end{bmatrix} = \left[\tilde{a}_{kij} \right] i, j = 1, 2, \ldots, n \tag{7.2}$$

where, $\tilde{a}_{kij} = (a_{kij}^l, a_{kij}^m, a_{kij}^r)$.

The average characteristic value $\tilde{a}_{ij} = (a_{ij}^l, a_{ij}^m, a_{ij}^r)$ of the kth triangular fuzzy number $\tilde{a}_{kij} = (a_{kij}^l, a_{kij}^m, a_{kij}^r)$ $(k = 1, 2, \ldots, K)$ determined by the kth expert through comparison between p_i and p_j is given by (Equation 7.3),

$$l_{ij} = \min_k \{a_{kij}^l\}; \, m_{ij} = \frac{1}{K} \sum_{k=1}^{K} a_{kij}^m; r_{ij} = \max_k \{a_{kij}^r\} \tag{7.3}$$

Then the triangular fuzzy number comparison matrices of the first grade and second grade evaluation indices can be obtained through the evaluation of different experts and is shown in Tables 7.34–7.39.

Table 7.34 Fuzzy number comparison matrix A_0.

Index	p_1	p_2	p_3	p_4	p_5	p_6
p_1	(1, 1, 1)	(0.333, 1.318, 3)	(0.333, 1.318, 3)	(2, 4.727, 6)	(1, 2.818, 5)	(1, 4.818, 8)
p_2	(0.333, 0.759, 3)	(1, 1, 1)	(0.333, 0.864, 3)	(2, 4.182, 7)	(1, 2.455, 5)	(1, 4.455, 9)
p_3	(0.333, 0.759, 3)	(0.333, 1.157, 3)	(1, 1, 1)	(2, 4.364, 8)	(1, 2.636, 6)	(1, 4.727, 9)
p_4	(0.167, 0.212, 5)	(0.143, 0.239, 0.5)	(0.125, 0.229, 0.5)	(1, 1, 1)	(0.25, 0.47, 1)	(0.333, 1.864, 3)
p_5	(0.2, 0.355, 1)	(0.2, 0.407, 1)	(0.167, 0.379, 1)	(1, 2.128, 4)	(1, 1, 1)	(1, 2.909, 5)
p_6	(0.125, 0.208, 1)	(0.111, 0.224, 1)	(0.111, 0.212, 1)	(0.333, 0.536, 3)	(0.20, 0.344, 1)	(1, 1, 1)

Table 7.35 Fuzzy number comparison matrix A_1.

Index	p_{11}	p_{12}
p_{11}	(1.0, 1.0, 1.0)	(1.0, 1.418, 3.0)
p_{12}	(0.333, 0.705, 1.0)	(1.0, 1.0, 1.0)

Table 7.36 Fuzzy number comparison matrix A_2.

Index	p_{21}	p_{22}	p_{23}	p_{24}
p_{21}	(1.0, 1.0, 1.0)	(0.5, 1.318, 4.00)	(1.0, 2.818, 6.0)	(1.0, 1.318, 2.0)
p_{22}	(0.25, 0.759, 2.0)	(1.0, 1.0, 1.0)	(1.0, 2.455, 4.0)	(1.0, 1.157, 2.0)
p_{23}	(0.167, 0.355, 1.0)	(0.25, 0.407, 1.0)	(1.0, 1.0, 1.0)	(0.2, 0.379, 1.0)
p_{24}	(0.5, 0.759, 1.0)	(0.5, 0.864, 1.0)	(1.0, 2.636, 5.0)	(1.0, 1.0, 1.0)

Table 7.37 Fuzzy number comparison matrix A_3.

Index	p_{31}	p_{32}	p_{33}	p_{34}
p_{31}	(1.0, 1.0, 1.0)	(0.5, 0.759, 2.0)	(0.333, 0.407, 3.0)	(1.0, 2.218, 6.0)
p_{32}	(0.5, 1.318, 2.0)	(1.0, 1.0, 1.0)	(0.5, 0.759, 2.0)	(2.0, 2.636, 7.0)
p_{33}	(0.333, 2.457, 3.0)	(0.5, 1.318, 2.0)	(1.0, 1.0, 1.0)	(2.0, 2.909, 8.0)
p_{34}	(0.167, 0.470, 1.0)	(0.143, 0.379, 0.5)	(0.125, 0.344, 0.5)	(1.0, 1.0, 1.0)

Table 7.38 Fuzzy number comparison matrix A_4.

Index	p_{41}	p_{42}
p_{41}	(1.0, 1.0, 1.0)	(1.0, 2.128, 5.0)
p_{42}	(0.20, 0.47, 1.00)	(1.0, 1.0, 1.0)

Table 7.39 Fuzzy number comparison matrix $\mathbf{A}_5$.

Index	P_{51}	P_{52}	P_{53}
P_{51}	(1.0, 1.0, 1.0)	(0.167, 0.355, 1.0)	(0.125, 0.344, 0.5)
P_{52}	(1.00, 2.818, 6.00)	(1.0, 1.0, 1.0)	(0.5, 0.759, 1.0)
P_{53}	(2.0, 2.636, 7.0)	(1.0, 1.318, 2.0)	(1.0, 1.0, 1.0)

For the evaluation index comparison matrix, the Fuzzy Hierarchy Method (Fuzzy-AHP) was used to determine the weight vector for each evaluation index. The analysis process is as follows (Equations 7.4–7.7).

Step 1. Fuzzy comparison matrix for evaluation index A_i

$$\tilde{A} = \begin{bmatrix} \tilde{a}_{11} & \cdots & \tilde{a}_{1j} & \cdots & \tilde{a}_{1n} \\ \vdots & \vdots & \vdots & \vdots & \vdots \\ \tilde{a}_{i1} & \cdots & \tilde{a}_{ij} & \cdots & \tilde{a}_{in} \\ \vdots & \vdots & \vdots & \vdots & \vdots \\ \tilde{a}_{n1} & \cdots & \tilde{a}_{nj} & \cdots & \tilde{a}_{nn} \end{bmatrix} = \begin{bmatrix} \tilde{a}_{ij} \end{bmatrix} i, j = 1, 2, \ldots, n \tag{7.4}$$

The calculation method for the fuzzy value $\tilde{S}_i$ is

$$\tilde{S}_i = \left(S_i^l, S_i^m, S_i^r\right) = \sum_{j=1}^{n} \tilde{a}_{ij} \div \left(\sum_{i=1}^{n}\sum_{j=1}^{n} \tilde{a}_{ij}\right) - \sum_{j=1}^{n} \tilde{a}_{ij} \otimes \left(\sum_{i=1}^{n}\sum_{j=1}^{n} \tilde{a}_{ij}\right)^{-1} \quad i - 1, 2, \ldots, n \tag{7.5}$$

where

$$\sum_{j=1}^{n} \tilde{a}_{ij} = \left(\sum_{j=1}^{n} l_{ij}, \sum_{j=1}^{n} m_{ij}, \sum_{j=1}^{n} r_{ij}\right) \tag{7.6}$$

and

$$\left(\sum_{i=1}^{n}\sum_{j=1}^{n} \tilde{a}_{ij}\right)^{-1} = \left[\sum_{i=1}^{n}\left(\sum_{j=1}^{n} l_{ij}, \sum_{j=1}^{n} m_{ij}, \sum_{j=1}^{n} r_{ij}\right)\right]^{-1} = \left(\frac{1}{\sum_{i=1}^{n}\sum_{j=1}^{n} l_{ij}}, \frac{1}{\sum_{i=1}^{n}\sum_{j=1}^{n} m_{ij}}, \frac{1}{\sum_{i=1}^{n}\sum_{j=1}^{n} r_{ij}}\right) \tag{7.7}$$

The triangular fuzzy number comparison matrix $\tilde{A}$ can be converted into $\tilde{S} = \left[\tilde{S}_1, \tilde{S}_1, \ldots, \tilde{S}_i, \ldots, \tilde{S}_n\right]^T$.

Step 2. Comparing fuzzy numbers of vector $\tilde{S}$

The comparison principle for fuzzy numbers is as follows.

If $\tilde{S}_1 = (S_1^l, S_1^m, S_1^r)$ and $\tilde{S}_2 = (S_2^l, S_2^m, S_2^r)$ are triangular fuzzy numbers, the probability of $\tilde{S}_1 \geq \tilde{S}_2$ is defined using the triangular fuzzy function as follows (Equation 7.8):

$$V\left(\tilde{S}_1 \geq \tilde{S}_2\right) = \sup_{x \geq y}\left\{\min\left[\mu_{S_1}(x), \mu_{S_2}(y)\right]\right\}. \tag{7.8}$$

The calculation formula is (Equation 7.9)

$$V\left(\tilde{S}_1 \geq \tilde{S}_2\right) = hgt\left(\tilde{S}_1 \cap \tilde{S}_2\right) = \mu_{S_2}(d) = \begin{cases} 1 & S_2^m \geq S_1^m \\ \dfrac{S_1^l - S_2^r}{(S_2^m - S_2^r) - (S_1^m - S_1^l)} & \text{otherwise} \\ 0 & S_1^l \geq S_2^r \end{cases} \tag{7.9}$$

Then the fuzzy numbers in vector $\tilde{S}$ can be compared with each other in the formula above.

Step 3. Determining the initial weight vector of triangular fuzzy number comparison matrix $\tilde{A}$

The initial weight vector of fuzzy number $\tilde{S}_g$ in vector $\tilde{S}$ can be calculated by the following formula (Equation 7.10)

$$w'\left(\tilde{S}_g\right) = V\left(\tilde{S}_g \geq \tilde{S}_1, \tilde{S}_2, \dots, \tilde{S}_k, \dots, \tilde{S}_n\right) = \min V\left(\tilde{S}_g \geq \tilde{S}_k\right) \tag{7.10}$$
$$k = 1, 2, \dots, n; \quad \text{and} \quad k \neq g$$

Then using Equation 7.11, the initial weight vector of the triangular fuzzy number comparison matrix is

$$\mathbf{W}' = \left[w'\left(\tilde{S}_1\right), w'\left(\tilde{S}_2\right), \dots, w'\left(\tilde{S}_n\right)\right]^{\mathrm{T}} \tag{7.11}$$

Step 4. Initial weight vector of normalised comparison matrix $\tilde{A}$

The normalised vector of initial weight vector is found using Equations 7.12 and 7.13 as follows

$$\mathbf{W} = \left[w\left(A_1\right), w\left(A_2\right), \dots, w\left(A_n\right)\right]^{\mathrm{T}} \tag{7.12}$$

where

$$w\left(A_i\right) = w'\left(\tilde{S}_i\right) \Big/ \sum_{i=1}^{n} w'\left(\tilde{S}_i\right) \tag{7.13}$$

The weight vectors of the cavern group overall risk evaluation model before construction can be obtained by the method above. The results are presented in Table 7.40 and Figure 7.75.

Table 7.40 Weight vector results.

Weight vector	Weight vector values
W_0	[0.2101, 0.2028, 0.2060, 0.1072, 0.1597, 0.1142]
W_1	[0.5937, 0.4063]
W_2	[0.298, 0.275, 0.157, 0.270]
W_3	[0.271, 0.291, 0.323, 0.116]
W_4	[0.662, 0.338]
W_5	[0.316, 0.684]

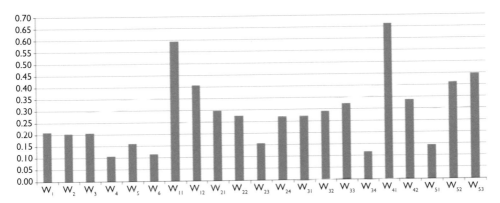

Figure 7.75 Weight vectors of different indices.

7.5.2.3 Determining the overall risk frequency

Based on the membership degree vector R_i and weight vector W_i of the overall risk evaluation index, the overall risk frequency for each section can be obtained finally through the fuzzy arithmetic procedure. The calculation process is as follows (Equations 7.14 & 7.15).

$$[N_0] = [W_0] \circ \begin{Bmatrix} R_1 \\ R_2 \\ R_3 \\ R_4 \\ R_5 \\ R_6 \end{Bmatrix} = [W_0] \circ \begin{Bmatrix} W_1 \circ N_1 \\ W_2 \circ N_2 \\ W_3 \circ N_3 \\ W_4 \circ N_4 \\ W_5 \circ N_5 \\ W_6 \circ N_6 \end{Bmatrix} \tag{7.14}$$

where, $\circ$ stands for the fuzzy operator, operator $M(\wedge, \vee)$ can be used, and the process is as follows.

$$b_j = \bigvee_{i=1}^{m} (w_i \wedge u_{ij}) = \max_{1 \le i \le m} \{\min(w_i, u_{ij})\}, \quad i = 1, 2, 3, 4, 5; j = 1, 2, \dots, m \tag{7.15}$$

Table 7.41 Risk loss evaluation standard for hydropower cavern group projects.

Level		A	B	C	D	E
Severity degree		Slight	Large	Serious	Very serious	Disaster
Value		0–5	5–10	10–15	15–20	20–25
Casualty S_1	Construction people	Slight injury 1 person	Serious injury 1 person or slight injury 2–10 persons	Dead 1–2 persons or serious injury 2–9 persons	Dead 3–9 persons or serious injury more than 10 persons	Dead more than 10 persons
	Other people	Slight injury 1 person	Slight injury 2–10 persons	Serious injury 1 person or slight injury more than 10 persons	Serious injury 2–9 persons	Dead more than 1 persons
Property damage S_2	Project	Less than 1 million	1 million–5 million	5 million–10 million	10 million–50 million	More than 50 million
	Other	Less than 100 thousand	100 thousand–500 thousand	500 thousand–1 million	1 million–2 million	More than 2 million
Schedule delay S_3	Long term	Less than 1 m	1 m–3 m	3 m–6 m	6 m–12 m	More than 12 m
	Short term	Less than 10 d	10 d–30 d	30 d–60 d	60 d–90 d	More than 90 d

Note: The 'dead person' includes 'missing person', the unit of property damage is Chinese yuan, and the delay is in m (month) and d (day).

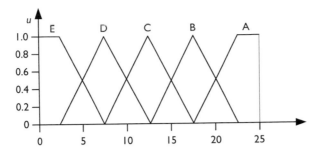

Figure 7.76 Risk loss membership functions.

Consequence		Disastrous	Severe	Serious	Considerable	Insignificant
Frequency	Grade	V	IV	III	II	I
Very likely	V	Unacceptable	Unacceptable	Unacceptable	Unwanted	Unwanted
Likely	IV	Unacceptable	Unacceptable	Unwanted	Unwanted	Acceptable
Occasional	III	Unacceptable	Unwanted	Unwanted	Acceptable	Acceptable
Unlikely	II	Unwanted	Unwanted	Acceptable	Acceptable	Negligible
Very unlikely	I	Unwanted	Acceptable	Acceptable	Negligible	Negligible

Figure 7.77 Risk decision matrix.

Then the cavern group overall risk assessment results $[N_0] = [u_{0\mathrm{I}} \quad u_{0\mathrm{II}} \quad u_{0\mathrm{III}} \quad u_{0\mathrm{IV}} \quad u_{0\mathrm{V}}]$ for each section can be obtained. The overall risk level can be decided according to the principle of maximum membership degree.

7.5.2.4 Determining overall risk consequence

The overall risk consequence for the large cavern groups includes the three factors of casualties, property damage and schedule delay. About six construction management experts are invited to suggest the damage values according to Table 7.41. The mean value is regarded as the calculation value. Then the overall risk consequence level value can be determined with the different weight values of the three factors. Finally, the overall risk consequence level is obtained via the membership function, which is shown in Figure 7.76.

7.5.2.5 Overall risk control analysis

Based on the overall risk frequency and overall risk consequence, the overall risk level of the analysed area can be obtained through the risk decision matrix, which is shown in Figure 7.77.

7.5.3 Method for assessment and mitigation of local risk for a large hydropower cavern group before construction

The overall risk assessment of different sections of large hydropower cavern groups can be implemented before construction, when the assessment area of each section is large. For the main powerhouse, its dimensions may be 25–30 m × 60–80 m × 25–35 m. Although the result is approximate and only used for preliminary estimation, the risk assessment can also provide a decision basis for making an overall construction plan and optimising the supporting design. But for the local areas, this result is not enough. The local risk types and their risk events can be recognised according to Figure 7.78 and the local risk assessment result can be used to determine the optimal design area and excavation control zone.

The local risk assessment process is based on the overall risk assessment result, shown in Figure 7.79. The local risk is also affected by cavern group size, structural surfaces, rock characteristics, *in situ* stress, groundwater and the excavation method. The main types of cavern group local risk are rockburst, surrounding rock unloading and splitting, deformation instability and collapse. The rockburst risk assessment can be performed based on the microseismic monitoring information, rock burst tendency index RVI and intelligent optimisation algorithms. The methodology is not discussed here in detail because it has been described in Chapter 6. The rock mass collapse instability risk assessment can be performed based on the fracture occurrence, block theory and reliability theory. Then the block stability and failure probability can be calculated based on the existing research results, while for the surrounding rock unloading, splitting and deformation instabilities, there is no suitable evaluation method and so the numerical analysis method is used. These types of local risk will be evaluated as a single type, i.e., excavation unloading deformation local risk.

7.5.3.1 Large deformation local risk assessment model before construction

The failure modes in the surrounding rock after cavern group excavation are mainly determined by cavern size, space effect, *in situ* principal stress values and directions, fault occurrence and filling, rock uniaxial compressive strength, tensile strength, joint occurrence and its spatial distribution, excavation method, and so on. So how do we comprehensively consider the effects of these various factors when evaluating the caverns for excavation stability as the key problem in engineering construction? We propose the dynamic feedback analysis method for the **overall** assessment—which can suggest the most suitable excavation method and support parameters—because only large faults can be considered in the numerical analysis model. But the **local** risk assessment method can be based on the numerical analysis model in order to reflect the effects of local faults, joints, rock types, blasting excavation method, and so on. The strength–stress ratio for the different units, which can be obtained via the numerical analysis model, can be used to reflect the comprehensive influence of the cavern excavation effects, large faults, stress field and rock character. The large deformation local risk assessment process for the cavern group is outlined in Figure 7.80.

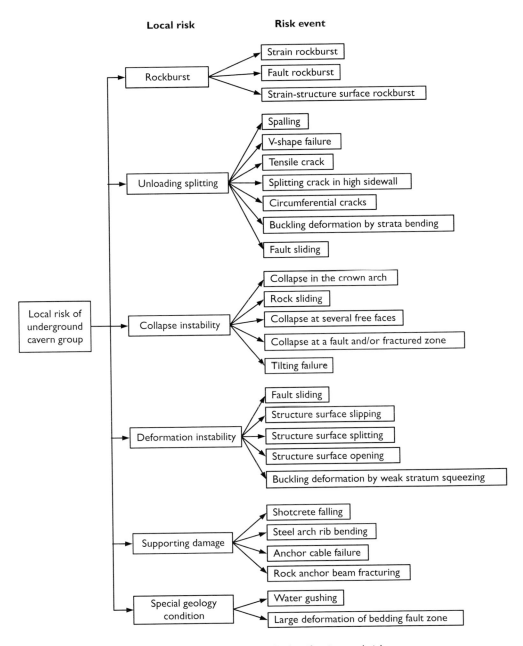

Figure 7.78 Cavern group local risk classification and risk events.

Based on the three-dimensional numerical analysis results for the cavern group peripheral rock, the local risk frequency grade evaluation model of the analysis units can be seen in Figure 7.81. In addition, the effects of the strength–stress ratio affected by local stress concentration, the relaxation depth affected by excavation unloading,

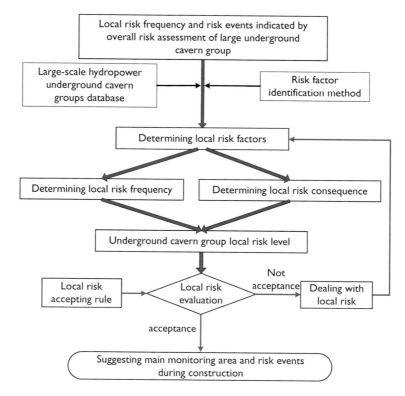

Figure 7.79 Large underground cavern group local risk assessment process.

and the perturbations caused by local faults and joints may be highlighted. The special geological conditions mainly include gas, karst, bedding fault zones and columnar joints. If any of these factors are present, they should of course be considered.

7.5.3.2 Index membership degree determining method

7.5.3.2.1 Area division standard for local risk assessment

Firstly, the local risk assessment range should be determined when the large deformation local risk is evaluated based on the numerical analysis results, due to the fact that there is an influencing boundary depth of excavation unloading, and there are also influencing ranges and degrees of the faults and joints on unloading and deformations. The determination method for the local risk assessment range can be seen in Figure 7.82. Secondly, the local risk frequency grade of the different analysis units in the assessment range can be calculated through the three-dimensional numerical analysis results and selection standards of the index values. The local risk frequency grade distribution can be obtained, but the area outside the assessment range cannot be analysed. Thirdly, the local risk consequence can be determined and the risk controlling measures can be suggested according to the local risk level.

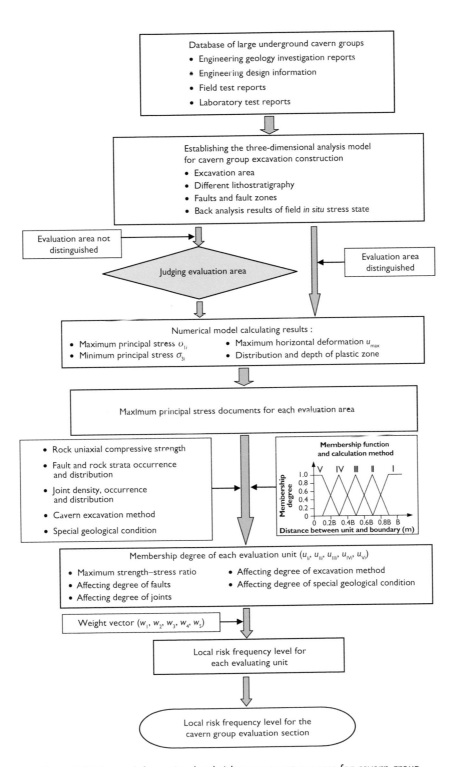

Figure 7.80 Large deformation local risk assessment process for cavern group.

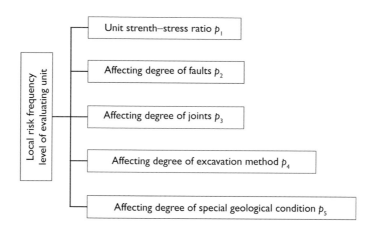

Figure 7.81 Local risk frequency grade evaluation model for cavern group.

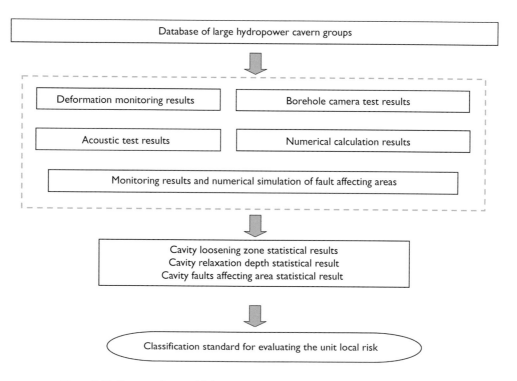

Figure 7.82 Process for establishing the large deformation local risk assessment.

According to statistical analysis results for the large hydropower underground cavern groups, the local risk assessment area of large deformation caused by excavation unloading is divided into four regions, as can be seen in Figure 7.83, and the associated division standard is shown in Table 7.42. In particular, the rock masses

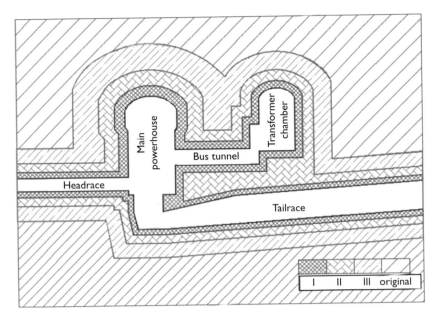

Figure 7.83 Schematic diagram of local risk assessment areas.

Table 7.42 Partition standard of large deformation local risk assessment.

Evaluation region	Region I large unloading	Region II large unloading	Region III weak unloading	Region IV original rock mass
Depth from excavation face (m)	0.1 B or 3 m	0.1 B–0.2 B or 3–6 m	0.2 B–0.4 B or 6–12 m	>0.4 B or 12 m

Note: B represents the span of the underground opening.

in regions I to IV should be taken into account, while the rock masses outside this range do not need to be considered. According to the existing research results, the influence of faults on the deformation caused by excavation unloading, the testing results of unloading relaxation depths, and the influence region of unloading fractures is considered in order to determine the partition standard, the reasons for which are described as follows.

1   Influence of faults

When the fault thicknesses are less than 1 m, the effect caused by the different thicknesses on the stress and deformation of the surrounding rock after excavation is relatively small and can be ignored. In addition, if the fault does not intersect the whole cavern group and the distance to the excavation face is greater than the cavern span, then the effect of the fault on the stress and deformation of the surrounding rock is very small, regardless of whether the fault is located in the vault, spandrel (the region between the arch and sidewall) or sidewall. Moreover, based on the displacement monitoring results for the existing large cavern groups, the deformation of the surrounding rock is mainly in the rock mass between the excavation boundary and one cavern span depth into the rock.

2   Test results for unloading relaxation depth

The sonic wave test results obtained after excavation show that the depth of most of the cavern loosened zones is less than 5 m. The loose rock depth caused by severe unloading, moderate unloading and weak unloading is less than one cavern span, which can be demonstrated by the monitoring results of the Ertan, Dagangshan, Pubugou, Xiluodu and Jinping I hydropower stations.

3   Effect area of unloading fractures

For small-scale hydropower stations, such as Shisanling, Guangxu, Jiangya and Dachaoshan, there are one or two ring cracks around the bus tunnel near the main powerhouse downstream side within a 2 m range while, for the large-scale hydropower stations, such as Longtan, Ertan and Laxiwa, there are several ring cracks around the bus tunnel near the main powerhouse downstream side within an 8 m range. Their width is usually greater than 10 mm, but 15 mm for Laxiwa.

Based on the above studies, the analysis areas are divided into three regions to evaluate the effect of the structural surface deformation and unloading fracturing caused by the excavation of large cavern groups. The surrounding rock deformation and fracturing are most serious in region I (Figure 7.83). In addition, some small blocks will collapse in region I. In region II, the deformation may be substantial if there are structural surfaces or fault zones present and the *in situ* stress is high, noting that the distance is more than 5 m from the free face to region II. There is no unloading cracking or just a small amount in the surrounding rock in region III. For Laxiwa and Jingping II, the distribution range of the ring crack in the bus tunnel is 10–15 m deep from the free face, which is less than the cavern span. The sonic wave testing results for the Pubugou hydropower station surrounding rock indicate that the unloading crack distribution range is less than the main powerhouse span. Therefore, the local risk assessment range was determined as one cavern span, which can satisfy the engineering analysis request. Even for the Jinping I hydropower station, which has the largest deformation of the high sidewall, the loosening range is not greater than one powerhouse span. Thus, the local risk assessment range, which is one cavern span depth from the free face, is suitable.

7.5.3.2.2   *Determination method and evaluation standard for the strength–stress ratio*

The maximum principal stress, σ_1, of the different elements of a typical analysis section can be selected from the numerical simulation results for the large cavern group, and the strength–stress ratio, σ_{ci}/σ_1, of each element can be calculated by determining its uniaxial compressive strength, σ_{ci}. Then the local risk frequency grade for each element can be determined according to Table 7.43, where I represents the smallest risk frequency grade, and V the greatest risk frequency grade. Their membership functions can be seen in Figure 7.84.

Table 7.43   Evaluation level and values of the rock strength–maximum principal stress ratio.

Local risk frequency level	I	II	III	IV	V
Rock strength–stress ratio, σ_{ci}/σ_{1i}	>7	7–4	4–2	2–1	<1

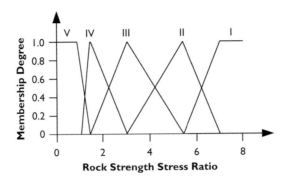

Figure 7.84 Membership function for the rock strength–maximum principal stress ratio.

The following factors should be carefully observed when determining the element strength–stress ratio: (1) The supporting effects of bolt, anchor cable and shotcrete should be considered, not including large faults, lithology of the strata or *in situ* stress, when determining the element maximum principal stress. (2) The maximum principal stress after excavation can be used to calculate the element strength–stress ratio. (3) If the local risk for the cavern group without support must be evaluated, then the support effect should not be considered in the three-dimensional numerical analysis.

7.5.3.2.3   Determination method and evaluation standard for local fault effect degree

The effect range of a major fault on cavern group stability is generally large, especially when the fault dip angle is low, and when the intersection angle between its strike direction and cavern axis is less than 40°, and it is located in the arch crown or sidewall. For example, seven faults at the Jinping II hydropower station, namely F_{16}, F_{21}, F_{24}, F_{25}, F_{36}, F_{65} and F_{68}, were considered when establishing the three-dimensional numerical model, which is shown in Figure 7.85. However, according to the geological investigation report, there are 20 faults. Among them, 11 faults and lithological interfaces, which may affect the cavern group local stability, have not been considered in the three-dimensional numerical model, see Figure 7.86. Therefore, all the faults potentially affecting the cavern group local stability should be considered in order to evaluate its deformation local risk.

In order to establish the evaluation method which can account for most faults, the influence rules for the fault location and occurrence on the stability of the surrounding rock should first be considered. Then, the influence rules for the fault's effect

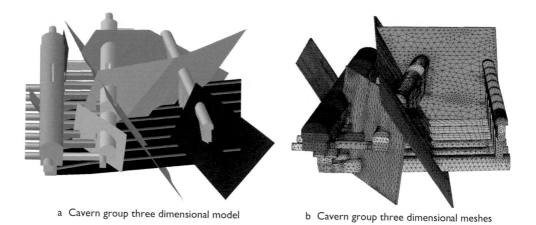

a Cavern group three dimensional model   b Cavern group three dimensional meshes

Figure 7.85 Three-dimensional model of the Jinping II cavern group with the main faults indicated by the planes.

Figure 7.86 Modelled faults affecting the local stability of the Jinping II cavern group.

on the surrounding rock stability in different stress fields and rock types should be analysed. According to existing engineering experience and numerical analysis results, the influences are shown as follows.

1   Influence of fault location on the surrounding rock stability

According to numerical analyses, when the fault is horizontal and near the crown, and the distance between the fault and the free face is the same as the cavern diameter, then the displacement at the crown mid-point increases by about 10% compared with the 'no fault' calculation model, while the distributions of the tensile stress zone and plastic zone are similar. When the distance is half the cavern diameter, then the crown midpoint displacement increases by about 40%, the tensile stress zone increases by 200% and the plastic zone increases by 100%. When the distance is 0.2 of the cavern diameter, then the crown midpoint displacement increases by about 300%, the tensile stress zone and plastic zone respectively increase by 500% and 100–200% times. The depth of the plastic zone increases by 250%.

When the fault is near the spandrel and the distance between the fault and free face is the same as the cavern diameter, then the displacement and distributions of the tensile stress zone and plastic zone are similar to the 'no fault' calculation model. When the distance is half the cavern diameter the tensile stress zone changes only slightly; the plastic zone increases by 200% times in comparison. When the distance is a fifth of the cavern diameter, then the displacements of the crown midpoint and sidewall near the faults all increase by about 100%, and the tensile stress area and plastic zone respectively increase by 200% and 200–300% times in comparison. The depth of the plastic zone increases by 300%.

When the fault is vertical and near the sidewall and the distance between the fault and free face is the same as the cavern diameter, the displacement and distributions of tensile stress zone and plastic zone are similar to the 'no fault' calculation model. When the distance is 0.5 times, the sidewall displacement increases by about 10%, and the tensile stress zone and plastic zone change little. When the distance is 0.2 times, the sidewall displacement increases by about 900%, and the tensile stress zone and plastic zone respectively increase 100% and 200–300% in comparison. The depth of the plastic zone increases by 200%.

In summary, when the fault is located near the cavern floor, its influence on the cavern stability is not significant, but the influence on cavern stability is significant when the fault is located near the spandrel or sidewall, and especially near the crown. Through analysis of the deformation, tensile stress distribution and plastic zone in the cavern surrounding rock after excavation, the influence of the fault on cavern stability is significant when it is located near the crown and spandrel within 1.0 times the cavern span, and near the sidewall within 0.5 times the cavern span.

2   Influence of fault thickness on surrounding rock stability

When the fault is located near the cavern crown and sidewall, and the distances between the fault and free face are 0.2 and 0.5 times the cavern span, the numerical analysis results indicate the following. When the fault thickness increases from 0 to 0.02 times the cavern span, then the displacement changes at the crown and sidewall are less than 5%, and the plastic zone changes little. When the thickness increases to 0.2 times, then the crown midpoint deformation increases by 90%. Therefore, the influence of the fault thickness from several tens of mm to hundreds of mm on the surrounding rock can be ignored.

3   Influence of a fault in different stress fields on the surrounding rock stability

The influence rules for faults located in different positions and with different lateral stress coefficients and fault–boundary distances on the surrounding rock stability are described as follows. When the fault is located near the crown, the influence of the fault on crown displacement and stress field decreases with an increase of the lateral stress coefficient, which is significant when the distance between the fault and free boundary is less than 0.5 times the cavern span. The sidewall displacement increases with the increase of lateral stress coefficient. When the fault is located near the spandrel, then the influence of the fault on vault displacement decreases with an increase in the lateral pressure coefficient, while it increases for a sidewall fault. These results can be seen in Figures 7.87(a) and (b). The influence of a fault on the stress field first increases and then decreases with increasing lateral stress, and the inflection point for the lateral stress coefficient is about 2.0. The plastic zone increases with the lateral stress coefficient, and, when it is less than 3.0, the plastic zone only increases by 25%. When the fault is located near the sidewall, the influence areas related to the fault are the crown and sidewall. The crown displacement decreases with an increase in the lateral stress coefficient, and the sidewall displacement increases. These results can be seen in Figures 7.87(c) and (d). The influence of the fault on the stress field also first increases then decreases along with the increasing lateral stress coefficient, and the inflection point for the lateral stress coefficient is also about 2.0. The plastic zone increases with the lateral stress coefficient, and when it is less than 3.0, the plastic zone only increases by 100%.

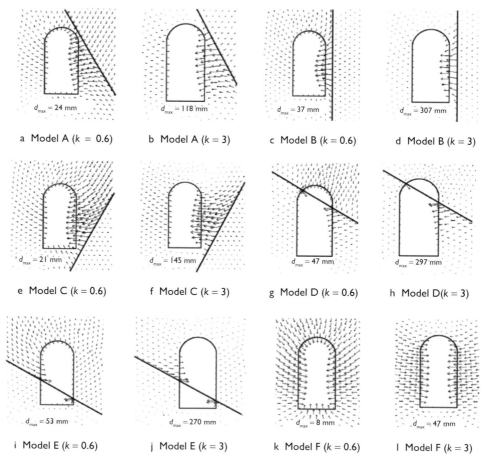

Figure 7.87 Displacement vector diagram for cavern analysis models (Huang *et al.*, 2009). *k* is the lateral stress coefficient.

4   Influence of the rock mass on the surrounding rock stability

When the fault is located near the cavern crown and the rock mass quality is grade II or III, the crown mid-point displacement will increase by about 85% compared with the 'no fault' situation, while the displacement of the sidewalls will decrease by 30–35% and the tensile stress area will increase by 4–5 times. When the rock mass quality is grade IV, then the displacement of the cavern crown will increase by about 120%, and the displacement of the sidewall will increase by about 10%.

When the fault is located near the spandrel and the rock mass quality is grade II or III, then the crown mid-point displacement will increase by about 37% compared with the 'no fault' situation, the displacement of sidewalls will increase by about 33%, and the tensile stress area in the grade III surrounding rock will increase by twice. When the rock mass quality is grade IV, the displacement at the cavern crown will increase by about 45%, the displacement of the sidewalls will increase by about 50%, and the tensile stress area will increase by four times.

When the fault is located near the sidewall and the rock mass quality is grade II or III, then the displacement of the cavern crown and sidewall without a fault will decrease by 10–15%, while the displacement of the sidewall with a fault will increase by 11 times. The tensile stress area in the grade III surrounding rock will increase by 500%. When the rock mass quality is grade IV, then the displacement of the sidewall without a fault will decrease by about 20%, the displacement of the sidewall with fault will increase by about 6 times, and the tensile stress area will increase by 250%.

If the rock mass quality is grade II, then the plastic zone will change little when there is a fault and it is near the crown and spandrel, while the plastic zone will increase by 100% when the fault is near to the sidewall. If the rock mass quality is grade III, then the plastic zone will increase by 100% when the fault is near the crown, and 50% when it is near the spandrel and sidewall. If the rock mass quality is grade IV, then the plastic zone will change little, regardless of where the fault is located.

5   Influence of fault occurrence on surrounding rock stability

There are some differences between the stresses at the cavern crown, floor, right and left sidewall, with different intersection angles and dip angles of the fault. In general, the maximum principal stress decreases and the minimum principal stress changes little with increase in the intersection angle between the fault surface and cavern axis. The influence of a low dip angle fault on the surrounding rock stress is small, while it is large when the fault is steep.

6   Influence of fault on surrounding rock stability in different areas

The surrounding rock deformation in the area between the fault and excavation surface is mainly influenced by the fault. If the fault is inclined towards the excavation area, then the deformations are mainly produced in the hanging wall, as can be seen in Figures 7.87(e) and (f). If the fault is inclined towards the surrounding rock inside, then the deformations are mainly produced in the footwall, as can be seen in Figures 7.87(a) and (b). If the fault intersects the opening, then the deformation near the intersecting area is the largest, as can be seen in Figures 7.87(g), (h), (i) and (j). The deformation direction can also be seen in these Figures for different situations, e.g., the radial deformations at the crown and the horizontal deformation at the sidewall. The lateral stress coefficient also influences the deformation direction and values; when it is 3, the deformation direction gradually rotates upwards and its value also increases significantly, as can be seen in Figures 7.87(k) and (l).

The fault location, distance from fault to excavation face, fault occurrence, *in situ* stress and rock mass quality are mainly considered in determining the influence degree, as can be seen in Tables 7.44 and 7.45. The membership functions can be seen in Figure 7.88.

Table 7.44 Fault influence model analysis partitions.

Unit influence degree partition	A	B	C
Ratio of distance from unit to excavation face (d_1) and from unit to fault (d_2)	<0.5	0.5–2.0	>2.0

Table 7.45 Faults influence degree values.

Intersection angle between fault strike direction and cavern axis	60–90°				30–60°				30°			
Fault dip angle	>70°	45–70°	20–45°	20°	>70°	45–70°	20–45°	20°	>70°	45–70°	20–45°	20°
Unit values A	0	−2	−5	−10	−2	−5	−10	−12	−5	−10	−12	−12
B	−1	−3	−3	−5	−3	−8	−6	−6	−8	−11	−8	−6
C	−2	−5	−2	0	−5	−10	−2	0	−10	−12	−5	0

Note: All values are for lateral stress coefficient less than 1.5 and the rock mass quality grade II. If the lateral stress coefficient is greater than 1.5 and less than 3, each of the values should be reduced by 2. If the lateral stress coefficient is equal to or greater than 3.0, then the values should be reduced by 3. If the rock mass quality is grade III, then the values should be reduced by 1. If the rock mass quality is grade IV, then the values should be reduced by 2. The minimum for the corrected values should be −12.

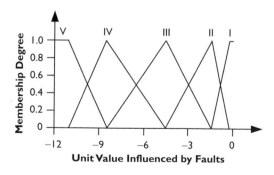

Figure 7.88 Membership values of the unit influence degree caused by faults.

7.5.3.2.4 Determining method and evaluation standard of joints and fracture effect degree

The occurrence of joints, bedding planes and separations, fractures and the relation between these and the cavern axis are all considered in determining the effect degree. If there are many sets of joints and fractures, then the effect degree should be obtained by means of their weights, which can be determined according to their assessment values. The area influenced by the joints and fractures should be determined first, then the assessment values can be obtained according to the analysis areas I, II and III. The standard can be seen in Table 7.46 and its membership degree can be seen in Figure 7.89.

7.5.3.2.5 Determining method and evaluation standard of construction method

The presplit blasting and smooth blasting technology are widely used because a relatively smooth excavation face can be obtained using these techniques. The presplit crack, which allows release of part of the rock stress in advance, forms more easily with 'middle cutting' first. In fact considering the field excavation test results from the Ertan and Pubugou hydropower station powerhouses which are in rock masses with high *in situ* stress, it is also known that the presplit fracture could not be obtained

Table 7.46 Model analysis unit influence degree values caused by fractures.

Intersection angle between joint strike direction and cavern axis	60–90°				30–60°				30°			
Joint dip angle	>70°	45–70°	20–45°	<20°	>70°	45–70°	20–45°	<20°	>70°	45–70°	20–45°	<20°
Unit values III region	0	−2	−5	−10	−2	−5	−10	−12	−5	−10	−12	−12
II region	−1	−3	−3	−5	−3	−8	−6	−6	−8	−11	−8	−6
I region	−2	−5	−2	0	−5	−10	−2	0	−10	−12	−5	0

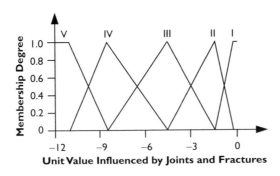

Figure 7.89 Membership degree of unit influence degree caused by joints and fractures.

without 'middle cutting' first, despite the fact that the explosive density used at the sites was greater than normal. If the excavation process involves 'middle cutting' first, followed by expanding excavation and smoothwall blasting, then the excavation effect will be improved. If the presplit blasting is first performed outside the protective layer followed by 'middle cutting', such as at the Jinping I and Jinping II hydropower stations, then the effect of the presplit crack will also not be ideal. Therefore, 'middle cutting' first is important for both the presplit blasting and smooth wall blasting excavation techniques. For the 2nd to 4th layers of the Guandi hydropower station main powerhouse, the excavation process involved 'middle cutting' first, followed by expanding excavation with two side protective layers, and finally presplitting along with the protective layer side and smoothwall blasting. For its 5th to 7th layers and transformer chamber, the excavation process involved 'middle cutting' first, followed by expanding excavation with two side protective layers, and finally smoothwall blasting. The blasting excavation height was less than 6 m. The monitored results indicated that the excavation effect was good.

The rock unloading and the relaxation depth for excavations in rock having a high *in situ* stress is greater than the typical stress situation, especially near the cavern intersection area. Thus, the tunnels which intersect with the sidewalls should be excavated first, and the sidewalls later—which reduces the adverse influence of blasting on the high sidewall stability.

The blasting methods used for the existing hydropower underground cavern groups in China were analysed by Lu *et al.* (2011). According to the authors' results, the influence degree of the cavern group excavation on the surrounding rock local risk can be determined according to the blasting excavation sequence and height, as shown in Table 7.47 and Figure 7.90.

7.5.3.2.6 Determination method and evaluation standard for special geological conditions

The influence degree for the special geological conditions on the surrounding rock stability can be determined according to the existing engineering cases, as shown in Table 7.48 and Figure 7.91.

Table 7.47 Evaluation standard for blasting excavation method influence degree.

Index description	Values	Grade
Most acceptable excavation sequence and blasting method, excavation height less than 8 m.	20–25	I
Most acceptable excavation sequence and blasting method, excavation height between 8 m and 10 m.	15–20	II
Reasonable excavation sequence and blasting method, excavation height between 8 m and 10 m.	10–15	III
Reasonable excavation sequence and blasting method, excavation height more than 10 m.	5–10	IV
Not reasonable excavation sequence and blasting method, excavation height more than 10 m.	0–5	V

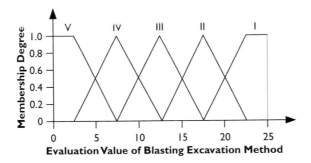

Figure 7.90 Membership function for cavern group blasting excavation methods.

Table 7.48 Evaluation standard for the special geological distribution influence degree.

Grade	I	II	III	IV	V
Influence of special geological conditions	Very little	Little	Medium	Large	Very large
Values	20–25	15–20	10 15	5–10	0–5

Note: The values should be determined according to adverse factors, such as karst, bedding fault zones and large faults.

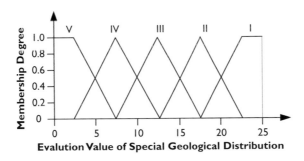

Figure 7.91 Membership function for special geological conditions.

Table 7.49 Weight vector for the evaluation indices.

Weight vector	Weight values
W'_0	[0.242, 0.233, 0.227, 0.167, 0.131]

The weights of the evaluation indices can be calculated using the fuzzy hierarchy method; the results are given in Table 7.49.

For the local risk assessment of the analysis area, the grid array should be first established, the geological conditions and *in situ* stress components may then be determined through the engineering geological and rock stress conditions. Next, the membership degree of each evaluation unit can be obtained using the method described above. Following this, the local risk frequency for each unit can be calculated using the fuzzy calculation method with the weight vector. The local risk occurrence possibility level is divided into five grades, labelled as I, II, III, IV and V, with grade V representing the maximal local risk. Finally, the 'local risk occurrence possibility grade distribution nephogram (cloud)' for the cavern group analysed area can be obtained, and the appropriate local risk controlling measures can be suggested.

7.5.4   Method for assessment and mitigation of local risk for a large hydropower cavern group during construction

The local risk assessment during construction should be performed based on the overall risk and local risk assessment results before construction, the excavation–revealed geological information, field monitoring, and dynamic feedback analysis results. The preliminary assessment results and control measures can be verified, and the modified risk control measures can be suggested for the current excavation layer and excavated area. The local risk assessment and control model can be divided into three types, as follows:

1   Current layer excavation of cavern group → Local risk assessment of excavated area → Risk assessment results accepted without reinforcing/supporting → Next layer excavation. The process is shown in Figure 7.92.
2   Current layer excavation of cavern group → Local risk assessment of excavated area → risk assessment results unaccepted and needing reinforcing support → Local risk assessment after reinforcement → Risk assessment results accepted and then the next layer excavation. The process is shown in Figure 7.93.
3   Current layer excavation of cavern group → Local risk assessment of excavated area → Risk assessment results unaccepted and needing reinforcing/supporting → Local risk assessment after reinforcement → Risk assessment results still unaccepted and needing further reinforcing/supporting → Local risk assessment after second reinforcement → Risk assessment results accepted and then next layer excavation. The process is shown in Figure 7.94.

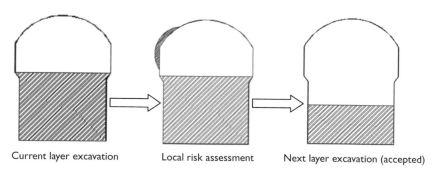

Current layer excavation   Local risk assessment   Next layer excavation (accepted)

Figure 7.92 Dynamic regulation type I for cavern group local risk.

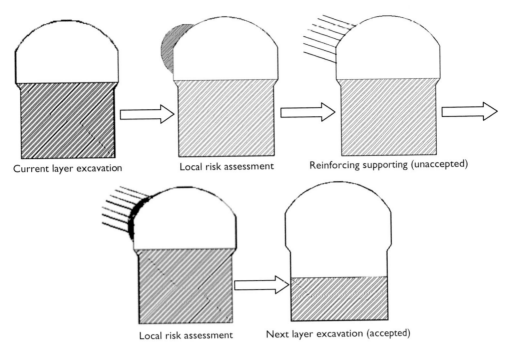

Current layer excavation   Local risk assessment   Reinforcing supporting (unaccepted)

Local risk assessment   Next layer excavation (accepted)

Figure 7.93 Dynamic regulation type II for cavern group local risk.

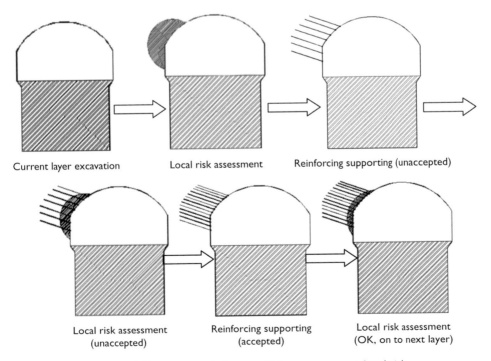

Current layer excavation   Local risk assessment   Reinforcing supporting (unaccepted)

Local risk assessment   Reinforcing supporting   Local risk assessment
(unaccepted)   (accepted)   (OK, on to next layer)

Figure 7.94 Dynamic regulation type III for cavern group local risk.

Table 7.50 Risk level and its accepted distribution depth.

Risk level	I	I–II	I–III	I–IV	V
Depth	≤3 m	≤6 m	≤9 m	≤15 m	—

The local risk level distribution contour map can then be drawn, where I stands for the maximum risk level and V stands for the minimum risk level. The risk control measures are suggested through the depths of different risk levels, as shown in Table 7.50.

7.6 ILLUSTRATIVE EXAMPLE: ASSESSMENT AND MITIGATION OF RISK FOR THE UNDERGROUND POWERHOUSE AT JINPING II HYDROPOWER STATION, CHINA

7.6.1 Epistemic uncertainty analysis

7.6.1.1 Geological setting

The Jinping II hydropower station is located in the southeastern region of the Song-fan-ganzi fold system; since the Mesozoic era, it underwent the Yinzhi and Yanshan movements, and especially the Himalayan movement, and formed a series of thrust faults, overturned strata, recumbent folds and mountain peaks. The north side of the site has a high, steep slope. The slope is at an angle of 50–70° at an altitude of 1330–1600 m, and is located at the intersection of steep bedding-parallel joints and NWW steep and gently inclined joints. At the altitude range 1600–1700 m, the slope has an angle of 35–45° and a steep bank at an altitude of 1700 m. Above an altitude of 1700 m, the slope angle is around 35°.

The sedimentary strata are part of the middle Triassic Yan-tang formation (T_{2y}) and Quaternary system (Q). T_{2y}^4 is a greyish-green, striped, micaceous marble with thickness of about 400 m. $T_{2y}^{5-(1)}$ is an ash black, fine marble and a white coarse grained marble. $T_{2y}^{5-(2)}$ is a grey-white, coarse grained marble containing H_2S, with a thickness of about 400 m. T_{2y}^6 is an ash black, moderately thin, layered, argillaceous limestone, at altitudes 1870 m and above, with a thickness of 350 m. The Quaternary system (Q) deposits consist of slope wash and associated accumulations.

The geological structure at the site is similar to the regional structure, has developed NNE structures, and is subject to a NWW–SEE *in situ* stress field. The strata strike is about NNE. Meanwhile, the site has also been subjected to ancient *in situ* stress fields, resulting in other structures.

There are five sets of fractures (dip direction and dip): (1) N10°W–N30°E, NE–NW∠73°–85°; (2) N60°–80°W, SW∠20°–40°, ∠70°–90°; (3) N40°–60°E, SE∠20°–40°; (4) EW, S20°–40°; and (5) N70°–85°E, SE∠60°–80°.

The underground powerhouse dimensions (length × width × height) are 352.4 m × 28.5 m × 28.5 m, equipped with eight 600 MW hydro-generator units, at an altitude of 1317 m. The transformer chamber size is 374.6 m × 19.8 m × 19.8 m (length × width × height), 45 m downstream of the main powerhouse.

7.6.1.2   Rock stress

The *in situ* stress has been tested at different locations (Figure 7.95) and at different depths using various methods. The results indicate that the three principal stresses change with the test location depth and some uncertainty exists about their variation.

1   The field test results indicate that the maximum principal stress, σ_1, near the powerhouse area is in the range 10.1–22.9 MPa, the σ_2 range is 7–20 MPa, and the σ_3 range is about 6–14 MPa. The principal stress values increase with depth, and are affected by the local structural surfaces in some areas.
2   The measured trend of the maximum principal stress is N120–160°E, which is similar to the zone's maximum principal stress direction NWW, while the minimum principal stress direction is NEE.
3   The maximum principal stress plunge angle is small, ranging from 3° to 56°, and its mean angle is around 28°, which is close to the slope angle.
4   The measured *in situ* stress is greater than that caused by the rock mass gravity, and there is considered to be a large residual tectonic stress in the rock mass.

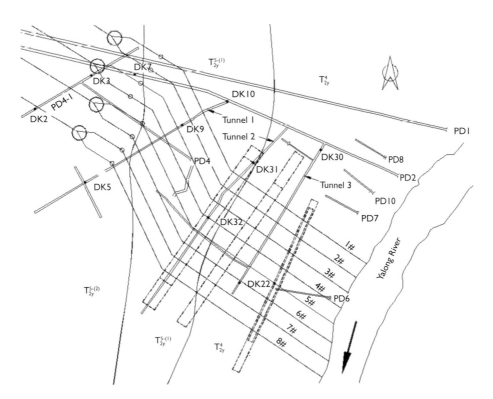

Figure 7.95 In situ stress measurement locations in the Jinping II cavern group area.

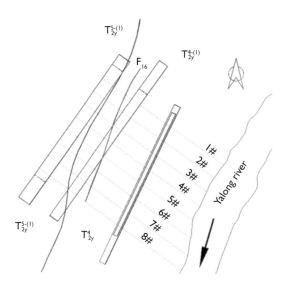

Figure 7.96 Jinping II hydropower station cavern group layout.

7.6.1.3 Hydrology

The site belongs to the eastern solution hydrogeology element (III) which has a width of 2.5–4.5 km and occurs in the middle Triassic Yantang formation. Shallow underground water occurs after rain, with a large flow during the rainy season. Deep underground water is mainly fracture or karst water, with a large head of 200–250 m, and there is little change with the seasons.

7.6.1.4 Specific project location

The main powerhouse is located in the T_{2y}^5 thick layer marble and avoids the influence of fault F_{16}, as can be seen in Figure 7.96. According to the intersection angle between the structural surface and the maximum principal stress direction and cavern axis direction, N35°E is selected as the powerhouse axis direction. It is located in steep strata with a S–N strike direction, on one side of a fold without a large fault in the vicinity. The maximum principal stress range is 10.6–16.8 MPa, with mean trend S43°E. The overlying rock thickness at the underground powerhouse is in the range 231–327 m.

7.6.1.5 Excavation and support method

According to the preliminary design analysis and similar engineering excavation experience, the excavation scheme and sequence for the Jinping II hydropower station cavern group are shown in Figure 7.97 and Table 7.51.

The support requirements for the Jinping II excavations are shown as follows:

1   Crown: Pre-stressed rockbolts: $T = 120$ kN; $\phi 28$ mm, $L = 5$ m; $\phi 32$ mm, $L = 7$ m; @1.5 × 1.5 m; steel fibre reinforced shotcrete with mesh reinforcement: thickness 150 mm.

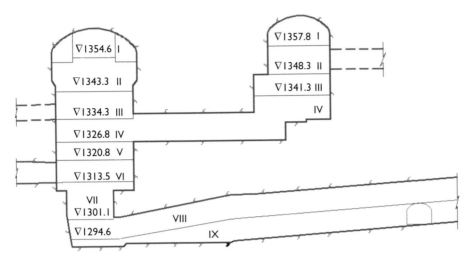

Figure 7.97 Underground cavern group excavation scheme.

Table 7.51 Underground cavern group excavation sequence.

Excavation stage nos.	Excavation layers for the powerhouse	Excavation layers for the transformer chamber	Other excavations
I	I		
2	II	I	
3	III	II	Bus tunnel
4	IV, VIII	III, IV	
5	V, VIII	IV	Upper layer of tailrace tunnel
6	VI, IX		Lower layer of tailrace tunnel
7	VII		

2   Upstream sidewall: Pre-stressed cable anchors: $T = 2000$ kN; $L = 45$ m; @4.5 × 4.5 m; mortared rockbolts: $\phi28$ mm, $L = 5$ m; grouted rockbolts: $\phi32$ mm, $L = 5$ m; @1.5 × 1.5 m; steel fibre reinforced shotcrete with mesh reinforcement: thickness 120 mm.

3   Downstream sidewall: mortared rockbolts: $\phi28$ mm, $L = 5$ m; grouted rockbolts: $\phi32$ mm, $L = 7$ m; @1.5 × 1.5 m; steel fibre reinforced shotcrete with mesh reinforcement: thickness 120 mm.

7.6.2   Assessment and mitigation of overall risk before construction

7.6.2.1   Assessment

The overall risk of the Jinping II underground cavern group should be evaluated before construction in order to optimise the excavation and support parameters. Based on the geological investigation report, each unit section will be evaluated, and the division of their evaluation areas is shown in Figure 7.98. During the overall risk assessment, the assessment results of the upstream sidewall, downstream sidewall and vault should be regarded as parts of the main powerhouse or transformer chamber section, which can be calculated with the proposed method described earlier. Their mean value is used as the section assessment result. The overall risk assessment results for the Jinping II underground cavern group can be seen in Figure 7.99. The overall risk levels for the main powerhouse unit sections #2, #3, #4 and #8 and transformer chamber unit sections #3 and #7 are high. During construction, the large deformation risk, collapse risk and anchor cable failure risk should be observed.

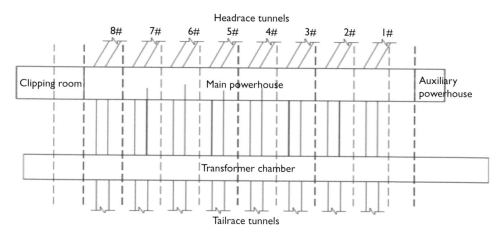

Figure 7.98 Overall risk assessment sections in the underground cavern group (clipping room ≡ assembly bay).

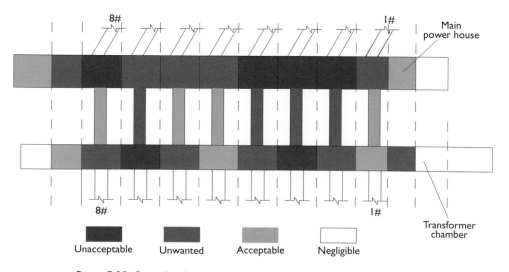

Figure 7.99 Overall risk assessment results for the Jinping II cavern group.

7.6.2.2 Risk mitigation measures

According to the cavern group overall risk assessment result, some monitoring and control measures need to be suggested in order to deal with the high risk sections—as follows:

1   Monitoring sections should be reasonably arranged, dynamic feedback analysis should be performed in a timely manner based on the monitoring information.
2   The spandrel and sidewall of the upstream and downstream sides should be reinforced with anchor cables and bolts.
3   The excavation height should be reduced, and the support should be installed early, especially for the fracture developed zone and fault affected area.
4   The anchor cable locked load near the area severely affected by the faults, *in situ* stress and fractures should be reduced in order to prevent the anchor cables from overloading.
5   The concrete layer near the fracture developed and rock mass broken area should be sprayed on as soon as possible, and the mesh space should be reduced accordingly.
6   The cavern intersection area should be reinforced with bolts, bolt piles or anchor cables.

7.6.3 Assessment and mitigation of local risk before the construction

7.6.3.1 Assessment

The #4 unit section is selected for local risk analysis before construction. According to the preliminary investigation report, the maximum principal stress is 12–24 MPa, mean value of the rock's dry uniaxial compressive strength is 95 MPa, mean value of the wet uniaxial compressive strength is 85 MPa, the rock mass quality grade is II. The fault distribution is shown in Figure 7.100. The joints can be divided into four

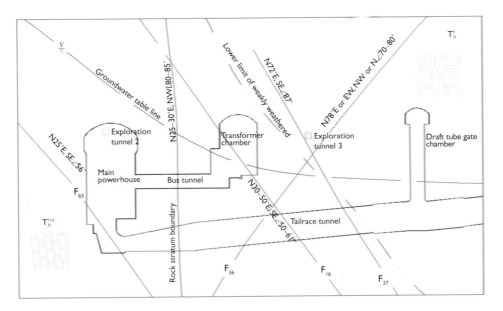

Figure 7.100 Geological section for the 4# unit, (F ≡ Fault).

main groups: (1) N63°W, NE∠81°; (2) N5°W, SW∠88°; (3) N60°W, SW∠87°; and (4) N55°E, SE∠36°. The excavation schemes of vault, rock anchor beam and sidewall are respectively AEP - II, BEP - I and WEP - V. The distributions of the underground cavern group plastic zone and maximum and minimum principal stress can be seen in Figures 7.101–7.103.

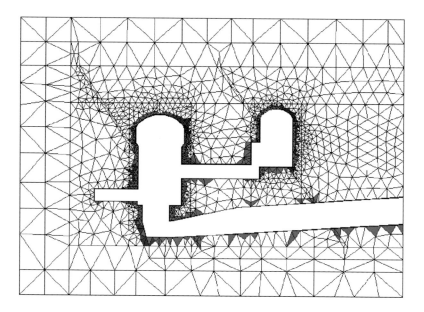

Figure 7.101 Plastic zone distribution for the 4# unit section.

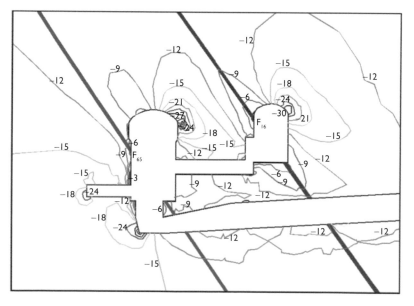

Figure 7.102 Maximum principal stress contour map of 4 # unit section (the regional maximum principal stress is acting approximately from top left to bottom right in this section).

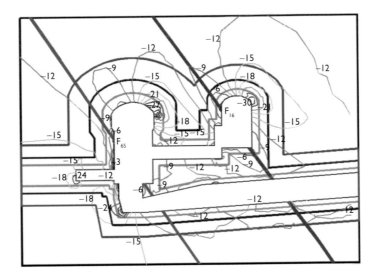

Figure 7.103 Maximum principal stress contour partition map for 4 # unit section.

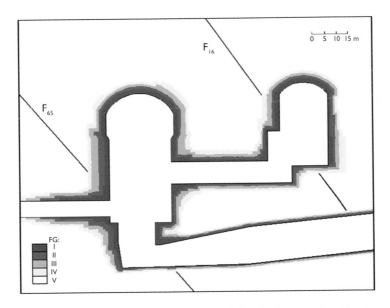

Figure 7.104 Local risk occurrence probability level distribution map for 4# unit section.

The distribution map of local risk occurrence probability level of the unit #4 section can be seen in Figure 7.104. The local risk levels for the main powerhouse upstream sidewall affected by fault F_{65} and transformer chamber upstream sidewall affected by fault F_{16} are the maximum levels; thus, these areas should be observed closely during excavation.

7.6.3.2 Risk mitigation measures

According to similar projects and previous construction experience, the measures to reinforce the surrounding rock are as follows:

1   The vault area can be reinforced by adding extra shotcrete layer thickness and steel arch ribs.
2   The broken rock mass or fracture developed area can be reinforced with high pressure grouting.
3   The pre-stressed ($T = 80$ kN or 120 kN) hollow grouted anchors with different lengths ($L = 6$ m, 7 m, 8 m, 9 m) can be used to increase the support parameters.
4   The unbonded pre-stressed anchor cables with different spacings (3.5 m × 3.5 m, 3.5 m × 4 m, 4 m × 4 m or 4.5 m × 4.5 m, etc.), different lengths ($L = 20$ m, 25 m, 30 m, 35 m or 40 m, etc.), and different pre-stresses ($T = 1000$ kN, 1200 kN, 1750 kN or 2000 kN) can be used to increase the support parameters.

7.6.4 Aleatory uncertainty analysis

7.6.4.1 Estimation of geological conditions at different layers

According to the cavern group preliminary investigation data, the main fault distributions of each unit section are shown in Table 7.52.

The geological conditions for each excavation layer revealed are shown as follows:

1   Main powerhouse first layer (I) excavated
The surrounding rock is weak, weathered marble. The main faults revealed are faults F_{56}–F_{63}, F_{68}, F_{77}, F_{78}, F_{80}–F_{85} with faults F_{80}–F_{85} being first exposed. The strike direction of steep fractures near the main powerhouse is N5–30°E, and that of the flat fractures is NE. Most of them have been lightly squeezed and filled with rock debris, and their aperture is more than 200 mm.

2   Main powerhouse second layer (II) and transformer chamber, first layer (I) excavated upstream sidewall: The strike direction of fault F_{65} is parallel with the main powerhouse axis, and its orientation has changed from N25°E SE∠56° to N26–35°E

Table 7.52 Main faults through different unit sections.

Unit section	Main faults
1#	F_{16}, F_{21}, F_{37}, F_{41}, F_{56}, F_{58}, F_{68}
2#	F_{16}, F_{37}, F_{41}, F_{56}, F_{58}
3#	F_{16}, F_{36}, F_{37}, F_{56}, F_{65}
4#	F_{16}, F_{36}, F_{37}, F_{65}
5#	F_{16}, F_{36}, F_{37}, F_{65}, F_{79}
6#	F_{16}, F_{25}, F_{60}
7#	F_{16}, F_{24}, F_{25}, F_{61}, F_{77}, F_{78}
8#	F_{16}, F_{24}, F_{25}, F_{61}, F_{62}, F_{78}

SE∠30–45°. One set of joints parallel with the main powerhouse axis is exposed, their aperture is 30–80 mm, and tracelength is 100 m. There are some areas with developed broken rock volumes with joints.

Downstream sidewall: Steep fault F_{68} affects the downstream sidewall stability, and its width is 30–50 mm. Its length from unit #1 to #2 is changed to #1 to #5. There are parallel joints developed along with the fault, and the maximum width of the parallel joint zone is more than 2 m. The occurrence of fault F_{78} has changed from N70°E NW∠62° to N85°E NW∠50°, and there are joint zones with occurrence SN W84°. There are also some areas with developed broken rock masses and joints.

There are faults F_{35}, F_{21} and F_{16} and one extrusion fracture zone in the transformer chamber surrounding rock. The right sidewall of the transformer chamber between 0+005–015 is affected by the hanging wall of fault F_{21}. Fault F_{16} is exposed at the vault near 0+100.00, the intersection angle with the cavern axis is 20°, and the width of the failure depth is 150–500 mm.

3 Excavation of main powerhouse layer III and transformer chamber layer II

One set of steep structural surfaces was revealed at the powerhouse upstream sidewall near 0+135, their strike direction being parallel with the powerhouse axis, their length is between 10 m and 20 m, they develop along with fault F_{65}, and their mechanical properties are poor. There are also horizontal joints from 0+170 to the vice powerhouse. Some rock mass collapsed due to fault F_{68} at the downstream sidewall. There are also large numbers of small faults, joints and local failure depths.

There are steep structural surfaces in the upstream and downstream sidewalls of the transformer chamber, and some rock blocks have collapsed as a result of their presence. The surrounding rock crossed by fault F_{16} was severely damaged. The faults and joints revealed include F_{16}, P_{16}, P_{17}, P_{19}, P_{20}.

The rock mass quality of the bus tunnel is good, with no large faults.

4 Excavation of main powerhouse layer IV and transformer chamber layers III and IV

Upstream sidewall of powerhouse: The fault fracture zone was revealed at 20–30 points, its width is usually about 50 mm, the maximum width being 300 mm. The horizontal fracture zone F_{65-1} is revealed between 0+065–0+135. There are many joints along the horizontal fault F_{65}.

Downstream sidewall of powerhouse: There is a horizontal fault between units #4–6.

There are large numbers of joints in the transformer chamber surrounding rock, and its rock mass quality is poorer than that of the powerhouse.

5 Excavation of main powerhouse layers V and VI

Fault F_{65-1} was revealed in the downstream sidewall of layer V of the powerhouse. Fault F_{65-2} was revealed at the upstream sidewall between units #5–8 and downstream sidewall between units #6–8. Its strike direction parallels the powerhouse axis, with mean dip angle less than 10°, width about 400 mm, and its character is poor.

6 Excavation of cavern group

The surrounding rock quality between the water-collecting well and pit #3 is poor. There is a fracture zone Pz_{17} near the water–collecting well. Fault F_{81} and fracture zone Pz_{19} were revealed near #2.

7.6.4.2 Estimation of three dimensional stress field

Note that the convention here is that compressive stresses are negative.

According to the field measuring results before construction, the maximum principal stress σ_1 magnitude is about -10.1 to -22.9 MPa, trend range N120–160°E, plunge range 3°–56°, with mean value 28°. The intermediate principal stress σ_2 magnitude is -7 to -20 MPa. The minimum principal stress σ_3 range is -6 to -14 MPa, and its plunge is NEE.

The analysis results from the numerical method and the test results can be seen in Table 7.53. From the results of unit sections #8 and #4, the maximum principal stress σ_1 range is about -14 to -16 MPa, trend range is N105–140°E, plunge range is 30° to 55°, with mean value 28°. The intermediate principal stress σ_2 range is -9 to -11 MPa, and plunge range is 20–30°. The minimum principal stress σ_3 range is -5 to -8 MPa, plunge range 40–55°, and trend range N25–40°E. The location of the principal stress zones and the surrounding rock damage is shown in Figure 7.105.

Through comparison analysis, the minimum value of the calculated maximum principal stress is shown to be 27.3% greater than the measurement result. The mean plunge angle of the measured maximum principal stress is about 28°, whereas the calculated result is 42.5°. There is thus large uncertainty between the test results and the actual values—which is often the case with the *in situ* stress values because of the wide spread of the *in situ* values caused by perturbations in the stress field introduced by the variety of discontinuities in the rock mass.

7.6.4.3 Local water variations

According to the survey data, most parts of exploration tunnel 2 are dry, with only some locations having seepage water. Some water may run out along the fault, but its quantity is not high. There is dripping water near fault F_{16} at the transformer chamber after excavation, and some areas have seepage water, as shown in Figures 7.106 and 7.107.

Table 7.53 Comparison of measured *in situ* stress values and the calculated values. (Negative values of the principal stresses indicate compression.)

Test points		σ_x/MPa	σ_y/MPa	σ_z/MPa	τ_{xy}/MPa	τ_{yz}/MPa	τ_{xz}/MPa
1	Measured	−7.03	−6.44	−9.53	−0.29	−0.36	−3.14
	Calculated	−6.05	−7.34	−8.76	−0.34	−0.48	−3.63
3	Measured	−11.37	−6.11	−7.12	0.79	0.25	−1.72
	Calculated	−12.53	−6.11	−6.73	0.66	0.31	−1.33
6	Measured	−8.85	−10.94	−12.12	2.14	1.93	−3.33
	Calculated	−8.26	−9.88	−10.75	2.15	1.63	−3.12
7	Measured	−11.74	−13.21	−9.63	−2.51	0.84	1.37
	Calculated	−11.49	−12.02	−9.9	−1.57	0.69	1.96
8	Measured	−12.18	−9.18	−11.04	−0.59	0.82	0.54
	Calculated	−11.86	−10.15	−10.18	−0.57	0.66	0.73
10	Measured	−10.28	−7.38	−6.94	1.46	0.4	−0.07
	Calculated	−8.98	−8.24	−7.24	1.47	0.3	−0.08

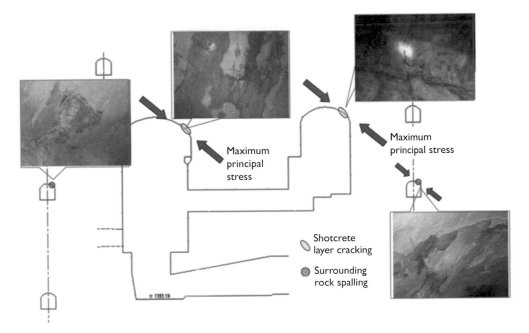

Figure 7.105 Surrounding rock damage and *in situ* stress at the Jinping II underground cavern group.

Figure 7.106 Dripping water on the rock surface.

Figure 7.107 Seepage water near the rock anchor beam.

7.6.4.4 Mechanical behaviour of the rock mass after excavation and in the long term

Five monitoring sections, i.e., S_1–S_5, were selected according to the Jinping II cavern group engineering geological conditions, as shown in Figure 7.108. The multi-point displacement meters, anchor cable load meters and rockbolt stress meters were installed as indicated in Figures 7.109–7.113.

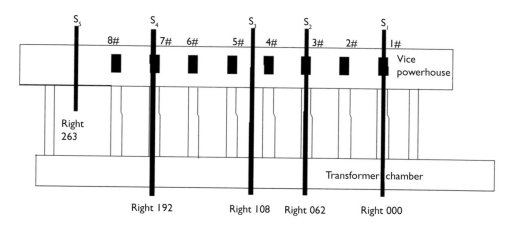

Figure 7.108 Cavern group monitoring sections.

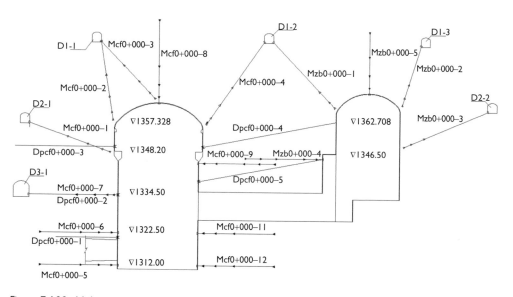

Figure 7.109 Multi-point displacement meters and anchor cable load meters in monitoring section S_1.

The relation between the measured deformation of S_1 by multiple position extensometer Mcf0+000 and the excavation process is shown in Figure 7.114. The anchor cable load monitoring curve at 0+009 upstream is shown in Figure 7.115.

In order to test the failure depth of the surrounding rock after cavern excavation, acoustic wave testing holes were arranged in the 0+270 and 0+186 sections, as shown in Figure 7.116. The test results indicated that the failure depth was about 1.4–2.9 m after the first layer was excavated. The depth in different sections after all the layers were excavated can be seen in Table 7.54.

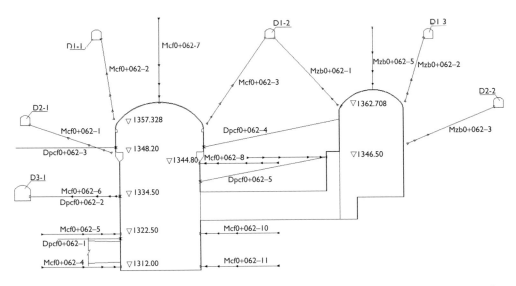

Figure 7.110 Multi-point displacement meters and anchor cable load meters in monitoring section S$_2$.

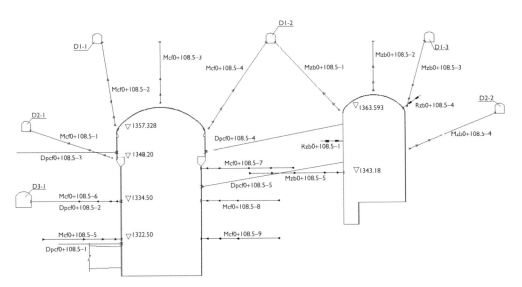

Figure 7.111 Multi-point displacement meters, anchor cable load meters and rockbolt stress meters in monitoring section S$_3$.

In the main powerhouse and transformer chamber, the number of anchor cables with bearing load greater than the designed value is more than 30%, and the number of anchor cables with more than 120% of the designed bearing load is 19%, as shown in Figure 7.117. These cables are mostly located near the stress

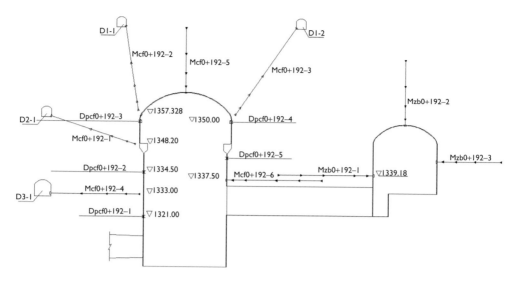

Figure 7.112 Multi-point displacement meters and anchor cable load meters in monitoring section S$_4$.

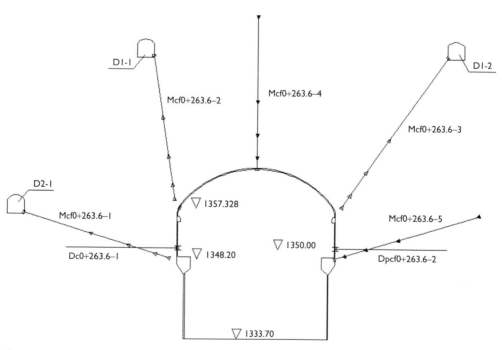

Figure 7.113 Multi-point displacement meters and anchor cable load meters in monitoring section S$_5$.

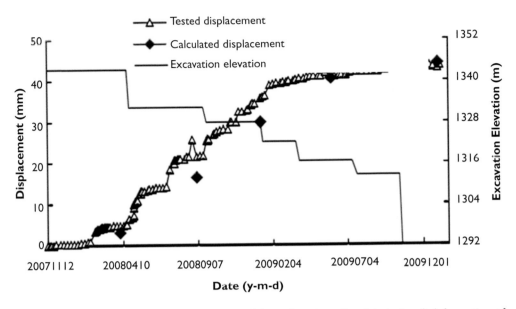

Figure 7.114 Comparison between the measured (tested) and predicted (calculated) deformation of the Mcf0+000 with excavation progress.

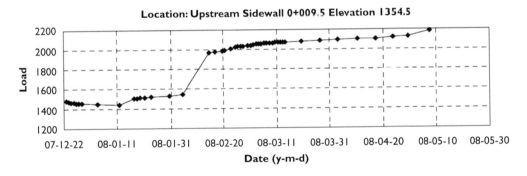

Figure 7.115 Anchor cable load monitoring curve at 0+009 upstream.

concentration area, near the powerhouse downstream spandrel and at the bottom of the transformer chamber upstream sidewall. These loads were controlled by reinforcement.

The number of bolts with stress lower than 100 MPa is 52.8% and the number of bolts with stress more than 300 MPa is 8.7%, as shown in Figure 7.118.

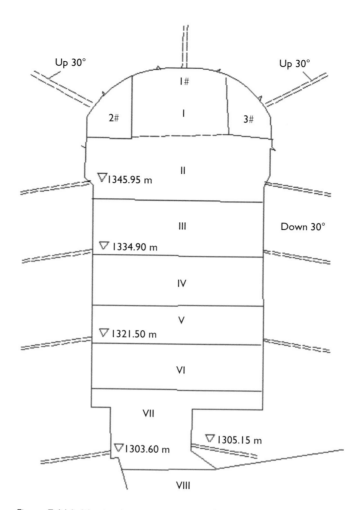

Figure 7.116 Monitoring arrangement for measuring the failure depth in the surrounding rock as a result of the progressive excavation process.

Table 7.54 Testing results for the failure depth in the surrounding rock.

Location of measurement boreholes	Section 0+000 (m)	Section 0+062 (m)	Section 0+124 (m)
Vault	3.0	3.0	3.0
Upstream skewback	1.8	1.8	2.0
Downstream skewback	2.6	2.8	2.4
Upstream sidewall	2.0	3.2	2.2
Downstream sidewall	1.6	2.4	3.0

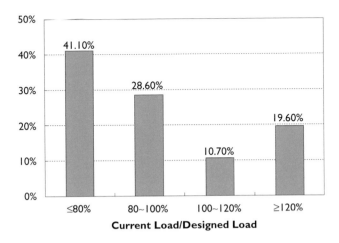

Figure 7.117 Powerhouse and transformer chamber anchor cable load distribution.

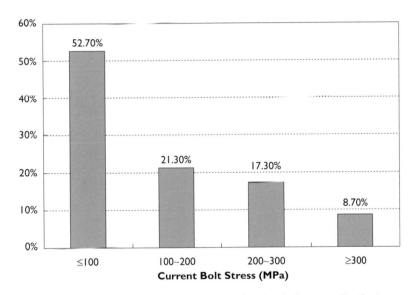

Figure 7.118 Powerhouse and transformer chamber bolt stress distribution.

7.6.5 Assessment and mitigation of local risk during construction

7.6.5.1 Construction of the main powerhouse layer I

The layer I excavation progress of the main powerhouse is shown in Figure 7.119.

During the layer I excavation, a fracture zone, along with fault F_{68}, was revealed in the downstream sidewall between the left section 0+037.00 and right section 0+018.00. Apart from a rock block collapsing near the left section 0+032, which was affected by

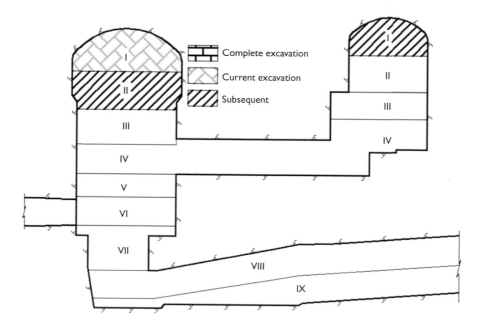

Figure 7.119 Excavation of the first layer of the main powerhouse.

the steep fault F_{68}, there was no other local instability. The faults and fractures (such as F_{81}, F_{84}, F_{56}, F_{61}, F_{77}, P_3 and P_{10}) crossing through the powerhouse vault were stable, and, as indicated by the field monitoring results, there was no influence on the surrounding rock deformation nor any supporting structure damage. There was seepage water on the sprayed concrete surface of the faults and fracture zones.

Fault F_{85} and two long structural surfaces (NE80 dip angle) were revealed at section S_3, and the occurrence of fault F_{85} is N30°–40°E SE∠10°–15°. Some rock blocks collapsed during the first layer excavation. The deformation at Mcf108.5-2 increased by 19 mm in one week, and its deformation rate was 2.35 mm/d, which reached the warning level of the surrounding rock deformation management standard, and was greater than the calculated value.

The local risk assessment for the surrounding rock at section S_3 was performed according to the dynamic feedback analysis results with the geological conditions revealed—and the rock mass was not reinforced after evaluation. The analysis results can be seen in Figure 7.120. The surrounding rock local risk level near fault F_{85} was high, and the areas of grade I and II were large; their depths exceeded the security range, indicating that they should be reinforced based on the original support scheme. The reinforcing support measures for the section S_3 are as follows: increasing one row unbonded pre-stressed anchor cable at upstream spandrel, $T = 2000$ kN, $L = 20$ m.

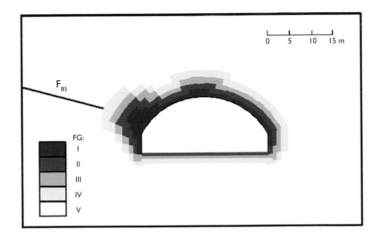

Figure 7.120 Local risk assessment result for the section S₃ after excavation.

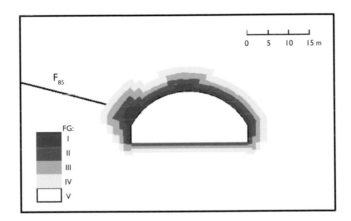

Figure 7.121 Local risk assessment result for section S₃ after reinforcement.

The local risk assessment result after reinforcement is shown in Figure 7.121. The depth of risk grade I area is less than 3 m; that of risk grade areas I and II is less than 5 m; and that of risk grade areas I–III is less than 8 m. According to the local risk assessment standard, the surrounding rock local risk level of section S₃ after reinforcement is acceptable.

The monitoring results of Mcf108.5-2 at section S₃ show that the deformation value suddenly increased by 19 mm on September 29, 2007, which was caused by fault F_{85}. After reinforcement, its deformation speed was less than 0.1 mm/w, and the total deformation was less than 30 mm after six months. The deformation monitoring results shown in Figure 7.122 indicate that the local risk was controlled after reinforcement.

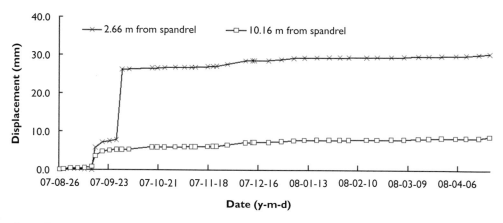

Figure 7.122 Deformation monitoring curve of Mcf0+108.5–2 at upstream abutment of section S₃.

7.6.5.2 Construction of main powerhouse layer II and transformer chamber layer I

The excavation progress of the layer II construction of the main powerhouse and layer I construction of the transformer chamber can be seen in Figure 7.123.

1   Right section 0+000–0+009 (S₁)

According to the engineering geology exploration report and geological information revealed during the layer I construction, the local risk of the section S₁ was evaluated based on the dynamic feedback analysis results with the original support scheme. The results can be seen in Figures 7.124 and 7.125. The local risk grade near fault F_{65} in the powerhouse upstream sidewall was high, and the local risk grade I and II areas affected by fault F_{68} in the powerhouse downstream sidewall was also large, from the abutment to the rock anchor beam. Local reinforcement measures should be carried out to prevent the surrounding rock from collapsing or undergoing large deformation during the subsequent excavation.

Many fractures near fault F_{65} were revealed at the section S₁, and some rock blocks collapsed along the fracture surface. The monitored displacement of the surrounding rock, obtained by the multi-point displacement meter Mcf0+000–4 at the downstream abutment of the right section 0+000, was about 15 mm. This exceeded the warning value of 12 mm in the deformation management standard for the powerhouse downstream sidewall surrounding rock. There was one horizontal crack with sprayed concrete near the abutment, the length of which was 8–10 m.

The reinforcement measures proposed are shown as follows:

1   Add four pressure dispersed and unbonded, pre-stressed, anchor cables at EL1353.10 of the powerhouse upstream, $T = 2000$ kN, $L = 20$ m; consolidation grouting at the influence zone of fault F_{65}.

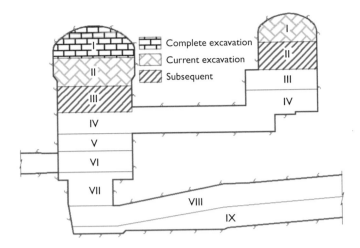

Figure 7.123 Excavation progress in the underground cavern group.

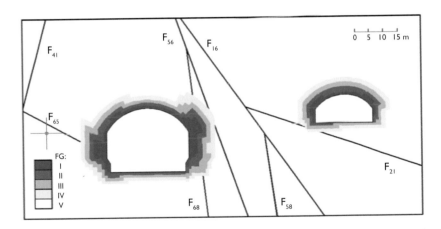

Figure 7.124 Local risk assessment result for section S_1 after construction.

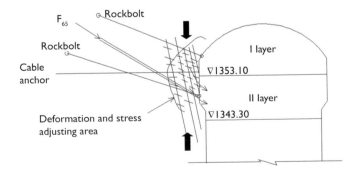

Figure 7.125 Indication of the surrounding rock failure near fault F65.

2 Anchor cables at EL1350 of the powerhouse upstream and downstream sidewall constructed as soon as possible, cable design load increased to 2000 kN.
3 Add pressure dispersed and unbonded, pre-stressed, anchor cables at the powerhouse downstream sidewall from the left section 0+58 to the right section 0+8, $T = 1750$ kN, $L = 28.5$ m.

The local risk assessment result of section S_1 after reinforcement is shown in Figure 7.126. The depth of the risk grade I area is less than 3 m, and that of the risk grade I and II areas is less than 6 m. According to the local risk assessment standard, the surrounding rock local risk level of section S_1 after reinforcement is acceptable.

The monitored results from the multi-point displacement meter and anchor load at the upstream sidewall of section S_1, shown in Figures 7.127 and 7.128, indicate that the displacement increased quickly during the second layer excavation, then slowly. The displacement 1.43 m into the sidewall did not change, and the displacement 3.93 m into the sidewall became small and close to zero. The reinforcement was not changed until the third layer excavation at 2008.04.17. The anchor monitored

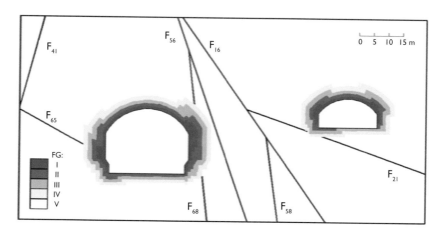

Figure 7.126 Local risk assessment result for section S_1 after reinforcement.

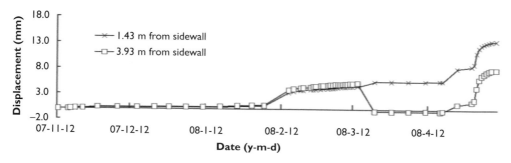

Figure 7.127 Monitored deformation curve from the multi-point displacement meter Mcf0+000–1 results for the upstream sidewall EL1357.558 of section S_1.

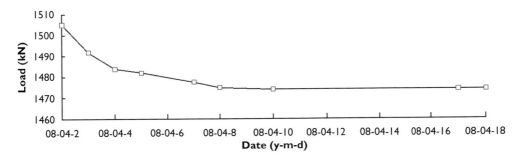

Figure 7.128 Monitored anchor load curve of Dpcf0+000–3 in the upstream sidewall EL1350.0 of section S₁.

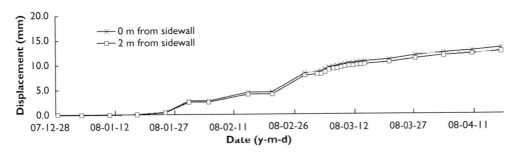

Figure 7.129 Monitored deformation curve of multi-point displacement meter Mcf0+000–4 in the downstream sidewall EL 1357.328 of section S₁.

load was 1500 kN and did not change suddenly. These monitoring results show that the local risk was controlled by the reinforcement, and that the deformation of the upstream rock anchor beam was small.

The monitored deformation curve from the multi-point displacement meter Mcf0+000–4 (shown in Figure 7.129) indicates that the deformation speed was high during the second layer excavation, then it reduced after local reinforcement measures were implemented. The deformation of the surrounding rock near the downstream rock anchor beam was controlled.

2   Right section 0+108.5 (S₃)

According to the geological information revealed during the first layer construction and the dynamic feedback analysis results, the local risk for the section S₃ (right section 0+108.5) was evaluated, and the results are included as Figure 7.130. It can be observed that the local risk grade affected by fault F₈₅ in the powerhouse upstream sidewall was high, and the local risk of the transformer chamber vault and downstream abutment affected by fault F₁₆ and fracture zones P₁₇ and P₁₈ was also high. Rock block collapse and/or large deformations may occur, and the surrounding rock should be reinforced, especially the downstream abutment of the transformer chamber.

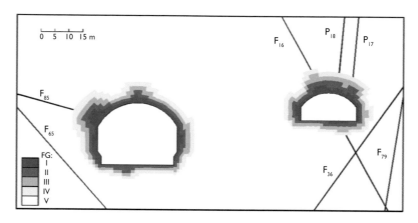

Figure 7.130 Local risk assessment result for section S₃ after construction.

During the process of excavation, the load in the anchor cable at the main powerhouse section S_3 (right section 0+108.500) increased rapidly, at close to the design value of 1800 kN. Fault F_{16} and the fracture zones with P_{17} and P_{18} were exposed at the transformer chamber vault. The intersection angle between the fault strike direction and cavern axis is 20° and its dip direction is facing the downstream side. The width of the fracture zones is 50–150 mm, and the affected zone width is about 2–4 m, both being filled with weathered breccias or schist.

The reinforcement measures proposed are shown as follows:

1   Add one row of unbonded pre-stressed anchor cables at the powerhouse upstream spandrel, $T = 2000$ kN, $L = 20$ m.
2   Add a steel arch rib @ 2 m near the area affected by fault F_{16}; pre-stressed bolts connected with arch rib, $\phi32$, 1.5 m @ 2.0 m, $L = 7$ m, $T = 120$ kN; spray fibre concrete.
3   Add pre-stressed hollow grouting bolts near fault F_{16}, $\phi32$, $L = 7$ m, $T = 120$ kN, and ordinary mortar bolts, $\phi32$, $L = 5$ m.

The local risk assessment result for section S_3 after reinforcement is shown in Figure 7.131. The depth of the risk grade I area is less than 3 m, and that of risk grade I and II areas is less than 6 m. The surrounding rock local risk level of section S_3 after reinforcement is acceptable.

The monitored anchor load results for the upstream sidewall of section S_3, shown in Figure 7.132, indicate that the displacement increased quickly before reinforcement and then reduced under the locked load, keeping the load at 1500 kN. The monitored anchor stresses at the sites (shown in Figure 7.133), 2 m and 5 m from the transformer chamber downstream abutment, were less than 60 MPa. So, the local risk for different areas in the section S_3 was reduced.

3   Right section 0+192 (S_4)

According to the geological information revealed during the layer I construction and the dynamic feedback analysis results, the local risk for section S_4 was evaluated,

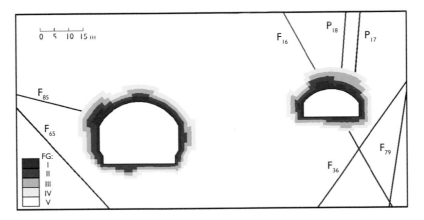

Figure 7.131 Local risk assessment result for section S₃ after reinforcement.

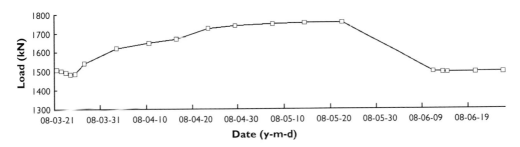

Figure 7.132 Monitored anchor load curve of Dpcf0+108.5–3 at upstream sidewall EL1350.0 of the section S₃.

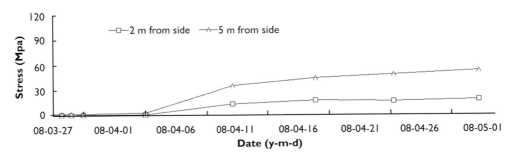

Figure 7.133 Monitored bolt stress curve for Rzb0+108.5–1 at downstream abutment EL1363.6 of section S₃.

and the results can be seen in Figure 7.134. Some rock blocks affected by faults F_{61}, F_{77} and F_{83} near the powerhouse vault will fall, and large deformations may occur near the powerhouse upstream sidewall and transformer chamber downstream abutment.

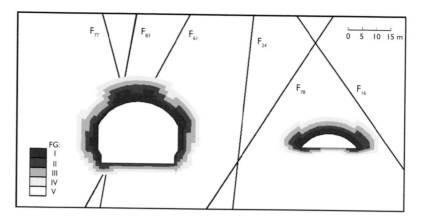

Figure 7.134 Local risk assessment result for the section S_4 after construction.

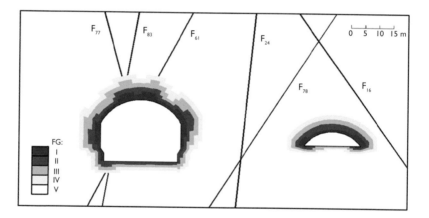

Figure 7.135 Local risk assessment result for section S_4 after reinforcement.

The monitored deformation of section S_4 at Mcf0+192–1 was about 20 mm during the second layer excavation, which was equal to the warning value of the deformation management standard for the powerhouse surrounding rock. The reason for the large deformation is that there is one set of steep strata (spacing 0.05–0.1 m) and multiple sets of structural surfaces (N16°W NE∠41°, EW S∠40°, N71°E NW∠60°).

The reinforcement measures proposed are as follows:

1   Add expansion shell and bolt grouting pre-stressed bolts in the upstream sidewall, $\phi32$, $L = 9$ m, $T = 80$ kN.
2   Add a row of pressure dispersed anchor cables at EL1353.25, $T = 2000$ kN, @3 m, $L = 25$ m, plunge angle upwards at 15°.

The local risk assessment results for section S_4 after reinforcement are shown in Figure 7.135. The depth of the risk grade I area is less than 3 m and that of the risk grade II area is less than 6 m. According to the local risk assessment standard, the surrounding rock local risk level of section S_4 after reinforcement is acceptable.

The monitored results from the multi-point displacement meter Mcf0+192–1 at EL1356.577 and section S_4, shown in Figure 7.136, indicate that the displacement increased from 1 to 20 mm during the layer II excavation. After reinforcement, the deformation speed reduced, and the load monitoring results of the anchor cable set after the second layer excavation, shown in Figure 7.137, indicate that the local risk was reduced by the reinforcement.

4   Right section 0+263 (S_5)

The fault F_{78} and bedding P_2 were exposed at section S_5 during the layer II excavation. The local risk assessment result obtained via the dynamic feedback analysis results can be seen in Figure 7.138, indicating that there is a rock block falling risk or large deformation risk near the vault affected by fault F_{78} and in the upstream sidewall affected by bedding P_2.

The monitored results from the multi-point displacement meter Mcf0+263.6–1 at section S_5, shown in Figure 7.139, indicate that the displacement increased from 12.8 to 21.4 mm in one week, and the displacement and deformation speed exceeded the warning values of the surrounding rock deformation management standard.

Reinforcement measures: The system support should be performed as soon as possible, and monitored and reinforced with pre-stressed bolts and anchor cables.

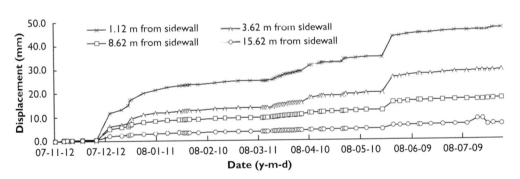

Figure 7.136 Monitored deformation curve from multi-point displacement meter Mcf0+192–1 at upstream sidewall EL1356.577 of section S_4.

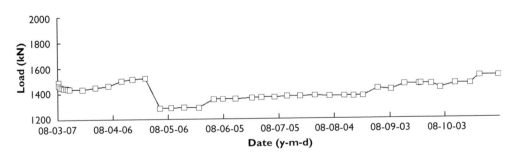

Figure 7.137 Monitored anchor load curve of Dpcf0+192–3 at EL1350.0 in section S_4.

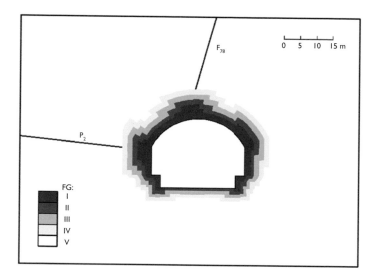

Figure 7.138 Local risk assessment result for the section S_5 after construction.

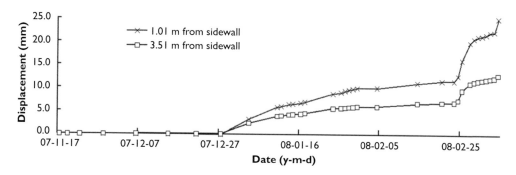

Figure 7.139 Monitored deformation curve from multi-point displacement meter Mcf0+263.6–1 at the upstream sidewall EL1348.2 of section S_5.

The local risk assessment result for the section S_5 after reinforcement is shown in Figure 7.140. The depth of the risk grade I area is less than 3 m, and that of the risk grade II area is less than 6 m. According to the local risk assessment standard, the surrounding rock local risk level of the section S_5 after reinforcement is acceptable.

The monitored results from the multi-point displacement meter Mcf0+263.6–1 at section S_5, shown in Figure 7.139, indicate that the displacement speed tended to be stable after 2008.02.27, and the load monitoring result for anchor cable Dpcf0+263.6–1, shown in Figure 7.141, was about 1470 kN. These results indicate that the local risk was reduced by the reinforcement.

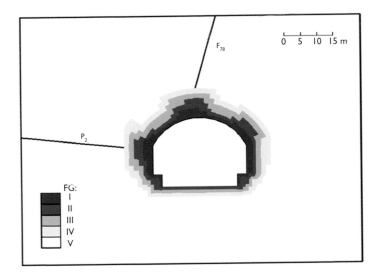

Figure 7.140 Local risk assessment result for the section S_5 after reinforcement.

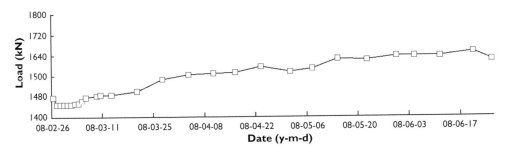

Figure 7.141 Monitored anchor load curve for Dpcf0+263.6–1 at upstream sidewall EL1350.0 of section S_5.

7.6.5.3 Construction of main powerhouse layer III and transformer chamber layer II

The excavation progress for the construction of layer III of the main powerhouse and layer II of the transformer chamber can be seen in Figure 7.142.

1   Right section 0+000 (S_1)

Fault F_{21} was revealed during the second layer construction and the local risk of section S_1 (right section 0+000) was evaluated, the results being included as Figure 7.143. The local risk grade near the powerhouse upstream rock anchor beam was high; the local risk, rock block collapse or large deformation, affected by fault F_{68} at the powerhouse upstream sidewall was high; and the local risk of the transformer chamber upstream as affected by fault F_{21} was also high.

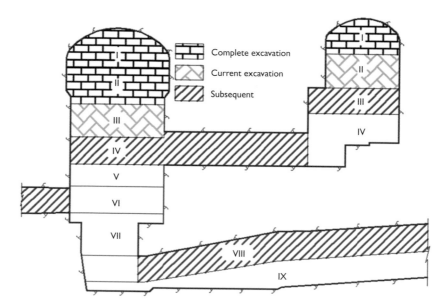

Figure 7.142 Excavation progress image for the underground cavern group.

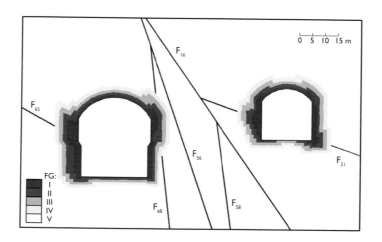

Figure 7.143 Local risk assessment result for the section S_1 after construction.

The maximum monitored deformation of the section S_1 was 27 mm during the third layer excavation. Some rock blocks collapsed due to fault F_{68}. There was a 10 m long horizontal crack in the sprayed concrete layer near the right section 0+002, as shown in Figure 7.144.

The anchor cable monitoring loads at Dpcf0+002.682–3 (EL1350.0) near the powerhouse upstream sidewall significantly increased to 2240 kN (shown in Figure 7.145), making them greater than the designed values. The loads may continue to increase in the subsequent excavation.

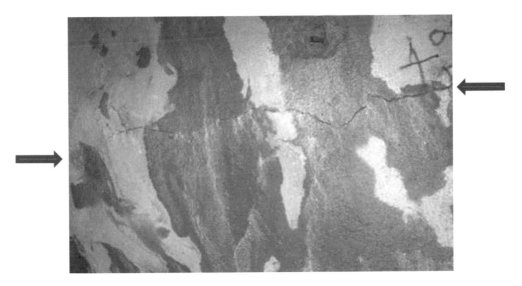

Figure 7.144 Crack in the sprayed concrete layer near the 0+002 section downstream.

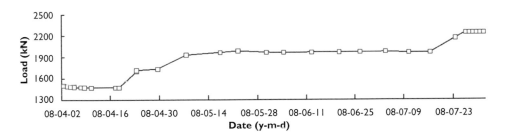

Figure 7.145 Monitored anchor load curve from Dpcf0+002.682 at the upstream sidewall EL1350.0 of the 0+002.682 section.

The reinforcement measures proposed are shown as follows.

1   Add two rows of load dispersed anchor cables ($T = 1750$ kN, $L = 28$ m) near the downstream abutment between the left section 0+010 and right section 0+022.

2   Change the spacing of the anchor cables, between the powerhouse and transformer chamber at EL1343.5 between the left section 0+33.8 and right section 0+035, from 4.5 m to 3.0 m.

3   Add pre-stressed bolts ($\phi32$, $T = 80$ kN, $L = 9$ m) in the downstream sidewall and abutment between the left section 0+030 and right section 0+035.

4   Add two anchor points of pressured dispersed anchor cables at EL1350.0 near Dpcf0+002.682–3, $T = 2000$ kN, $L = 20$ m, and fault F_{65}, affecting the zone consolidation grouting.

The local risk assessment result for the section S$_1$ after reinforcement is shown in Figure 7.146. The depth of the risk grade I area is less than 3 m, and that of the risk grade I and II areas is less than 6 m. Therefore, the surrounding rock local risk level of the section S$_1$ after reinforcement is acceptable.

The monitored results for the displacement and anchor cable load near the powerhouse upstream rock anchor beam indicated that the deformation was less than the warning value of 25 mm (shown in Figure 7.147), and the anchor cable load was about 2200 kN (shown in Figure 7.148). It is thus shown that the local displacement risk and anchor cable overloading risk were reduced by the reinforcement.

After the powerhouse downstream sidewall was reinforced with bolts and anchor cables, the monitored result from the multi-point displacement meter reduced from 24 mm to 15 mm (shown in Figure 7.149), and the anchor cable load was about 1770 kN (shown in Figure 7.150). It is thus shown that the local displacement risk and anchor cable overloading risk have been reduced by the reinforcement.

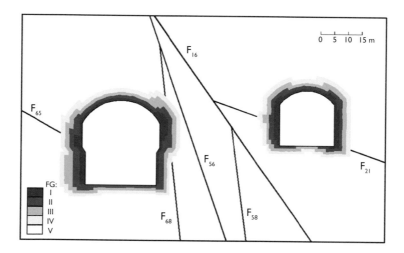

Figure 7.146 Local risk assessment result for the section S$_1$ after reinforcement.

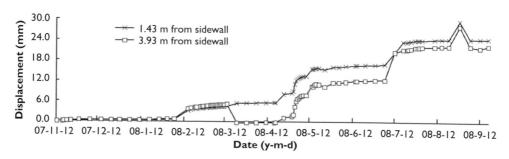

Figure 7.147 Monitored deformation curve of multi-point displacement meter Mcf0+000–1 in the upstream sidewall EL1357.558 of section S$_1$.

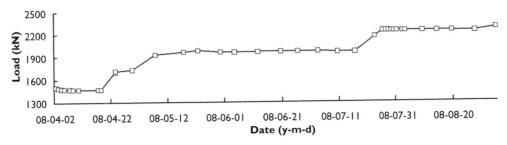

Figure 7.148 Monitored anchor load curve from Dpcf0+002.682–3 in the upstream sidewall EL1350.0 of right section 0+002.682.

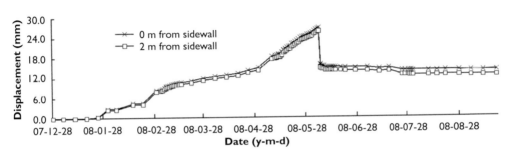

Figure 7.149 Monitored deformation curve from multi-point displacement meter Mcf0+000–4 in the downstream sidewall EL1357.328 of section S_1.

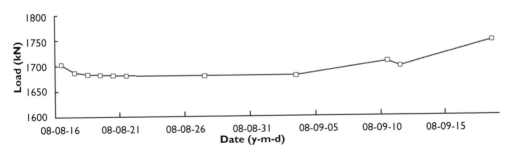

Figure 7.150 Monitored anchor load curve for Dpcf0+002.682–4 in downstream sidewall EL1350.0 of right section 0+002.682.

The displacement monitoring results from the transformer chamber downstream, shown in Figure 7.151, indicate that its deformation increased suddenly during the second layer excavation, but it changed little after being supported by bolts and sprayed concrete. Therefore the original support scheme is successful in ensuring that the surrounding rock is stable.

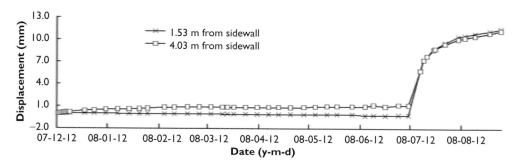

Figure 7.151 Monitored deformation curve from multi-point displacement meter Mzb0+000–3 in the downstream sidewall EL1353.18 of the left section 0+001.

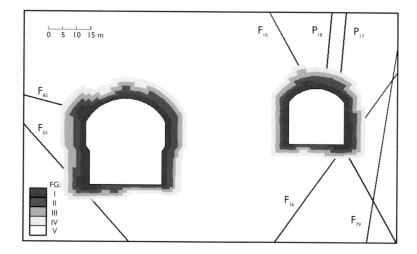

Figure 7.152 Local risk assessment result for the unit #5 section after layer III had been excavated.

2 Right 0+80–0+142 section

Taking into account the fault F_{65} revealed during the excavation of the right section 0+80–0+142 and the dynamic feedback analysis results, the local risk of the unit #5 section was evaluated, and the results can be seen in Figure 7.152. The local risk grade near the powerhouse upstream sidewall affected by fault F_{65} was high and the local risk grade near the transformer chamber downstream sidewall was also high. These should be reinforced—otherwise the surrounding rock will collapse or the anchor cables will exceed their design loads.

During the excavation, the surrounding rock above the upstream rock anchor beam was reinforced by bolts and anchor cables, but some failed due to the cracks between EL1348.5 and EL1350, as shown in Figure 7.153.

The bolt stress monitoring results (shown in Figure 7.154) demonstrate that the stress in the bolt at the downstream abutment increased rapidly and that the maximum stress was 240 MPa, caused by fault F_{16} and the fractured zones P_{17} and P_{18} at EL1340–EL1348 near the transformer chamber downstream section 0+068–0+130.

Strutural surface cracking

Shotcrete layer collapsing

Figure 7.153 Damaged rock near the upstream rock anchor beam.

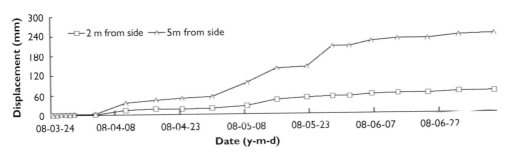

Figure 7.154 Monitored stress increase of R∠b0 bolt+108.5–4 at the transformer chamber downstream abutment EL1363.6 of right section 0+108.5.

The reinforcement measures proposed are as follows.

1   Add pre-stressed bolts in rock near to the anchor beam with a down dip angle of 5–10°.
2   Reduce the row spacing of the anchor cables at the powerhouse upstream EL1339.0–1330.0 from 4.5 to 3.0 m, and change the anchor cable locking loads from 1750 kN to 1050 kN, or from 2000 kN to 1200 kN, i.e., 60% of the designed load.
3   Add two rows of tensioned anchor cables (T = 1750 kN, L = 25 m @ 4 m) near the transformer chamber downstream between the right sections 0+85 and 0+125 and tensioned bolts (ϕ32, T = 80 kN, L = 9 m, 1 m × 1 m).

The local risk assessment result for the unit #5 section after reinforcement is shown in Figure 7.155. The depth of risk grade I area is less than 3 m, and that of the risk grade I and II areas is less than 6 m. The surrounding rock local risk level is acceptable.

The displacement monitoring results (shown in Figure 7.156) and anchor cable load (shown in Figure 7.157) in the upstream sidewall during excavation of layer III indicate that the displacement speed was high and the anchor cable load increased

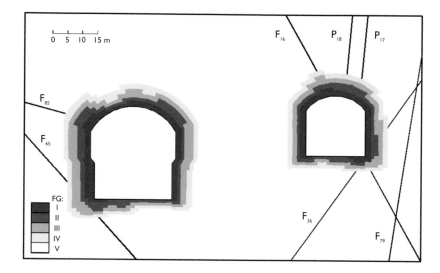

Figure 7.155 Local risk assessment result for the unit #5 section after reinforcement.

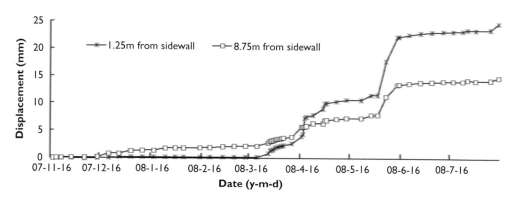

Figure 7.156 Monitored deformation plot of multi-point displacement meter Mcf0+108.5–1 at upstream sidewall EL1357.004 of section S_3.

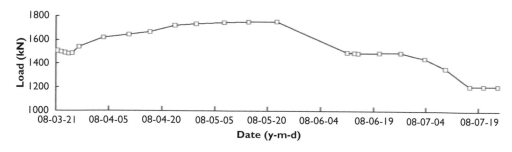

Figure 7.157 Monitored anchor load plot for Dpcf0+108.5–3 at upstream sidewall EL1350.0 of section S_3.

during the excavation of layer III. Its local risk grade was high. After reinforcement, the displacement remained at 24 mm and the anchor cable load reduced to 1200 kN. It can thus be shown that the local displacement risk and anchor cable overloading risk were reduced by the reinforcement.

The bolt stress monitoring results (shown in Figure 7.158) show that the bolt stress at the downstream abutment reduced from 240 MPa to 140 MPa after reinforcement, and the local risk was controlled.

3   Right section 0+263.6 (S_5)

The monitored displacement of the powerhouse upstream sidewall increased suddenly, then retained a low increase during the excavation of the upper half of layer III, as shown in Figure 7.159. The local risk of the right section 0+263.6 was evaluated, and the results can be seen in Figure 7.160. The local risk grade for the upstream sidewall is high.

The reinforcement measures proposed are as follows: add seven mortar bolts (ϕ32, $L = 9$ m) and 32 pre-stressed bolts (ϕ32, $L = 9$ m, $T = 80/120$ kN).

The local risk assessment result for section S_5 after reinforcement of the upper half of the layer III is shown in Figure 7.161. The risk grade I area decreased significantly but its depth near the upstream and downstream rock anchor beam was still large, at about 3.5 m. The local risk was controlled, but this area should be monitored in the future.

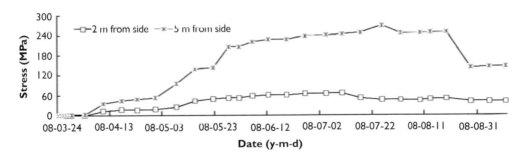

Figure 7.158 Monitored stress plot for Rzb0 bolt+108.5–4 at transformer chamber downstream abutment EL1363.6 of section S_3.

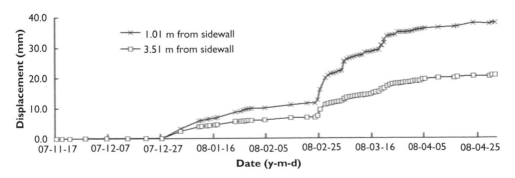

Figure 7.159 Monitored deformation plot for multi-point displacement meter Mcf0+263.5–1 at upstream sidewall EL1348.2 of section S_5.

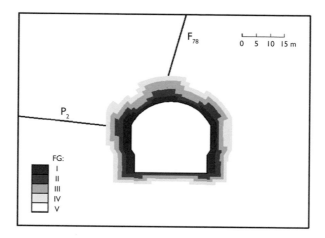

Figure 7.160 Local risk assessment result for section S$_5$ after layer III upper half part excavation.

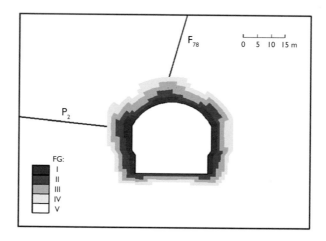

Figure 7.161 Local risk assessment result for section S$_5$ after layer III upper half part reinforcement.

The monitored displacement of the powerhouse upstream sidewall also increased suddenly, then continued to increase at a lower rate during the excavation of the lower half of layer III, as shown in Figure 7.159. The local risk of the right section 0+263.6 was also evaluated, and the results can be seen in Figure 7.162.

The reinforcement measures proposed are shown as follows: reduce the second row anchor cable space to 3 m between the right sections 0+255.62 and 0+273.62 and their locked loads; add two pre-stressed anchor cables ($T = 1750$ kN, $L = 20$ m).

The local risk assessment result for section S$_5$ after reinforcement of the lower half of layer III is shown in Figure 7.163. The monitoring displacement results, shown in Figure 7.164 and remaining at 37 mm, indicate that its local risk was controlled.

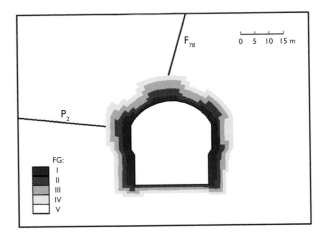

Figure 7.162 Local risk assessment result for the section S$_5$ after layer III lower half part excavation.

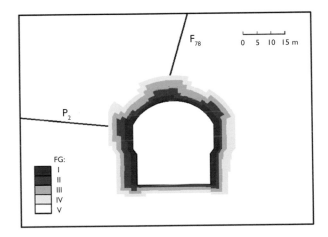

Figure 7.163 Local risk assessment result for the section S$_5$ after layer III lower half part reinforcement.

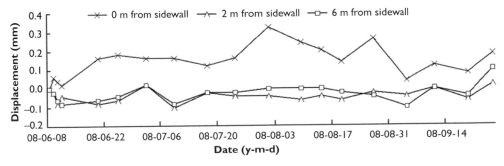

Figure 7.164 Monitored deformation curve for multi-point displacement meter Mcf0+263.5–1B (added later) at upstream sidewall EL1348.2 of section S$_5$.

7.6.5.4 Construction of main powerhouse layer IV and transformer chamber layer III

The excavation progress for the construction of layer IV of the main powerhouse and layer III of the transformer chamber can be seen in Figure 7.165.

1 Right section 0+008–0+032

According to the revealed geological condition and dynamic feedback analysis results, the local risk of monitoring section S_1 was evaluated after construction of layer IV of the main powerhouse, and the results can be seen in Figure 7.166. The local risk

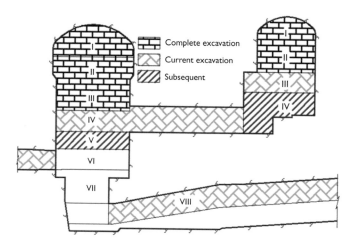

Figure 7.165 Excavation progress for layer IV construction of main powerhouse and layer III construction of transformer chamber.

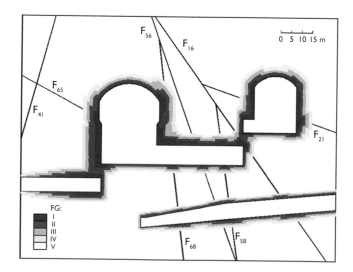

Figure 7.166 Local risk assessment result for section S_1 after layer IV construction of main powerhouse.

grades at the powerhouse upstream sidewall, intersections of bus tunnel with the powerhouse downstream sidewall and transformer chamber upstream sidewall were high, and the local risk of rock block collapse or large deformation should be avoided.

The surrounding rock collapsed more than once during the excavation of layer IV for the main powerhouse, which was affected by faults F_{68} and F_{16}. The monitored deformation of this section was also large. The load in the anchor cable tension meter Dpcf0+006–1 at the powerhouse downstream sidewall was 2590 kN, as shown in Figure 7.167. Sprayed concrete cracks first appeared in this area.

The surrounding rock stability of the transformer chamber upstream sidewall was poor, which was affected by the steeply dipping fault F_{16}, fault F_{21} and a large fracture zone. In addition, water seeped out along the soft structural surface.

The reinforcement measures proposed are shown as follows:

1   Add one row of stress dispersed anchor cables ($T = 2000$ kN, $L = 28$ m @ 4 m) near the powerhouse upstream rock anchor beam and one row of pressure dispersed anchor cables ($T = 1750$ kN, $L = 28$ m @ 4 m) at EL1354 of the powerhouse downstream between the left section 0+008 and the right section 0+032.
2   Add one row of stress dispersed anchor cables ($T = 1750$ kN, $L = 28$ m @ 4 m) at EL1352 of the transformer chamber upstream between the right section 0+5–0+80 and pre-stressed bolts ($\phi32$, $T = 80$ kN, $L = 8$ m).
3   Reduce the spacing of the anchor cables between the powerhouse and transformer chamber to 3 m.
4   Add pre-stressed bolts near the fault revealed in the tailrace tunnel.

The local risk assessment result for section S_1 after reinforcement is shown in Figure 7.168. The area of risk grade I reduced significantly and satisfies the engineering construction request.

The monitored results for the displacement at the powerhouse upstream sidewall indicate that the deformation increase was first high and then low during construction of layer IV (from 20 September 2008); finally it remained at about 11 mm, as shown in Figure 7.169, thus the local risk was controlled.

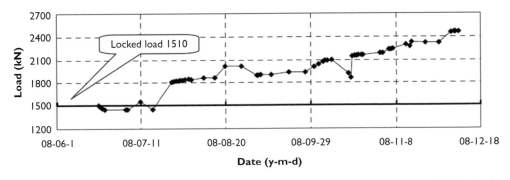

Figure 7.167 Load increase for Dpcf0+006–1 at powerhouse downstream sidewall EL1357.3 of right section 0+006.

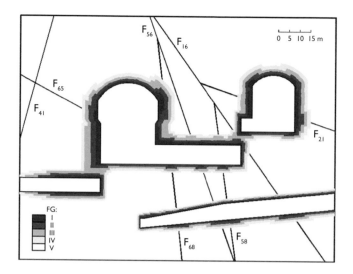

Figure 7.168 Local risk assessment result for the section S₁ after reinforcement.

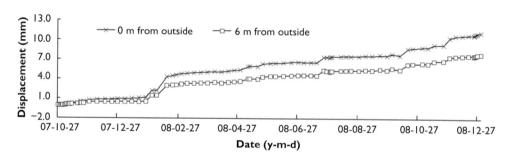

Figure 7.169 Monitored deformation curve from multi-point displacement meter Mcf0+000–2 at powerhouse upstream sidewall EL1357.328 of right section 0+000.

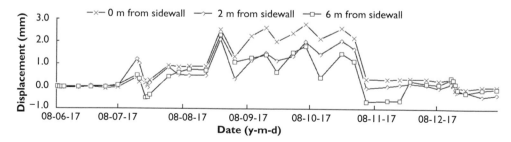

Figure 7.170 Monitored deformation changes for multi-point displacement meter Mcf0+000–9 at powerhouse downstream sidewall EL1344.8 of right section 0+000.

The monitored results of the displacement and anchor cable load at the power-house downstream sidewall indicate that the increased deformation was large and the anchor cable load was greater than 1700 kN, as shown in Figures 7.170 and 7.171. After reinforcement, the load reduced to 1100 kN and the deformation was stable. The local risk was reduced by the reinforcement.

The monitored results for displacement at the transformer chamber upstream and downstream sidewall (shown in Figures 7.172 and 7.173) indicate that the

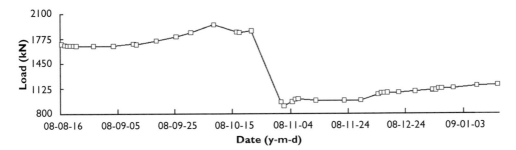

Figure 7.171 Load variations for Dpcf0+000–4 at powerhouse downstream sidewall EL1350.0 of right section 0+002.682.

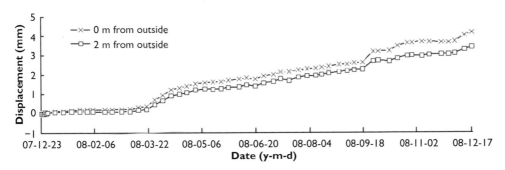

Figure 7.172 Monitored deformation curve for multi-point displacement meter Mzb0+000–1 at transformer chamber upstream sidewall EL1362.708 of section S$_1$.

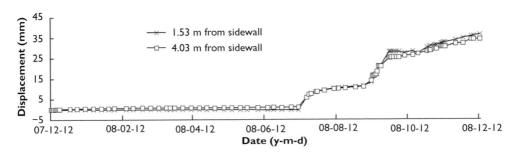

Figure 7.173 Monitored deformation curve for multi-point displacement meter Mzb0+000–3 at transformer chamber downstream sidewall EL1353.18 of left section 0+001.

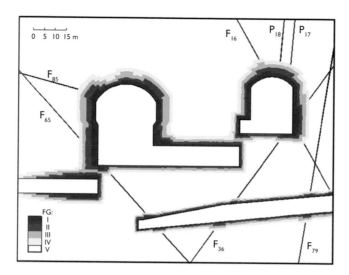

Figure 7.174 Local risk assessment result for the unit #5 section after layer IV construction of main powerhouse.

deformation change was high and its maximum was higher than the warning value of 30 mm, especially at the downstream sidewall. The displacement local risk was reduced by the reinforcement.

2   Right section 0+108.5–0+192

According to the revealed geological condition and dynamic feedback analysis results, the local risk for the unit #5 section was evaluated after the construction of layer IV of the main powerhouse; the results can be seen in Figure 7.174. The local risk grades at the powerhouse upstream sidewall, upstream bus tunnel intersection and transformer chamber downstream sidewall were high, and the local risk is rock block collapse or large deformation.

Some local surrounding rock collapsed at the upstream sidewall between the right sections 0+108.5 and 0+192 during excavation of layer IV of the main powerhouse, which was affected by the steep dip bedding, fault F_{65} and joints. This also happened at the intersection between the powerhouse and high pressure pipe, and there were some rock cracks at the intersection between the powerhouse and bus tunnel, and some vertical sprayed concrete cracks near the upstream and downstream rock anchor beam, as shown in Figures 7.175 and 7.176.

Some sprayed concrete crazed and collapsed at the transformer chamber downstream spandrel between right section 0+85 and 0+125, and the steel grid at the right section 0+115 bent significantly.

The reinforcement measures proposed are as follows.

1   Add drain holes and pre-stressed bolts at the powerhouse downstream abutment and discharge the water in the drainage gallery.
2   Reduce the spacings of anchor cables at the powerhouse upstream (EL1339.0– 1330.0) to 3 m and their locked load to the design value of 60%, which is

Figure 7.175 Sprayed concrete bulging outwards.   Figure 7.176 Rock cracking at cavern intersection.

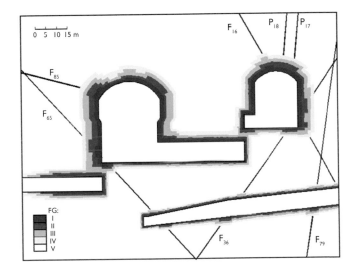

Figure 7.177 Local risk assessment result for unit 5# section after reinforcement.

1750 kN, anchor cable lock tonnage is 1050 kN, 2000 kN cable lock tonnage is 1200 kN; and the mortar anchors are replaced by pre-stressed bolts, and implement consolidation grouting near the fault F_{65} influence area.

3   Add one row of stress dispersed anchor cables ($T = 2000$ kN, $L = 20$ m) and pre-stressed bolts ($\phi32$, $T = 120$ kN, $L = 7$ m) at EL1340–EL1348 of the transformer chamber downstream between the right sections 0+068–0+130.

4   Add pre-stressed bolts near the fault revealed in the tailrace tunnel.

The local risk assessment result for section S_3 after reinforcement is shown in Figure 7.177. The area of risk grade I reduces significantly, the depth of the risk grade I area being less than 3 m, and that of the risk grade I and II areas less than 6 m, which satisfies the engineering construction request.

The monitored results for displacement at the powerhouse upstream sidewall (as shown in Figures 7.178 and 7.179) indicate that the deformation near the rock anchor beam increased during construction of layer IV; remained at about 41 mm, then increased by 20 mm. The displacement at the middle section of the upstream

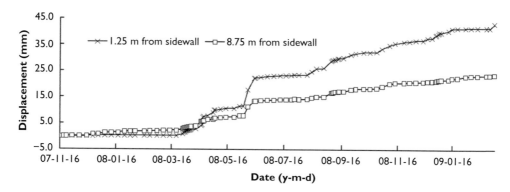

Figure 7.178 Monitored deformation curve for multi-point displacement meter Mcf0+108–1 at powerhouse upstream sidewall EL1357.004 of section S_3.

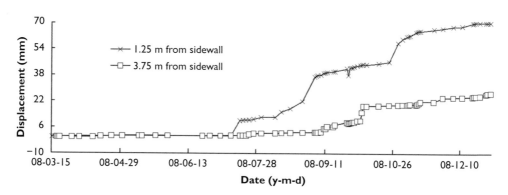

Figure 7.179 Monitored deformation curve for multi-point displacement meter Mcf0+110.5–6 at powerhouse upstream sidewall EL1334.5, right section 0+110.5.

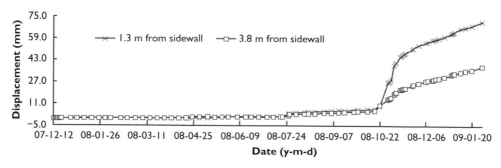

Figure 7.180 Monitored deformation curve for multi-point displacement meter Mzb0+108.5–4 at transformer chamber upstream sidewall EL1361.518 of section S_3.

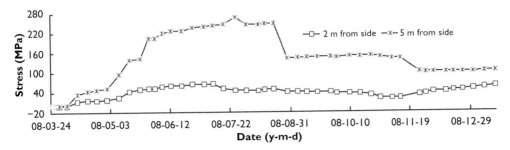

Figure 7.181 Monitored stress data for Rzb0+108.5–4 at downstream abutment EL1363.6 of section S₃.

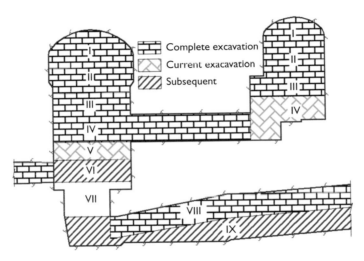

Figure 7.182 Excavation progress for layer V construction of the main powerhouse and layer IV construction of the transformer chamber.

sidewall increased from 10 mm to 70 mm, which was affected by fault F_{65}. After reinforcement, the local risk was controlled.

The displacement at the transformer chamber downstream abutment first increased quickly, then remained stable after reinforcement, as shown in Figure 7.180. The bolt stress also reduced, as shown in Figure 7.181. In this way, the local risk was reduced by the reinforcement.

7.6.5.5 Construction of layer V of the main powerhouse

The excavation progress for the construction of layer V of the main powerhouse and layer IV of the transformer chamber can be seen in Figure 7.182.

1   Right section 0+000–30.00

According to the revealed geological conditions and dynamic feedback analysis results, the local risk of monitoring section S_1 was evaluated after construction of layer V of the main powerhouse, and the results can be seen in Figure 7.183. The local risk grades at the powerhouse downstream sidewall and intersection between the transformer chamber upstream sidewall and bus tunnel were high, and the local risk of rock collapse or large deformation should be avoided.

The surrounding rock collapsed more than once during excavation of layer IV of the transformer chamber (Figure 7.184), which was affected by faults F_{21} and F_{16}, and the overbreak was serious (Figure 7.185). There were cracks in the sprayed concrete at both ends of the bus tunnel.

The reinforcement measures proposed are shown as follows:

1   Reduce the spacing of anchor cables near the intersection between the power-house downstream sidewall and bus tunnel to 3 m and the spacing of the pre-stressed bolts.

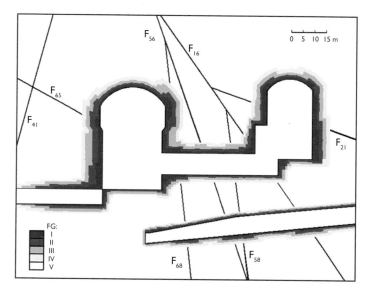

Figure 7.183 Local risk assessment result for unit #1 section after layer V construction of the main powerhouse.

Figure 7.184 Rock mass collapse near fault F_{16}.

Figure 7.185 Overbreak near fault F_{21}.

2   Add two rows of stress dispersed anchor cables ($T = 2000$ kN, $L = 20$ m @ 4 m) at the transformer chamber upstream sidewall bottom and pre-stressed bolts ($\phi 32$, $L = 8$ m); fill the overbreak area with concrete.

3   Add one row of stress dispersed anchor cables ($T = 1750$ kN, $L = 28$ m @ 4 m) at EL1352 of the transformer chamber upstream between the right section 0+5–0+80 and pre-stressed bolts ($\phi 32$, $T = 80$ kN, $L = 8$ m).

4   Add pre-stressed bolts near the intersection between the high pressure pipe and powerhouse upstream sidewall.

The local risk assessment result for the unit #1 section after reinforcement is shown in Figure 7.186. The area of risk grade I reduced significantly and the depth of the risk grade I and II areas is less than 6 m, which satisfies the engineering construction request.

The monitored results for the displacement and anchor cable at the powerhouse sidewalls (shown in Figures 7.187 and 7.188) indicate that the deformation near the

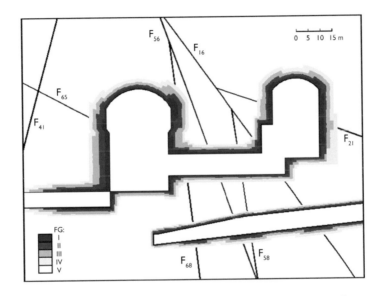

Figure 7.186 Local risk assessment result for unit #1 section after reinforcement.

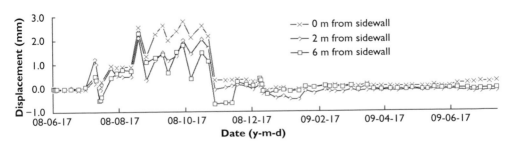

Figure 7.187 Monitored deformation curve for multi-point displacement meter Mcf0+000–9 at powerhouse downstream sidewall EL1344.8 of section S_1.

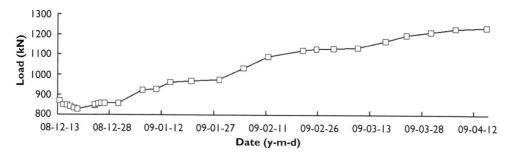

Figure 7.188 Load change results for Dpcf0+000–2 at powerhouse upstream sidewall EL1334.5 of section S$_1$.

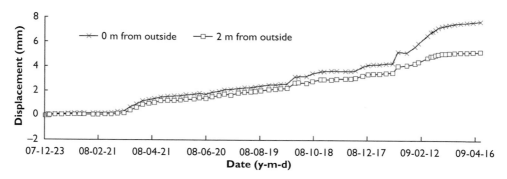

Figure 7.189 Monitored deformation curve for multi-point displacement meter Mzb0+000–1 at transformer chamber upstream sidewall EL1362.708 of section S$_1$.

downstream rock anchor beam was stable and did not change, and the anchor cable load in the middle area of the upstream sidewall increased quickly and then remained at about 1200 kN. Therefore, the influence of the construction of layer V to the stability of the sidewalls was small, and the local risk could be controlled by the original supporting scheme.

The displacement at the transformer chamber upstream abutment first increased quickly, then reduced after reinforcement (shown in Figure 7.189), which indicated that the surrounding rock stress state was changed and that the local risk grade was acceptable.

2   Right section 0+263.6 (S$_5$)

The load measured by the anchor cable tension meter DPcf0+263.6–2 at the downstream sidewall of the monitored section S$_5$ exceeded 2100 kN (shown in Figure 7.190), which was installed in February 2009, and the displacement at the downstream sidewall increased constantly and did not converge to a stable value, as shown in Figure 7.191. According to the revealed geological condition and the dynamic feedback analysis results, the local risk grade at the powerhouse downstream sidewall was high, as shown in Figure 7.192.

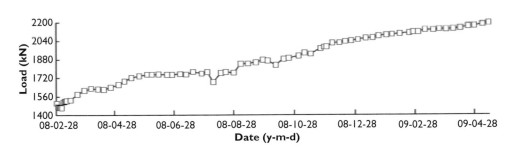

Figure 7.190 Load results for Dpcf0+263.6–2 at powerhouse downstream sidewall EL1350.0 of section S_5.

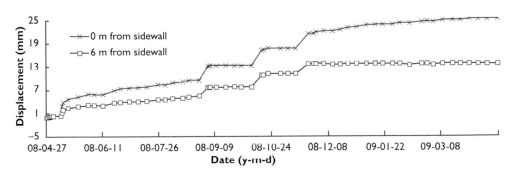

Figure 7.191 Monitored deformation by multi-point displacement meter Mcf0+263.6–5 at powerhouse downstream sidewall EL1348.2 of section S_5.

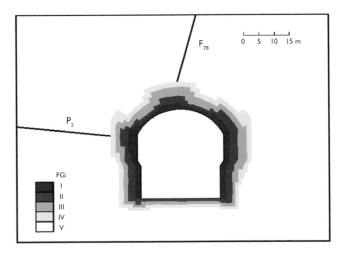

Figure 7.192 Local risk assessment result for monitored section S_5 after lower III layer construction of the main powerhouse.

The reinforcement measures proposed are as follows: Add one row of stress dispersed anchor cables (T = 2000 kN, L = 20 m @ 3 m) at the powerhouse upstream sidewall EL1353.25 between the right section 0+232.6–294.62.

The local risk assessment result for section S_5 after reinforcement of layer III is shown in Figure 7.193. Its local risk was controlled: the anchor cable load remained at 2400 kN (as shown in Figure 7.194); and the maximum deformation was 32 mm (as shown in Figure 7.195) and did not increase.

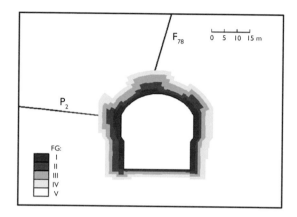

Figure 7.193 Local risk assessment result for monitored section S_5 after reinforcement.

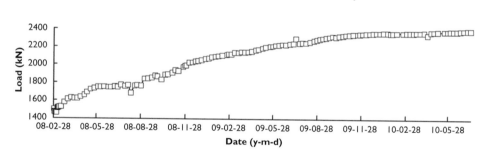

Figure 7.194 Load increase for Dpcf0+263.6–2 at powerhouse upstream sidewall EL1350.0 of section S_5.

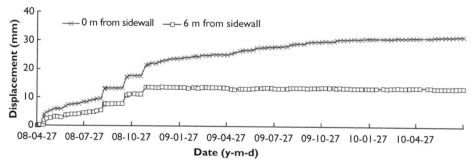

Figure 7.195 Monitored deformation curve for multi-point displacement meter Mcf0+263.6–5 at the powerhouse downstream sidewall EL1348.2 of section S_5.

7.6.5.6   Construction of layers VI, VIII and IX of the main powerhouse

The excavation progress for the construction of layers VI, VIII and IX of the main powerhouse can be seen in Figure 7.196.

According to the revealed geological condition and dynamic feedback analysis results, the local risk of the unit #1 section was evaluated after construction of layers VI, VIII and IX of the main powerhouse, Figure 7.197. It can be observed that the

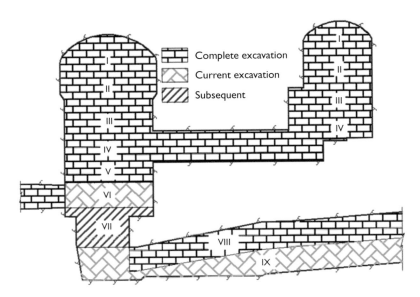

Figure 7.196 Excavation progress for the layers VI, VIII and IX of the main powerhouse.

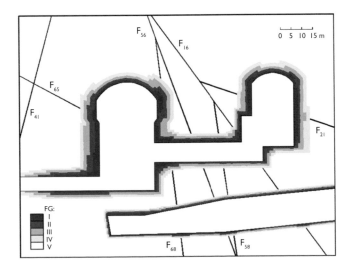

Figure 7.197 Local risk assessment result for the unit #1 section after layers VI, VIII and IX construction of the main powerhouse.

local risk grades at the intersection between the upstream sidewall and high pressure pipe and between the downstream sidewall and bus tunnel were high, and the local risk of rock collapse or large deformation should be avoided.

There were some sprayed concrete cracks and rock cracks at the intersection area; so these areas should be reinforced, for which the measures proposed are shown as follows.

1   Reduce the spacing of the anchor cables to 3 m on the downside of the intersection between the powerhouse downstream sidewall and bus tunnel.
2   Add one row of stress dispersed anchor cables ($T = 1750$ kN, $L = 20$ m @ 4 m) at EL1316.50 in both sidewalls.
3   Add pressured bolts ($\phi 32$, $T = 80$ kN, $L = 8$ m, 1.5 m × 1.5 m).

The local risk assessment result of the unit #1 section after reinforcement is shown in Figure 7.198. The area of the risk grade I is reduced significantly and the depth of the risk grade I and II areas is less than 6 m, which satisfies the engineering construction request.

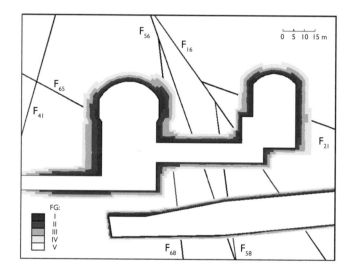

Figure 7.198 Local risk assessment result for unit 1# section after reinforcement.

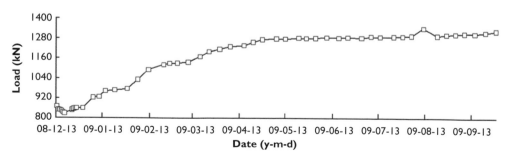

Figure 7.199 Load changes by Dpcf0+000–2 at the powerhouse upstream sidewall EL1334.5 of section S₁.

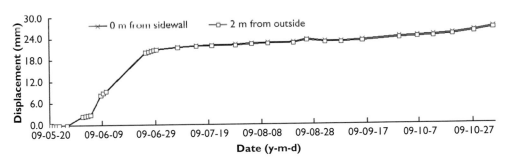

Figure 7.200 Monitored deformation curve of multi-point displacement meter Mcf0+000–11 at powerhouse downstream sidewall EL1322.5 of section S_1.

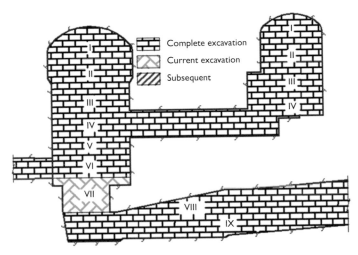

Figure 7.201 Excavation progress for layer VII construction of the main powerhouse.

The monitored results for the displacement and anchor cables at the powerhouse sidewalls (shown in Figures 7.199 and 7.200) during construction of layers VI, VIII and IX of the main powerhouse indicate that the deformation at the downside of the downstream bus tunnel increased quickly, its maximum was 20 mm, but it did not change after reinforcement. The anchor cable load was also controlled. Therefore, the local risk could be controlled by the reinforcement supporting scheme.

7.6.5.7 Construction of layer VII of the main powerhouse

The excavation progress for the construction of layer VII of the main powerhouse can be seen in Figure 7.201.

1   Right section 0+0.00 (S_1)
According to the revealed geological condition and dynamic feedback analysis results, the local risk for section S_1 has been evaluated after construction of layer VII of the

main powerhouse, and the results can be seen in Figure 7.202. The local risk grades for the upside of the machine stable and intersection between the powerhouse downstream sidewall and tailrace tunnel were high, and the rock collapse local risk should be avoided.

Some rock blocks collapsed near the intersection during construction of layer VII. Therefore, these areas should be reinforced, for which the measures proposed are shown as follows.

1   Add stressed bolts and fill with concrete near the rock collapse area.
2   Add stressed bolts at the sides of machine stable, $\phi 32$, $T = 80$ kN, $L = 8$ m, 1.5 m × 1.5 m.
3   Reduce the spacing of the anchor cables to 3 m in the downside area of the intersection between the powerhouse downstream sidewall and bus tunnel.

The local risk assessment result for the unit #1 section after reinforcement is shown in Figure 7.203. The region of risk grade I has reduced significantly and the depths of the risk grade I and II areas are less than 6 m, which satisfies the engineering construction requirement.

The monitored results for displacement and anchor cable at the powerhouse sidewalls (shown in Figures 7.204–7.207) during construction of layer VII of the main powerhouse indicate that the deformation of the powerhouse and transformer chamber surrounding rock was stable and changed little. Therefore, the local risk was controlled by the reinforcement supporting scheme.

2   Right section 0+108.5 (S$_3$)
According to the revealed geological conditions and dynamic feedback analysis results, the local risk for the section S$_3$ was evaluated after construction of layer VII of

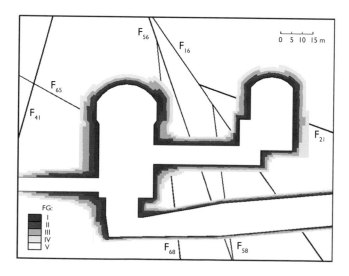

Figure 7.202 Local risk assessment result for section S$_1$ after layer VII construction of the main powerhouse.

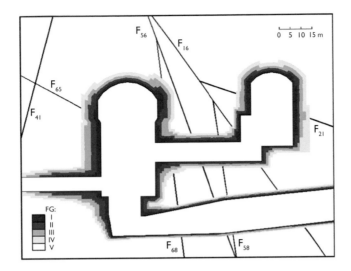

Figure 7.203 Local risk assessment result for section S₁ after reinforcement.

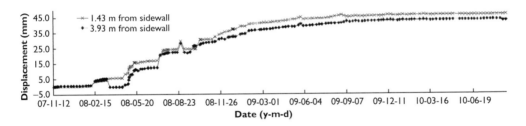

Figure 7.204 Monitored deformation curve for multi-point displacement meter Mcf0+000–1 results at the powerhouse upstream sidewall EL1357.558 of section S₁.

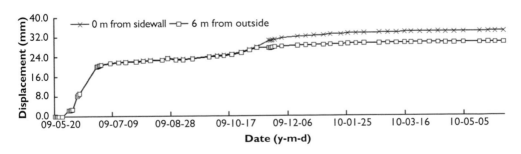

Figure 7.205 Monitored deformation curve for multi-point displacement meter Mcf0+000–11 results at the powerhouse downstream sidewall EL1322.5 of section S₁.

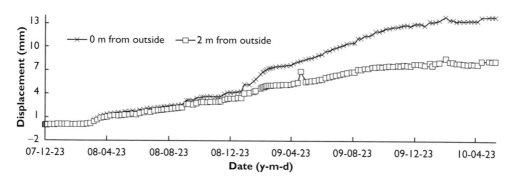

Figure 7.206 Monitored deformation curve for multi-point displacement meter Mzb0+000–1 results at the transformer chamber upstream sidewall EL1362.708 of section S_1.

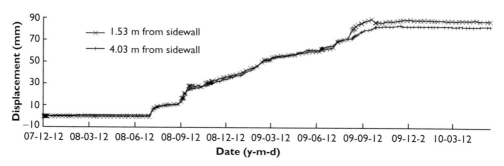

Figure 7.207 Monitored deformation curve for multi-point displacement meter Mzb0+000–3 results at the transformer chamber upstream sidewall EL1353.18 at left section 0+001.

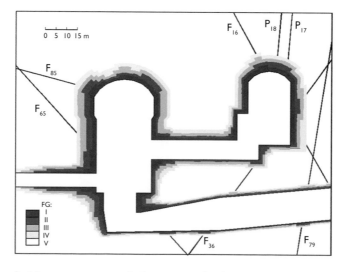

Figure 7.208 Local risk assessment result for section S_3 after layer VII construction of the main powerhouse.

the main powerhouse, with the results being included as Figure 7.208. The local risk grades at the upside of the machine stable and intersection between the powerhouse downstream sidewall and tailrace tunnel were high, and the rock collapse local risk should be avoided.

Some rock blocks collapsed near the intersection and the upside as affected by fault F_{65} during construction of layer VII. Therefore, these areas should be reinforced, for which the proposed measures are as follows: add stressed bolts at the sides of machine stable, $\phi 32$, $T = 80$ kN, $L = 8$ m, 1.5 m × 1.5 m.

The local risk assessment result for the unit #5 section after reinforcement is shown in Figure 7.209. The area of risk grade I reduced significantly and the depth of the risk grade I and II areas is less than 6 m, which satisfies the engineering construction requirement.

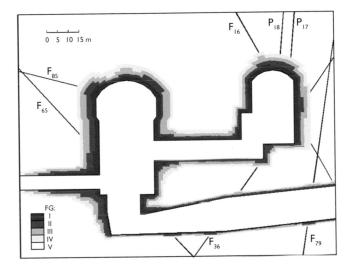

Figure 7.209 Local risk assessment result for section S_3 after reinforcement.

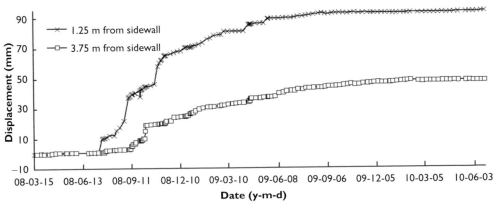

Figure 7.210 Monitored deformation curve for multi-point displacement meter Mcf0+108–6 results at the powerhouse upstream sidewall EL1334.5 of section S_3.

The monitored results for the displacement and anchor cables at the power-house sidewalls (shown in Figures 7.210–7.211) during construction of layer VII of the main powerhouse indicated that the maximum deformation at the powerhouse upstream sidewall was 93 mm, as affected by fault F_{65}; the maximum deformation at the powerhouse downstream sidewall was 52 mm, as affected by excavation unloading; the maximum deformation at the transformer chamber upstream side-wall was 37 mm, as affected by excavation unloading (Figure 7.212); and the maxi-mum deformation at the transformer chamber downstream sidewall was 92 mm, as affected by faults F_{16} and F_{36} and fracture zones P_{17} and P_{18} (Figure 7.213); finally, they all remained stable. Therefore, the local risk was controlled by the reinforcement supporting scheme.

7.6.5.8   Construction of different types of tunnel

According to the revealed geological condition and dynamic feedback analysis results, the local risk of other tunnels was also evaluated during the construction of the Jinping II hydropower underground cavern group, and appropriate measures were proposed to reduce the local risk as follows.

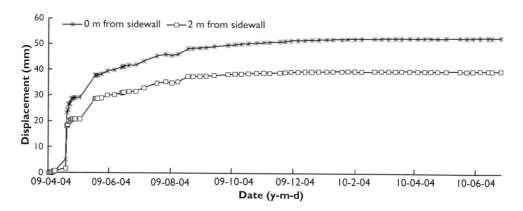

Figure 7.211 Monitored deformation curve for multi-point displacement meter Mcf0+108–9 results at the powerhouse downstream sidewall EL1322.5 of section S_3.

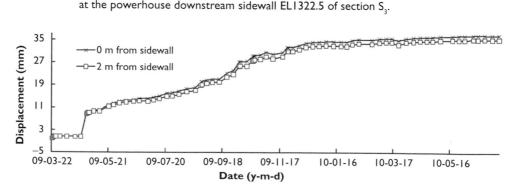

Figure 7.212 Monitored deformation curve for multi-point displacement meter Mzb0+108.5–5 results at the transformer chamber upstream sidewall EL1343.18 of section S_3.

1   Some sprayed concrete cracks were found in the area 3–4 m away from the #1 construction tunnel entrance near the erecting bay on 16 April 2008, as shown in Figure 7.214. It is a temporary tunnel, so the local risk was not dealt with and only monitored. No more cracks appeared later.

2   There was rock block collapse along fracture N70°W/SW∠85° and fault F_{21} at the left side of the intersection between the transformer chamber and #3 bus tunnel, as shown in Figure 7.215. The surrounding rock quality at the right-hand side was also poor due to fracture zone Pz17. Pre-stressed bolts (ϕ32, $L = 8$ m, $T = 120$ kN) were used to reinforce the surrounding rock, and after construction the rock mass was stable.

3   Some surrounding rock bulged outward and collapsed near the high-pressure pipe exit, especially the #6 pipe. The different failure types are shown in Figure 7.216, and their statistical occurrences are shown in Figure 7.217. The surrounding rock near the powerhouse upstream sidewall was severely damaged.

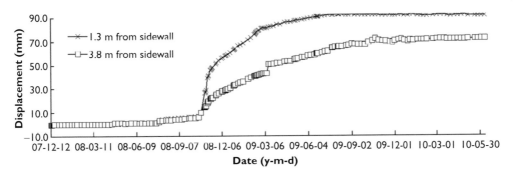

Figure 7.213 Monitored deformation curve of multi-point displacement meter Mzb0+108.5–4 results at the transformer chamber downstream sidewall EL1361.518 of section S_3.

Figure 7.214 Sprayed concrete cracks near #1 construction tunnel entrance.

Figure 7.215 Rock block collapse near #3 bus tunnel entrance.

a Spraved concrete and surrounding rock collapse   b Surrounding rock collapse

c Rock bedding opening   d Sprayed concrete spalling   e Tunnel type changed

Figure 7.216 Failure types of high pressure pipes.

The spacing of the bolts was reduced and pre-stressed bolts were used to reinforce the surrounding rock. The monitoring results showed that the local risk was controlled.

4   There were several ring cracks at the #3 and #4 bus tunnels near the power-house downstream sidewall during construction of layers V and VI, as shown in Figure 7.218. The spacing of the anchor cables near the intersection was reduced to 3 m, and more pre-stressed bolts were used. The monitoring results showed that there were no new cracks.

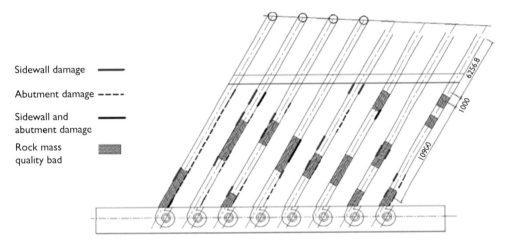

Sidewall damage ──────

Abutment damage ── ── ──

Sidewall and abutment damage ──────

Rock mass quality bad ▨▨▨

Figure 7.217 Statistical results for different failure types in the high pressure pipe exit tunnels.

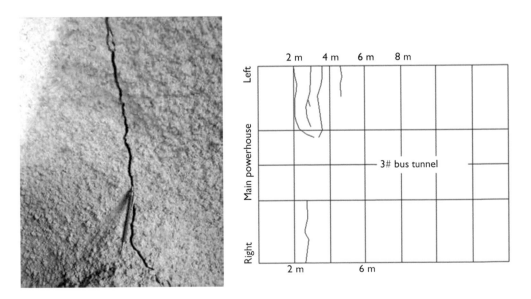

Figure 7.218 Sprayed concrete cracks at 3# bus tunnel.

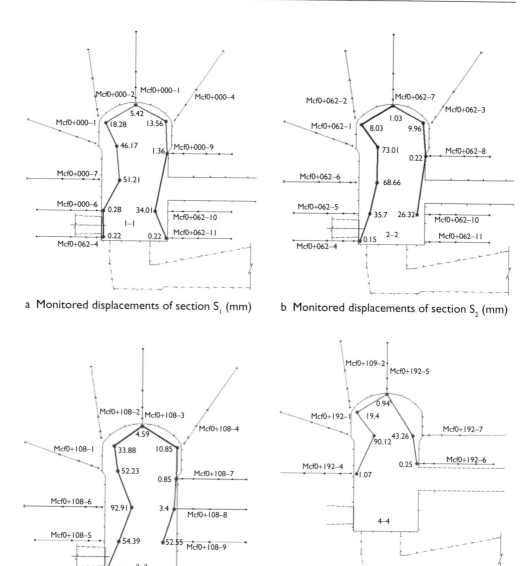

a Monitored displacements of section S$_1$ (mm)

b Monitored displacements of section S$_2$ (mm)

c Monitored displacements of section S$_3$ (mm)

d Monitored displacements of section S$_4$ (mm)

Figure 7.219 Monitored displacement for different sections (until September 2010).

7.6.5.9   Overall evaluation of the complete construction and final design

The underground cavern group was excavated by December 2009. The monitoring results (up until September 2010) indicated that the deformation at the vault was small, and the deformation at the powerhouse upstream sidewall was larger than that at the downstream sidewall. The maximum deformation was in the area between the centre of the upstream sidewall and rock anchor beam, which was 93 mm as monitored by Mcf0+108–6 at the right section 0+108.5, while the deformation at the downstream sidewall bottom was smaller, at 53 mm, as monitored by Mcf0+108–9 in the same section. The deformation distribution characters were similar, as shown in Figure 7.219.

The test results for the powerhouse displacements are shown in Table 7.55 until July 2009. It can be observed that the displacements were relatively low, the range being from 1.0 m to 3.6 m. The maximum was found at the powerhouse downstream sidewall EL1346.0 on the right side of the 0+186.00 section. The amount at the upstream sidewall was larger than that at the same elevation of the downstream sidewall. The test results for the transformer chamber are shown in Table 7.56: the movement at the vault was largest, at 3 m, and that at the sidewalls was about 2–3 m.

The loads of about 30% of the anchor cables in the powerhouse and transformer chamber exceeded their design values, and the anchor cables with more than 20% load accounted for 20%. The overloaded anchor cables were replaced or reinforced with other cables, and so the surrounding rock was effectively reinforced by the final set of anchor cables. The stress in most of the monitored bolts was lower than the design value. The bolts with a stress of less than 100 MPa accounted for 53%, and those with stress greater than 300 MPa accounted for 9%.

The monitoring results above demonstrate that dynamic feedback analysis and local risk assessment can be used to reduce the construction risk, and ensure that the surrounding rock is stable.

7.6.6   Important points

7.6.6.1   Optimisation of bench height of layers II and III, and the excavation procedure for layers IV–IX

The excavation scheme was adjusted when layer III was excavated, as shown in Table 7.57. The excavation process is shown in Figure 7.220.

Table 7.55 Test results for powerhouse displacements (m).

Section	1#	2#	3#	4#	5#	6#	7#	8#	9#
Right 0+000	2.8	2.0	3.0	2.4	1.4	2.8	1.8	2.0	1.2
Right 0+062	2.0	1.8	1.0	2.4	1.0	1.2	1.2	2.0	1.8
Right 0+124	1.6	2.2	3.2	2.8	1.6	3.4	3.4	1.6	1.2
Right 0+186	1.4	3.0	2.8	2.4	3.6	3.2	2.2	2.4	2.0

Table 7.56 Test results for transformer chamber displacements (m).

Section	1#	2#	3#	4#	5#
Right 0+000	2.6	3.0	1.8	2.0	1.6
Right 0+062	2.8	3.0	1.8	3.2	2.4
Right 0+124	2.4	3.0	2.0	3.2	3.0

Table 7.57 Powerhouse excavation scheme before and after adjustment.

Layer	Original scheme		Optimisation scheme	
	Elevation	Height (m)	Elevation	Height (m)
Layer IV	1334.3–1326.8	7.5	1332.3–1327.8	4.5
Layer V	1326.8–1320.8	6.0	1327.8–1323.3	4.5
Layer VI	1320.8–1313.5	7.3	1323.3–1314.3	9.0
Layer VII	1313.5–1301.1	12.4	1314.3–1302.1	12.2
Layer VIII	1301.1–1294.6	6.5	1302.1–1294.1	8.0
Layer IX	1294.6–1292.1	2.5	1294.1–1292.1	2.0

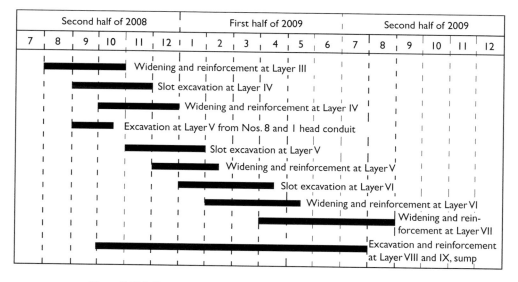

Figure 7.220 Excavation progress for the adjusted excavation scheme.

The differences between the original excavation scheme and adjusted excavation scheme are as follows:

1   the layer IV and V excavation heights of the adjusted scheme were lower than those of the original scheme; while those of layers VI and VIII became larger;
2   the middle sections of layers IV and V could be excavated at the same time;
3   the layers III–VI and VIII could be excavated at the same time; and
4   the layer VI excavation height was 9 m.

The surrounding rock deformations, plastic zones and anchor cable loads were compared through numerical analyses of the two types of excavation scheme, and the results are as follows.

1   Deformation characteristics
There were no obvious differences between the final deformations of the two excavation schemes. The deformation caused by the excavation of layer VI of the adjusted scheme was greater than that of the original scheme due to its larger excavation height. The deformation caused by the excavation of layers IV and V of the adjusted scheme was smaller than that of the original scheme due to its smaller excavation height.

2   Plastic zone characteristics
The plastic zone volume was similar, the maximum difference being 1.5%.

3   Anchor cable load characteristics
The differences in the anchor cable loads of the two schemes were also small.

7.6.6.2   More than ten local warnings and reinforcement improved the main powerhouse and transformer chamber

A summary of the excavation local risks and the risk management measures for the underground powerhouse and transformer chambers can be found in Tables 7.58 and 7.59 respectively.

7.6.6.3   Support reinforcement for different types of tunnel

The local risk and local risk management measures for other tunnels in the Jinping II cavern group are included as Table 7.60.

Table 7.58 Excavation local risk and risk management measures for the powerhouse.

Local risk warning	Local risk management measures
Rock block collapse at vice powerhouse downstream abutment and left side 0+56	Adding mortar bolts ($L = 6$ m) and hollow grouted bolts ($L = 9$ m) in the area affect ed by fault F_{68}. Adding two rows of pre-stressed anchor cables ($T = 2000$ kN, $L = 20$ m) between left side 0+58 and right side 0+8 sections.
Monitored deformation at S3 section increasing suddenly	Monitoring and system support timely implemented.
Anchor cable load over standard load in powerhouse upstream of the right side 0+009 section	System support timely implemented and addition of four root anchor cables ($T = 2000$ kN, $L = 20$ m).
Large rock deformation at right side of 0+263.6 section	Reducing anchor cable spacing to 3 m and adding two root anchor cables and pre-stressed bolts ($\phi 32, L = 9$ m, $T = 120$ kN)
Anchor cable over standard load	Reducing the anchor cable locked load, full-length bonded anchor cable replaced by stress dispersed anchor cable; plunge angle changed into down dip 5°.
Large deformation at right side 0+192 section	Adding pre-stressed bolts ($\phi 32, L = 9$ m, $T = 80$ kN).

(Continued)

Table 7.58 (*Continued*)

Local risk warning	Local risk management measures
Large deformation and large anchor cable load in upstream sidewall	Reducing anchor cable spacing to 3 m, bolt spacing at EL1342.0–1344.0 m changed to 1.0 m × 1.0 m; adding pre-stressed bolts ($\phi32, L = 9$ m, $T = 120$ kN) at upstream EL1355.5–1349.5 m between right side 0+070–0+110; bolt spacing at upstream EL1338.0–1342.5 m between right side 0+033.8–0+045 changed from 1.5 m × 3 m to 1 m × 2 m; adding pre-stressed bolts ($\phi32, L = 9$ m, $T = 80$ kN) at upstream EL1327.0–1334.5 m between right side 0+051.4–0+139.9.
Large deformation at upstream sidewall	Reducing anchor cable spacing to 3 m; rock mass grouting near faults F_{16} and F_{65}.
Anomaly deformation and supporting damage at downstream abutment of right side 0+000 section	Adding pre-stressed bolts and anchor cables near downstream abutment.
Stability problem of upstream sidewall during layer III excavation	Adding pre-stressed bolts and testing loosened rock depth; adding anchor cables; adding acoustic wave test hole at El1346 of rock anchor beam.
Stability of upstream sidewall near fault F_{65}	Consolidation grouting near fault F_{65}.
Surrounding rock safety monitoring	Adding one monitoring section system at the right side 0+140.
Powerhouse sidewall sprayed concrete cracking	Adding a row of anchor cables on the upside of the upstream and downstream rock anchor beam and drain hole.
Rock mass stability of end wall of the erecting bay	Adding anchor cables and monitoring.
Powerhouse deformation monitoring data analysis	Adding a row of dispersed stress anchor cables ($T = 1750$ kN, $L = 20$ m) at downstream sidewall EL1316.5; adding pre-stressed hollow grouting bolts near fracture zone.
Cracking of foundation pit rock mass	Adding bolts and anchor cables between the pits.

7.6.6.4 *Overall evaluation of the complete construction process and final design*

The various test data, statistical analysis results, cavern group excavation dynamic feedback analysis results and dynamic risk assessment results indicated that the surrounding rock deformation was not large (given the sizes of the excavations) and achieved a stable state after excavation. The bolt stresses were lower than the required safety standard values. Some anchor cable loads exceeded the design values, but they were controlled after reinforcement, or changed. The local risk grades for some regions were high, but they were acceptable. The loosened rock depth was similar to other projects, and satisfied the engineering construction requirements. Therefore, the dynamic local risk assessment based on dynamic feedback analysis and its management measures are confirmed as useful, if not essential, in reducing the surrounding rock failure possibility.

Table 7.59 Excavation local risk and risk management measures for the transformer chamber.

Local risk	Local risk management measures
Fault F_{16} revealed at right side 0+100, fault 0.5 m wide and influence zone 2–4 m wide.	Adding pre-stressed bolts at vault; adding steel arch rib and grouting near fault F_{16}.
Sprayed concrete cracking and collapse at downstream abutment between right side 0+085–0+125 m; steel arch rib bending significantly at right side 0+115 m.	Adding two rows of pre-stressed anchor cables at EL1363.5 and 1365.5; adding pre-stressed hollow grouting bolts between EL1361.77–1366.7.
Faults and fracture zones revealed at transformer chamber upstream EL1341–EL1352 of right side 0+5–0+80.	Adding pre-stressed bolts and anchor cables.
Fault F_{16} and compressive zone P_{17} and P_{18} revealed at transformer chamber downstream EL1340–EL1348 of right side 0+068–0+130.	Adding pre-stressed bolts and anchor cables.
Steep dip and flat structural surfaces revealed at powerhouse downstream sidewall EL1352 of right side 0+035–0+055 section.	Adding mortared bolts with plate.
Rock block falling down at transformer chamber EL1340–1349 m between right side 0+100 section and left side 0+93.8 section.	Adding anchor cables and mortared bolts.
Steep dip bedding revealed at transformer chamber upstream sidewall between 7# and 8# bus tunnel.	Adding pre-stressed anchor cable ($T = 1750$ kN, 3 m $\times$ 4.5 m).

Table 7.60 Local risk and local risk management measures for other tunnels in the Jinping II complex.

Local risk	Local risk management measures
Bus tunnel ring crack near powerhouse downstream sidewall.	Adding a row of pre-stressed anchor cables ($L = 20$ m, $T = 1750$ kN) at bus tunnel downside of powerhouse downstream sidewall (EL1316.5 m); adding loosened rock mass region depth testing hole at the bus tunnel downside of right side 0+108.5 section.
Rock mass at high pressure pipe damaged seriously.	Adding supporting steel arch ribs.
Rock mass damaged seriously, affected by steep dip structural surfaces and low dip structural surfaces at high pressure pipe.	Adding pre-stressed bolts ($\phi 32, L = 8$ m) and steel arch rib at 6# and 8# high pressure pipe exit areas.
Compressive fracture zone P_{19} and steep dip structural surfaces revealed at upstream sidewall of collection well; weathered fracture zone P_{17} and low dip structural surfaces revealed at downstream sidewall.	Adding pre-stressed bolts ($\phi 32, L = 9$ m) and mortar bolts ($\phi 32, L = 9$ m, 1.5 m $\times$ 1.5 m).

7.7　CHAPTER SUMMARY

In this Chapter, we have recorded the progressive construction of the Jinping II hydro-power cavern group and the associated implementation of a risk assessment procedure incorporating remedial actions. The content of the Chapter is summarised through the following key points.

1　A database of large-scale, hydropower cavern groups has been compiled, including 60 projects with spans of more than 20 m. The engineering geology investigation data, excavation revealed information, excavation scheme, supporting parameters, failure type and degree of damage, reinforcement measures and field monitoring information have been systematically analysed and presented. In addition, the corresponding qualitative and quantitative relations, such as those between geological condition, excavation scheme, support/reinforcement parameters, surrounding rock failure modes, reinforcement measures and field monitoring information, have also been analysed and explained.

2　The influence of the geological conditions on large underground cavern group design and construction is significant. The uncertainty of geological setting, rock stress, hydrology, properties of the rock mass, specific project location and excavation, support method and their influence on engineering construction have been analysed through the cases described. There is a large degree of subjective uncertainty because the geological information obtained before construction is not sufficient for design and construction. This uncertainty can be reduced through more investigation, monitoring during construction and associated evaluation methods.

3　There is heterogeneity present in rock masses which causes uncertainty. This uncertainty has been analysed through study of geological variations, rock stress variations, local water variations and the mechanical behaviour of the rock mass after excavation and in the long-term. This uncertainty cannot be avoided, only reduced.

4　The main characteristics of hydropower large-scale cavern groups are large-scale, intersection of excavations, excavation in layers and dynamic feedback optimisation design. The construction risk assessment methods with three stages have been proposed according to the geological information mastered before and after excavation. Firstly, the overall risk of different sections is assessed according to the geological survey information obtained before construction. Secondly, the local risk of typical sections can be assessed according to the geological survey information and numerical analysis results before construction. Finally, the local risk can assessed according to the geological information revealed during construction together with the dynamic feedback analysis results. The associated assessment methods have been proposed.

5   The overall risk of different sections of the Jinping II hydropower large cavern group has been assessed based on the analysis of uncertainties in the engineering geology conditions. Sections or areas with high risk grade have been determined. Then the local risk has been assessed according to the geological information revealed and field monitored information during construction of the seven layers. Local risk management measures have been suggested and verified through the field monitoring results of deformation, bolt stresses and anchor cable loads. In addition, local risk warning information and management measures for different tunnels have been analysed.

* * * * *

An outline of 24 cavern risk events during hydropower project construction is provided in Appendix A. The outline includes a brief description and photograph of each risk event, together with the necessary remedial actions.

Chapter 8

Concluding remarks

It will be recalled that the stimulus for this book was generated by (a) the 2007–2011 work of the Design Methodology Commission of the International Society for Rock Mechanics (ISRM), which resulted in our previous book "Rock Engineering Design", and (b) through the further encouragement provided by Professor E.T. Brown of Golder Associates in Australia who recommended that the logical follow-on subject should be "Rock Engineering Risk". This current book is the result, following the Design Methodology Commission's work during the period 2011–2015. The members of the Commission are listed in the Acknowledgements section at the beginning of the book.

Indeed, it has been a long journey through this book—from the Frontispiece flowchart, through the seven preceding Chapters, to these concluding remarks. We have introduced the rock engineering risk subject, discussed uncertainty and risk, including the concepts of epistemic and aleatory uncertainties, explained how the Rock Engineering Systems approach enables significant reductions in epistemic uncertainty, outlined the idiosyncrasies of rock fractures and *in situ* rock stress, considered how the design of an underground radioactive waste repository is approached, and presented the two major, detailed case examples of the construction of long, deep tunnels and a hydropower cavern complex. Throughout this journey, the Frontispiece flowchart has been the governing flowchart, ensuring that an overall risk reduction approach has been continually borne in mind, however much detail has been involved. Also, 'sub-flowcharts' have been developed, viz., for guiding epistemic and aleatory uncertainty analyses for the tunnel and cavern construction examples in Chapters 6 & 7.

We have thus provided the overall framework for developing a risk reduction approach to underground rock engineering and demonstrated its use through the major case examples, but what about the details? How can one check that the necessary procedures have been followed in any particular case? What is the auditing procedure? In our previous book, "Rock Engineering Design", we provided a suite of Protocol Sheets which enabled, through an independent auditing procedure, a mechanism for checking that the details of the necessary design work have been carried out satisfactorily. This procedure is preferably used contemporaneously with the design work so that any necessary corrections can be made, but it can also be used afterwards in a forensic mode. From all the discussion and conclusions throughout the book, it is evident that a set of detailed Risk Reduction Protocol Sheets is required to ensure that

a) there is a suitable framework established for all the manifold factors involved and b) that the necessary auditing procedure is in place so that the work can be checked step-by-step. Let us hope that these Protocol Sheets can be developed soon.

Meanwhile, we also hope that you have enjoyed reading the book and that you will be able to contribute to the risk reduction approaches—whether this be in civil/mining engineering practice, research or teaching.

John A. Hudson and Xia-Ting Feng,
January 2015

Cavern risk events during construction

Given the many and varied adverse events that occurred during the construction of the Jinping II hydropower caverns, plus knowledge of risk events that have occurred in other schemes, we now provide a reference list of the risk events that can occur during construction for hydropower schemes, although most of the occurrences also potentially apply to other underground construction work. The photographic illustrations are all taken from the construction activities at the Jinping II project and other Chinese schemes.

The subjects covered in the following text are:

1 Tilting failure
2 Slippage along bedding surfaces
3 Spalling
4 Unloading and rock structure
5 Strain rockburst
6 Fault rockburst
7 V-shaped failure
8 Tensile cracking
9 Rock block sliding
10 Rock block collapse
11 Circumferential cracks in tunnel's intersection with high cavern sidewalls
12 Splitting cracks in high sidewalls
13 Collapse at cavern intersections
14 Collapse at a fault or fractured zone
15 Fault sliding
16 Buckling deformation by squeezing of a weak stratum
17 Buckling by strata bending
18 Bedding plane splitting
19 Joint or bedding plane opening
20 Deep Excavation Damaged Zone (EDZ) damage
21 Supporting system failure
22 Anchor cable overload
23 Large deformation at the rock anchor beam
24 Severe water/mud inflow in karst regions

1   Tilting failure

The surrounding rock becomes loose and cracked, then tilting failure will occur if it is caused by reverse and steeply dipping bedding surfaces and fractures, as shown in Figure A1. This risk can be avoided or reduced by timely support using bolts and/or anchor cables.

2   Slippage along bedding surfaces

If there is one steeply dipping bedding surface or fault/joint revealed at the sidewall and the shear stress is greater than its shear strength after excavation, the upper rock mass can slip along the bedding surface. The risk management measures are as follows: make the intersection angle between the structural surface strike direction and cavern axis direction larger, where possible; support with anchor cables having a suitable dip angle; reinforce with steel arch ribs; use grouting for surrounding rock consolidation.

There was clear cracking and slipping along the bedding surfaces and rock blocks falling down during the excavation of the 3rd layer of the Penshui powerhouse. The width of cracks along the bedding surface was 10–20 mm, which increased with further excavation, and the final opening width was 50–100 mm. Rock blocks also collapsed during the excavation of the 5th layer, as shown in Figure A2.

Figure A1   Tilting failure of powerhouse upstream sidewall.

Figure A2   Rock mass opening and slipping along the bedding surface in the powerhouse downstream sidewall, Penshui powerhouse, China.

3   Spalling

Spalling occurs in brittle rock containing high *in situ* stress and which is concentrated around openings. It often occurs at the spandrel and sidewall. The risk management measures are: reduce the main stress concentration by arranging the cavern axis to be parallel/sub-parallel with the major horizontal principal stress component; reduce the concentrated tangential stress at the surrounding rock surface by suitable blasting methods; spray the concrete with steel net or fibres in a timely manner; use pre-stressed bolts.

The shotcrete on the downstream sidewall of the Jinping I powerhouse and transformer chamber broke away from the surrounding rock, and fresh rock could be seen. Some rock mass volumes split with an obvious inward bulge, as shown in Figure A3(a). Sometimes the rock mass may be compressed into debris, as shown in Figure A3(b).

Spalling also occurred at the downstream skewback of the Jinping II powerhouse and transformer chamber and the outside skewback of the high-pressure pipes and bus tunnels, as shown in Figure A4. Spalling usually occurred several hours after excavation, the rock width being 200–300 mm, and the maximum was 700 mm.

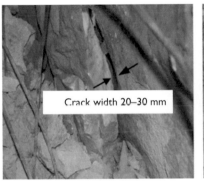

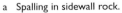

| a   At transformer chamber downstream skewback. | b   At transformer chamber downstream haunch. |

Figure A3 Typical spalling of surrounding rock at the Jinping II hydropower project.

| a   Spalling in sidewall rock. | b   Spalling at 1# tail tunnel outside vault. |

Figure A4 Typical spalling of surrounding rock, Jinping II project.

4   Unloading and rock structure

The rock mass unloaded and opened 10–70 mm along the bedding structure and the location of the drilling holes changed after excavation, as shown in Figure A5. This usually occurred at the upstream sidewall of the powerhouse and transformer chamber and inside the high-pressure pipes and bus tunnels.

5   Strain rockburst

A strain rockburst occurs when the elastic strain energy in the rock mass caused by stress concentration is larger than its storage capacity. If the elastic strain energy is larger than the damage energy, rock can be violently ejected. This failure usually occurs at the cavern vault, abutment, sidewall and intersections, as shown in Figure A6.

The risk management measures are as follows: make the intersection angle between the maximum principal stress direction and the cavern axis direction small; excavate with smooth wall blasting; support with steel net and shock resistant bolts. A strain rockburst occurred at the downstream spandrel during excavation of the 1st layer, as shown in Figure A7. The surrounding rock was reinforced with pre-stressed hollow grouted bolts (ϕ 32, $L = 6$ m, $T = 120$ kN).

Drilling hole location changed

Figure A5 Typical unloading and surrounding rock structure.

Figure A6 Moderate rockburst, drainage gallery.   *Figure A7* Slight rockburst, downstream spandrel.

6   Fault rockburst

Fault rockbursts can occur in stiff rock masses containing high *in situ* stress if there is one stiff fracture or large joint and it releases a large amount of energy after excavation. This failure usually occurs at the spandrel or sidewall where the stress is concentrated, as shown in Figure A8. The risk management measures are as follows: keep away from the large fracture or excavate firstly with a pilot tunnel; consolidation grouting; advance support and shock resistant bolts.

7   V-shaped failure

In a stiff, brittle rock mass with high *in situ* stress and asymmetrical loading, spalling can occur, or a V-shape type failure. This usually occurs at the spandrel and intersections, as shown in Figure A9. The risk management measures are as follows: reduce the asymmetrical load by modifying the cavern axis direction (where possible); reduce the disturbance to the surrounding rock by controlling the blasting methods; support with pre-stressed bolts or steel arch ribs; reduce the concentration of stress by drilling holes and cutting grooves.

Figure A8 Strong fault rockburst in the bus tunnel.

Figure A9 V-shape failure in auxiliary tunnel.

A V-shaped failure occurred at the left side spandrel of the Jinping II high pressure pipe because the rock mass stress was high, as shown in Figure A10. It was reinforced with steel arch ribs and bolts.

8   Tensile cracking

Tensile cracking may occur when the complete surrounding rock at the high sidewall becomes significantly broken and the rock stress is changed. This usually occurs at the sidewall with loosened rock or the step area bulging outwards, as shown in Figure A11.

The risk management measures are as follows: reduce the asymmetrical load by altering the cavern axis (where possible); reduce the disturbance to the surrounding rock by controlling the blasting methods; support with pre-stressed bolts and adjust their plunge angle; support with anchor bars at the step area bulging outwards.

Tensile cracking occurred at the sidewall of pits #1–4, and there were horizontal and vertical cracks in the sidewalls of pits #2 and 3, and some rock blocks collapsed, as shown in Figure A12. The surrounding rock was first filled with concrete, then

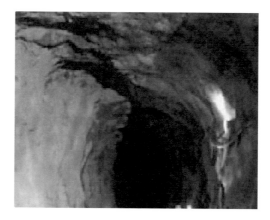

Figure A10 V-shape failure in high pressure pipe.

Figure A11 Tensile cracking at the erection bay sidewall.

Figure A12 Surrounding rock failure process at the 3# pit sidewall.

Figure A13 Local risk control at 3# pit sidewall.

reinforced with vertical anchor bars and bolts and three rows of anchor cables at different elevations, as shown in Figure A13.

9   Rock block sliding

The minimum number of faces that a rock block can have is four: a tetrahedron. Thus, for a rock block to form, a minimum of three fracture sets are required in addition to the excavation surface. If the resisting force is lower than the sliding force after excavation, then the rock block can slide along one or two surfaces. This type of failure usually occurs at the spandrel or sidewall, as shown in Figure A14. The risk management measures are shown as follows: reduce the disturbance to the surrounding rock by controlling the blasting methods; support with pre-stressed bolts or anchor cables; support with anti-sliding piles.

The unstable rock block formed by one steeply dipping fracture zone and one set of planar joints with EW strike direction without support collapsed during the excavation of the 2nd layer of the draft tube gate chamber, as shown in Figure A15. This area was first filled with concrete, then reinforced with two rows of anchor bars (3 ϕ 28 mm, $L = 8$ m). The surrounding rock was stable after reinforcement.

Figure A14 Rock wedge sliding at rock anchor beam.

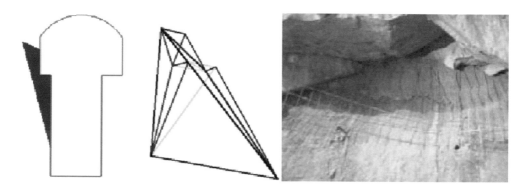

Figure A15 Rock block sliding at upstream sidewall of draft tube gate chamber.

10   Rock block collapse

The moving rock block formed by three different sets of structural surfaces and free surfaces may slide and fall if its own weight is high enough. This failure usually occurs at the vault with different structural surfaces present, as shown in Figure A16. The risk management measures are as follows: reduce the disturbance to the surrounding rock by controlling the blasting methods, support with pre-stressed bolts or anchor cables or steel arch ribs.

11   Circumferential cracks in tunnels intersecting high cavern sidewalls

If the normal unloading is severe and the vertical stress concentration is significant after the high sidewall has been formed in the rock mass, and the excavation causes the normal restraint to be released, then circumferential cracks are produced. This risk usually occurs at the bus tunnel, high pressure pipe or transportation tunnel near the sidewalls. The risk management measures are as follows: excavate the tunnels first

Figure A16 Rock block falling at vault of draft tube gate chamber.

Figure A17 Circumferential cracks at 1# construction tunnel invert.

and then the sidewalls; provide support in a timely manner at the cavern intersection and sidewall; support with bolts or anchor cables or anchor bars.

Circumferential cracks appeared at the intersection of the #1 construction tunnel and sidewall, as shown in Figure A17. They were reinforced with anchor cable ($L = 20$ m, $T = 1000$ kN). The monitoring results showed that the surrounding rock was stable.

12   Splitting crack in the high sidewall

When unloading occurs in different directions at the cavern intersection in a brittle rock mass, then splitting crack(s) may appear in the high sidewall. The risk management measures are as follows: excavate the tunnels first and then the sidewalls, plus interval excavating; provide the support in a timely manner at the cavern intersection and sidewall; support with bolts or anchor cables.

Vertical splitting cracks appeared at two sides of the bus tunnel near the transformer chamber upstream sidewall after excavation, as shown in Figure A18. They were reinforced with 20 m long anchor cables, after which the surrounding rock became stable.

Figure A18 Splitting crack at the transformer chamber upstream sidewall.

Figure A19 Collapse at the intersection of high pressure pipe and sidewall.

13 Collapse at cavern intersection

If there are several structural surfaces at the cavern intersection, collapse may occur easily after excavation without support, as shown in Figure A19.

The risk management measures are shown as follows: excavate the tunnels first, then the sidewalls; reduce the disturbance to the surrounding rock by controlling the blasting methods; provide support in a timely manner at the cavern intersection and sidewall; support with steel arch ribs; use advance consolidation grouting.

Rock collapse occurred at the intersection of the #3 bus tunnel and transformer chamber upstream sidewall; its volume was about 12 m^3, and the bolts had been damaged, as shown in Figure A20. Through the field survey, it was found that faults F16 and F21 and fracture zones Pz7 and Pz16–Pz18 were revealed near the intersection. The surrounding rock was reinforced with two rows of anchor cables at EL1334.5 m and 1331.5 m. The damaged area was filled with concrete and reinforced with 8 m long pre-stressed bolts.

Figure A20 Collapse at 3# bus tunnel intersection.

Figure A21 Collapse at a fault.

14   Collapse at a fault or fractured zone

If there are faults or fractured zones, and their strength is low and their stability poor, then collapse is likely to occur after excavation. This risk will occur near the fault or fractured zone or area of developed joints, as shown in Figure A21.

The risk management measures are as follows: make the intersection angle between the structural surface strike direction and cavern axis direction large; reduce the disturbance to the surrounding rock by controlling the blasting methods; provide support in a timely manner with advance consolidation grouting or bolts; support with bolts or anchor cables; replace the rock mass with reinforced concrete.

The width of fault F16 revealed at the transformer chamber was large and filled with shale, and its mechanical character was poor. After excavation, the concentrated stress field caused fractured rock mass damage, then collapse occurred, as shown in Figure A22. The surrounding rock was reinforced with steel arch ribs, pre-stressed bolts and consolidation grouting. The overbreak area was filled with concrete and reinforced with anchor cables.

15 Fault sliding

If the intersection angle between the fault strike direction and cavern axis is small and the *in situ* stress is high, then the hangingwall and footwall of the fault will move due to the excavation unloading. This risk usually occurs at the sidewall, as shown in Figure A23.

The risk management measures are shown as follows: make the intersection angle between structural surface strike direction and cavern axis direction large; provide support in a timely manner with bolts or anchor cables or anchor bars; consolidation grouting.

16 Buckling deformation by squeezing of a weak stratum

The rock stress may reach the rock mass yield strength at an area of concentrated stress, weathered fracture zone, local soft rock, or clay interlayer zone after excavation, following which the soft rock mass will extrude. This risk usually occurs at the sidewall base having a soft interlayer and other weak zones where there is concentrated stress.

Figure A22 Collapse at fault F16 fracture zone.

Figure A23 Fault sliding at draft tube gate chamber.

Figure A24 Fracture zone bulging outward at powerhouse abutment.

The risk management measures are as follows: excavate the soft rock mass first; reduce the disturbance to the surrounding rock by controlling the blasting methods; provide support in a timely manner with bolts or anchor cables; replace the rock mass with reinforced concrete.

There are several fracture zones revealed at the downstream abutment between the vice powerhouse and unit #1 section, their mechanical character is poor, their stress concentration is the maximum, and the underground water reduces their shear strength, so that shear failure and expansion occur easily. Some shotcrete bulged outwards and spalled at the downstream abutment, finally bulging 400 mm outwards, as shown in Figure A24. The damaged shotcrete was cleaned and reinforced with steel net and shotcrete.

17 Buckling deformation by strata bending

If the tangential stress in a thin layer of the rock mass is greater than its strength, then it is easily broken off. For the complete rock mass, it splits first, then bulges outwards under highly concentrated stress. This risk usually occurs at the sidewall base in a steeply dipping, thinly layered rock mass, or at the vault where there is a horizontal structural surface.

The risk management measures are as follows: reduce the disturbance to the surrounding rock by controlling the blasting methods; make the intersection angle between the structural surface strike direction and cavern axis direction large; provide support in a timely manner with bolts or anchor cables, or steel arch ribs or grouting.

It was found that the shotcrete came away from the rock mass surface after cleaning at the Jinping I powerhouse downstream arch crown (EL1670–1671 m) at the 0 + 132 – 0 + 185 m section. The rock failure types were splitting and bedding opening, the rock plate widths were 100–200 mm, and they were parallel with the excavation surface, as shown in Figure A25. This risk usually occurs at the vault or arch crown of the high pressure pipe and bus tunnel.

Figure A25 Buckling deformation by strata bending at 0 + 150 m.

Figure A26 Bedding plane splitting in the sidewall.

18 Bedding plane splitting

If the joint dip angle is large and the intersection angle between its strike direction and the cavern axis is small, and the tangential stress concentrates significantly after normal unloading, then bedding planes can split. This risk usually occurs at the sidewall base or the cavern side.

The risk management measures are shown as follows: make the intersection angle between structural surface strike direction and cavern axis direction large; provide support in a timely manner with bolts or anchor cables or steel arch ribs or consolidation grouting.

For one occurrence, the mean dip angle of the rock bedding is 80°, the intersection angle between its strike direction and cavern axis is about 15°, and the rock mass is also affected by fault F65, so that bedding plane splitting occurred, as shown in Figure A26. The area was reinforced with pre-stressed bolts and anchor cables.

19 Joint or bedding plane opening

If the intersection angle between structural surfaces strike direction and unloading direction is large, then the joints or bedding planes will loosen and open because of

Figure A27 Joint loosening and opening in the transformer chamber sidewall.

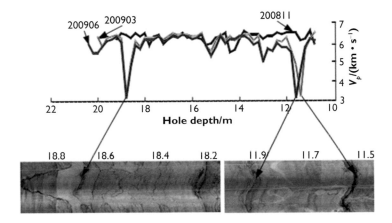

Figure A28 Test results of borehole wave velocity and borehole camera photos.

the unloading or local tensile stress action. This risk usually occurs at the sidewall central area, cavern intersection or step area, as shown in Figure A27.

The risk management measures are shown as follows: reduce the disturbance to the surrounding rock by controlling the blasting methods; support with bolts or consolidation grouting.

20 Deep Excavation Damaged Zone (EDZ) with crack

The EDZ increases as a function of the unloading caused by excavation, and new cracks appear, as can be seen in Figure A28. If the *in situ* stress is high and there are many joints or fractures, then the EDZ depth may be large. This can be avoided or reduced by using pre-stressed bolts or anchor cables.

21 Supporting system failure

If the rock mass quality is poor and the *in situ* stress is high, then the supporting system may be damaged by the pressure caused by the failure of the surrounding rock. For the Jinping I powerhouse, the shotcrete crazed and the steel arch ribs bent in

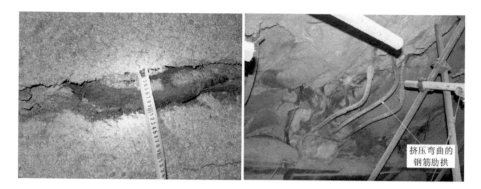

Figure A29 Damage to support system.

Figure A30 Cracks at the top of the upstream rock anchor beam.

the downstream arch area, the crack width was 50–100 mm, and their lengths were 2–20 m, as shown in Figure A29. They were reinforced with fresh shotcrete and pre-stressed bolts.

22 Anchor cable overload

If the deformation of the surrounding rock continues to increase after it has been supported with anchor cables, then the associated deformation and tensile stress of the anchor cables will increase, and the rock deformation will be limited. However, if the anchor cable load is large enough, then it will be damaged and anchor cable overload will occur. The main reasons for this are the high *in situ* stress and low rock mass strength, with time-dependent deformation. This risk usually occurs at both side-walls. It can be avoided or reduced by using a low locked load in the anchor cables, which may be 60–70% or 50–60% of the design load, depending on the severity of the overloading.

23 Large deformation at the rock anchor beam

If the rock mass quality near the rock anchor beam is poor, and the *in situ* stress is high, and there are steeply dipping rock structural surfaces, then the deformation can be large and shotcrete cracks may appear, so that the rock anchor beam cannot be used. This risk can be reduced by using a suitable blasting method and advance bolts and consolidation grouting; otherwise, the beam should be constructed later.

For the Guandi hydropower station, there were several cracks at the rock anchor beam (as shown in Figure A30); the maximum deformation at the upstream side was 34 mm, and the relative deformation of both sides was 58 mm; thus, the gantry crane could not be used.

24 Severe water/mud inflows in karst regions

In karst areas, gushing water and/or mud may occur during excavation; in particular, the water pressure can be high. This risk can be reduced by advance exploration and monitoring and suitable excavation methods. For one project, the underground cavern group was affected by karst systems KW51 and W84, several incidences of gushing water and mud occurred, and the inflow water volume exceeded 10,000 stere (i.e., m^3).

The Chinese 'Basic Quality' (BQ) system for rock mass classification

B1 INTRODUCTION

This Chinese BQ unified rock classification system[1] has been developed to aid in the evaluation of the stability of engineering rock masses and to provide rock characterisation assistance for design and construction in rock engineering. This Standard is applicable to rock mass classification for all kinds of rock engineering. The engineering classification of a rock mass should be conducted in a way in which qualitative and quantitative evaluations are combined together, and in two steps: firstly, the determination of the basic rock mass quality; and, secondly, to determine rock mass classes according to the requirements of different kinds of rock engineering.

B2 TERMINOLOGY AND SYMBOLS

B2.1 Terminology

Rock engineering: The kinds of surface and underground engineering in which the rock mass serves as a foundation or environment for the construction project.

Engineering rock mass: The rock mass in the area influenced by rock engineering, including underground engineering, the foundations for industrial and civil construction, dam foundations and rock slopes.

Rock mass basic quality: Includes the basic properties inherent in a rock mass and on which the stability of the engineering rock mass depends. The rock mass basic quality is defined by the rock solidity and rock mass integrity.

Structural plane (discontinuity): Fracture plane or plane in a rock mass that is liable to fracture, such as a bedding plane, joint, fault, schistosity, etc. —also called discontinuity.

---

1 This BQ Chinese Standard has been translated from the original Chinese language so some of the nuances in the rock descriptions may have been lost. Thus, if necessary for detailed use of the BQ system, we recommend referring to the original Chinese text. This English version has been included to give the reader an impression of the BQ system because it has been referred to in several chapters of this book. ("Standard for Engineering Classification of Rock Masses", The National Department of Technical Monitorial Affairs and The Ministry of Construction, PRC, GB 50218–94.) See also Section B11.

Rock mass volumetric joint count (J_v): The number of joints (structural planes) in a unit volume of rock mass.

Point load strength index ($I_{s(50)}$): Point load strength of a diametrically loaded cylindrical specimen having a diameter of 50 mm.

Stand-up time for an underground rock mass: The time for a rock mass to withstand any kind of failure without support.

Initial stress field: The rock stress field in the natural condition, resulting from gravity and tectonics, also called the 'natural stress field'.

B2.2 Symbols

Table B1 List of symbols and their meaning.

No.	Symbol	Meaning
1	γ	Unit weight of the rock
2	R_c	Saturated uniaxial compressive strength of intact rock
3	$I_{s(50)}$	Point load strength index
4	E	Elastic modulus of the rock mass
5	v	Poisson's ratio of rock mass
6	ϕ	Friction angle of a rock mass or structural plane
7	c	Cohesion of a rock mass or structural plane
8	K_v	Intactness index for a rock mass
9	J_v	Volumetric joint count for a rock mass
10	K_1	Corrective coefficient for the influence of underground water
11	K_2	Corrective coefficient for the influence of the orientation of the main weak structural planes
12	K_3	Corrective coefficient for the influence of the initial stress state
13	f_0	Basic value of the bearing capacity of a rock foundation
14	η	Reduction coefficient for the influence of the rock foundation configuration
15	BQ	Rock mass basic quality index
16	$[BQ]$	Corrected rock mass basic quality index

B3 CLASSIFICATION PARAMETERS FOR THE ROCK MASS BASIC QUALITY

B3.1 Classification parameters and the method of their determination

The rock mass basic quality should be determined by two parameters, i.e., rock solidness and rock mass integrity. Both of these should be determined by two methods: namely, qualitative classification and quantitative indices.

B3.2 Qualitative classification of rock mass solidity

Rock solidity should be qualitatively classified according to Table B2.

During qualitative classification of the rock solidity, the weathering degree of the rock should be determined according to Table B3.

Table B2 Qualitative classification of rock solidity.

Class name		Qualitative evaluation	Representative rock type
Strong Rock	Hard Rock	When hammered, clear sound with rebound and shock to hand, difficult to fragment. After being immersed in water, no significant absorption effect noticed	Unweathered and slightly weathered rock, such as granite, syenite, diorite, diabase, basalt, andesite, gneiss, quartz schist, siliceous slate, quartzite, conglomerate with siliceous cement, quartz sandstone, siliceous limestone, etc.
	Less Hard Rock	When hammered, fairly clear sound, with slight rebound and shock to hand, fairly difficult to fragment. After being immersed in water, a slight absorption effect is noticed	1 Weakly weathered hard rock 2 Unweathered to slightly weathered rock, such as welded tuff, dolomite, marble, slate, limestone, sandstone with calcareous cement, etc.
Weak Rock	Less Soft Rock	When hammered, unclear sound without rebound, fairly easy to fragment. After being immersed in water, can be scratched by nail.	1 Strongly weathered hard rock 2 Weakly weathered, less hard rock 3 Unweathered to slightly weathered rocks, such as: tuff, phyllite, sandy mudstone, marl, argillaceous sandstone, siltstone, shale, etc.
	Soft Rock	When hammered, dull sound without rebound, but with surface depression, easy to fragment. After being immersed in water, can be broken off by fingers.	1 Strongly weathered hard rock 2 Weakly to strongly weathered, less hard rock 3 Weakly weathered, less soft rock 4 Unweathered mudstone, etc.
	Very Soft Rock	When hammered, dull sound without rebound, but with deep surface depression, can be crushed by hand. After being immersed in water, can be kneaded into a ball	1 Any kind of completely weathered rock 2 Any kind of 'semi-rock'

Table B3 Classification of rock weathering degree.

Weathering degree	Description of weathered rock texture
Unweathered	Unaltered, fresh rock
Slightly weathered	Colour and lustre of minerals basically unaltered; parts of the joint surfaces stained by ferrous and manganese substances
Weakly weathered	Partly decomposed, colour and lustre of minerals clearly altered, weathered minerals noticed in joints
Strongly weathered	Mostly decomposed, colour and lustre of minerals clearly altered, feldspar, mica etc. weathered into secondary minerals
Completely weathered	Completely decomposed, most mineral assemblages, except quartz, weathered into soils

Table B4 Qualitative classification of rock mass integrity.

Class name	Quantitative description of structural planes		Interconnection degree of the main structural plane sets	Type of main structural plane sets	Corresponding structural type of rock mass
	Number of joint sets	Mean spacing, m.			
Intact	I to 2	>1.0	Well-interconnected or partially interconnected	Joints, fractures, bedding planes.	Integral or macro-bedded structure
Mostly intact	I to 2	>1.0	Poorly interconnected	Joints, fractures, bedding planes.	Massive or thick-bedded structure
	2 to 3	1.0 to 0.4	Well interconnected or partially interconnected		Massive structure
Partially fractured	2 to 3	1.0 to 0.4	Poorly interconnected	Joints, fractures, bedding planes, minor faults.	Cracked-massive or thickly-bedded structure
	≥3	0.4 to 0.2	Well-interconnected Partially interconnected		Mosaic-cataclastic Medium-bedded to thin-bedded structure
Fractured	≥3	0.4 to 0.2	Poorly interconnected	All kinds of discontinuities	Cracked-massive structure
		≤0.2 Very low	Partially interconnected or badly interconnected		Cataclastic structure
Very fractured	Many fractures		Very poorly interconnected	All kinds of discontinuities	Loose structure

Note: The mean spacing refers to the mean spacing of the main structural planes (I to 2 sets).

Table B5 Classification of the degree of interconnectedness of the structural planes.

Interconnection degree	Characteristics of the structural planes
Well interconnected	Aperture width less than I mm, unfilled Aperture width I to 3 mm, with siliceous or ferruginous cement Aperture width greater than 3 mm, with rough joint surfaces and siliceous cement
Partially interconnected	Aperture width I to 3 mm, with calcareous or argillaceous cement Aperture width larger than 3 mm, with rough joint surfaces and ferriferous or calcareous cement
Poorly interconnected	Aperture width I to 3 mm, with planar surfaces, and argillaceous or argillaceous-calcareous cement Aperture width larger than 3 mm, mostly filled with argillaceous material or debris
Very poorly interconnected	With argillaceous fillings possibly mixed with debris, the thickness of which is larger than the undulations of the structural planes' surfaces

Table B6 Relation between R_c and qualitatively classified rock solidity.

R_c (MPa)	>60	60–30	30–15	15–5	<5
Rock solidity	Hard rock	Not so hard rock	Not so soft rock	Soft rock	Very soft rock

Table B7 Correspondence of K_v and J_v.

J_v (joint count/m³)	<3	3–10	10–20	20–35	>35
K_v		>0.75	0.75–0.55	0.55–0.35	0.35–0.15	0.15

Table B8 Relation between K_v and the qualitatively classified rock integrity.

K_v	>0.75	0.75–0.55	0.55–0.35	0.35–0.15	0.15
Rock mass integrity	Intact	Less intact	Less fractured	Fractured	Very fractured

B3.3   Qualitative classification of rock mass integrity

The rock mass integrity should be qualitatively classified according to Table B4.

The interconnection degree of the structural planes should be determined by the characteristics of the structural planes according to Table B5.

B3.4   Determination and classification of quantitative indices

For the quantitative index of rock solidity, the saturated uniaxial compressive strength of the rock (R_c) should be adopted. Measured data for R_c should be used. In the case when measured data for R_c cannot be obtained, the value converted from the measured point load strength index ($I_{s(50)}$) can be used, via Equation B1:

$$R_c = 22.82 \, I_{s(50)}^{0.75} \tag{B1}$$

The corresponding relation between the saturated uniaxial compressive strength of the rock (R_c) and the qualitatively classified rock solidness can be determined in accordance with Table B6.

For the quantitative index of rock mass integrity, the intactness index of the rock mass (K_v) should be adopted. Measured data of K_v should be used. In the case when measured data are not available, the value of K_v can be determined through its correspondence to the volumetric joint count for the rock mass (J_v), in accordance with Table B7.

The corresponding relation between the intactness index for the rock mass (K_v) and the qualitatively classified rock integrity can be determined in accordance with Table B8.

The quantitative indices K_v and J_v should be measured in accordance with the following Section B6.

B4   CLASSIFICATION OF ROCK MASS BASIC QUALITY

B4.1   Determination of the rock mass basic quality class

The class of rock mass basic quality should be determined through a combination of the qualitative characteristics and the rock mass basic quality index (BQ), in accordance with Table B9.

In the case of inconsistency between the rock mass basic quality class determined by the qualitative characteristics and the class determined by the Basic Quality index (BQ), the class should be re-determined by further analysis of both approaches, with further tests being conducted if necessary.

B4.2   Qualitative characteristics of the basic quality and the basic quality index

The qualitative characteristics of the rock mass basic quality should be determined by a combination of the rock solidity and the rock mass integrity, obtained in accordance with Tables B2 and B4.

The rock mass basic quality index (BQ) should be calculated using the classification parameters R_c in MPa units and K_v via Equation B2.

$$BQ = 90 + 3R_c + 250K_v \tag{B2}$$

Table B9   Classification of the rock mass basic quality.

Basic quality class	Qualitative characteristics of the rock mass basic quality	Rock mass Basic Quality index (BQ)
I	Hard rock, intact rock mass	>550
II	Hard rock, not so intact rock mass Not so hard rock, intact rock mass	550–451
III	Hard rock, not so fractured rock mass Not so hard rock or inter-bedding of hard and soft rock, not so intact rock mass Not so soft rock, intact rock mass	450–351
IV	Hard rock, fractured rock mass; Not so hard rock, less fractured to fractured rock mass Not so soft rock or inter-bedding of hard and soft rock, with predominantly soft rock, not so intact to not so fractured rock mass; Soft rock, intact to not so intact rock mass.	350–251
V	Not so soft rock, fractured rock mass; Soft rock, not so fractured to fractured rock mass; All kinds of very soft rock and all kinds of very fractured rock mass	≤250

Note: The following restrictions should be followed when using Equation B2.

1   If $R_c > 90K_v + 30$, then let $R_c = 90K_v + 30$. These values of R_c and K_v should be substituted into Equation B2 to calculate the value of BQ;

2   If $K_v > 0.04R_c + 0.4$, then let $K_v = 0.04R_c + 0.4$. These values of K_v and R_c should be substituted into Equation B2 to calculate the value of BQ.

B5   ENGINEERING CLASSIFICATION FOR A ROCK MASS

B5.1   General rules

The class for the rock mass basic quality, determined in accordance with Table B9, may be used as the rock mass class in the preliminary engineering classification.

For detailed classification of a rock mass, necessary corrective parameters should be taken into consideration. Depending on the features of different rock masses, these are: the state of underground water, the initial rock stress state, the relative orientation of the project's structural axis or strike with the occurrences of the main weak discontinuity sets. For slopes, the influence of surface water should also be taken into consideration.

When no measured data are available, the initial stress state in the rock mass may be evaluated according to the thickness of the overburden/depth of excavation, topography and morphology, history of geological tectonic movements, the main lineaments and any special phenomena which occur during excavation, such as rockbursting, core discing, etc., in accordance with Section B7 of this Standard.

The influences of the following factors on the rock mass class should be considered, if they become significant factors affecting rock mass stability: expandibility and solubility of the rock mass, and the existence of large weak structural planes with a fairly high degree of continuity in the rock engineering volume.

In the preliminary classification of the rock mass, the physical and mechanical parameters of the rock mass may be adopted in accordance with Table B13 in Section B8.1 of this Standard. The peak values of shearing strength of structural planes may be adopted according to rock solidness and interconnection degree of structural planes in accordance with Table B14 in Section B8.2 of this Standard.

B5.2   Engineering rock mass classification

During detailed classification of a rock mass and if one or more of the cases below should occur, correction should be made to the rock mass Basic Quality index (BQ) and the class of rock mass should be determined with the corrected value in accordance with Table B9.

1   The existence of underground water.
2   The stability of the rock mass is affected by weak structural planes, one of which is dominant.
3   The existence of high initial stress phenomena, as indicated in Table B12 of Section B7 in this Standard.

The corrected rock mass Basic Quality index (BQ) for the rock mass may be calculated according to the following Section B9.

Table B10 The basic value of bearing capacity for a rock foundation (f_0).

Rock mass class	I	II	III	IV	V
f_0 (MPa)	>7.0	7.0–4.0	4.0–2.0	2.0–0.5	0.5

Table B11 The reduction coefficient for the influence of the specific rock foundation configuration (η).

Rock foundation configuration	Planar type	Inward type	Outward type	Stepwise type
Slope of rock surface (°)	0–10	10–20	10–20	Height of steps 5 m
η	1.0	0.9	0.8	0.7

Note: The case in which the structural planes are dipping towards the slope surface of a rock foundation is termed the 'outward type' and the opposite case is termed the 'inward type'.

For an underground excavation with a span equal to or less than 20 m, the already-estimated rock mass class should be appropriately adjusted if the actual stand-up time does not agree with the stand-up time for that class of rock mass, as indicated in Section B10 of this Standard.

For detailed classification of a rock mass for large or special underground projects, other Standards can also be used so that the results can be compared in order to determine the rock mass class more effectively.

The rock mass class for the foundations of civil structures should be determined in accordance with Table B9.

The bearing capacity of a rock foundation for civil structures can be determined according to the following rules.

1   The basic value of the bearing capacity on a rock foundation for various classes of rock mass (f_0) can be determined according to Table B10.
2   When the influence of the rock foundation configuration is to be considered, the standardised value of bearing capacity for rock foundations (f_k) can be determined according to Equation B3:

$$f_k = \eta f_0 \tag{B3}$$

3   The reduction coefficient for the influence of the specific rock foundation configuration (η) can be adopted according to Table B11.

During detailed classification of a rock mass for a slope, according to different heights of the slope, correction should be made according to the following factors: underground water, surface water, initial stress field, spatial combination of structural planes, relation between the orientations of the structural planes and the slope surfaces.

B6   ESTABLISHING THE K_v AND J_v INDICES

B6.1   The K_v index

For determination of the intactness index for a rock mass (K_v), the longitudinal wave velocity should be measured at representative points and sections, chosen from the different rock types. The longitudinal wave velocity in intact rock specimens taken from the same rock mass should also be measured. The value for the intactness index of the rock mass should be calculated according to Equation B4:

$$K_v = (V_{pm}/V_{pr})^2 \tag{B4}$$

In which V_{pm} is the longitudinal wave velocity of the rock mass in km/s, and V_{pr} is the longitudinal wave velocity for the intact rock, also in km/s.

B6.2   The J_v index

For the determination of the volumetric joint count for the rock mass (J_v), the number of joints (discontinuities) should be investigated at representative outcrops or the excavation face, chosen from the different rock types. Besides the joint sets, joints having a length more than 1 m should be counted separately as well. Joints which have been re-cemented by siliceous, ferrous and calcareous fillings should not be included.

The area of each sampling location should not be less than 2×5 m². The value J_v should be calculated using Equation B5:

$$J_v = S_1 + S_2 + \cdots\cdots + S_n + S_k \tag{B5}$$

where J_v is the volumetric joint count for the rock mass (number of joints/m³), $S_1 - S_n$ are the numbers of joints intersected in a 1 m perpendicular scanline for each joint set, and S_k is the number of through going joints in a 1 m³ rock mass.

B7   PRELIMINARY ASSESSMENT OF THE ROCK STRESS FIELD

In the case where there is a lack of *in situ* test data, a preliminary assessment of the rock stress field can be made on the basis of geological investigation results by the following methods.

1   For an isolated mountain area (or butte) with a gentle landform, the vertical stress for the preliminary rock stress field can be considered as the gravity induced stress. The horizontal stress should then not be greater than the value of $\gamma H v/(1 - v)$, where γ is the rock density in kN/m³, H is the depth to the underground location in m, and v is Poisson's ratio.

2   The direction of the maximum principal stress can be assessed on the basis of tectonics. The most recent tectonic system can be determined according to the superposition relation of the tectonic events, with emphasis on the first order. If the vertical stress is a gravity stress and is one of the principal stresses, the greater horizontal principal stress could be given a value of $(0.8-1.2)\gamma H$ or greater.

Table B12 Main phenomena occurring during rock excavation in a highly stressed region.

Stress state	Main phenomena	R_c/σ_{max}
Very high stress	1 Hard rock: During excavation rockbursts occasionally occur. Rock fragments can be expelled. The sidewalls of the excavation peel off. New fissures and cracks can be seen. The resultant cavity may lose its intended shape. Similarly, for open pit excavations, rock peeling can occur affecting the pit shape.	4
	2 Soft rock: Rock discing often occurs. During excavation, the sidewalls peel and large displacements can occur over a long time period. It can be difficult to establish the intended cavity shape. For open pit excavations, there can be upheavals and rock peeling. It may be difficult to establish the intended pit geometry.	
High stress	1 Hard rock: During excavation, rockbursts may occur. Some peeling-off from the sidewalls and rock falls may be noted. There will be some new fracturing. The cavity may be shaped badly. For open pit excavation, rock peeling may occasionally occur. The pit shape may not be ideal.	4–7
	2 Soft rock: Core discing occurs. During excavation, the sidewalls displace over a prolonged period. The cavity may be shaped badly. For open pit excavation, upheaval occurs and the pit may be shaped badly.	

Note: σ_{max} is the estimated maximum normal stress component in the plane perpendicular to the excavation axis.

3   At a depth greater than 1000 m, the preliminary stress field tends to a hydro-static stress state. At a depth greater than 1500 m, the stress field can be taken as hydrostatic.

4   In a gorge area, starting from the slope surface to the inner rock mass of the mountain, the stress field can be divided into three zones, i.e., the stress relief zone, the stress concentration zone and the zone having a stabilised stress state. The horizontal extent of the area in which the stress state is influenced by the gorge landform is generally one to three times the gorge width. The direction of the maximum principal stress for the rock mass on both sides of the gorge is, in general, parallel to the valley, but at depth below the valley bottom it tends to be horizontal and perpendicular to the valley axis.

5   Wherever core discing or a rockburst has occurred, the likelihood of high rock stress should be considered. Then, the stress state can be assessed in accordance with Table B12.

B8   PHYSICAL AND MECHANICAL PARAMETERS OF THE ROCK MASS AND DISCONTINUITIES

B8.1   Rock mass parameters

The physical and mechanical parameters of the rock mass may be estimated from Table B13.

Table B13 Physical and mechanical parameters for the rock mass.

Class of rock mass: basic quality	Unit weight of rock γ (kN/m³)	Peak value of shear strength		Elastic modulus E (GPa)	Poisson's ratio v
		Internal friction angle ø (°)	Cohesion c (MPa)		
I	>26.5	>60	>2.1	>33	0.2
II		60–50	2.1–1.5	33–20	0.2–0.25
III	26.5–24.5	50–39	1.5–0.7	20–6	0.25–0.3
IV	24.5–22.5	39–27	0.7–0.2	6–1.3	0.3–0.35
V	<22.5	<27	<0.2	<1.3	>0.35

Table B14 The shear strengths of discontinuities.

No.	The solidity of the rock and the interconnectedness of the discontinuities	Internal friction angle ϕ (°)	Cohesion c (MPa)
1	Hard rock, well-interconnected	>37	>0.15
2	Hard rock to less hard rock, reasonably interconnected; Strong soft rock, well-interconnected	37–29	0.15–0.1
3	Hard rock to less hard rock, poorly interconnected; Stronger soft rock to soft rock, reasonably interconnected	29–19	0.10–0.06
4	Softer hard rock to strong soft rock, badly interconnected to very poorly interconnected; Soft rock, poorly interconnected; Argillaceous interfaces in soft rock	19–13	0.06–0.03
5	Less hard rock and all kinds of soft rock, very poorly interconnected; Argillaceous intercalations in soft rock	<13	0.03

B8.2   Discontinuity parameters

The friction and cohesion of discontinuities may be estimated from Table B14.

B9   CORRECTED VALUE OF THE ROCK MASS BASIC QUALITY INDEX

The corrected rock mass basic quality index, [BQ], may be calculated using Equation B6.

$$[BQ] = BQ - 100(K_1 + K_2 + K_3) \tag{B6}$$

where BQ is the rock mass basic quality index, K_1 is the correction coefficient for the influence of underground water, K_2 is the correction coefficient for the influence of the orientation of the main weak structural planes, and K_3 is the correction coefficient for the influence of the initial stress state.

Table B15 Correction coefficient for the influence of underground water, K_1.

State of the underground water outflow	BQ >450	450–350	350–250	250
Wet or dripping	0	0.1	0.2–0.3	0.4–0.6
Pressure ≤0.1 MPa or with a runoff ≤10 l/min	0.1	0.2–0.3	0.4–0.6	0.7–0.9
Pressure >0.1 MPa or with a runoff >10 l/min	0.2	0.4–0.6	0.7–0.9	1.0

Table B16 Correction coefficient for the influence of the orientation of the main weak structural planes, K_2.

Occurrence of the main weak structural planes and the relation with the excavation axis	Angle between the orientation of the main weak structural planes and axis of excavation: 30° Their inclination: 30–75°	Angle between the orientation of the main weak structural planes and axis of excavation: 60° Their inclination: >75°	Other cases
K_2	0.4–0.6	0–0.2	0.2–0.4

Table B17 Correction coefficient for the influence of the initial stress state, K_3.

Initial stress state	BQ >550	550–450	450–350	350–250	250
Very high stress region	1.0	1.0	1.0–1.5	1.0–1.5	1.0
High stress region	0.5	0.5	0.5	0.5–1.0	0.5–1.0

The values of K_1, K_2, K_3 may be determined in accordance with Tables B15 to B17. In the case when the actual states are not described in these Tables, the correction coefficient takes the value zero. If the value of [BQ] becomes negative, the situation should be treated as a special case.

B10 STAND-UP TIME FOR AN UNDERGROUND ROCK MASS

The stand-up time can be estimated from Table B18.

Table B18 Stand-up time for an underground rock mass.

Class of rock mass	Stand-Up Time
I	Excavation span ≤20 m will remain in a stable state over the long term, small rock pieces may fall occasionally, no significant rockfall.
II	Excavation span 10–20 m will remain in a stable state on the whole, rock pieces may fall and/or there could be small rockfalls occurring locally.
	Excavation span 10 m will also remain in a stable state over the long term, small rockfalls may occur occasionally.
III	Excavation span 10–20 m will remain stable from several days to a month, small and medium rockfalls occurring locally.
	Excavation span 5–10 m, will remain in a stable state for several months, block displacements and/or small to medium rockfalls occurring locally.
	Excavation span 5 m will generally remain in a stable state.
IV	Excavation span >5 m is generally not capable of remaining in a stable state. Loosening causing deformation, small rockfalls occurring over several days to several months, followed by medium to large rockfalls. These are mainly loosening failure of the roof at shallow depths and plastic flow and squeezing failure at greater depths.
	Excavation span ≤5 m is capable of remaining in a stable state for only several days to a month.
V	Not stable without support and/or reinforcement

Note: Height of small rockfall 3 m, or volume 30 m³; height of medium rockfall 3–6 m, or volume 30–100 m³; Height of large rockfall >6 m, or volume >100 m³.

B11   ACKNOWLEDGEMENTS

The Chief Compiling Institutions for the BQ Standard as reported in this Appendix B were the Changjiang River Scientific Research Institute, the Changjiang Water Resources Committee, and the Ministry of Water Resources. The Compiling Institutions were the Northeastern University, the 4th Design and Research Institute of the General Staff, the Southwest Department of the China Academy of Railway Sciences, the Ministry of Railways China Institute of Geotechnical Investigation and Surveying, and the Ministry of Construction.

The main authors are Dong Xuesheng, Lin Yunmei, Liu Fuzheng, Xing Nianxin, Wang Shichun, Su Yibing, Li Yunlin, Li Zhaoquan, Zhang Kecheng, and Xu Fu-an.

* * * *

Note: The original version of this Standard was written in Chinese. It was then translated into English by Chinese colleagues and that translation was then polished by John A. Hudson. Thus, it is possible that, in this English version, some of the nuances of the Chinese Standard may have been 'lost in translation'.

References and bibliography

Aaltonen, I., Lahti, M., Engström, J., Mattila, J., Paananen, M., Paulamäki, S., Gehör, S., Kärki, A., Ahokas, T., Torvela, T. & Front, K. (2010) *Geological model of the Olkiluoto site, Version 2.0*. Posiva, Working Report 2010-70.

AFTES. (2002) Recommandations relatives à la méthode convergence-confinement. (Association française des travaux en souterrain). AFTES Groupe de travail n7 (anime´ par Panet, M. avec la collaboration de Bouvard, A., Dardard, B., Dubois, P., Givet, O., Guilloux, A., Launay, J., Nguyen, MinhDuc., Piraud, J., Tournery, H., & Wong, H.). *Tunnels et Ouvrages Souterrains*, 170, 79–89.

Agüero, A., Pinedo, P., Simón, I., Cancio, D., Moraleda, M., Trueba, C. & Pérez-Sánchez, D. (2008) Application of the Spanish methodological approach for biosphere assessment to a generic high-level waste disposal site. *Science of the Total Environment*, 403, 34–58.

Ali, K.M. & Hasan, K. (2002). Rock mass characterization to indicate slope instability at Bandarban, Bangladesh: a Rock Engineering Systems approach. *Environ Eng Geosci*, 8, 105–119.

Anderson, E.M. (1942) *The Dynamics of Faulting and Dyke Formation*. Edinburgh: Oliver and Boyd.

Andersson, J., Ahokas, H., Hudson, J.A., Koskinen, L., Luukkonen, A., Löfman, J., Keto, V., Pitkänen, P., Mattila, J., Ikonen, A.T.K., Ylä-Mella, M. (2007) *Olkiluoto Site Description 2006. Part 1*. Posiva: Report number: 2007–3. (Download available from www.posiva.fi)

Andersson, J.A., Munier, R., Ström, A., Söderbäck, B., Almén, K.E. & Olsson, L. (2004) *When is there sufficient information from the site investigations?* SKB Report R-04-23, Stockholm, Sweden. (Download available from www.skb.se)

Andersson, J.C. (2007) *Äspö Pillar Stability Experiment, Final report: Rock mass response to coupled mechanical thermal loading*. Svensk Kärnbränslehantering AB. Report number: TR-07-01. (Download available from www.skb.se)

Andersson, J.C. (2007). *Rock mass response to coupled mechanical thermal loading. Äspö Pillar Stability Experiment, Sweden*. Royal Institute of Technology (KTH), Stockholm: PhD thesis. (Download available from www.kth.se)

Andrieux, P. & Hadjigeorgiou, J. (2008) The destressability index methodology for the assessment of the likelihood of success of a large-scale confined destress blast in an underground mine pillar. *Int J Rock Mech Min*, 45, 407–421.

Andrieux, P., Brummer, R., Liu, Q., Mortazavi, A. & Simser, B. (2003) Large-scale panel destress blast at Brunswick Mine. *Bull Can Inst Min Metall Pet*, 96, 78–87.

Arnold, P (1993) *The development of a rock engineering methodology using a system perspective*. PhD Thesis, Imperial College, University of London, London.

AS/NZS—Standards Australia & Standards New Zealand. (2009) AS/NZS ISO 31000:2009. *Risk management—Principles and guidelines*. Sydney: Standards Australia & Standards New Zealand.

Aughenbaugh, J.M. & Paredis, C.J.J. (2006) The value of using imprecise probabilities in engineering design. *J Mech Design*, 128, 969–979.

Avila, R. & Moberg, L. (1999) A systematic approach to the migration of 137Cs in forest ecosystems using interaction matrices. *J Environ Radioactiv*, 45, 271–282.

Aydan, O., Akagi, T. & Kawamoto, T. (1993) The squeezing potential of rocks around tunnels: theory and prediction. *Rock Mech Rock Eng*, 26, 137–163.

Bakalowicz, M. (2005) Karst groundwater: a challenge for new resources. *Hydrogeol J*, 13, 148–160.

Barla, G. (1995) Squeezing rocks in tunnels. *ISRM News Journal*, 3/4, 44–49.

Barla, G. (2001) Tunnelling under squeezing rock conditions. In: *Eurosummer-School in Tunnel Mechanics, Innsbruck*. Berlin: Logos Verlag Berlin GmbH. pp. 169–268.

Barla, G. (2010) Innovative tunneling construction method to cope with squeezing at the Saint Martin La Porte access adit (Lyon-Turin base tunnel). In: Vrkljan, I. (ed.), *Rock Engineering in Difficult Rock Conditions-Soft Rocks and Karst: Proceedings of ISRM Regional Symposium Eurock2009, 29–31 October 2009, Dubrovnik, Croatia*. London: Taylor & Francis Group. pp. 15–24.

Barla, G., Barla, M., Bonini, M.& Debernardi, D. (2007) Lessons learned during the excavation of the Saint Martin La Porte access gallery along the Lyon-Turin Base tunnel. In: Schneider, E., John, M. & Brandner, R. (eds.), *BBT 2007, Internationales Symposium Brenner Basistunnel und Zulaufstrecke, 2007, Innsbruck, Austria*. Innsbruck: Innsbruck University Press. pp. 45–52.

Barla, G., Bonini, M. & Semeraro, M. (2011) Analysis of the behaviour of a yield control support system in squeezing rock. *Tunn Undergr Sp Tech*, 26, 1, 146–154.

Barton, N., Lien, R. & Lunde, J. (1974) Engineering classification of rock masses for the design of tunnel support. *Rock Mech Rock Eng*, 6, 189–236.

Baynes, F.J. (2010) Sources of geotechnical risk. *Q J Eng Geol Hydroge*, 43, 321–331.

Beck, D.A. & Brady, B.H.G. (2002) Evaluation and application of controlling parameters for seismic events in hard-rock mines. *Int J Rock Mech Min*, 39, 633–642.

Bedi, A. (2013) *A proposed framework for characterising uncertainty and variability in rock mechanics and rock engineering*. PhD thesis, Imperial College London.

Bedi, A. & Harrison, J.P. (2012) Dealing with epistemic uncertainty in fractured rock masses. In: Stille H. et al. (eds): *Eurock'12-Rock Engineering and Technology for Sustainable Underground Construction; Proceedings of the ISRM International Symposium Eurock'12, May 28–30, 2012, Stockholm, Sweden*. ISRM.

Bedi, A. & Harrison, J.P. (2013a) Characterisation and propagation of epistemic uncertainty in rock engineering: a slope stability example. In: Kwasniewski, M. & Lydzba, D.(Eds). *Eurock 2013: Rock mechanics for resources, energy and environment, Proceedings of the ISRM International Symposium Eurock'13, 23–26 September, Wroclaw, Poland*. London: Taylor & Francis Group.

Bedi, A. & Harrison, J.P. (2013b). A comparison of Bayesian techniques and non-probabilistic models in rock engineering design. In: Pyrak-Nolte, L.J., Chan, A. Dershowitz, W., Morris, J. & Rostami, J. (eds.) *Proceedings 47th US Rock Mechanics/Geomechanics Symposium 2013, 23–26 June 2013, San Francisco, California, USA*. Alexandria, Virginia, USA: American Rock Mechanics Association. Volume 1 of 4, pp. 3035–3044.

Benardos, A.G. & Kaliampakos, D.C. (2004a) A methodology for assessing geotechnical hazards for TBM tunneling—illustrated by the Athens Metro, Greece. *Int J Rock Mech Min*, 41, 987–999.

Benardos, A.G., Kaliampakos, D.C. (2004b). Modelling TBM performance with artificial neural networks. *Tunn Undergr Sp Tech*, 19, 597–605.

Bernaud, D. (1991) *Tunnels profonds dans les milieux viscoplastiques: approches expérimentale et numérique*. PhD thesis, Paris, France: Ecole Nationale des Ponts et Chaussées.

Bernaud, D. & Rousset, G. (1996) The new implicit method for tunnel analysis. *Int J Numer Anal Meth Geomech*, 20, 673–690.

Billig, B., Gipperich, C., Wulff, M. & Schaab, A. (2008) Ausbausysteme für den maschinellen Tunnelbau in druckhaftem Gebirge. In: *Taschenbuch für den Tunnelbau 2008*. Essen: Verlag Glückauf GmbH, pp. 223–262.

Boldini D., Lackner R., Mang H.A. (2003): The role of an advanced shotcrete constitutive model for reliable predictions of ground-shotcrete interaction in tunneling. In *Proceedings 10th Congress of the International Society for Rock Mechanics , Johannesburg, South Africa*. Johannesburg: South African Institute for Mining and Metallurgy. pp. 125–130.

Bond, A. & Harris, A. (2008) *Decoding Eurocode7*. London: Taylor & Francis.

Bonnini, M. & Barla, G. (2012) The Saint Martin la Porte access adit (Lyon–Turin Base Tunnel) revisited. *Tunn Undergr Sp Tech*, 30, 38–54.

Brown, E.T. (2012) Risk assessment and management in underground rock engineering—an overview. *J Rock Mech Geotech Eng*, 4(3), 193–204.

Brown, E.T. & Booth, A. (2009) Risk management. In: Read, J. & Stacey, P. (eds.) *Guidelines for Open Pit Design*. Melbourne: CSIRO Publishing. pp. 381–400.

Budetta, P., Santo, A. & Vivenzio, F. (2008). Landslide hazard mapping along the coastline of the Cilento region (Italy) by means of a GIS-based parameter rating approach. *Geomorphology*, 94, 340–352.

Budil, A., Höllrigl, M. & Brötz, K. (2004) Strenger Tunnel-Gebirgsdruck und Ausbau. *Felsbau*, 22(1), 39–43.

Cai, B., Deng, Z.W. & Wu, G.Z (2012) Stability evaluation on rockslide zone of Dagangshan underground powerhouse. *Yangtze River*, 43, 33–35.

Cai, J., Zhao, J., Hudson, J.A. & Wu, X. (1996) Using neural networks in Rock Engineering Systems for cavern performance auditing. In: Barla, G. (ed.) *EUROCK 96—Prediction and Performance in Rock Mechanics and Rock Engineering:Proc. Eurock '96, ISRM International Symposium, 2–5 September 1996, Turin, Italy*. Rotterdam: Balkema, ISBN: 905410843 6. pp. 965–972.

Cai, J.G., Zhao, J. & Hudson, J.A. (1998) Computerization of Rock Engineering Systems using neural networks with an expert system. *Rock Mech Rock Eng*, 31, 135–152. DOI: 10.1007/s006030050015.

Cai, J.G., Bian, H.Y., Zhao, J. & Feng. X.T. (2000) Rock engineering system approach in tunnelling and underground excavation. In: Zhao, J., Shirlaw, J.N. & Krishnan, R. (eds). *Tunnels and Underground Structures: Proc. ICTUS 2000. Singapore*. Rotterdam: Balkeema. pp. 343–346.

Cai, M., Kaiser, P.K., Uno, H., Tasaka, Y. & Minami, M. (2004) Estimation of rock mass deformation modulus and strength of jointed hard rock masses using the GSI system. *Int J Rock Mech Min*, 41, 3–19.

Cancelli, A., & Crosta, G. (1994) Hazard and risk assessment in rockfall prone areas. In: Skipp, B.O. (ed.), *Risk and Reliability in Ground Engineering*. Springfield: Thomas Telford. pp. 177–190.

Cantieni, L. & Anagnostou, G. (2009) The interaction between yielding supports and squeezing ground. *Tunn Undergr Sp Tech*, 24, 309–322.

Carranza-Torres, C. & Fairhurst, C. (2000) Application of the convergence–confinement method of tunnel design to rock masses that satisfy the Hoek–Brown failure criterion. *Tunn Undergr Sp Tech*, 15, 187–213.

Carter, T.G. (2011) Himalayan ground conditions challenge innovation for successful TBM tunnelling. In: *Risk Management in Tunnelling Proc. Hydrovision India 2011*, Session 5c.

Castaldini, D., Genevois, R., Panizza, M., Puccinelli, A., Berti, M. & Simoni, A. (1998) An integrated approach for analysing earthquake-induced surface effects: a case study from the Northern Apennines. Italy. *J Geodyn*, 26 (2–4), 413–441.

Castro, L.A.M., Carter, T.G. & Lightfoot, N. (2009) Investigating factors influencing fault-slip in seismically active structures. In: Diederichs, M. & Grasselli, G. (eds.) *ROCKENG09: Proceedings of the 3rd CANUS Rock Mechanics Symposium, Toronto, May 2009.*

Ceryan, N. & Ceryan, S. (2008) An application of the interaction matrices method for slope failure susceptibility zoning: Dogankent Settlement Area (Giresun, NE Turkey). *B Eng Geol Environ*, 67, 375–385.

Cesano, D., Olofsson, B. & Bagtzoglou, A.C. (2000) Parameters regulating groundwater inflows into hard rock tunnels: a statistical study of the Bolmen Tunnel in southern Sweden. *Tunn Undergr Sp Tech*, 15, 153–165.

Chen, B.R., Feng, X.T., Xiao, Y.X., *et al.* (2010) Acoustic emission test on damage evolution of surrounding rock in deep-buried tunnel during TBM excavation. *Chinese J Rock Mech Eng*, 29, 1562–1569. (in Chinese)

Chen, B.R., Feng, X.T., Zeng, X.H., Xiao, Y.X., Zhang, Z.T., Ming, H.J. & Feng, G.L. (2011) Real-time microseismic monitoring and its characteristic analysis during TBM tunneling in deep-buried tunnel. *Chinese J Rock Mech Eng*, 30, 275–283. (in Chinese)

Chen, B.R., Feng, X.T., Ming, H.J., Zhou, H., Zeng, X.H., Feng, G.L. & Xiao, Y.X. (2012) Evolution law and mechanism of rockburst in deep tunnel: time-delayed rockburst. *Chinese J Rock Mech Eng*, 31, 561–569. (in Chinese)

Chen, D.F. (2012) *Risk estimation of rockburst based on neural network*. MS thesis. Shenyang: Northeastern University. (in Chinese)

Chen, G.Q., Feng, X.T., Zhou, H., Chen, B.R., Huang, S.L. & Zhang, C.Q. (2007) Numerical analysis of the long-term stability of the seepage tunnel in Jinping II hydropower station. *Rock and Soil Mechanics*. 28, 417–422.

Chen, L. & Huang, H.W. (2005) Risk analysis of rock tunnel engineering. *Chinese J Rock Mech Eng*, 24, 110–115.

Chen, W. & Ruan, H.N. (2007) Numerical modeling of tunnel excavation and lining under high initial stress and high underground water pressures at Jinping mountain. In: Barták, J., Hrdina, G., Romancov, G. & Zlámal, J (eds.) *Underground Space—The 4th Dimension of Metropolises: Proceedings of the World Tunnel Congress 2007 and 33rd ITA/AITES Annual General Assembly, Prague, 5–10 May, 2007.* London: Taylor & Francis Group. Chapter 75, DOI: 10.1201/NOE0415408073.ch75.

Cheng, W., Wang, W., Huang, S. & Ma, P. (2013) Acoustic emission monitoring of rockbursts during TBM-excavated headrace tunneling at Jinping II hydropower station, *J Rock Mech Geotech Eng*, 5, 486–494.

Chen, Y., Zhang, X.C., Wang, H.R. & Zhang, N.T. (2013) Influence of soft rock factor and time dependence in deep tunnel. *Appl Mech Mater*, (353–356) 1625–1629.

Chen, Z.A., Sun, Z.L., Peng, S.B. & Xi, Q.X. (2000) *Hydropower Engineering in China (Engineering Geology Series)*. Beijing, China: China Electric Power Press.

Chinese Railway Tunnel Guidelines (2008) *Code for design on tunnel of railway*. China Railway Publishing House.

Clark, G.T. & Borst, A. (2002) Addressing risk in Seattle's underground. *PB Network*. 1, 34–37.

Company Standard of Hydrochina Corporation (2012) Q/HYDROCHINA 009-2012. *Underground Powerhouse Design Guideline of Hydropower Station*. Beijing, China: Hydrochina Corporation.

Condor, J. & Asghari, K. (2009) An alternative theoretical methodology for monitoring the risks of $CO2$ leakage from wellbores. *Energy Procedia* 1, 2599–2605.

Cook, N.G.W. (1965) A note on rockbursts considered as a problem of stability. *J S Afr I Min Metall*, March, 515–523.

Cook, N.G.W., Hoek, E., Pretorius, J.P.G., Ortlepp, W.D. & Salamon, M.D.G. (1966) Rock mechanics applied to the study of rockbursts. *J S Afr I Min Metall*, May, 506–515.

Corbetta, F. (1990) *Nouvelles méthodes d' étude des tunnels profonds-calculs analytiques et numériques.* PhD thesis, Ecole des Mines de Paris.

Corkum, A.G. & Martin, C.D (2007) Modeling a mine-by test at the Mont Terri rock laboratory, Switzerland. *Int J Rock Mech Min*, 44, 846–859.

Cornet, F.H. & Burlet D. (1992) Stress field determinations in France by hydraulic tests in boreholes. *J Geophys Res, Solid Earth, 1978–2012*, 97, B8, 11829–11849. DOI: 10.1029/90JB02638.

Cosgrove, J.W. & Hudson, J.A. (2015). *Structural Geology and Rock Engineering.* London: Imperial College Press.

Dai, Y. (2009) Geological features and treatment of mud rushing in Yunwu Mountain tunnel. *Railway Engineering*, 10, 33–35. (in Chinese)

Dalgic, S. (2002) Tunneling in squeezing rocks, the Bolu tunnel, Anatolian motorway, Turkey. *Eng Geol*, 67, 73–96.

Damjanac, B. & Fairhurst, C. (2010) Evidence for a long-term strength threshold in crystalline rock. *Rock Mech Rock Eng*, 43, 1–19.

Davies, J.P., Clarke, B.A., Ockleston, G.O., Whiter, J.T. & Cunningham, R.J. (1999) Application of interaction matrices to the problem of sewer collapse. In: *Proceedings of the 11th European Sewage and Refuse Symposium, Liquid Wastes Section, 4–6 May 1999, Munich, Germany.*

Day, M.J. (2004) Karstic problems in the construction of Milwaukee's deep tunnels. *Environ Geol*, 45, 859–863.

Debernardi, D. & Barla, G. (2009) New viscoplastic model for design analysis of tunnels in squeezing conditions. *Rock Mech. Rock Eng*, 42, 259–288.

Detoumay, E. (1986) Elastoplastic model of a deep tunnel for a rock with variable dilatancy. *Rock Mech. Rock Eng*, 19, 99–108.

Dong, F.T., Song, H.W., Guo, Z.H., Lu, S.M. & Liang, S.J. (1994) Roadway support theory based on broken rock zone. *Journal of China Coal Society.* 19, 22–32. (in Chinese)

Dubois, D. & Guyonnet, D. (2011) Risk-informed decision-making in the presence of epistemic uncertainty. *Int J Gen Syst*, 40 (2), 145–167.

Dunn, J. (2004) *Auditing Theory and Practice.* Int. Book Dist., and 2nd Edition, 2006, London: Prentice Hall.

Durrheim, R.J., Jager, A.J., Klokow, J.W. & Booyens, D. (1995) Back analysis to determine the mechanism and risk of rockbursts—3 case histories from south African gold mines. In: *Proc. 26th Int. Conf. of Safety in Mines Research Institutes, Central Mining Institute, Katowice, Poland.* (5), pp. 41–56.

Dwivedi, R.D., Singh, M., Viladkar, M.N. & Goel, R.K. (2013) Prediction of tunnel deformation in squeezing grounds. *Eng Geol*,161, 55–64.

Ehrbar H. & Schoch Keller S. (2006) Geologische Risiken und Massnahmenplanung am Beispiel des Teilabschnitts Sedrun, In: Löw, S. (ed.) *Geologie und Geotechnik der Basistunnels am Gotthard und Lötschberg, Proceedings of GEAT05, 26–28 September, 2005, ETH, Zürich.* In English.

Einstein, H.H. (1996) Risk and risk analysis in rock engineering. *Tunn Undergr Sp Tech*, 11, 141–155.

Einstein, H.H., Dudt, J.-P., Halabe, V.B. & Descoeudres, F. (1992) *Decision Aids in Tunneling; Principle and Practical Application.* Monograph, prepared for the Swiss Federal Office of Transportation.

Einstein, H. H., Chiaverio, F. & Köppel, U. (1994) Risk analysis for the Adler Tunnel. *Tunnels and Tunnelling*, 26 (11) 28–30.

Einstein H.H., Sousa, R.L., Karam, K., Manzella, I. & Kveldsvik, V. (2010) Rock slopes from mechanics to decision making. In: Zhao, J. Labiouse, V. Dudt, J.P. & Mathier, J.F. (eds.) *Rock Mechanics in Civil and Environmental Engineering: Proceedings European Rock*

Mechanics Symposium, EUROCK 2010, 15–18 June 2010, EPFL Lausanne, Switzerland. Leiden: CRC Press/Balkema, pp. 3–13.

Eng, T., Hudson, J.A., Stephansson, O., Skagius, K. & Wiborgh, M. (1994) *Scenario development methodologies.* Swedish Nuclear Fuel and Waste Management, Stockholm, Sweden. SKB Technical Report 94–28.

Environ. Geol. (2009) Tsang, C.-F. (Guest ed.) Special Issue on the DECOVALEX THMC Project (Safety Assessment of Nuclear Waste Repositories). *Environ Geol,* 57, 1217–1390. ISSN: 0943-0105 (Print) 1866–6299 (Online)

Eskesen, S.D., Tengborg, P., Kampmann, J. & Veicherts, T.H. (2004) Guidelines for tunneling risk management: International Tunneling Association working group No.2. *Tunn Undergr Sp Tech,* 19, 217–237.

Evans, K.F., Engelder, T. & Plumb, R.A. (1989) Appalachian Stress Study: 1. A detailed description of *in situ* stress variations in Devonian shales of the Appalachian Plateau. *J Geophys Res-Sol Ea, (1978–2012),* 94 (B6), 7129–7154. Published online: 20 Sep 2012—DOI: 10.1029/JB094iB06p07129.

Evans, K.F., Oertel, G. & Engelder, T. (1989) Appalachian Stress Study: 2. Analysis of Devonian shale core: Some implications for the nature of contemporary stress variations and Alleghanian Deformation in Devonian rocks. *J Geophys Res-Sol Ea, (1978–2012),* 94 (B6), 7155–7170. Published online: 20 Sep 2012—DOI: 10.1029/JB094iB06p07155.

Fan, Q.X. & Wang, Y.F. (2011) A case study of rock mass engineering of underground powerhouse at Xiluodu hydropower station. *Chinese J Rock Mech Eng,* 30, 2986–2993.

Faramarzi, F., Ebrahimi Farsangi, M. & Mansouri, H. (2012) An RES-based model for risk assessment and prediction of backbreak in bench blasting. *Rock Mech Rock Eng,* Online August 2012: 1–11.

Feng, W.K., Huang, R.Q. & Li, T.B. (2012) Deformation analysis of a soft-hard rock contact zone surrounding a tunnel. *Tunn Undergr Sp Tech,* 32, 190–197.

Feng, X.T. (2000) *Introduction of Intelligent Rock Mechanics.* Beijing, China: Science Press.

Feng, X.T. & Hudson, J.A. (2003) The ways ahead for rock engineering design methodologies. *Int J Rock Mech Min,* 41, 255–273.

Feng, X.T. & Hudson, J.A. (2010) Specifying the required information for rock engineering modelling and design. *Int J Rock Mech Min Sci* 47, 179–194.

Feng, X.T. & Hudson, J.A. (2011) *Rock Engineering Design.* London: CRC Press, Taylor & Francis.

Feng, X.T. & Wang, L.N. (1994) Rock-burst prediction based on neural networks. *Trans Nonferrous Met Soc China,* 4, 7–14.

Feng, X.T., Chen, B.R., Ming, H.J., *et al.* (2012) Evolution law and mechanism of rockbursts at deep tunnels: immediate rockburst. *Chinese J Rock Mech Eng,* 31, 433–444. (in Chinese)

Feng, X.T., Zhang, C.Q., Li, S.J., Qiu, S.L., *et al.* (2013) *Dynamic design method for deep tunnels in hard rock.* Beijing, China: Science Press.

Feng, Y. (1990) Research on the tunnel support of soft rock in China. *Ground Pressure and Strata Control,* 7, 1–5. (in Chinese)

Ferentinou, M.D. & Sakellariou, M.G. (2007) Computational intelligence tools for the prediction of slope performance. *Comput Geotech,* 34, 362–384.

Ferentinou, M.D., Hasiotis, T. & Sakellariou, M.G. (2012). Application of computational intelligence tools for the analysis of marine geotechnical properties in the head of Zakynthos canyon, Greece. *Comput Geosci,* 40, 166–174.

Franciss, F.O. (1997) *Weak rock tunnelling.* Boston: AA Balkema Press.

Frough, O. & Torabi, S.R. (2013) An application of rock engineering systems for estimating TBM downtimes. *Eng Geol,* 157, 112–123.

Gao, Z.Z. (2009) The application of the beam geological advance prediction system for the TBM construction of the Jinping diversion tunnel. *Railway Construction Technology.* 11, 65–67.

Gay, N.C. & Ortlepp, W.D. (1979) Anatomy of a mining-induced fault zone. *Geol Soc Am Bull, Part I*, 90, 47–58.

Ge X. & Hou, M. (2011) A new 3D in-situ rock stress measuring method: borehole wall stress relief method (BWSRM) and development of geostress measuring instrument based on BWSRM and its primary applications to engineering. *Chinese J Rock Mech Eng*, 30, (11) 2161–2180.

Ge, Y.H., Ye, Z.H., Li, S.C., Lu, W. & Zhang, Q.S. (2009) Risk measurement and control of water inrush into Qiyue mountain tunnel. In: Shi, Y. *et al.* (eds.) *Cutting-Edge Research Topics on Multiple Criteria Decision Making: Proceedings 20th International Conference, MCDM 2009, June 21–26 2009, Chengdu/Jiuzhaigou, China,*. Berlin: Springer-Verlag. 35, pp. 523–530.

Geo, H.K. (2012) *Catalogue of Notable Tunnel Failure Case Histories*. Geotechnical Engineering Office, Hong Kong . Available from: http://www.cedd.gov.hk/eng/publications/geo/doc/HK%20NotableTunnel%20Cat.pdf.

Gioda, G. (1982) On the non-linear 'squeezing' effects around circular tunnels. *Int J Num Anal Meth Geomech*, 6, 21–46.

Gioda, G. & Cividini, A. (1996) Numerical methods for the analysis of tunnel performance in squeezing rocks. *Rock Mech Rock Eng*, 29, 171–193.

Glcick, J. (1987) *Chaos: Making a New Science*. Viking Penguin. (ISBN: 0670811785)

Goel, R.K., Jethwa, J.L. & Paithankar, A.G. (1995a) An empirical approach for predicting ground conditions for tunnelling and its practical benefits. In: Daemen, J.J.K. & Schulz, R.A. (eds.) *Proc. 35th US Symposium on Rock Mechanics,1995, Univ Reno, Nevada, USA*, pp.431–435.

Goel, R.K., Jethwa, J.L. & Paithankar, A.G. (1995b). Indian experiences with Q and RMR systems. *Tunn Undergr Sp Tech*, 10, 97–109.

Gong, F.Q., Li, X.B. (2007) A distance discriminant analysis method for prediction of possibility and classification of rockburst and its application. *Chinese J Rock Mech Eng*, 26, 1013–1018. (in Chinese)

González, Nicieza C., Álvarez, Vigil A.E., Menéndez, Diaz A., González, Palacio C. (2008) Influence of the depth and shape of a tunnel in the application of the convergence–confinement method. *Tunn Undergr Sp Tech*, 23, 25–37.

Goricki, A., Schubert, W., Steidl, A. & Vigl, L. (2002) Geotechnical risk assessment as the basis for cost estimates in tunnelling. *Felsbau* 20, (5) 24–40.

Goricki, A., Schubert, W. & Riedmueller, G. (2004) New developments for the design and construction of tunnels in complex rock mass. *Int J Rock Mech Min*, 41, 1–6.

Goricki, A. Button, E.A., Schubert, W., Poetsch, M. & Leitner, R. (2005) The influence of discontinuity orientation on the behaviour of tunnels. *Felsbau* 23/5, VGE, 12–18.

Graziani, A., Boldini, D. (2012) Influence of hydro-mechanical coupling on tunnel response in clays. *J Geotech Geoenviron*, 138, 415–418.

Gu, M.C. (2001) Research on rockburst in Qinling railway tunnel. *Research on Water Resources and Hydropower* 3/4, 19–26. (in Chinese)

Gu, Y.C., Zheng, D.J., Guo, H.Z. & He, X.F. (2008) Regression analysis of 3D initial geostress field for dam site of Xiaowan hydropower station. *Rock Soil Mech*. 29, 1015–1020. (in Chinese)

Guan, B. (2003) *Key Points In Tunnel Designing*. Beijing: China Communications Press. (in Chinese)

Guan, B. & Zhao, Y. (2011) *Construction Technology in Tunnels with Soft and Weak Surrounding Rocks*. Beijing: China Communications Press. (in Chinese)

Guntli, P., Weber, A., (2009) Einspurtunnel Ost km 118.835–120.039, Einspurtunnel West km 218.855–220.062 Tavetscher Zwischenmassiv Nord und ClavanievZone, *Geologischer Schlussbericht, Geologie, Geotechnik, Hydrogeologie*, Sieber Cassina + Handke AG,

Bericht Nr. GR 360B-2, 26 Juni 2009. [Railway tracks, *Geological Survey Final Report*, Report No. GR 360B-2, Sieber Cassina + Handke AG, 26 June 2009.] (in German)

Gunzberger, Y. & Cornet, F.H. (2007) Rheological characterization of a sedimentary formation from a stress profile inversion. *Geophys J Int*, 168, 402–418.

Guo, J. & Du, X. P. (2007) Sensitivity analysis with mixture of epistemic and aleatory uncertainties. *AIAA J*, 45, 2337–2349.

Hagros, A., Johansson, E., & Hudson, J.A. (2007) *Time dependency in the mechanical properties of crystalline rocks: a literature survey*. Posiva Oy, Finland. Working Report 2008-68.

Hakala, M., Hudson, J.A., Harrison, J.P. & Johansson, E. (2008). *Assessment of the potential for rock spalling at the Olkiluoto site*. Posiva Oy, Finland. Posiva Working Report 2008-83. (Download available from www.posiva.fi)

Han, R.G. (1987) *New Austrian tunnelling method in underground engineering*. Beijing: Science Press. (in Chinese)

Han, R.R., Zhang, J.H., Zhang, X., Jiang, F., Li J., Zhang, E.B. & Hou, P. (2009) Study of prediction on feedback calculation of excavated monitor to the underground powerhouse of Xiluodu. *Journal of Shandong University (Engineering Science)*. 39, 140–144.

Harrison, J.P. (2002) Pragmatism, fracture geometry, and rock engineering. In: *Stiftelsen Svensk Bergteknisk Forskning: Bergmekanikdag*. Stockholm, Sweden. pp. 1–18.

Harrison, J.P. (2012) Rock engineering, uncertainty and Eurocode 7: implications for rock mass characterisation. In: *Proc. XIV Ciclo di Conferenze di Meccanica e Ingegneria delle Rocce: Nuovi Metodi di Indagine Monitoraggio e Modellazione degli Ammassi Rocciosi., 21–22 Nov. 2012, Torino. Italy*.

Harrison J.P. (2014) Eurocode 7 and rock engineering: current problems and future opportunities. In: Alejano, L.R., Perucho, A., Olalla, C. & Jiménez, R. (eds.) *Rock engineering and rock mechanics: structures in and on rock: Proceedings, ISRM European Regional Symposium Eurock 2014, 26–28 May 2014, Vigo, Spain*. Leiden: CRC Press/Balkema. p.1531e6.

Harrison, J.P. & Hudson, J.A. (2000) *Engineering Rock Mechanics: Illustrative Worked Examples*. Oxford: Elsevier. (Also published in Chinese by Science Press of Beijing, 2009.)

Harrison, J.P. & Hudson, J.A. (2006) Comprehensive hazard identification in rock engineering using interaction matrix mechanism pathways. In: *50 Years of Rock Mechanics—Landmarks and Future Challenges: Proc. 41st U.S. Symposium on Rock Mechanics (USRMS), Golden Rocks 2006, 17–21 June 2006, Golden, CO., USA*. Alexandria, Virginia, USA: American Rock Mechanics Association.

Harrison, J.P. & Hudson J.A. (2009) Incorporating parameter variability in rock mechanics analyses: fuzzy mathematics applied to underground rock spalling. *Rock Mech Rock Eng*, 43, 219–224.

Harrison, J.P., Hudson J.A. & Carter J.N.C. (2007). Is there a relation between the *in situ* principal stress magnitudes in rock masses? In: *Proc. 1st Canadian-US Rock Mechanics Symposium*, Vancouver, Canada.

He, J.D., Xie, H.Q., Wang, Q.Z. & Xiao, M.L. (2009) Inversion analysis of initial geostress in dam site of Guandi hydropower project. *J Geotech Eng*, 31, 166–171. (in Chinese)

He, M.C., Jing, H.H.& Sun, X.M. (2002) *Engineering Mechanics of Soft Rock*. Beijing: Science Press.

He, M.C., Yang, J. & Qi, G. (2007) Optimized design and its application of coupling support for soft rock roadway at great depth. *J Liaoning Tech Univ*, 26, 40–42.

Heidbach, O., Tingay, M., Barth, A., Reinecker, J., Kurfess, D. & Müller, B. (2008) *The World Stress Map*. database release 2008, doi:10.1594/GFZ.WSM.Rel2008, 2008.

Hijazo Ramiro, T. & González de Vallejo, L.I. (2012) In-situ stress amplification due to geological factors in tunnels: the case of Pajares tunnels, Spain. *Eng Geol*, 137–138, 13–20. ISSN 0013-7952.

Hill, S.E. & Rosenbaum, M.S. (1998) Assessing the significant factors in a rock weathering system. *Q J Eng Geol*, 31, 85–94.

Hisatake, M. (2003) Effects of steel fiber reinforced high-strength shotcrete in a squeezing tunnel. *Tunn Undergr Sp Tech*, 18, 197–204.

Hoek, E. (1990) Estimating Mohr–Coulomb friction and cohesion values from the Hoek–Brown failure criterion. *Int J Rock Mech Min* 27, 227–229.

Hoek, E. (1991) When is a design in rock engineering acceptable? In: Wittke, W. (ed.) *Proceedings of the 7th Congress, International Society for Rock Mechanics, September 1991, Aachen, Germany*. Rotterdam: A.A. Balkema. Vol. 3, pp. 1485–1497.

Hoek, E. (2001) Big tunnels in bad rock. *J Geotech Geoenviron*, 127, 726–740.

Hoek, E. & Brown, E.T. (1980) *Underground Excavations in Rock*. London: Institution of Mining and Metallurgy.

Hoek, E. & Guevara, R. (2009) Overcoming squeezing in the Yacambú-Quibor tunnel, Venezuela. *Rock Mech Rock Eng*, 42, 389–418.

Hoek, E. & Marinos, P. (2000) Predicting tunnel squeezing problems in weak heterogeneous rock masses. Parts 1 and 2. *Tunnels Tunnell Int*. 32, 34–36 and 45–51.

Hoek, E. & Marinos, P.G. (2009) Tunnelling in overstressed rock In: Vrkljan, I. (ed.), *Rock Engineering in Difficult Rock Conditions-Soft Rocks and Karst: Proceedings of ISRM Regional Symposium Eurock2009, 29–31 October 2009, Dubrovnik, Croatia*. London: Taylor & Francis Group. pp. 49–60.

Hoek, E., Kaiser, P.K., Bawden, W.F. (1995) *Support of underground excavations in hard rock*. Rotterdam: Balkema.

Hoek, E., Carranza-Torres, C., Diederichs, M.S., & Corkum, B. (2008) Integration of geotechnical and structural design in tunneling. In: *Proceedings University of Minnesota 56th Annual Geotechnical Engineering Conference, 29 February 2008, Minneapolis*. pp. 1–53.

Hou, D.Q., Feng, M., Liao, C.G., Xin, X.L. & Zang, H.Y. (2012) Engineering practice of anchor cables supporting to large-scale underground caverns. *Chinese J Rock Mech Eng*, 31, 963–972.

Hsiao, F.Y., Wang, C.L. & Chern, J.C. (2009) Numerical simulation of rock deformation for support design in tunnel intersection area. *Tunn Undergr Sp Tech*, 24, 14–21.

Hu, B., Feng, X.T., Huang, X.H. (2005) Regression analysis of initial geostress field for left bank high slope region at Longtan hydropower station. *J Rock Mech Eng*, 24, 4055–4064. (in Chinese)

Huadong Engineering Corporation Limited of Hydrochina Corporation. (2005) *Feasibility Study Report of Jinping II Hydropower Station along Yalong River—Engineering Geology (second volume)*. Hangzhou, China.

Huang, D. (2007) *Study on Unloading Deformation Mechanism and Stability of Excavating Surrounding Rock Mass of Large Underground Caverns*. PhD thesis of Chengdu University of Technology, Chengdu, China.

Huang, D., Huang, R.Q. & Zhang, Y.X. (2009) Analysis on influence of fault location and strength on deformation and stress distribution of surrounding rocks of large underground openings. *J Civil Arch Environ Eng*, 31, 68–73.

Huang H. W. (2006) State-of-the-art of the research on risk management in construction of tunnel and underground works. *Chinese J Undergr Sp Eng*, 2 (1), 13–20. (in Chinese)

Huang, Q.X., Deng, J.H., Su, P.Y., Wang, D.K. & You, P. (2011) Displacement characteristics analysis of surrounding rock in underground powerhouse chambers at Pubugou hydropower station during construction. *Chinese J Rock Mech Eng*, 30, 3032–3042.

Huang, R.Q., Huang, D., Duan, S.H., & Wu, Q. (2011) Geomechanics mechanism and characteristics of surrounding rock mass deformation failure in construction phase for underground powerhouse of Jinping I hydropower station. *Chinese J Rock Mech Eng* 30, 23–35.

Huang, R.Q., Huang, J., Ju, N.P. & Li, Y.R. (2013) Automated tunnel rock classification using rock engineering systems. *Eng Geol*, 156, 20–27.

Huang, S.L., Feng, X.T., Zhang, C.Q. (2008) A new generalized polyaxial strain energy strength criterion of brittle rock and examination of polyaxial tests. *Chinese J Rock Mech Eng*, 27, 124–134.

Hudson, J.A. (1989). *Rock Mechanics Principles in Engineering Practice*. London: Butterworths.

Hudson, J.A. (1991) Atlas of rock engineering mechanisms. Part 1—Underground excavations. *Int J Rock Mech Min*, 28, 523–526.

Hudson, J.A. (1992) Atlas of rock engineering mechanisms. Part 2—Slopes. *Int J Rock Mech Min*, 29, 157–159.

Hudson, J.A. (1992) *Rock Engineering Systems: Theory & Practice*. New York: Ellis Horwood.

Hudson, J.A. (2010) Stresses in rock masses: a review of key points. In: Vrkljan, I. (ed.), *Rock Engineering in Difficult Rock Conditions-Soft Rocks and Karst: Proceedings of ISRM Regional Symposium Eurock2009, 29–31 October 2009, Dubrovnik, Croatia*. London: Taylor & Francis Group. pp. 61.

Hudson, J.A. & Cosgrove, J.W. (1997) Integrated structural geology and engineering rock mechanics approach to site characterisation. *Int J Rock Mech Min*, 34, 3–4, paper 136.

Hudson, J.A, & Feng, X.T. (2010). Technical auditing of rock mechanics modelling and rock engineering design. *Int J Rock Mech Min*, 47, 877–886.

Hudson, J.A. & Feng, X.T. (2010). Variability of *in situ* rock stress. In: *Proc. 5th Int. Symp*. In Situ *Rock Stress, August, 2010, Beijing, China*.

Hudson, J.A. & Harrison, J.P. (1992) A new approach to studying complete rock engineering problems. *Q J Eng Geol*, 25, 93–105.

Hudson, J.A. & Harrison, J.P. (1997) *Engineering Rock Mechanics: An Introduction to the Principles*. Oxford: Elsevier Ltd.

Hudson, J.A. & Hudson, J.L. (1993) Establishing the potential behavioural modes of rock engineering systems by computer simulation of interaction matrix energy flux. *Int J Rock Mech Min*, 30, 457–468.

Hudson, J.A. & Lamas, L. (eds.) (2012). *ISRM 50th Year Anniversary Commemorative Book— 1962–2012*. Portugal: International Society for Rock Mechanics (ISRM).

Hudson, J.A., Stephansson, O., Andersson, J., Tsang, C.-F. & Jing, L. (2001) Coupled T–H–M issues relating to radioactive waste repository design and performance. *Int J Rock Mech Min*, 38, 143–161.

Hudson, J.A., Cornet, F.H. & Christiansson, R. (2003) ISRM suggested methods for rock stress estimation—Part 1: Strategy for rock stress estimation. *Int J Rock Mech Min*, 40, 991–998.

Hudson, J.A., Stephansson, O. & Andersson, J. (2005) Guidance on numerical modelling of thermo-hydro-mechanical coupled processes for performance assessment of radioactive waste repositories. *Int J Rock Mech Min*, 42, 850–870.

Hudson, J.A., Cosgrove, J.W. & Johansson, E. (2008) *Estimating the mechanical properties of the brittle deformation zones at Olkiluoto*. Posiva Oy, Finland. Posiva Working Report 2008-67. Downloadable from www.posiva.fi.

Hudson, J.A., Bäckström, A., Rutqvist, J., Jing, L., Backers, T., Chijimatsu, M., Christiansson, R., Feng, X-T., Kobayashi, A., Koyama, T., Lee, H-S., Neretnieks, I., Pan, P. Z., Rinne, M. & Shen, B.T. (2009) Characterising and modelling the excavation damaged zone in crystalline rock in the context of radioactive waste disposal. *Journal of Engineering Geology*, 57, 1275–1297.

Isaksson, T. & Stille, H. (2005) Model for estimation of time and cost for tunnel projects based on risk evaluation. *Rock Mech Rock Eng*, 38, 373–398.

ISO/GUIDE (2009), *ISO/GUIDE 73:2009-11, Risk management vocabulary*. Vernier, Geneva Switzerland: International Organization for Standardization.

Itasca (2010) 3DEC. http://www.itascacg.com/3dec/overview.php.

Ivars, D.M. & Hakami, H. (2005) *Effect of a sub-horizontal fracture zone and rock mass heterogeneity on the stress field in Forsmark area—A numerical study using 3DEC*. Stockholm: Svensk Kärnbränslehantering AB, SKB Report 05-59.

Jafar, K.H., Kourosh, S., Bahram, R, Jamal, R. & Hadi, B. (2010) Risk assessment based selection of rock TBM for adverse geological conditions using Fuzzy-AHP. *B Eng Geol Environ*, 69, 523–532.

Jethwa, J.L., Singh, B. & Singh, B. (1984) Estimation of ultimate rock pressure for tunnel linings under squeezing rock conditions—a new approach. In: Brown, E.T. & Hudson, J.A. (eds.) *Design and performance of underground excavations in rock: Proceedings of the Symposium of the International Society for Rock Mechanics, 3–4 Sept 1984, Cambridge, UK*. London: British Geotechnical Society. pp. 231–238.

Jiang, J. (2006) Mechanism and countermeasures of water-bursting in railway tunnel engineering. *China Railway Science*, 27, 76–82. (in Chinese)

Jiang, Q. (2010) *Study on model and stability of surrounding rock of large underground caverns under high geo-stress condition*. PhD thesis of Institute of Rock & Soil Mechanics, Chinese Academy of Sciences, Wuhan, China.

Jiang, Q., Feng, X.T., Chen, G.Q. (2008) Study on constitutive model of hard rock considering surrounding rock deterioration under high geostresses. *Chinese J Rock Mech Eng*, 27, 144–152.

Jiang, Q., Feng, X.T., Su, G.S., & Chen, G.Q. (2010) Stability analysis of large underground caverns in Laxiwa hydropower plant under high crustal stress. *J Hydro Eng*, 29, 132–140.

Jiang, Q., Feng, X.T., Xiang, T.B. & Su, G.S. (2010) Rockburst characteristics and numerical simulation based on a new energy index: a case study of a tunnel at 2,500 m depth. *B Eng Geol Environ*, 69, 381–388.

Jiang, Q., Chen, J.L., Feng, X.T., Chen, J., Cui, J., Wan, X.B. & Hou, J. (2013) Failure format and interactive mechanism of pre-stressed thru-anchor cable in a large underground caverns. *Rock Soil Mech*, 34, 2271–2279.

Jiang, Y.Z., Xu, W.Y., Wang, R.H. & Wang, W. (2008) Numerical analysis of the long-term stability of large-scale underground cavern of hydropower station. *Rock Soil Mech*, 29, 52–58.

Jiao, Y. & Hudson, J.A. (1995) The fully-coupled model for Rock Engineering Systems. In: Stephansson, O. (ed.) Special Issue: Thermal-Hydro-mechanical Coupling in Rock Mechanics. *Int J Rock Mech Min*, 32, (5) 491–512.

Jiao, Y.Y., Zhang, X.L., Zhao J. & Liu Q.S. (2007) Viscous boundary of DDA for modeling stress wave propagation in jointed rock. *Int J Rock Mech Min*, 44, 1070–1076.

Jimenez, R. & Recio, D. (2011) A linear classifier for probabilistic prediction of squeezing conditions in Himalayan tunnels. *Eng Geol*, 121, 101–109.

Jing, L. (2003) A review of techniques, advances and outstanding issues in numerical modelling for rock mechanics and rock engineering. *Int J Rock Mech Min*, 40, 283–353.

Kaiser, P.K. & Cai, M. (2012) Design of rock support system under rockburst condition. *J Rock Mech Geotech Eng*, 4, 215–227.

Kaiser, P.K., Tannant, D.D. & McCreath, D.R. (1996) *Canadian Rockburst Support Handbook*. Sudbury Ontario: Geomechanics Research Centre, Laurentian University.

Kaiser, P.K., Diederichs, M.S., Martin, C.D., Sharp, J. & Steiner, W. (2000) Underground works in hard rock tunnelling and mining. In: *Proceedings GeoEng 2000, 19–24 November 2000, Melbourne, Australia*. Melbourne: Technomic Publishing Co. pp. 841–926.

Kampmann, J., Eskesen, S.D. & Summers, J.W. (1998) Risk assessment helps select the contractor for the Copenhagen Metro System. In: Ferreira, A.A. & Negro, A. (eds.) *Tunnels and*

Metropolises: Proceedings of the World Tunnel Congress '98, 25–30 April 1998, Sao Paulo, Brazil. pp. 123–128.

Kawatani, T., Tezuka, H., Morita, R. & Shimaya, S. (1999) Tunnel construction with a large-scale TBM in a collapse-prone poor rock. In: *Challenges for the 21st century, Proceedings ITA World Tunnel Congress '99, June 1999 Oslo.* Rotterdam: A.A. Balkema. Vol. 2, pp. 895–901.

Khaled M.A. & Khaled, H. (2002) Rock mass characterization to indicate slope instability at Bandarban, Bangladesh; a rock engineering systems approach. *Environ Eng Geosci,* 8, 105–119.

KhaloKakaie, R. & Zare Naghadehi, M. (2012) Ranking the rock slope instability potential using the interaction matrix (IM) technique; a case study in Iran. *Arab J Geosci 5,* 263–273, DOI: 10.1007/s12517-010-0150-1.

Khanlari, G., Meybodi, R.G. & Mokhtari, E. (2012) Engineering geological study of the second part of water supply Karaj to Tehran tunnel with emphasis on squeezing problems. *Eng Geol,* 145–146, 9–17.

Kidybinski, A. & Dubinski, J. (1990) *Strata Control in Deep Mines.* Rotterdam: A.A. Balkema Press.

Kim, M.-K., Yoo, Y.-I., Jae-Joon Song, J.-J. (2008). Methodology to quantify rock behavior around shallow tunnels by Rock Engineering Systems. *Geosystem Engineering,* 11, 2, 37–42.

Kimura, F., Okabayashi, N. & Kawamoto, T. (1987) Tunnelling through squeezing rock in two large fault zones of the Enasan Tunnel II. *Rock Mech Rock Eng,* 20, 151–166.

Kishida, K., Hosoda, T. & Tomita, A. (2007) Long term tunnel stability in soft rock considering the influence of permeability using a coupled analysis. In: Eberhardt, E., Stead, D. & Morrison, T. (eds.) *Meeting Society's Challenges and Demands: Proceedings of the 1st Canada-US Rock Mechanics Symposium, 27–31 May 2007, Vancouver, Canada.* London: Taylor & Francis Group. pp. 419–425.

Kong, W.K. (2011) Water ingress assessment for rock tunnels: a tool for risk planning. *Rock Mech Rock Eng 44,* 755–765.

Kovári, K. (2005) Method and device for stabilizing a cavity excavated in underground construction. *US Patent Appl. 200550191138.*

Kovari, K. & Staus, J. (1996) Basic considerations on tunnelling in squeezing ground. *Rock Mech Rock Eng,* 29, 203–210.

Ladanyi B. (1974) Use of long-term concept in determination of ground pressure on tunnel linings. In: Advances *in Rock Mechanics, Proc. 3rd Congress Int. Soc. Rock Mech., 1–7 September 1974, Denver, Colorado, USA.* Washington DC: National Academy of Sciences. Vol. II, Part B, pp. 1150–1156.

Latham J-P & Lu Ping (1999) Development of an assessment system for the blastability of rock masses. *Int J Rock Mech Min,* 36, 41–55.

Laubscher, D.H. (1990) A geomechanics classification system for the rating of rock mass in mine design. *J S Afr I Min Metall,* 90, 257–273.

Lee, M.F., Mollison, L.J., Mikula, P. & Pascoe, M. (2006). *In situ* rock stress measurements in Western Australia's Yilgarn Craton. In: Lu, M., Li, C.C., Kjørholt, H. & Dahle, H. (eds.), In situ *Rock Stress, Proc. Int. Symp. 19–21 June 2006, Trondheim, Norway.* London: Taylor & Francis Group. pp. 35–38.

Lemaitre, J. & Chaboche, J.L. (1996) Mécanique des matériaux solides. *Dunod,* Paris: Wiley. pp. 253–341.

Li, B.Y. (1999) "Down Three Zones" in the prediction of the water inrush from coal bed floor aquifer theory, development and application. *Journal of Shandong Mining Institute.* 8, 11–18. (in Chinese)

Li, G.F., He, M.C., Zhang, G.F. & Tao, Z.G. (2010) Deformation mechanism and excavation process of large span intersection within deep soft rock roadway. *Mining Science and Technology*. 20, 28–34.

Li, G.L. & Wu, S.H. (2011) Analysis on phased safety monitoring of underground power house in construction period. *Yangtze River*. 42, 59–63.

Li, J.H., Wu, W.F. & Li, J.C. (2013) Control and monitoring of rock mass engineering of super large underground powerhouse cavern groups at Xiluodu hydropower station. *Chinese J Rock Mech Eng*, 32, 8–14.

Li, N., Sun, H.C., Yao, X.C. & Shi, G.B. (2008) Cause analysis of circumferential splits in surrounding rock of busbar tunnels in underground powerhouses and reinforced measures. *Chinese J Rock Mech Eng*, 27, 439–446.

Li, S.C., Xue, Y.G., Zhang, Q.S., *et al.* (2008) Key technology study on comprehensive prediction and early-warning of geological hazards during tunnel construction in high-risk Karst areas. *Chinese J Rock Mech Eng*, 27, 1297–1307.

Li, S.J., Feng, X.T., Zhang, C.S., *et al.* (2010) Testing on formation and evolution of TBM excavation damaged zone in deep-buried tunnel based on digital panoramic borehole camera technique. *Chinese J Rock Mech Eng*, 29, 1106–1112. (in Chinese)

Li, S.J., Feng, X.T., Li, Z.H., *et al.* (2011) In-situ experiments on width and evolution characteristics of excavation damaged zone in deeply buried tunnels. *Science China Technology Sciences*, 54, 167–174.

Li, S.J., Feng, X.T. & Li, Z.H. (2012) Evolution of fractures in the excavation damaged zone of a deeply buried tunnel during TBM construction. *Int J Rock Mech Min*, 55, 125–138.

Li, S.J., Feng, X.T., Li, Z.H., *et al.* (2012) *In situ* monitoring of rockburst nucleation and evolution in the deeply buried tunnels of Jinping II hydropower station. *Eng Geol*, 137–138, 85–96.

Li, X. & Li, Y. (2014) Research on risk assessment system for water inrush in the karst tunnel construction based on GIS: Case study on the diversion tunnel groups of the Jinping II Hydropower Station. *Tunn Undergr Sp Tech*, 40, 182–191.

Li, Y. (2009) *Study on the stability induced by stepped excavations and rheological effect of underground cavern groups in high in situ stress areas*. PhD thesis of Shandong University, Jinan, China.

Li, Z.G. (1997) Study on the surrounding rock damage of Ertan hydropower station underground cavern group. *Water Power*, 50–53.

Li, Z.H., Li, S.J., Feng, X.T., Zhou, H., Chen, B.R. & Zhang, C.Q. (2011) Characteristics and formation mechanism of core discing in deep rock mass. *Chinese J Rock Mech Eng*, 30, 2254–2266.

Li, Z.K., Zhou, Z., Tang, X.F., Liao, C.G., Hou, D.Q., Xing, X.L., Zhang, Z.Z., Liu, Z.G. & Chen, Q.H. (2009) Stability analysis and considerations of underground powerhouse caverns group of Jinping I hydropower station. *Chinese J Rock Mech Eng*, 28, 2167–2175.

Li, Z.K., Zhou, Z., Xu, Q.J., Liao, C.G., Yin, H.L., Zhang, Z.Z., Cheng, L.J., Liu, Z.G. & Guo, K. (2010) Implementation and development of AI back analysis for the underground powerhouse of Jinping-I hydropower project. *J Hydro Eng*, 29, 177–183.

Likhitruangslip, V. & Ioannou, P.G. (2004) Risk-sensitive decision support system for tunnel construction. In: Yegian, M.K. & Kavazanjian, E. (eds.) Geotechnical engineering for transportation projects: Proceedings of Geo-Trans 2004, 27–31 July 2004, Los Angeles, California, USA. Geotechnical special publication, No.126. Reston, Virginia: American Society of Civil Engineers.

Lin, H.I. & Lee, C.H. (2009) An approach to assessing the hydraulic conductivity disturbance in fractured rocks around the Syueshan tunnel, Taiwan. *Tunn Undergr Sp Tech*, 24, 222–230.

Liu, N., Zhu, W.S., Xin, X.L. (2008) Back regression analysis on the initial geostress field of the Shuangjiangkou hydropower station. *J Shandong U (EngSci)*, 38, 121–126. (in Chinese)

Liu, N., Zhang, C.S. & Chu, W.J. (2011) Simulating time-dependent failure of deep marble with particle flow code. *Chinese J Rock Mech Eng* 30, 1989–1996. (in Chinese)

Liu, N., Zhang, C.S., Wu, X.M., *et al.* (2013) Time-dependent failure of deep-buried marble in Jinping II hydropower station. *Advances in Science and Technology of Water Resources.* 33, 63–67. (in Chinese)

Liu, Q.S. & Zhang, H. (2003) Study on stability and support of rock masses surrounding deep coal-mine roadway. *Chinese J Rock Mech Eng*, 22, 2195–2220.

Ljunggren, C., Chang, Y., Janson, T. & Christiansson, R. (2003) An overview of rockstress measurement methods. *Int J Rock Mech Min*, 40, 975–989.

Lombardi, G. (1973) Dimensioning of tunnel linings with regard to construction Procedure. *Tunnels and Tunneling.* 5, 340–435.

Lombardi, G. & Panciera, A. (1997) Problems with TBM & linings in squeezing ground. *Tunnels and Tunneling*, 29, 54–56.

Lu, B., Wang, J.M., Ding, X.L., Wu, A.Q., Duan, S.H. & Huang, S.L. (2010) Study of deformation and cracking mechanism of surrounding rock of Jinping I underground powerhouse. *Chinese J Rock Mech Eng*, 29, 2429–2441.

Lu, P. & Latham J-P. (1994) A continuous quantitative coding approach to the interaction matrix in rock engineering systems based on grey systems approaches. In: *Proc. 7th Int. Cong. of IAEG, 5–9 Sep 1994, Lisbon, Portugal.* Rotterdam: A.A. Balkema. pp. 4761–4770.

Lu, W.B., Geng, X., Chen, M., Shu, D.Q. & Zhou, C.B. (2011) Study of selection of excavation procedure and contour blasting method for deep underground powerhouse. *Chinese J Rock Mech Eng*, 30, 110–115.

Lu, Y. (2009) *Karst in China*. Higher Education Press, China.

Lunardi, P. (2000) The design and construction of tunnels using the approach based on the analysis of controlled deformation in rocks and soils. *Tunnels and Tunnelling Int., special supplement, ADECO—RS Approach,* May.

Malan, D.F. (1998) *An investigation into the identification and modelling of time dependent behaviour of deep level excavations in hard rock.* PhD thesis, University of the Witwatersrand, Johannesburg, South Africa.

Malan, D.F. & Spottiswoode, S.M. (1997) Time-dependent fracture zone behaviour and seismicity surrounding deep level stoping operations. In: Gibowicz, S.J. & Lasocki, S. (eds.) Proc. *4th Int. Symp. Rockbursts & Seismicity in Mines, August 1997, Krakow, Poland.* Rotterdam: A.A. Balkema. pp. 173–177.

Martin, C.D., Tannant, D.D., Yazici, S. & Kaiser, P.K. (1999) Stress path and instability around mine openings. In: *Vouille, G. & Berest, P. (eds.) Proceeding of the 9th ISRM Congress on Rock Mechanics, Paris, 1999.* Rotterdam: A.A. Balkema. Vol. 1, pp. 311–315.

Masoud Sadeghi & Rasouli Vamegh. (2011) Application of rock engineering systems in evaluation of stability of underground excavations. *Q Amirkabir J Civil Environ Eng*, 43 (1), 89–95.

Masset, O. & Loew, S. (2010), Hydraulic conductivity distribution in crystalline rocks, derived from inflows to tunnels and galleries in the Central Alps, Switzerland. *Hydrogeology Journal*, 18, 863–891.

Mavroulidou, M., Hughes, S.J. & Hellawell, E.E. (2004) A qualitative tool combining an interaction matrix and a GIS, to map vulnerability to traffic induced air pollution. *J Environ Manage*, 70, 283–289.

Mavroulidou, M., Hughes, S.J. & Hellawell, E.E. (2007) Developing the interaction matrix technique as a tool assessing the impact of traffic on air quality. *J Environ Manage*, 84, 513–522.

Mazzoccola, D.F. & Hudson, J.A. (1996) A comprehensive method of rock mass characterization for indicating natural slope instability. *Q J Eng Geol*, 29, 37–56.

Mazzoccola, D.F., Millar, D.L. & Hudson, J.A. (1997) Information, uncertainty and decision making in site investigation for rock engineering. *Geotech Geol Eng*, 15, 145–180.

Meguid, M.A. & Rowe, R.K. (2006) Stability of D-shaped tunnels in a Mohr–Coulomb material under anisotropic stress conditions. *Can Geotech J*, 2006, 43, 273–281.

Meng, L.B., Li, T.B., Jiang, Y., Wang, R. & Li, Y.R. (2013) Characteristics and mechanisms of large deformation in the Zhegu mountain tunnel on the Sichuan-Tibet highway. *Tunn Undergr Sp Tech*, 37, 157–164.

Meng, Z.P., Li, G.Q. & Xie, X.T. (2012) A geological assessment method of floor water inrush risk and its application. *Eng Geol*, 143–144, 51–60.

Millar, D.L. (2008). *Parallel distributed processing in rock engineering systems*. PhD thesis, Imperial College London.

Millar, D.L. & Hudson, J.A. (1993) Rock engineering system performance monitoring using neural networks. In: *Artificial Intelligence in the Minerals Sector, Proceedings of an Institution of Mining and Metallurgy Conference, April, 1993, Nottingham*.

Millar, D.L. & Hudson, J.A., (1994) Performance monitoring of rock engineering systems using neural networks. *T I Min Metall A*, 103, A3–A16.

Millar, D.L., Jiao, Y., Arnold, P.N. & Hudson, J.A. (1994) A rock engineering systems approach to risk assessment. In: *Proc. IMM Conference: Risk in Materials Extraction Industry, 22–24 March 1994, Exeter, U.K.* London: Thomas Telford.

Mitri, H.S., Hassani, F.P. & Kebbe, R. (1993) A strain energy approach for the prediction of rockburst potential in underground hard rock mines. *In: Proc. First Can. Symp. on Numerical Modelling Applications in Mining and Geomech, 1993, McGill University.* pp. 228–239.

Mo, Y. & Zhou, X. (2008) Dynamic monitoring and simulation analysis of surrounding rock deformation of tunnel in karst region. *Chinese J Rock Mech Eng*, 27 (S2), 3816–3821. (in Chinese)

Molag, M. & Trijssenaar-Buhre, I.J.M. (2006) Risk Assessment Guidelines for Tunnels. In: *Safe & Reliable Tunnels, Innovative European Achievements, Second International Symposium, Lausanne, Switzerland.* pp. 1–10.

Mosayebi, A., Shahriar, K. & Taki, R. (2004) Determination of Kuh Sefid fault zone in Cheshmeh Langan tunnel using TSP. In: *Proc. 6th Iranian Tunneling Conference, Tehran, Iran.*

Moulton, B.G., Cass, D.T., Nowak, D.E. & Poulin, R.M. (1995) Tunnel boring machine concept for converging ground. In: *Proc. Rapid excavation and tunnelling conference, San Francisco, USA.* Littleton: SME Inc. pp. 509–523.

Movahednejad, A.E. (2008) Deep tunnelling in fault zone case study in Kuhrang 3 water conveyance tunnel. In: Kanjlia, V.K., Ramamurthy, T., Wahi, P.P. & Gupta, A.C. (eds.) *Underground Facilities for Better Environment and Safety, Proceedings of the World Tunnel Congress 2008, 22–24 September, 2008, Agra, India.* India: Central Board of Irrigation and Power. pp. 1119–1123.

Naghadehi, M.Z., Jimenez, R., KhaloKakaie, R. & Jalali, S.-M.E. (2011) A probabilistic systems methodology to analyze the importance of factors affecting the stability of rock slopes. *Eng Geol.* 118, 82–92.

Nakano, R. (1974) On the design of water tunnels in relation with the type and magnitude of rock load with special references to the mechanism and prediction of squeezing welling rock pressure. *B Natl Res Inst Agr Eng*, Ministry of Agriculture and Forestry (Japan). 12, 89–142.

Nathanail, C.I., Earle, D.A. & Hudson, J.A. (1992) A stability hazard indicator system for slope failures in heterogeneous strata. In: Hudson, J.A. (ed.) *Proceedings of EUROCK '92 International Symposium of the International Society for Rock Mechanics on Rock Characterization. 14–17 September 1992, Chester, UK.* London: Thomas Telford. pp. 111–116.

Nguyen, Minh, D. & Corbetta, F. (1992), New methods for rock—support analysis of tunnels in elasto-plastic media. In: Kaiser, P.K. & McCreath, D.R. (eds) *Rock support in mining and underground construction.* Rotterdam, Netherlands: A.A. Balkema. ISBN: 90 5410 044 3. pp. 83–90.

Nguyen, Minh D. & Guo, C. (1996) Recent progress in convergence confinement method. In: Barla, G. (ed.) *EUROCK 96—Prediction and Performance in Rock Mechanics and Rock Engineering: Proc. Eurock '96, ISRM International Symposium, 2–5 September 1996, Turin, Italy.* Rotterdam: Balkema, ISBN: 905410843 6. pp. 855–860.

Nývlt, O., Prívara, S. & Ferkl, L. (2011) Probabilistic risk assessment of highway tunnels. *Tunn Undergr Sp Tech*, 26, 71–82.

OECD. (2013). *Radioactive waste management: underground research laboratories (URL).* NEA/RWM/R(2013)2. OECD, Nuclear Energy Agency.

Olivier, M. (2011) *Transient tunnel inflow and hydraulic conductivity of fractured crystalline rocks in the Central Alps (Switzerland).* Doctoral and Habilitation Theses. ETH, Switzerland.

Oreste, P. (2003) A procedure for determining the reaction curve of shotcrete lining considering transient conditions. *Rock Mech Rock Eng*, 36, 209–236.

Ortlepp, W.D. (1978) The mechanism of a rockburst. In: *Proc. of the 19th US Rock Mechanics Symposium, University of Nevada, Reno.* pp. 476–483.

Ortlepp, W.D. (2000) Observation of mining-induced faults in an intact rock mass at depth. *Int J Rock Mech Min*, 37, 423–436.

Palmström, A. (1995) *RMi-a rock mass characterization system for rock engineering purposes.* PhD thesis, Oslo University, Norway.

Palmstrom, A. & Stille, H. (2007) Ground behaviour and rock engineering tools for underground excavations. *Tunn Undergr Sp Tech*, 22, 363–376.

Palmström, A., & Stille, H. (2010) *Rock Engineering.* London: ICE Publishing, Thomas Telford Ltd.

Pan, J.Z. & He, J. (2000) *Large Dams in China: A Fifty-Year Review.* Beijing, China: China Water & Power Press.

Panet, M. (1979) Time-dependent deformations in underground works. In: *Proceedings of the 4th Congress of the International Society for Rock Mechanics, 2–8 Sept 1979, Montreux, Switzerland.* Rotterdam: A.A. Balkema. pp. 279–290.

Panet, M. (1995) *Le calcul des tunnels par la méthode convergence–confinement.* Paris: Presses de l'École Nationale des Ponts et Chaussées.

Panet, M. (1996) Two case histories of tunnels through squeezing rocks. *Rock Mech Rock Eng*, 29, 155–164.

Panthi, K.K. (2006) *Analysis of engineering geological uncertainties related to tunnelling in Himalayan rock mass conditions.* PhD thesis, Department of Geology and Mineral Resources Engineering. Norwegian University of Science and Technology (NTNU), Norway.

Panthi, K.K. (2011) Assessment on stress induced instability in a tunnel project of the Himalaya. In: Qihu Qian &Yingxin Zhou (eds.) *Harmonizing Rock Engineering and the Environment, Proceedings of the 12th ISRM International Congress on Rock Mechanic., 18–21 October 201,1 Beijing China.* Leiden, The Netherlands: CRC Press/Balkeema, Taylor & Francis Group. ISBN: 9780415804448. pp. 1777–1782.

Panthi, K.K. & Nilsen B. (2007) Uncertainty analysis of tunnel squeezing for two tunnel cases from Nepal Himalaya. *Int J Rock Mech Min*, 44, 67–76.

Pariseau, W.G. (2007) *Design Analysis in Rock Mechanics.* London: Taylor & Francis.

Pellet, F. (2009) Contact between a tunnel lining and a damage susceptible viscoplastic medium. *Comput Model Eng Sci*, 52, 279–295.

Peng, J.S. (1998) The rockburst and it protection of Ertan hydropower station underground engineering. *Water Power*, pp. 39–40.

Pennington, T.W. & Richards, D.P. (2010) Understanding uncertainty: assessment and management of geotechnical risk in tunnel construction. In: Proc. *GeoRisk 2011: Risk Assessment and Management in Geoengineering, June 26–28, 2011, Atlanta, Georgia, USA.* Reston, Viginia, USA: American Society of Civil Engineers (ASCE). ISBN: 98-1-61839-019-6. pp. 552–558.

Pers, K. Skagius, K., Södergren, S., Wiborgh, M., Hedin, A., Morén, L., Sellin, P., Ström, A., Pusch, R. & Bruno, J. (1999) *Identification and structuring of process.* SKB Technical Report TR-99-20, Sweden, plus Appendices. Available from www.skb.se.

Perzyna, P. (1966) Fundamental problems in viscoplasticity. *Advances in Applied Mechanics, Academic Press.* 9, pp. 243–377.

Phienwej, N., Thakur, P.K. & Cording, E.J. (2007) Time-dependent response of tunnels considering creep effect. *Int J Geomech*, 7, 296–306.

Price, N.J. & Cosgrove, J.W. (1990) *Analysis of Geological Structures.* Cambridge: Cambridge University Press, UK.

Priest, S.D. (1993) *Discontinuity Analysis for Rock Engineering.* London: Chapman & Hall.

Prigogine, I. & Stengers, I. (1985) *Order Out of Chaos: Man's New Dialogue with Nature,* London: Fontana.

Qiu, S.L. (2011) *Study on deformation and failure mechanism of deep-buried hard rock under loading and unloading conditions and rockburst vulnerability assessment methods.* Wuhan: Institute of Rock and Soil Mechanics, Chinese Academy of Sciences. (in Chinese)

Qiu, S.L., Feng, X.T., Zhang, C.Q. & Wu, W.P. (2011) Development and validation of rockburst vulnerability index (RVI) in deep hard rock tunnels. *Chinese J Rock Mech Eng.*

RAEng. (2014). *Thinking Like an Engineer: implications for education system.* UK Royal Academy of Engineering Report. downloadable from: www.raeng.org.uk/thinkinglikeanengineer

Rafiee, R. (2014) Development of a rock behaviour index around underground space using a rock engineering system. *J Geol Min Res,* 6, 46–56.

Rafiee R., Ataei, M., Khalokakaie, R., Jalali, S.M.E. & Sereshki, F. (2014) Determination and assessment of parameters influencing rock mass cavability in block caving mines using the probabilistic rock engineering system. *Rock Mech Rock Eng,* Published online. DOI 10.1007/s00603-014-0614-9.

Ramamurthy, T. & Arora, V.K. (1994) Strength predictions for jointed rocks in confined and unconfined states. *Int J Rock Mech Min,* 31, 9–22.

Ramoni, M. & Anagnostou, G. (2011) The interaction between shield, ground and tunnel support in TBM tunnelling through squeezing ground. *Rock Mech Rock Eng,* 44, 37–61.

Read, J. & Stacey, P. (2009) *Guidelines for Open Pit Slope Design.* Australia: CSIRO Publishing.

Reilly, J.J. (2003) The Relationship of Risk Mitigation to Management and Probable Cost. In: Saveur, J. (ed.) *(Re) Claiming the Underground Space, Proceedings of the 29th ITA World Tunneling Congress, Amsterdam, Netherlands.* Amsterdam, The Netherlands: A.A. Balkeema, Taylor & Francis. pp. 1035–1041.

Reilly, J.J & Brown, J. (2004) Management and control of cost and risk for tunneling and infrastructure projects. In: *Proceedings of World Tunnel Congress and 13th ITA Assembly,* Singapore. pp. 1–8.

Reyes, O. & Einstein, H.H. (1991) Failure mechanisms of fractured rock-a fracture coalescence model. In: Wittke, W. (ed.) *Proceedings of the 7th Congress, International Society for Rock Mechanics, September 1991, Aachen, Germany.* Rotterdam: A.A. Balkema. Vol. 1, pp. 333–340.

Rose, B., Andrieux, P.P., O'Connor, C.P. & Verreault, M., (2011) A systematic approach to rock mechanics challenges at Xstrata Zinc Brunswick mine. In: *Sainsbury, D., Hart, R., Detournay, C. & Nelson, M (eds), Continuum and Distinct element numerical modelling in Geomechanics, Proceedings of the 2nd International FLAC/DEM Symposium,*

14–16 February 2011, Melbourne, Australia. CD-ROM available directly from Itasca, Minneapolis, Itasca. pp. 49–57.

Ross-Brown, D.M. & Richards D.P. (2014) Challenges with empirical methods in the design of long-life mining excavations. In: *Proc. 1st ISRM Int. Conf. on Applied Empirical Design Methods in Mining, 9–11 June 2014 Lima, Peru.*

Rozos, D., Pyrgiotis, L., Skias, S. & Tsagaratos, P. (2008) An implementation of rock engineering system for ranking the instability potential of natural slopes in Greek territory: an application in Karditsa County. *Landslides*, 5, 261–270.

Rozos, D., Bathrellos, G.D. & Skillodimou, H.D. (2011) Comparison of the implementation of rock engineering system and analytic hierarchy process methods, upon landslide susceptibility mapping, using GIS: a case study from the Eastern Achaia County of Peloponnesus, Greece. *Environ Earth Sci*, 63, 49–63.

Ryder, J.A .(1988) Excess shear stress in the assessment of geologically hazardous situations. *J S Afr I Min Metall*, 88, 27–39.

Saaty, T.L. (1990) How to make a decision: the analytic hierarchy process. *Eur J Oper Res*, 48, 2–26.

Saaty, T.L. (2006) Rank from comparisons and from ratings in the analytic hierarchy/network processes. *Eur J Oper Res*, 168, 557–570.

Sabatakakis, N., Koukis, G., Vassiliades, E. & Lainas, S. (2012) Landslide susceptibility zonation in Greece. *Natural Hazards*, DOI 10.1007/s11069-012-0381-4 (first published on line, now on paper (2013) 65 (1), 523–543.)

Saeidi, O., Azadmehr, A. & Torabi, S.R. (2014) Development of a rock groutability index based on the Rock Engineering Systems (RES): A case study. *Indian Geotech J*, 44, 49–58.

Sakurai, S. (1993) Back analysis in rock engineering. In: Hudson, J.A. (ed.) *Comprehensive Rock Engineering*. Oxford: Pergamon Press. Vol. 4, pp. 543–569.

Sakurai, S. & Akutagawa, S. (1993) Some aspects of back analysis in geotechnical engineering. In: Ribeiro e Sousa, L. & Grossmann, N.F. (eds.): *Safety and environmental issues in rock mechanics: Proc. ISRM Symposium, EUROCK'93, 21–24 June 1993, Lisbon, Portugal.* Rotterdam: A.A. Balkema. pp. 1133–1140.

Sakurai, S., Akutagawa, S. & Tokudome, O. (1995) Back analysis of non-elastic strains based on the minimum norm solution. *J JSCE*, 517 (III-31), 197–202.

Salamon, M.D.G. (1970) Stability, instability and design of pillar workings. *Int J Rock Mech Min*, 7, 613–631.

Sandrone, F. (2008) *Analysis of pathologies and long term behavior of Swiss National Road tunnels.* Thesis EPFL no. 4019, École Polytechnique Fédérale de Lausanne, Lausanne.

Schubert, W. (1992) Erfahrungen bei der Durchörterung einer Großstörung im Inntaltunnel. *Felsbau* 11.

Schubert, W. (1996) Dealing with squeezing conditions in Alpine tunnels. *Rock Mech Rock Eng*, 29, 145–153.

Schubert, W., Golser, J. & Schwab, P. (1996) Improvement of support for squeezing rock. *Int J Rock Mech Min*, 33, 329A.

Schubert, W., Moritz, B. & Sellner, P. (2000) Tunnelling methods for squeezing ground. In: Ervin, M.C. (ed.) *GeoEng 2000; Proc. Intern. Symp. on Geotechnical and Geological Engineering, 19–24 November 2000, Melbourne, Australia.* CRC Press, Taylor & Francis Group. ISBN: 1-58716-067-6. Also on CD-ROM.

Schweiger, H.F., Pottler, R.K. & Steiner, H. (1991) Effects of seepage force on shotcrete lining of a large undersea cavern. In: *Proc. 7th Conf. of the Int. Association for Computer Methods and Advances in Geomechanics, April 1991, Cairns, Australia.* pp. 1503–1508.

Shalabi, F.I. (2005) FE analysis of time-dependent behaviour of tunnelling in squeezing ground using two different creep models. *Tunn Undergr Sp Tech*, 20, 271–279.

Shang, Y., Park, H-D. & Yang, Z. (2005) Engineering geological zonation using an interaction matrix of geological factors: an example from one section of Sichuan-Tibet Highway. *Geosci J*, 9, 375–387.

Shang, Y.J., Wang, S.J., Li, G.C. & Yang, Z.F. (2000) Retrospective case example using a comprehensive suitability index (CSI) for siting the Shisan-Ling power station, China. *Int J Rock Mech Min*, 37, 839–853.

Shao, J.Z. (2006) A discussion on mine water filling factors and flooding control measures in Huaibei mining area. *China Coal Field*. 18, 43–45. (in Chinese)

Sheorey, P.R. (1991) Experiences with application of the NGI classification to coal measures. *Int J Rock Mech Min*, 28, 27–33.

Shin, H.S., Kwon, Y.C., Jung, Y.S., Bae, G.J., & Kim, Y.G. (2009) Methodology for quantitative hazard assessment for tunnel collapses based on case histories in Korea. *Int J Rock Mech Min*, 46, 1072–1087.

Shin, J.H., Addenbrooke, T.I. & Potts, D.M. (2002) A numerical study of the effect of groundwater movement on long-term tunnel behavior. *Géotechnique*. 52, 391–403.

Shrestha, G.L. & Broch, E. (2008) Influences of the valley morphology and rock mass strength on tunnel convergence: with a case study of Khimti 1 headrace tunnel in Nepal. *Tunn Undergr Sp Tech*, 23, 638–650.

Shrestha, G.L. (2005) *Stress induced problems in Himalayan tunnels with special reference to squeezing*. Ph.D. thesis. Department of Geology and Mineral Resources Engineering. Norwegian University of Science and Technology. Trondheim.

Singh, B. & Goel, R.K. (1999) *Rock Mass Classification, a Practical Approach in Civil Engineering*. New York: Elsevier.

Singh, M., Singh, B. & Choudhari, J. (2007) Critical strain and squeezing of rock mass in tunnels. *Tunn Undergr Sp Tech*, 22, 343–350.

Skagius, K., Ström, A. & Wiborgh, M. (1995) *The use of interaction matrices for identification, structuring and ranking of FEPs in a repository system: application on the far-field of a deep geological repository for spent fuel*. Svensk Kärnbränslehantering AB, Stockholm, SKB Technical Report 95-22.

Skagius, K., Wiborgh, M., Ström, A. & Morén, L. (1997). Performance assessment of the geosphere barrier of a deep geological repository for spent fuel—the use of interaction matrices for identification, structuring and ranking of features, events and processes. *Nucl Eng Des*, 176, 155–162.

Smith, G.J. & Rosenbaum, M.S. (1993) Abandoned mine workings in chalk: approaches for appraisal and evaluation. *Q J Eng Geol Hydroge*, 26, 281–291. doi:10.1144/GSL. QJEGH.1993.026.004.04v.

Smith, G.M., Watkins, B.M., Little, R.H., Jones, H.M. & Mortimer, A.M. (1996) *Biosphere Modeling and Dose Assessment for Yucca Mountain*. Final Report to Electric Power Research Institute, Palo Alto, CA, USA. [Prepared by Quantisci]. TR-107190 3294-18.

Sousa, R.L. (2010) *Risk Analysis for Tunneling Projects*. PhD thesis of Cambridge MA, Massachusetts Institute of Technology.

Sousa, R.L. & Einstein, H.H. (2012) Risk analysis during tunnel construction using Bayesian networks: Porto Metro case Study. *Tunn Undergr Sp Tech*, 27, 86–100.

Special Issue: DECOVALEX (1995) (Stephansson, O. guest ed.) Thermo-hydro-mechanical coupling in rock mechanics. *Int J Rock Mech Min*, 32, (5) 387–535.

Special Issue: DECOVALEX II (2001) *Int J Rock Mech Min*, 38, (1) 1–162.

Special Issue: DECOVALEX III (2005) *Int J Rock Mech Min*, 42, (5 & 6) 593–610.

Special Issue: DECOVALEX 2011. (2013) Part 1, *J Rock Mech Geotech Eng*, 5, (1), 1–84; Part 2, *J Rock Mech Geotech Eng*, 5, (2), 84–155.

Special Issue: (2009) Tsang, C.-F. (Guest ed.) Special Issue on the DECOVALEX THMC Project (Safety Assessment of Nuclear Waste Repositories). *Environ Geol*, 57, 1217–1390. ISSN: 0943-0105 (Print) 1866-6299 (Online)

Spiegelhalter, D.J. (2010) Quantifying uncertainty. Contribution to *Handling uncertainty in science*. Meeting, 22 March 2010, Royal Society, UK. Audio recording downloadable from www.royalsociety.org.

Spiegelhalter, D.J. & Riesch, H. (2011) Don't know, can't know: embracing deeper uncertainties when analysing risks. *Philos T Roy Soc A, 369*, 4730–4750.

Standards Australia. (2009) AS/NZS ISO 31000: 2009. *Risk management—principles and guidelines*. Sydney: Standards Australia. See AS/NZ.

Steiner, W. (1996) Tunnelling in squeezing rocks: case histories. *Rock Mech Rock Eng, 29*, 211–246.

Stephansson, O. (1993) Rock stress in the fennoscandian shield. In: Hudson, J.A. (ed.) *Comprehensive Rock Engineering*. Oxford: Pergamon Press. Vol. 3, Chapter 17, pp. 445–459.

Stephansson, O. & Hudson, J.A. (1993) *SKI/SKB FEPs identification and characterization via the "rock engineering systems" approach*. Svensk Kärnbränslehantering AB, Stockholm. SKB Arbetsrapport: 93–96.

Stephansson, O. & Zang, A. (2012) ISRM suggested methods for rock stress estimation—Part 5: Establishing a model for the in-situ stress at a given site. *Rock Mech Rock Eng, 45*, 955–969. DOI: 10.1007/s00603-012-0270-x.

Stephansson, O. & Zang, A. (2015) ISRM Suggested Methods for Rock Stress Estimation—Part 5: Establishing a model for the *in situ* stress at a given site. In: Ulusay, R. (ed.) *The ISRM Suggested Methods for Rock Characterization, Testing and Monitoring 2007–2014*. pp. 187–201.

Stephansson, O., Jing, L. & Tsang, C-F. (eds.) (1996) *Coupled Thermo-Hydro-Mechanical Processes of Fractured Media*. Elsevier Book Series: Developments in geotechnical engineering, Vol.79. Amsterdam: Elsevier, ISBN 13: 978-0-444-82545-2.

Stephansson, O., Hudson, J.A. & Jing, L. (eds.) (2004) *Coupled Thermo-Hydro-Mechanical-Chemical Processes in Geosystems*. Vol. 2, Elsevier Geo-Engineering Book Series, London : Elsevier. ISBN: 008044525X, 9780080445250.

Sterpi, D. & Gioda, G. (2009) Visco-plastic behaviour around advancing tunnels in squeezing rock. *Rock Mech Rock Eng, 42*, 319–339.

Sturk, R., Olsson, L. & Johansson, J. (1996) Risk and decision analysis for large underground projects, as applied to the Stockholm ring road tunnels. *Tunn Undergr Sp Tech, 11*, 157–164.

Su, L.J. & Fu, C.H. (2011) Study on groundwater detection technology for construction of diversion tunnel of Jingping hydropower station II. *Water Resources and Hydropower Engineering*. 42, (3) 16–18, 34.

Sun, S.J., Wang, L., Zhang, D.F., Wang, H.P. (2011) Time-history response analysis of surrounding rock mass in process of excavating deep tunnel. *Journal of China Coal Society*. 36, 738–746.

Sunamura, T. (1992) *Geomorphology of Rocky Coasts*. Wiley–Blackwell, ISBN-10:0471917753; ISBN-13: 978-0471917755.

Šušteršič, J., Jovičič, V., Zajc, A. & Ercegovič, R. (2004) Evaluation of improvement in the bearing capacity of fibre reinforced shotcrete tunnel lining. In: Di Prisco, M., Felicetti. R. & Plizzari, G.A. (eds.) *Fibre Reinforced Concrete, BEFIB 2004: Proceedings of Sixth International RILEM Symposium, 20–22 September, 2004, Varenna, Italy*. F-92220 Bagneux – France: RILEM Publications S.A.R.L., Vol. 2, pp. 985–994.

Tan, T.K. (1982) The mechanical problems for the long-term stability of underground galleries. *Chinese J Rock Mech Eng 1*, 1–20.

Tan, T.K. (1993) The importance of creep and time-dependent dilatancy, as revealed from case records in China. In: Hudson, J.A. (ed.) *Comprehensive Rock Engineering*. Oxford: Pergamon Press, Vol. 3, pp. 709–744.

Tang, C.A. & Hudson, J.A. (2010) *Rock Failure Mechanisms—Explained and Illustrated.* London. UK: Taylor & Francis.

Tang, W.Z., Qiang, M.S., Duffield, C.F., Young, D.M. & Lu, Y.M. (2007) Risk management in the Chinese construction industry. *J Constr Eng M ASCE*, 133, 944–956.

Temel, T., Janssen, W. & Karimov, F. (2002a) Systems analysis by graph-theoretic techniques: assessment of institutional linkages in the agricultural innovation system of Azerbaijan. In: *Proceedings Xth EAAE Congress on Exploring Diversity in the European Agri-Food System', 28–31 August 2002, Zaragoza, Spain.*

Temel, T., Janssen, W. & Karimov. F. (2002b). *The Agricultural Innovation System of Azerbaijan: An Assessment of Institutional Linkages.* The Hague: International Service for National Agricultural Research. Country Report 64, May 1992.

Terzaghi, K. (1946) Rock defects and loads on tunnel supports. In: Proctor, R.V. & White, T.L. (eds.) *Rock Tunneling with Steel Support, Section 1.* Youngstown, Ohio: Commercial Shearing and Stamping Company.

The International Tunnelling Insurance Group. (2006) *A code of practice for risk management of tunnel works.* Available: http://www.munichre.com/publications/tunnel_code_of_practice_en.pdf.

Thompson, P.M. & Chandler, N.A. (2004) *In situ* rock stress determinations in deep boreholes at the underground research laboratory. *Int J Rock Mech Min*, 41, 1305–1316.

Thut, A., Nateropp, D., Steiner, P. & Stolz, M. (2006) Tunnelling in squeezing rock—yielding elements and face control. In: *Proceedings 8th International Conference on Tunnel Construction and Underground Structures, Ljubljana, Slovenia.* pp. 150–157.

Trifu, C.I. & Suorineni, F.T. (2009) Use of microseismic monitoring for rockburst management at Vale Inco mines. In: Tang, Chun'An. (ed.) *Proc. of 7th Intl. Symp. on Rockburst and Seismicity in Mines: Controlling Seismic Hazard and Sustainable Development of Deep Mines, 21–23 August 2009Dalian, China.* New York: Rinton Press. pp. 1105–1114.

Tsang, C-F., Bernier, F. & Davies, C. (2005) Geohydromechanical processes in the Excavation Damaged Zone in crystalline rock, rock salt, and indurated and plastic clays—in the context of radioactive waste disposal. *Int J Rock Mech Min*, 42, 109–125.

Tseng, D.J., Tsai, B.R. & Chang, L.C. (2001) A case study on ground treatment for a rock tunnel with high groundwater ingression in Taiwan. *Tunn Undergr Sp Tech*, 16, 175–183.

Ulusay, R. & Hudson, J.A. (2007) *The Complete ISRM Suggested Methods for Rock Characterisation, Testing and Monitoring: 1974-2006.* ISRM, Portugal.

Vahed Ghiasi, Husaini Omar, Bujang B. Kim Huat, Ratnasamy Muniandi, Zainuddin B. Md Yusof. (2011) Risk management overview of tunnels using numerical modeling. *Journal of Engineering, Design and Technology.* 9, 110–124. http://dx.doi.org/10.1108/17260531111121495.

Vahed Ghiasi, Husaini Omar, Jamal Rostami, Zainuddin B. Md. Yusoff, Samad Ghiasi, Bujang K. Huat & Ratnasamy Muniandy. (2011) Geotechnical and geological studies of NWCT tunnel in Iran focusing on the stabilization analysis and design of support: A case study. *Sci Res Essays*, 6, 79–97. DOI: 10.5897/SRE10.704.

Valley, B., Kaiser, P.K. & Duff, D. (2010) Consideration of uncertainty in modelling the behaviour of underground excavations. In: Van Sint Jan, M. & Potvin, Y. (eds.) *Deep Mining 2012: Proceedings of the 5th International Seminar on Deep and High Stress Mining, Santiago, Chile.* Perth: Australian Centre for Geomechanics. pp. 423–435.

Valli, J., Hakala, M. & Kuula, H. (2011) *Modelling of the* in situ *stress state at the Olkiluoto site, western Finland.* Posiva Oy, Finland. Posiva Working Report 2011-34.

Van Dorp, F., Egan, M., Kessler, J.H., Nilsson, S., Pinedo, P., Smith, G. & Torres, C. (1999) Biosphere modeling for the assessment of radioactive waste repositories; the development of a common basis by the BIOMOVS II reference biospheres working group. *J Environ Radioactiv*, 42, 225–236.

Velasco, H.R., Ayub, J.J., Belli, M. & Sansone, U. (2006) Interaction matrices as a first step toward a general model of radionuclide cycling: application to the 137Cs behavior in a grassland ecosystem. *J Radioanal Nucl Ch*, 268 (3), 503–509.

Vincenzi, V., Gargini, A., Goldscheider, N. (2009) Using tracer tests and hydrological observations to evaluate effects of tunnel drainage on groundwater and surface waters in the Northern Apennines (Italy). *Hydrogeology Journal*. 17, 135–150.

Vu, T.M., Sulem, J., Subrin, D. & Monin, N. (2012) Semi-analytical solution for stresses and displacements in a tunnel excavated in transversely isotropic formation with non-linear behaviour. *Rock Mech Rock Eng*, 46, 213–229. (03/2013) DOI 10.1007/s00603-012-0296-0.

Vu, T.M., Sulem, J., Subrin, D., Monin, N. & Lascols, J. (2013) Anisotropic closure in squeezing rocks: the example of Saint-Martin- la-Porte access gallery. *Rock Mech Rock Eng*, 46, 231–246.

Wang, G.D., Shi, Y.C., Liu, H.C., Kou, J-W., Shan, Z-G. & Zhou, C-H. (2006) Rock mass classification methods for deep buried tunnels. In: *IAEG2006: Engineering geology for tomorrow's cities: Proceedings 10th IAEG International Congress, 6–10 September 2006, Nottingham, UK*. CD: Paper number 574. The Geological Society of London, 2006.

Wang, H.N., Utilib, S. & Jiang, M.J. (2014) An analytical approach for the sequential excavation of axisymmetric lined tunnels in viscoelastic rock. *Int J Rock Mech Min*, 68, 85–106.

Wang, S.R., Liu, Z.W., Qu, X.H. & Fang, J.B. (2009) Large deformation mechanics mechanism and rigid-gap-flexible-layer supporting technology of soft rock tunnel. *China J Highway Trans*, 22 (6), 90–95.

Wang, W.D., Guo, J., Gang, F.L. & Chang, X.S. (2012) A subjective and objective integrated weighting method for landslides susceptibility mapping based on GIS. Environ Earth Sci, 65, 1705–1714.

Wang, X. & Wang, M. (2006) Analysis of mechanism of water inrush in Karst tunnels. In: Zhu, H., Zhang, F., Chin, C-T.& Zhang, D. (eds.) *GeoShanghai 2006: Underground Construction and Ground Movement: Proceedings GeoShanghai International Conference, 6–8 June 2006, Shanghai, China*. American Society of Civil Engineers. ISBN (print): 978-0-7844-0867-4. pp. 66–72.

Wang, X.C. & Ma, G.Y. (2002) Introduction to engineering project of water diversion from the upper Yangtze river into the upper Yellow river and its main engineering geological problems. *J Eng Geol* 10, 38–45.

Wang, X.H. (1998) The multi-parameters measuring method forecasting water inrush from working faces. *Coal Geology & Exploration*. 26, 36–39. (in Chinese)

Wang, X.M. (2013) *Study on Rock Fractures and Rock Blocks in Wudongde Dam Area*. PhD thesis of China University of Geosciences, Beijing, China.

Wang, Z-L., Shi Y-C., Shan Z-G.,, Xie D-Z., Wang J-X. & Li X-L. (2013) DBQ system: a classification method of soft wall rock for deep tunnels. *Journal of Geological Hazards and Environment Preservation*, 24 (1), 84–87.

Wei, J.B. & Deng J.H. (2010) Variation of excavation damaged zone and back analysis of large scale underground powerhouse with high geostress. *Rock Soil Mech*, 31, 330–336.

Wei, J.B., Deng, J.H., Wang, D.K., Cai, D.W. & Hu, J.Z. (2010) Characterization of deformation and fracture for rock mass in underground powerhouse of Jinping I hydropower station. *Chinese J Rock Mech Eng*, 29, 1198–1205.

Wenner, D. & Harrison, J. P. (1996) Techniques to cope with uncertain parameters in geomechanics on different levels of information. In: Barla, G. (ed.) *EUROCK 96—Prediction and Performance in Rock Mechanics and Rock Engineering: Proc. Eurock '96: ISRM International Symposium, 2–5 September 1996, Turin, Italy*. Rotterdam: Balkema, ISBN: 905410843 6. pp. 327–334.

Wiles, T.D., Marisett, S.D. & Martin, C.D. (1998) *Correlation between local energy release density and observed bursting conditions at Creighton mine.* Mine Modelling Report, Sudbury, Canada.

Wilson, A.H. (1980) A method of estimating the closure and strength of lining required in drivages surrounded by a yield zone. *Int J Rock Mech Min*, 17, 349–355.

Windsor, C.R. & Thompson, A.G. (1993) Rock reinforcement-technology, testing, design and evaluation. In: Hudson JA (ed) *Comprehensive Rock Engineering, Principles, Practice & Projects.* Oxford: Pergamon Press, Vol. 4, pp. 451–484.

Wu, J.Y., Zhu, H.C., Zhu, Y.S. (2013) Study cracking propagation behavior of marble and its influence in ground safety of the Jinping II project. In: *Continuum and Distinct Element numerical modeling in geomechanics, Proceedings of the 3rd International FLAC/DEM Symposium, 22–24 October 2013, Hangzhou, P.R. China.* Minneapolis: Itasca International Inc.

Wu, Q., Pang, W. & Wei, Y. (2006) The coupling technology of GIS and ANN of evaluation of vulnerability of groundwater inrush into coal mines from aquifers underlying the coal seams. *J Chin Coal Soc*, 3, 314–319.

Wu, Q., Liu, Y.Z., Liu, D.H. & Zhou, W.F. (2011) Prediction of floor water inrush: the application of GIS-Based AHP vulnerable index method to Donghuantuo coal mine, China. *Rock Mech Rock Eng*, 44, 591–600.

Wu, S.Y. & Wang, G. (2011) Rock mechanical problems and optimization for the long and deep diversion tunnels at Jinping II hydropower station. *J Rock Mech Geotech Eng*, 3, 314–328.

Wu, S.Y., Ren, X.H., Chen, X.R. & Zhang, J.X. (2005) Stability analysis and supporting design of surrounding rocks of diversion tunnel for Jinping hydropower station. *Chinese J Rock Mech Eng*, 24, 3777–3782.

Xiang, T.B. (2010) *Study on dynamic feed-back optimization design method of large underground powerhouse cavern group during construction period.* Master thesis of Institute of Rock & Soil Mechanics, Chinese Academy of Sciences, Wuhan, China.

Xiang, T.B., Feng, X.T., Jiang, Q. & Chen, J. (2012) Failure mode classification and control for surrounding rock of large-scale cavern group. In: Cai, M. (ed.) *Rock Mechanics: Achievements and Ambitions.* Boca Raton, Florida, USA: CRC Press, Taylor & Francis Group. pp. 527–533.

Xiao, P.X., Yin, J.M., Ai, K. & Jing, F. (2006) Geostress measurement and application analysis for Guandi hydropower plant's underground powerhouse. *Chinese J Undergr Sp. Eng*, 2, 895–898.

Xie, H.Q., Xiao, M.L. & He, J.D. (2008) Regression analysis of initial geostress for dam region of Jinping hydropower station. *J Yangtze River Sci Res Inst.* 25, 50–54. (in Chinese)

Xu, D.P., Feng, X.T., Cui, Y.J., Jiang, Q., Jiang, Y.L. & Huang, K. (2012) Shear behaviors of interlayer staggered zone at Baihetan hydropower station. *Chinese J Rock Mech Eng*, 31, 2692–2703.

Yang, D. & Zhang, P. (2008) Study on the rapid treatment method of mud gushing and landslide at shallowly covered karst section in Yunwu Mountain tunnel, *West-China Exploration Engineering*, 10, 192–194. (in Chinese)

Yang, Y. & Zhang, Q. (1998) The application of neural networks to Rock Engineering Systems (RES). *Int J Rock Mech Min*, 35, 727–745.

Yang, Y. & Zhang, Q. (1997) A hierarchical analysis for rock engineering using artificial neural networks. *Rock Mech Rock Eng*, 30, 207–222.

Yao, X.C., Li, N., Chen, L.J., Sun, H.C. & Jing, M.G. (2011) Back analysis of surrounding rock stability based on excavation process of underground powerhouse at Laxiwa hydropower station. *Chinese J Rock Mech Eng*, 30, 3052–3059.

Yassaghi, A., & Salari-Rad, H. (2005) Squeezing rock conditions at an igneous contact zone in the taloun tunnels, Tehran-Shomal freeway, Iran: a case study. *Int J Rock Mech Min*, 42, 95–108.

Yasuhara, H., Polak, A., Mitani, Y., Grader, A., Halleck, P., & Elsworth, D. (2006) Evolution of fracture permeability through fluid–rock reaction under hydrothermal conditions. *Earth Planet Sc Lett*, 244, 186–200.

Yasuhara, H., Kinoshita, N., Ohfuji, H., Lee, D.S., Nakashima, S., & Kishida, K. (2011) Temporal alteration of fracture permeability in granite under hydrothermal conditions and its interpretation by coupled chemo-mechanical model. *Appl Geochem*, 26, 2074–2088.

Zang, A. & Stephansson, O. (2010) *Stress Field of the Earth's Crust*. London, New York: Springer. ISBN: 978-1-4020-8443-0 (Print) 978-1-4020-8444-7 (Online); DOI 10.1007/978-1-4020-8444-7.

Zarei, H.R., Uromeihy, A. & Sharifzadeh, M. (2009) Identification and classification of hazards related to groundwater inflow in Sabzkuh tunnel. In: *Proceeding of 6th Iranian Engineering Geology and Environment Conference*. Tarbiat Modarres University, Tehran, Iran.

Zarei, H.R., Uromeihy, A. & Sharifzadeh, M. (2012) Identifying geological hazards related to tunneling in carbonate Karstic rocks–Zagros, Iran. *Arabian J Geosci*, 5, 457–464.

Zhang, C., Zhou, H., Feng, X.T. & Zhang, Z.H. (2008) Method for obtaining local stress field based on the interpolation and numerical simulation method. *Rock Soil Mech*, 29, 2016–2024. (in Chinese)

Zhang, C., Feng, X.T. & Zhou, H. (2012) Estimation of *in situ* stress along deep tunnels buried in complex geological conditions. *Int J Rock Mech Min*, 52, 139–162.

Zhang, C.Q., Zhou, H. & Feng, X.T. (2011) An index for estimating the stability of brittle surrounding rock mass: FAI and its engineering application. *Rock Mech Rock Eng*, 44, 401–414.

Zhang, C.S., Chu, W.J., Liu, N., Zhu, Y.S. & Hou, J. (2011) Laboratory tests and numerical simulations of brittle marble and squeezing schist at Jinping II hydropower station, China. *J Rock Mech Geotech Eng*, 3, 30–38.

Zhang, C.S., Liu, N., Zhu, H.C., Chu, W.J. & Wu, J.Y. (2013) Time-dependent behavior of crack propagation and evaluation of control effect of Jinping deep marble. *Chinese J Rock Mech Eng*, 32, 1964–1972.

Zhang, F., Zhou, S.D. & Guo, W. (2006) Layout and supporting design of Penshui hydropower station underground powerhouse. *Yangtze River*, 37, 35–36, 44.

Zhang, J.C. & Shen, B.H. (2004) Coal mining under aquifers in china: a case study. *Int J Rock Mech Min*, 41, 629–639.

Zhang, J.J. & Fu, B.J. (2008) Rockburst and its criteria and control. *Chinese J Rock Mech Eng*, 27, 2034–2042.

Zhang, J.X., Ren, X.H. & Yang, S. (2010) Study on transport of groundwater of Jinping cascade-II hydropower station with long diversion tunnel region. In: *Earth and Space: Proceedings 12th Biennial International Conference on Engineering, Construction, and Operations in Challenging Environments; and Fourth NASA/ARO/ASCE Workshop on Granular Materials in Lunar and Martian Exploration 14–17 March 2010. Honolulu, Hawaii, USA.* American Society of Civil Engineers. pp. 2247–2255. ISBN (print): 978-0-7844-1096-7; doi: 10.1061/41096(366)207.

Zhang, K., Zhou, H., Feng, X.T., Shao, J.F., Yang, Y.S. & Zhang, Y.G. (2010) Experimental research on elastoplastic coupling character of marble. *Rock and Soil Mechanics*. 31, 2425–2434. (in Chinese)

Zhang, L.Q., Yang, Z.F., Liao, Q.L. & Chen, J. (2004) An application of the Rock Engineering Systems (RES) methodology in rockfall hazard assessment on the Chengdu-Lhasa Highway, China. In: Hudson, J.A. & Feng, X.T. (eds.) *Proceedings of the ISRM SINOROCK 2004*

Symposium, May 2004, Three Gorges Dam, China. Paper 3B 11, *Int J Rock Mech Min*, 41, 833–838. CD-ROM. doi:10.1016/j.ijrmms.2004.03.144.

Zhang, M. (2010) *Tunnel Construction Technology in Karst Fault of Yiwan Railway.* Beijing: Science Press. (in Chinese)

Zhang, X.B. (2010) Collapse treatment for powerhouse arch-top of Dagangshan hydropower station. *Sichuan Water Power.* 29, 55–59.

Zhang, X.G. (2010) *Risk Assessment Guidelines of Highway Bridge and Tunnel Engineering Design.* Beijing, China: China Communications Press.

Zhang, Y. (2011) *Study of Geomechanics Mechanism and Characteristics of Surrounding Rock Mass Deformation Failure for Underground Powerhouse of Jinping I Hydropower Station.* Master thesis of Chengdu University of Technology, Chengdu, China.

Zhang, Y., Xiao, P.X., Ding, X.L., Ou, W.B., Lu, B., Liao, C.G. & Dong, Z.H. (2012) Study of deformation and failure characteristics for surrounding rocks of underground powerhouse caverns under high geostress condition and countermeasures. *Chinese J Rock Mech Eng*, 31, 228–244.

Zhao, X.J., Chen, B.R., Zhao, H.B., Jie, B.H. & Ning, Z.F. (2012) Laboratory creep tests for time-dependent properties of a marble in Jinping II hydropower station. *J Rock Mech Geotech Eng*, 4, 168–176.

Zhao, Y., Li, P.F. & Tian, S.M. (2013) Prevention and treatment technologies of railway tunnel water inrush and mud gushing in China. *J Rock Mech Geotech Eng*, 5, 468–477.

Zhao, Z.N., Feng, X.T., Chen, B.R., Feng, G. L. & Chen, T.Y. (2013) Study of relativity between rockburst and microseismic activity zone in deep tunnel. *Rock and Soil Mechanics*, 34, 491–497.

Zhou, H., Zhang, K., Feng, X.T., Shao, J.F. & Qiu, S.L. (2010) Elastoplastic coupling mechanical model for brittle marble. *Chinese J Rock Mech Eng*, 29, 2398–2409.

Zhou, H., Zhang, C., Li, Z., Hu, D. & Hou J. (2014) Analysis of mechanical behavior of soft rocks and stability control in deep tunnels. *J Rock Mech Geotech Eng*, doi: 10.1016/j.jrmge.2014.03.003.

Zhu, J.L., Tian, S.L. & Xie, C.S. (2013) Collapse treatment for powerhouse arch-top of Dagangshan hydropower station. *Yangtze River*, 44, 45–47.

Zhu, W.S., Qi, Y.P., Guo, Y.H. & Yang W.M. (2012) 3D Damage rheology analysis of deformation and fracture of surrounding rocks in Jinping I hydropower station underground powerhouse. *Chinese J Rock Mech Eng*, 31, 865–872.

Zhu, Z.Q., Sheng, Q., Zhang, Y.H. & Li, Y.F. (2013) Research on excavation damage zone of underground powerhouse of Dagangshan hydropower station. *Chinese J Rock Mech Eng*, 32, 734–739.

Zhuang, H. & Mu, J. (2009) The prevention and treatment of large cross section tunnel mud gushing in karst areas, *Railway Engineering* 6, 49–51. (in Chinese)

Zi, J.Q., Yang, G.P. Xie, G.H. & Xue, X.W. (2009) Karst Processing Technology of Goupitan Hydropower Station Underground Powerhouse. *Shuili Shuidian Shigong.* pp. 41–47.

Zoback, M.D., Barton, C.A., Brudy, M., Castillo, D.A, Finkbeiner, T., Grollimund, B.R., Moos, D.B., Peska, P., Ward, C.D. & Wiprut, D.J. (2003) Determination of stress orientation and magnitude in deep wells. *Int J Rock Mech Min*, 40, 1049–1076.

Zou, H.Y. & Xiao, M. (2010) Study of methodology for assessment of excavation disturbed zone of underground caverns. *Chinese J Rock Mech Eng*, 29, 513–519.

Colour plates

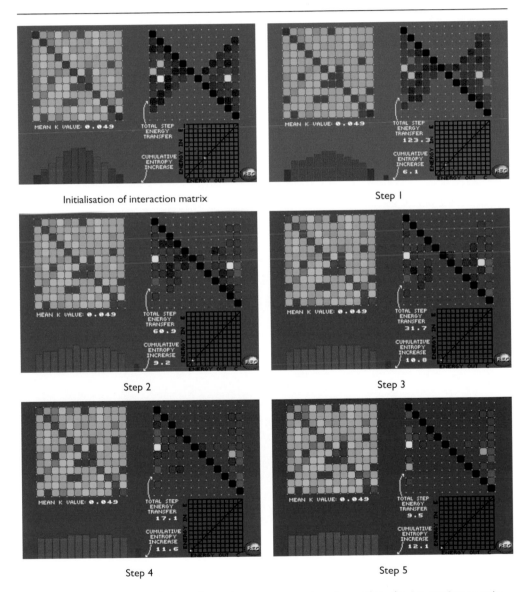

Initialisation of interaction matrix

Step 1

Step 2

Step 3

Step 4

Step 5

Figure 3.19 Computer simulation of step-by-step energy transfer within the interaction matrix—Case B: irregular attenuation.

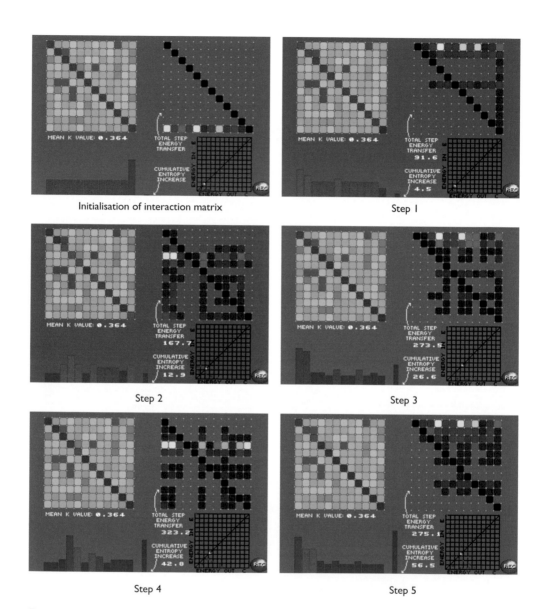

Initialisation of interaction matrix

Step 1

Step 2

Step 3

Step 4

Step 5

Figure 3.20 Computer simulation of step-by-step energy transfer within the interaction matrix—
Case C: chaotic behaviour when the energy transfer amounts are high.

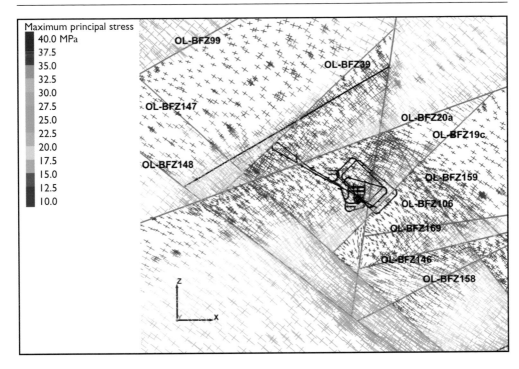

Figure 4.14a Maximum principal stresses in the horizontal cutting plane, depth level −150 m, from Valli et al. (2011). Scale: ~4 km laterally.

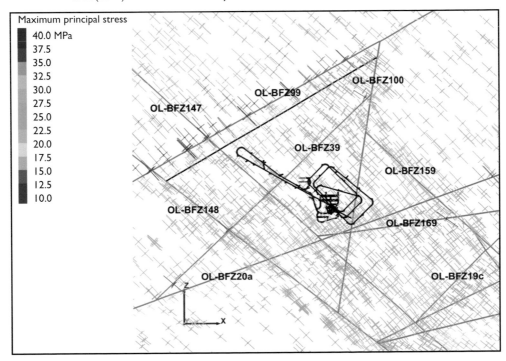

Figure 4.14b Maximum principal stresses in the horizontal cutting plane, depth level −420 m, from Valli et al. (2011). Scale: ~4 km laterally.

Figure 6.2 Failure of the support system and damage to the rock mass due to rockburst and spalling occurrence. (a), (b) and (c): broken support system in TBM tunnel; (d), (e) and (f): crushed rock masses.

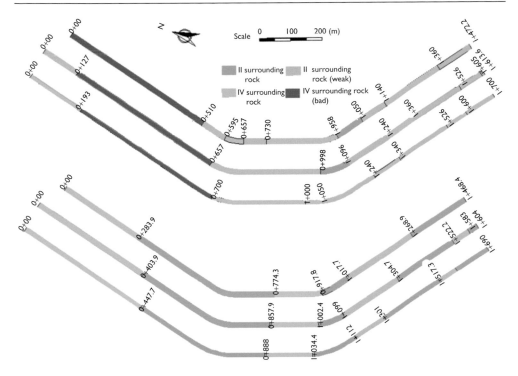

Figure 6.11 Difference between the estimated rock mass classification before construction (lower three tunnels) and the actual rock mass classification after excavation for the same tunnels (upper three tunnels), Baihetan Power Station, China.

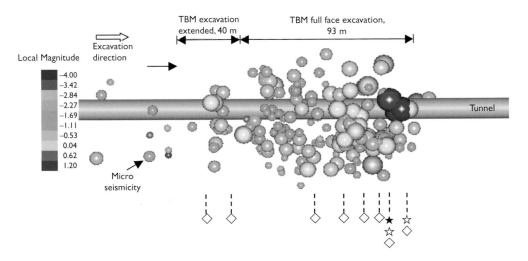

Figure 6.26c Comparison of microseismic events location and rockburst events during D&B pilot – TBM extension excavation and TBM full face excavation (◇: slight rockburst cases, ☆: moderate rockburst cases, ★: intense rockburst cases) (after Feng et al., 2013).

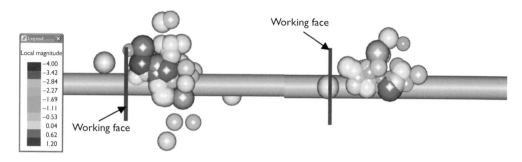

Figure 6.29b Location of microseismicity during September 6th–8th during which time the rockburst risk was not controlled (left part) and during September 9th–11th during which time the rockburst risk was controlled successfully (right part), (after Feng *et al.*, 2013).

Figure 6.79 High pressure flow point near the tunnel crown at chainage AK10+612 in Access Tunnel #A: (a) before treatment.